Handbook of
Ceramics and Composites

Volume 2:
MECHANICAL PROPERTIES AND
SPECIALTY APPLICATIONS

Handbook of Ceramics and Composites

edited by

Nicholas P. Cheremisinoff

Volume 1 Synthesis and Properties
Volume 2 Mechanical Properties and Specialty Applications

Other Volumes in Preparation

Handbook of
Ceramics and Composites

Volume 2:
MECHANICAL PROPERTIES AND
SPECIALTY APPLICATIONS

edited by

Nicholas P. Cheremisinoff

MARCEL DEKKER, INC.

New York • Basel • Hong Kong

Library of Congress Cataloging-in-Publication Data
(Revised for volume 2)

Handbook of ceramics and composites.

    Includes bibliographical references and indexes.
    Contents: v. 1. Synthesis and properties – v. 2.
Mechanical properties and specialty applications.
    1. Composite materials. 2. Ceramic materials.
I. Chereminisoff, Nicholas P.
TA418.9.C6H32 1990   620.1'18   90-2833
ISBN 0-8247-8005-1 (v. 1)
ISBN 0-8247-8006-X (v. 2)

MARCEL DEKKER, INC.
270 Madison Avenue, New York, New York 10016

Current printing (last digit):
10 9 8 7 6 5 4 3 2 1

PRINTED IN THE UNITED STATES OF AMERICA

Preface

The *Handbook of Ceramics and Composites* is a multivolume series devoted to reporting the latest technological advances in polymeric composites and blends, reinforced polymeric and composite materials, and ceramics of engineering importance. The first volume, published in 1990, contained overview discussions concerning the structural and physical properties of important classes of materials. This second volume contains twelve chapters covering topics ranging from physical properties testing and characterization to specialty composites.

The most common failure mode in laminated composite materials is interlaminar fracture, or delamination between plies of the laminate. Chapter 1 describes the experimental methods and associated data analysis techniques that are used to measure the resistance of polymer composite materials to interlaminar fracture (or fracture toughness).

The adoption of advanced fibrous composites by the aerospace industry triggered off a massive research effort. The ability to design a structure composed of fiber-reinforced plastics (FRP) must be matched by the ability to fabricate the component and knowledge of damage, failure, and fracture behavior in its service time. Chapter 2 continues the discussion of fabrication flaws with emphasis given to FRP.

Chapter 3 embraces the subject of improving fracture resistance. For epoxy resins, which this chapter is devoted to, fracture resistance can be achieved by the addition of rubbery or rigid second phase dispersions, or both, in an epoxy matrix. These multiphase epoxy systems are specialty polymeric materials for applications as high-performance adhesives and matrices for advanced fiber composites.

Chapter 4 covers the important phenomenon of fretting, which is caused by the relative cyclic sliding motion of two surfaces in intimate contact. This type of surface damage is caused by cyclic straining due to fatigue loading of one or both of the mating components. Surface damage induced by fretting (for example, microcracking) can accelerate fatigue failure development. Again, discussions in this chapter are slanted toward continuous fiber reinforced polymer composites, which continue to replace metals in an increasing number of engineering applications.

Polyurethanes have become an important class of polymers, particularly in their use with glass fiber reinforced composites. Chapter 5 covers the basic chemistry of polyurethanes; the reaction injection molding (RIM) process, which has become a major manufacturing method; and the physical properties of high-glass-transition-temperature polyurethanes and glass fibers-reinforced composites.

Hydraulic cement paste and concrete are considered composite materials because of their multiphase nature. There is enormous interest in improving the strength of concrete. Chapter 6 covers the methods of polymer impregnation and steel-fiber reinforcement, and their combination to improve the mechanical properties of concrete. Chapter 7 returns to the subject of manufacturing flaws with composites. Emphasis is given to the consequences of prepreg aging, variation of production conditions, hot/moist environmental conditioning, low-energy impact, and pin bearing, for various laminate systems and sandwich panels.

Chapter 8 covers the use of graphite/epoxy composites and the phenomenon of delamination that may develop during manufacturing due to nonoptimal curing or the introduction of interlaminar stresses and contamination.

Chapter 9 covers phenolic resin fibers, which are important in the production of ceramic and abrasive materials. This material has important uses ranging from aerospace construction to flame-proof and high-temperature-resistant materials.

Chapter 10 is intended as an introductory discussion of an old class of construction material that has regained interest in the composites family, namely, wood. This chapter describes the application of silane treatment as a means of improving the compatibility between wood fiber and a polymer matrix.

In the last two decades considerable attention has been given to the chemistry and physics of the mixed oxides. These materials have found application in a variety of technical fields. The chemical and physical characterization of these materials is a difficult task due to their complexity. Since these materials will be discussed in detail in subsequent volumes, Chapter 11 is devoted to their characterization. Mössbauer spectroscopy is highlighted in this review.

Finally, Chapter 12 is devoted to nuclear magnetic resonance spectroscopy, which has recently become an established and powerful technique for the study of high-performance ceramics. This chapter also serves as an introduction to high-refractory-ceramics silicon carbide, silicon nitride, and related materials.

This volume represents the efforts of 19 researchers and practitioners who devoted their time and labor to producing this work. Contributors are to be considered responsible for the statements and information reported in their respective chapter. Heartfelt gratitude is extended to each of these individuals. A special thanks is extended to the staff of Marcel Dekker, Inc. for their fine production of this series and to Marcel Dekker and Graham Garratt for coordination and direction.

Nicholas P. Cheremisinoff

Contents

Contributors

Mustafa Akay Department of Mechanical and Industrial Engineering, University of Ulster at Jordanstown, Newtownabbey, N. Ireland

Ghazala Anwar Centre for Solid State Physics, University of the Punjab, Lahore, Pakistan

Keith R. Carduner Scientific Research Laboratories, Ford Motor Company, Dearborn, Michigan

Stephen L. Dieckman Materials and Components Technology Division, Argonne National Laboratory, Argonne, Illinois

William A. Ellingson Materials and Components Technology Division, Argonne National Laboratory, Argonne, Illinois

Joseph E. Grady Structures Division, NASA/Lewis Research Center, Cleveland, Ohio

Güngör Gündüz Kimya Mühendisliği Bölümü, Orta Doğu Teknik Üniversitesi, (Middle East Technical University), Ankara, Türkiye (Turkey)

Galen R. Hatfield Washington Research Center, W. R. Grace and Company, Columbia, Maryland

Olaf Jacobs* Technical University, Hamburg, Germany

Eiichi Jinen Mechanical and Systems Engineering, Kyoto Institute of Technology, Kyoto, Japan

B. V. Kokta Center for Research in Pulp and Paper, University of Quebec at Three Rivers, Quebec, Canada

It-Meng Low Applied Physics Department, Curtin University of Technology, Perth, Australia

Yiu-Wing Mai Centre for Advanced Materials Technology, Department of Mechanical Engineering, University of Sydney, Sydney, Australia

---

* Current affiliation: Deutsche Airbus GmbH, Bremen, Germany

Abdul Mateen Institute of Chemical Engineering and Technology, University of the Punjab, Lahore, Pakistan

Svetozar Musić Laboratory for the Synthesis of New Materials, Rudjer Bošković Institute, Zagreb, Yugoslavia

R. G. Raj Center for Research of Pulp and Paper, University of Quebec at Three Rivers, Quebec, Canada

Saadat Anwar Siddiqi Centre for Solid State Physics, University of the Punjab, Lahore, Pakistan

Lin Ye Department of Engineering Mechanics, Xian Jiaotong University, Xian, Shaanxi, China

S. Y. Zhang Institute of Mechanics, Chinese Academy of Sciences, Beijing, China

1

Fracture Toughness Testing of Polymer Matrix Composites

Joseph E. Grady
Structures Division
NASA/Lewis Research Center
Cleveland, Ohio

INTRODUCTION

The most common failure mode in laminated composite materials is interlaminar fracture, or *delamination* between plies of the laminate. In this chapter we describe the experimental methods and associated data analysis techniques that are used to measure the resistance to interlaminar fracture, or *fracture toughness*, of polymer matrix composite materials. In the first section, a brief background in the use of energy methods to characterize fracture behavior in elastic solids is given. This serves as the basis for the explanations of fracture toughness tests and failure criteria presented in the next section. Material properties of polymers can vary considerably with loading rate. In the third section we describe the rate-dependent fracture behavior observed in polymer composites. In the final section, an overview is given of the variety of approaches used in the design of delamination-resistant composite materials.

FRACTURE MECHANICS BACKGROUND

This section is divided into two parts. In the first part, the criterion for interlaminar fracture is given. Next, the various modes of fracture are illustrated.

Fracture Criterion

An elastic body under an applied load satisfies the energy balance [1,2]

$$W = U + T + D \tag{1}$$

where

W = external work done by applied loads
U = strain energy
T = kinetic energy
D = energy dissipated through fracture

A fracture of the structure causes an increase in crack surface area, as shown in Figure 1. The fracture also causes changes in external work, strain energy, kinetic energy, and fracture energy such that the energy balance condition (1) remains satisfied:

$$\Delta W = \Delta U + \Delta T + \Delta D \tag{2}$$

If the *fracture resistance* of the material, R, is defined as the energy dissipated in the process of generating a unit of new crack surface area,

$$R = \frac{\Delta D}{\Delta A} \tag{3}$$

and the energy release rate, G, is defined as

$$G = \frac{\Delta W}{\Delta A} - \frac{\Delta U}{\Delta A} \tag{4}$$

then from (2)–(4) we have

$$G - R = \frac{\Delta T}{\Delta A}$$

so the rate at which the cracked structure gains kinetic energy is determined by the excess crack driving force $(G - R)$. Under static loading conditions, the kinetic energy due to fracture is negligible, and the energy balance at fracture can be taken as

$$G - R = 0$$

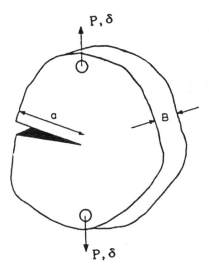

Figure 1 Cracked structure under load. (Adapted from Ref. 2.)

So fracture can occur when the energy release rate is equal to the fracture resistance of the material. If the critical energy release rate G_c is defined as the fracture resistance of the material, fracture occurs when the energy release rate reaches this critical value: when $G = G_c$. The critical value G_c is the *fracture toughness* of the material.

Compliance Method

Energy conservation principles can also be used [2] to express the energy release rate terms of the applied loading and elastic material properties of the cracked structure. This approach gives the result

$$G = \frac{P^2}{2B}\frac{dC}{da}$$

(5)

where a is the crack length and B is the crack width, as shown in Figure 1. The compliance, C, of the cracked structure is given by

$$C = \frac{\delta}{P}$$

where δ is the displacement produced at the loading point by the applied load, P.

Fracture Modes

A crack can propagate in any combination of the three modes shown in Figure 2. A mode I fracture is driven by the "crack-opening" action from the normal stress, σ_y, perpendicular to the crack plane. A shearing stress, σ_{xy}, parallel to the plane of the fracture will cause the crack to propagate with a mode II (in-plane shearing) type of deformation, and the shearing stress σ_{xz} will drive a mode III (out-of-plane shearing) crack extension. Experimental measurements have shown that the fracture resistance of most materials depends on the mode of loading at the crack tip. Therefore, separate material properties G_{Ic}, G_{IIc}, and G_{IIIc} are needed to characterize the fracture toughness of a particular material under different types of loading. We describe a series of test methods and data reduction procedures for measuring interlaminar fracture toughness of polymer matrix composites under mode I, mode II, and mixed mode I–mode II loading.

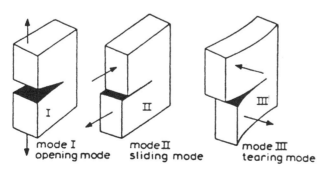

Figure 2 Fracture modes. (Adapted from Ref. 2.)

INTERLAMINAR FRACTURE TESTS AND DATA REDUCTION PROCEDURES

Mode I Loading

The double cantilever beam (DCB) specimen was originally used to measure the toughness of adhesive bonds between metals [3]. Since then it has been modified for use as a mode I interlaminar fracture toughness test specimen for composites. A typical composite DCB specimen is shown in Figure 3a. The delamination is usually initiated from an embedded notch by placing a thin nonadhesive film between two of the plies during layup [5]. This prevents those plies from bonding together in that area during the cure cycle, and creates an initial delamination, as shown in Figure 3.

If the two arms of the DCB specimen are considered to be beams cantilevered at the crack tip, the elastic compliance is given from simple beam theory as

$$C(a) = \frac{\delta}{P} = \frac{8a^3}{EBh^3} \tag{6}$$

where E is the flexural modulus of the two cantilevered arms [6] and the other parameters are as defined in Figure 3b. The energy release rate for the double cantilever beam is therefore given from (5) as

$$G_I = \frac{12P^2a^2}{EB^2h^3} \tag{7}$$

This is called the *beam model* of the DCB test specimen.

Crack Tip Compliance

Several modifications to the beam model have been used to model more closely the actual deformation of the DCB test specimen. These are outlined below. The first modification accounts for the flexural compliance at the crack tip.

The arms of the DCB test specimen are not rigidly clamped at the crack tip, as assumed in the beam model. In fact, there can be a rotation (θ) about the z-axis and a transverse

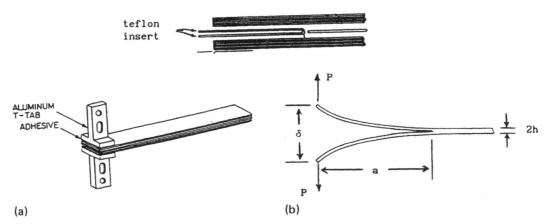

Figure 3 Double cantilever beam specimen: (a) test specimen geometry; (b) deformed shape. (Adapted from Ref. 4 and 5.)

displacement (δ) at the crack tip due to the compliant restraint. The assumption of a "fixed" boundary may therefore be more closely satisfied at some point ahead of the actual crack tip, where the rotation and displacement are negligible. In this case, an *effective crack length* [7] given by

$$a_{\text{eff}} = a + a^*$$

is used in (6) to account for the compliance at the crack tip. The correction term a^* is usually expressed in terms of the thickness, t, of the interply layer:

$$a^* = \beta t$$

where β is a constant chosen to match Eq. (6) with the measured compliance data. A value of $\beta = 0.37$ was used for DCB tests of adhesive bonds [7], but similar tests of a stiffer graphite–epoxy composite [8] found β to be negligibly small.

Shear Compliance

In highly orthotropic materials such as those considered here, there may be a significant amount of transverse shear deformation because the shear modulus is usually much lower than the in-plane moduli. When the shear compliance is considered, the total compliance of the DCB test specimen is given [6] by

$$C(a) = \frac{8a^3}{EBh^3}(1 + S)$$

where S is the shear correction factor:

$$S = \frac{3}{10}\frac{E}{G_{13}}\left(\frac{h}{a}\right)^2$$

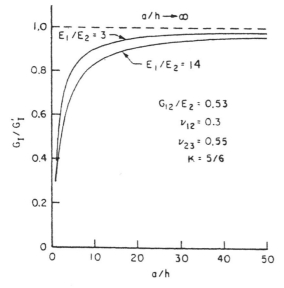

Figure 4 Energy release rate correction for beam theory. (Adapted from Ref. 9.)

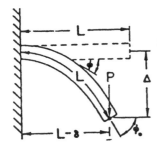

Figure 5 Large displacement of cantilevered beam. (Adapted from Ref. 10.)

and G_{13} is the transverse shear modulus, which is equal to the in-plane shear modulus, G_{12}, if the composite material is assumed to be transversely isotropic. In this case, the energy release rate is

$$G_I = \frac{12P^2}{EB^2h} \left[\left(\frac{a}{h} \right)^2 + \frac{1}{10} \frac{E}{G_{13}} \right]$$

(8)

The relative contribution of the shear term in (8) is expressed [6,9] in terms of the crack length and material orthotropy ratio, E_1/E_2. The shear correction increases the energy release rate by approximately 10% for a highly orthotropic material such as graphite–epoxy ($E_1/E_2 = 14$) when the crack length is relatively short ($a/h = 17$). As shown in Figure 4, the effects of shear compliance decrease for longer crack lengths and lower orthotropy ratios.

Large Displacements

If the compliance of the DCB specimen is too high, the linear elastic analysis described in this section may be inadequate to model the large displacements and rotations that occur under loading. A nonlinear elastic model of a cantilevered composite beam was therefore developed [10], as shown in Figure 5, in which arbitrarily large displacements were accounted for. The deformation of the cantilevered arms (and the resulting energy release rate) deviates significantly from that calculated using linear beam theory, as shown in Figure 6, for relatively large ($\delta/a > 0.3$) displacements.

Data Reduction Procedures: DCB Specimen

Load–displacement test data typical of a uniform double cantilever beam made from a brittle epoxy matrix composite are shown in Figure 7. Multiple cycles are shown of loading, stable crack propagation, and arrest, followed by elastic unloading. The unloading should always be elastic; that is, the displacement should return to zero when the load is removed. Permanent deformation of the test specimen would indicate that failure mechanisms other than interlaminar fracture have contributed to the energy dissipated by the structure during the loading cycle, and would therefore result in an erroneously high calculation of fracture toughness. In this section, several different techniques for determining the fracture toughness from load–displacement data like that shown in Figure 7 are presented.

Beam analysis method. The effects of transverse shear deformation and large displacements can be minimized by careful design of the DCB specimen, as discussed in

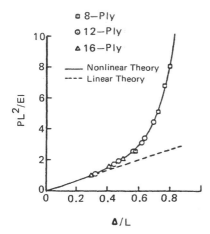

Figure 6 Nonlinear beam theory is required for $\delta/a \geq 0.3$. (Adapted from Ref. 10.)

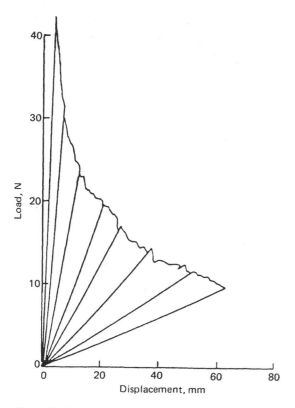

Figure 7 Typical results from a DCB test of a brittle (AS4/3501-6) matrix composite. (Adapted from Ref. 11.)

the following section. Assuming that these effects do not contribute significantly to the deformation of the test specimen, the simple beam model can be used to calculate the mode I fracture toughness, G_{Ic}, from load–displacement data such as those shown in Figure 7. This approach is called the *beam analysis method*. Combining Eqs. (6) and (7), we have [4]

$$G_I = \frac{3}{8B} \frac{P\delta}{a} \tag{9}$$

The mode I fracture toughness of the composite is therefore given by

$$G_{Ic} = \frac{3}{8B} \frac{P_c \delta_c}{a_c} \tag{10}$$

where P_c is the critical applied load that produces a separation δ_c of the cantilevered arms and causes the existing delamination, of length a_c, to extend. This expression is more useful than Eq. (7) for estimating fracture toughness from test data for two reasons. The flexural modulus, E, does not have to be estimated, either from test data or analysis, for Eq. (10), and the a^2 term in Eq. (7) is eliminated, thereby reducing the error in G_I for a given error in crack length measurement.

The fracture toughness can therefore be determined by measuring load, displacement, and crack length during a DCB test. A typical configuration for this test is shown in Figure 8.

Compliance calibration method. The empirical-based data reduction scheme described here was first used to measure the fracture toughness of unreinforced polymers [13] and later applied to composites [5,14]. In this approach the measured compliance is assumed to be of the more general form

$$C(a) = \frac{a^n}{K}$$

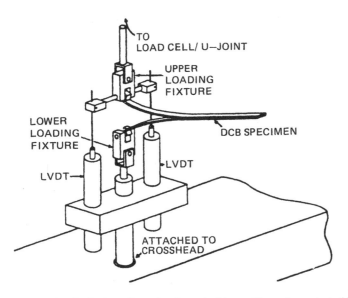

Figure 8 Typical configuration for a double cantilever beam test. (Adapted from Ref. 12.)

where the constants n and K are determined by curve fitting a plot of $\log(C)$ versus $\log(a)$. The energy release rate, from Eq. (5), is then

$$G_1 = \frac{nP\delta}{2Ba}$$

This empirical approach has an advantage in that it does not inherently assume that the beam bending model determines how the compliance varies with crack length. The effects of shear deformation, crack tip compliance, and/or large displacements will change the overall compliance of the test specimen and therefore will be accounted for automatically in the calculated curve-fit parameters, n and K.

Area method. A single load–crack extension–unload cycle for a brittle matrix composite DCB specimen is shown schematically in Figure 9. The applied load is increased until a load of P_1 is reached, which causes an existing crack (delamination) of length a_1 to extend. When the crack reaches a length a_2, its propagation is arrested. Due to the crack extension, the load drops to P_2 and the load point displacement increases to δ_2. The applied load is then removed. The area between the loading and unloading curves, designated ΔA, represents the decrease in stored strain energy caused by the crack extension. Given that

$$\Delta a = a_2 - a_1$$

is the distance the crack extended due to the load, then

$$G_c = \frac{\Delta A}{B \Delta a} \tag{11}$$

is the critical strain energy release rate required to cause crack extension; where $B\Delta a$ is the amount of new crack surface area generated by the crack extension. For the brittle fracture behavior depicted in Figure 9, Eq. (11) can be approximated [4] by

$$G_c = \frac{P_1\delta_2 - P_2\delta_1}{2B\Delta a} \tag{12}$$

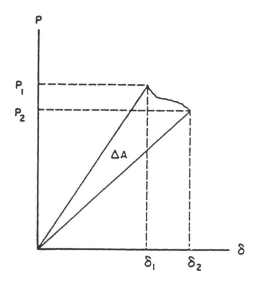

Figure 9 Single loading–unloading curve for a brittle material. (Adapted from Ref. 4.)

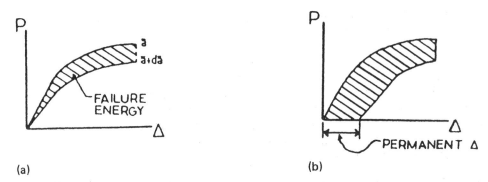

Figure 10 Nonlinear load–displacement behavior with inelastic deformation: (a) actual fracture energy; (b) area method calculation. (Adapted from Ref. 12.)

The advantage of this method is that fracture toughness is calculated from the difference in area between the two load–displacement curves (i.e, before and after crack extension). There is no requirement for the load–displacement curve to be linear. Therefore, Eq. (11) can be generalized for materials that have nonlinear elastic load–displacement curves, such as a composite material with a toughened (and therefore more ductile) matrix, like that shown in Figure 10. In this case, however, the simplification made in Eq. (12) for the linear case is not applicable, and a numerical integration scheme may be required to calculate the area between the loading and unloading curves.

***J*-integral method.**   The fracture energy released from a DCB specimen with a nonlinear load–displacement curve is shown in Figure 10a. If a permanent deformation were caused by the loading, the area method would overestimate the toughness of the material, as shown in Figure 10b. A different means of calculating the energy released due to fracture is therefore required when permanent deformation occurs in the material. The *J*-integral, defined in Ref. 15, is a measure of the energy available for crack extension and can be applied even when irreversible deformation occurs in the material. For the case of elastic deformation, the *J*-integral has the same value and meaning as the energy release rate; that is, $J = G$ for elastic deformation. An empirical method of calculating J for fracture toughness testing is given in Ref. 12 and outlined in Figure 11. In step 1, load–displacement

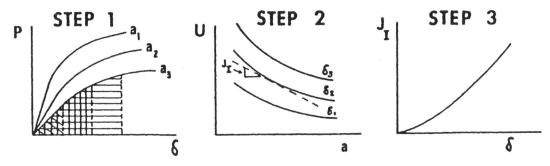

Figure 11 Experimental procedure for *J*-integral calculation. (Adapted from Ref. 12.)

curves are plotted for successive delamination lengths, $a_1 < a_2 < a_3$ as shown in the figure. The strain energy, U, is then calculated by integrating these curves:

$$U(\delta) = \int_0^\delta P(\zeta)\, d\zeta$$

and plotted as shown in the figure, using a curve fit to smooth the data. The J-integral is then calculated by differentiating the smoothed data

$$J = -\frac{1}{B}\frac{dU}{dA} \tag{13}$$

as shown in step 3. The fracture toughness of the material, J_{Ic}, is taken as the value of J from Eq. (13) evaluated at the critical displacement δ_c just before crack extension occurs.

Design Considerations: DCB Specimen

In this section, we provide some guidelines for the effective design of a mode I fracture toughness test and for a physically meaningful interpretation of the test results. Included are guidelines for choosing ply orientations and dimensions of the DCB test specimen based on the collective results presented above.

Hinged loading. If the load is applied to the test specimen through hinges adhesively bonded to the cantilevered arms as shown in Figure 12, the bending moment caused by overrestraining the arms is eliminated [16]. For this configuration, the effective delamination length (a) and specimen length (L) are defined from the location of the hinge, as shown in the figure.

Tapered width. The width-tapered DCB specimen geometry shown in Figure 13 was originally used in [18] and was designed to maintain a constant fracture load as the crack length increases. The compliance of the width-tapered DCB is given by Eq. (6); however, the width B of the specimen increases lineraly along the span;

$$B = ka$$

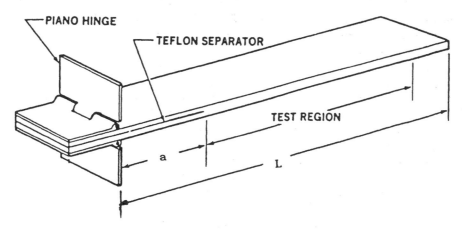

Figure 12 Hinged double cantilever beam. (Adapted from Ref. 16.)

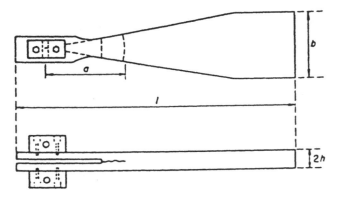

Figure 13 Width-tapered double cantilever beam. (Adapted from Ref. 17.)

where a denotes crack length and k is the taper ratio. The compliance of the test specimen is therefore given by

$$C(a) = \frac{8a^2}{Eh^3}\frac{1}{k}$$

and the energy release rate, from Eq. (5), is

$$G_1 = \frac{8P^2}{Eh^3}\left(\frac{1}{k}\right)^2$$

which is independent of crack length. The critical load, P_c, required to cause crack extension is therefore also independent of crack length. Similar results have been achieved [18,19] using DCB specimens tapered in the *thickness* direction.

Ply orientation. For several reasons, DCB test specimens are generally constructed from unidirectional 0-degree plies. High bending stiffness can minimize large deflections and the resulting analytical complexities they cause. In addition, the delamination will tend to follow the 0-degree plies [20] and grow in a self-similar manner at a 0/0 interface. This is not necessarily true in laminates with multiple-ply orientations [21], where the delamination may "wander" between several ply interfaces during loading and therefore violate the LEFM assumption of self-similar crack propagation. This can result in erroneously high measurements of apparent fracture energy [21]. To minimize the tendency for the delamination to wander between ply interfaces in specimens with multiple-ply orientations, symmetry should be used in the specimen design. The delamination should be located at the specimen midplane, and the layup should be symmetric about the midplane. In addition, the cantilevered arms of the DCB specimen should be symmetric about their respective midplanes, to avoid any twisting that would otherwise occur during loading [20] and to divert the crack path.

An additional factor to consider when choosing the ply orientations of a DCB test specimen is the effect of fiber bridging on the fracture resistance of toughened-matrix composites. When a delamination grows between two 0° plies, individual fibers can "bridge" the delamination, as shown in Figure 14, and significantly increase the apparent fracture toughness of the material [22–24]. This occurs in part because the fibers of similarly oriented plies can "nest" together and migrate into the neighboring ply when pressure

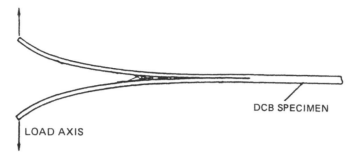

Figure 14 Fiber bridging in DCB specimen. (Adapted from Ref. 22.)

is applied during the cure cycle. The material shows an increasing resistance to fracture, or "*R*-curve" type of behavior, as the delamination grows and the bridged zone develops. Although this is a beneficial characteristic in practical applications of composite materials, it is undesirable in fracture toughness testing because the apparent fracture toughness of the material will vary with delamination length, fiber volume ratio, and test specimen geometry.

The fracture toughness data shown in Figure 15 were measured from a unidirectional C6000/Hx205 DCB specimen with the delamination at the 0/0 interface along the midplane [23]. The fiber bridging zone develops as the crack extends, tripling the initial fracture resistance of the material before reaching a plateau at a crack length of 100 mm. The fiber *nesting* across the delamination was then eliminated by orienting the two plies on either side of the delamination at small angles to each other such that the fibers from the

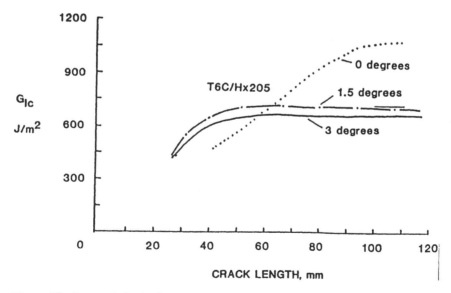

Figure 15 *R*-curve behavior in toughened-matrix composite due to fiber bridging. (Adapted from Ref. 23.)

two plies could not migrate across the ply interface during the cure cycle. Although the amount of fiber bridging (and the resultant *R*-curve behavior) is significantly reduced, neither one is completely eliminated, as shown in the figure. This was attributed [23] to the linking of the primary delamination, with secondary flaws originating near a neighboring ply interface within the crack tip *process zone*, as shown in Figure 16. This type of fiber bridging is most likely to occur in composites with toughened (and therefore more ductile) resins [26] because of the large process zone that develops near the delamination crack tip.

A significant decrease in *R*-curve behavior due to fiber bridging was noted when the thickness of the resin-rich interface layer was increased [24] and also when a thicker test specimen is used [27], which reduces the crack face separation and therefore decreases the amount of load that is supported by the bridging fibers.

From the discussion above, the following techniques can be used to estimate the actual mode I fracture toughness of a composite material that develops significant amounts of fiber bridging:

1. Reduce the amount of fiber nesting in the plies by either of the two alternate ply orientations discussed above; and/or

2. Use a thicker test specimen; and/or

3. Use the initial measured values of G_c before a significant amount of fiber bridging develops.

For brittle matrix composites, the mode I fracture toughness of the neat resin is a good estimate of the mode I interlaminar toughness of the composite [27], because of the small size of the process zone near the crack tip. The same is not true for the tougher resins, however [28].

Specimen thickness. The beam theory model assumes that the deformation of the DCB specimen is linear elastic. The thickness of the specimen should therefore be chosen such that the displacements of the cantilevered arms remain linear elastic throughout the entire

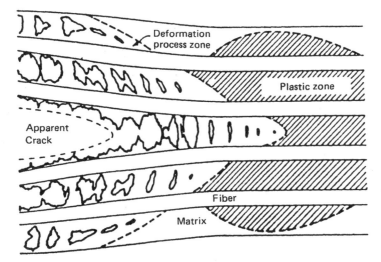

Figure 16 Crack tip process zone in toughened composite. (Adapted from Ref. 25.)

loading range, that is, until fracture occurs. As shown in Ref. 29, Eq. (6) and (7) can be combined to give

$$\frac{\delta_c}{a} = 4a\sqrt{\frac{G_{Ic}}{3Eh^3}}$$

where δ_c is the maximum displacement reached before crack extension. From Ref. 10, nonlinear displacements become significant for $\delta/a \geq 0.3$. To maintain linear elastic behavior to fracture, the test specimen must be designed such that

$$4a\sqrt{\frac{G_{Ic}}{3Eh^3}} \leq 0.3$$

Given a maximum allowable crack length (a_*) for the test, this can be accomplished by choosing h large enough. Solving for h gives

$$h^3 \geq \frac{16G_{Ic}a_*^2}{0.9E}$$

Thicker test specimens are therefore required for tougher materials, as shown in Figure 17.

Initial crack length. The effect of transverse shear varies with aspect ratio (a/h) and orthotropy ratio (E_1/E_2). Since crack extension increases the aspect ratio and therefore decreases the contribution of transverse shear deformation to the overall compliance of the test specimen, the shear compliance is greatest at the initial crack length. Therefore, the choice of a long enough initial delamination will ensure that the shear compliance is minimized over the entire range of crack lengths. In Figure 4, for example, it is shown that the simple beam theory would overestimate G_I by approximately 5% for a highly orthotropic material ($E_1/E_2 = 14$), with a crack-length aspect ratio (a/h) of 40. This corresponds to an initial notch length of approximately 2 in. for a 20-ply graphite–epoxy laminate.

Mode I fracture toughness measurements for a variety of different composites are given in Table 1 on page 39.

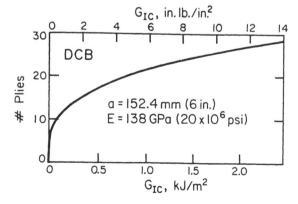

Figure 17 Minimum DCB specimen thickness required to maintain linear elastic behavior. (Adapted from Ref. 29).

Mode II Loading

The fracture resistance of composite materials is dependent on the mode of loading. Interlaminar fracture toughness under mode II loading is usually much greater than that under mode I, particularly for brittle epoxy matrix composites. A separate mode II test is therefore required to characterize interlaminar fracture behavior. In contrast to the case of mode I loading, there is no single test specimen geometry that is universally used to measure G_{IIc}. The two most commonly used tests for mode II fracture toughness measurements are therefore reviewed here. Mode II toughness measurements for a variety of different composites are given in Table 2 on page 43.

End-Notched Flexure Test

The end-notched flexure (ENF) specimen, shown in Figure 18, is the most frequently used specimen geometry to produce pure shear loading at the delamination crack tip. A compliance-based fracture mechanics approach is again used to express the mode II energy release rate in terms of the applied loading and the geometry of the test specimen. The elastic compliance was first derived in Refs. 41 and 42 using beam theory. For a unidirectionally reinforced laminate, the compliance is

$$C(a) = \frac{\delta}{P} = \frac{2L^3 + 3a^3}{8EBh^3} \tag{14}$$

where δ is the load-point displacement, $E = E_{11}$ is the modulus in the fiber direction, and the load P is applied at the midspan, as shown in Figure 18. Following the procedure outlined previously, the energy release rate, from Eq. (5), is

$$G_{II} = \frac{9P^2a^2}{16EB^2h^3} \tag{15}$$

where the energy release rate in this case is designated G_{II} because the crack is under pure shear loading. A finite element analysis was used in [43] to show that this "beam model" of the ENF specimen, Eq. (15), predicted G_{II} with less than 10% error for crack lengths $a/L \geq 0.5$.

Shear compliance. For large specimen aspect ratios (h/L) and highly orthotropic materials, the effect of shear deformation on the compliance of the ENF specimen is significant. When shear compliance is included, the overall compliance of the test specimen is [44]

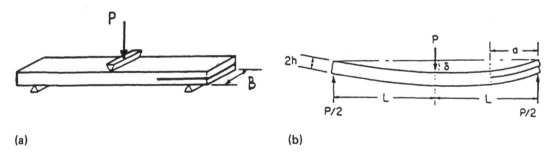

(a)　　　　　　　　　　　　　　　(b)

Figure 18 End-notched flexure specimen: (a) test specimen geometry; (b) deformed shape. (Adapted from Ref. 39 and 40.)

$$C(a) = \frac{2L^3 + 3a^3}{8EBh^3}(1 + S) \tag{16}$$

where S is the compliance due to shear deformation, given by

$$S = 2h^2 \frac{E_{11}}{G_{13}} \frac{1.2L + 0.9a}{2L^3 + 3a^3}$$

and G_{13} is the transverse shear modulus. The energy release rate is therefore

$$G_{II} = \frac{9P^2a^2}{16EB^2h^3}\left[1 + 0.2\frac{E_{11}}{G_{13}}\left(\frac{h}{a}\right)^2\right] \tag{17}$$

when shear deformation is accounted for.

Friction effects.   When the transverse load is applied to the ENF test specimen, the crack surfaces come into contact. Friction between the crack surfaces can retard crack growth by dissipating energy that would have otherwise been used as fracture energy. As the delamination extends and more fracture surface area is generated, this effect may become more pronounced. The reduced energy available for fracture is derived in [44] as

$$G_{II}(\mu) = G_{II} - \frac{3P^2\mu a}{4EB^2h^2} \tag{18}$$

where the last term is the energy dissipated through friction, and μ is the coefficient of friction between the contacting crack surfaces. Although the actual coefficient of friction is difficult to obtain, it was shown in Ref. 43 that frictional effects on the energy release rate are negligible for crack lengths $a/L \geq 0.5$ if $\mu \leq 0.3$.

Data Reduction Procedures: ENF Specimen

A typical load–displacement curve for an ENF test specimen made from a brittle matrix composite is shown in Figure 19. Several approaches are described here that can be used to determine G_{IIc} from this type of test data.

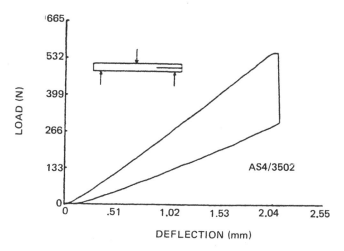

Figure 19   Linear elastic behavior of a brittle AS4/3502 ENF specimen. (Adapted from Ref. 40.)

Compliance calibration method. This approach is similar to the beam analysis method given earlier for the DCB specimen, and is described more thoroughly in Ref. 45. The compliance of the ENF specimen, given in Eq. (14), can be written as

$$C(a) = \left(\frac{3}{8EBh^3}\right)a^3 + \frac{2L^3}{8EBh^3}$$

$$= ma^3 + \text{constant}$$

(19)

where the parameter m is determined from a curve fit of the compliance versus crack length, as shown in Figure 20. Equations (14) and (15) can be combined to give

$$G_{II} = \frac{9P^2a^2}{2B(2L^3 + 3a^3)}C(a)$$

(20)

and the mode II fracture toughness is therefore

$$G_{IIc} = \frac{3P_c^2a^2}{2B}m$$

(21)

where P_c is the critical load that causes a delamination of length a to propagate and m is the effective measured value of the bracketed term in parentheses in Eq. (19), estimated from the slope of the plot in Figure 20. For a given experimental error in measurement of the delamination crack length, there will be more data scatter in the values of G_{IIc} calculated in this manner than there were for G_{Ic}, because of the a_c^2 term in the numerator of equation (21).

Modified compliance calibration method. A compliance calibration method that incorporates the effects of shear compliance is described in [47]. Equation (16) can be written as

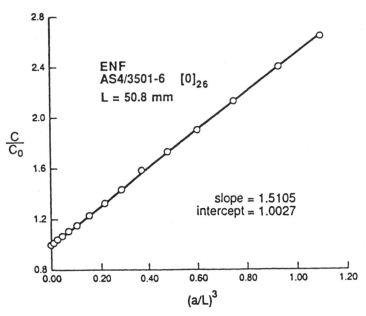

Figure 20 Compliance calibration curve for an AS4/3501-6 ENF specimen. (Adapted from Ref. 46.)

$$C(a) = \frac{2L^3}{8EBh^3} \left[1 + 1.2\gamma + 0.9\gamma\frac{a}{L} + 1.5\left(\frac{a}{L}\right)^3 \right] \qquad (22)$$

where $E = E_{11}$, and

$$\gamma = \left(\frac{h}{L}\right)^2 \frac{E_{11}}{G_{13}}$$

If C_0 is the compliance measured from an otherwise identical ENF specimen, but without a delamination,

$$C_0 = \frac{2L^3}{8EBh^3} (1 + 1.2\gamma) \qquad (23)$$

dividing (22) by (23) gives

$$\frac{C}{C_0} = A_0 + A_1\frac{a}{L} + A_3\left(\frac{a}{L}\right)^3 \qquad (24)$$

where the coefficients A_0, A_1, and A_3 are determined from a least-squares curve fit of the compliance versus crack length, normalized by the (constant) compliance C_0 of the uncracked ENF specimen. From Eq. (5), the energy release rate is

$$G_{II} = \frac{P^2C_0}{2BL} \left[A_1 + 3A_3\left(\frac{a}{L}\right)^2 \right] \qquad (25)$$

The fracture toughness, G_{IIc}, is calculated using Eq. (25) at the critical load $P = P_c$ just prior to crack extension.

Beam analysis method. Results of finite element analysis presented in Ref. 48 indicate that beam theory models will accurately predict the overall compliance of the ENF specimen if transverse shear effects are included. However, because they cannot account for the high shear stress near the crack tip, the beam theory models may underestimate G_{II} by 20 to 40%, depending on the test specimen dimensions and material orthotropy. Two nondimensional correction factors were derived to curve-fit the beam theory estimates of G_{II} to the finite element results:

$$\frac{G_{II}^{FE}}{G_{II}^{BT}} = \alpha + \beta \frac{E_{11}}{G_{13}} \left(\frac{h}{a}\right)^2$$

where the curve-fit parameters α and β are given in Ref. 48 for several typical test specimen dimensions, and the team theory estimate of G_{II} is determined from the measured compliance using Eq. (20).

Area method. A loading curve typical for an ENF test of a toughened-matrix composite is shown in Figure 21. Due to the nonlinear elastic deformation of the matrix, the area method should be used here to determine G_{IIc}. This approach was described earlier for data reduction of the DCB specimen.

Design Considerations: ENF Specimen

From the discussion presented above, several suggestions for the effective design of ENF test specimens can be determined. Variables that are considered include ply thickness, initial crack length, and specimen aspect ratio (a/L). In the cases considered here, a 0° unidirectional layup is assumed.

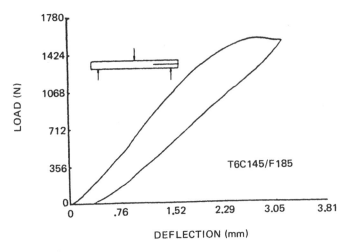

Figure 21 Nonlinear elastic behavior of a tough AS4/F-185 ENF specimen. (Adapted from Ref. 40.)

Initial crack length. If the data analysis method does not account for shear deformation, the dimensions of the test specimen should be chosen so as to minimize the shear contribution. As shown in Figure 22 and Eq. (17), the effect of shear compliance decreases with crack length. For highly orthotropic materials such as graphite–epoxy ($E_{11}/G_{13} \simeq 30$), shear compliance adds less than 2% to the apparent fracture resistance of the material for (a/h) $\geq$ 18. As shown in the figure, this would correspond to a crack length of approximately 2.5 cm for a 20-ply graphite–epoxy laminate.

Specimen thickness. Linear elastic material behavior and small deflection theory are assumed in the beam theory model of the ENF test specimen. The dimensions of the specimen must be chosen such that these assumptions will be valid. Based on the material linearity criterion, a minimum thickness requirement is derived in [44] as

$$h \geq \frac{L^2 G_{\mathrm{IIc}}}{a^2 \epsilon^2 E_{11}}$$

where ϵ is the maximum allowable strain at which the deformation is linear elastic.

Based on the small deflection criterion, a second thickness requirement is derived [44]:

$$h^3 \geq \frac{G_{\mathrm{IIc}}(L^2 + 3a^2)^2}{4E_{11}a^2(y')^2}$$

where y' is the maximum allowable slope, chosen so as to minimize the error induced by using the small deflection beam theory.

These two thickness requirements are used to determine the minimum number of plies to use for an ENF specimen. As shown in Figure 23, the small deflection requirement determines the minimum thickness for brittle materials, but the material linearity requirement determines the thickness for tougher materials.

Specimen length. In ENF tests, the specimen length is generally chosen such that a/L = 0.5, so that the crack tip is initially halfway between the loading point and the support

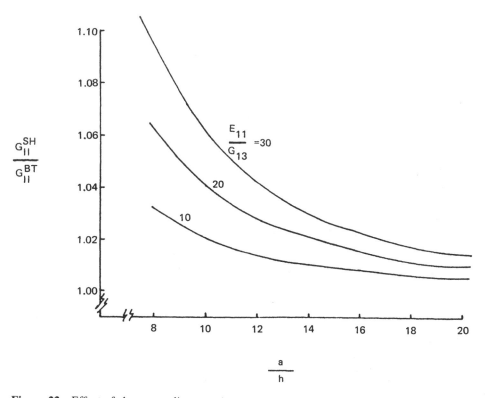

Figure 22 Effect of shear compliance on beam theory calculation of G_{II} from an ENF test.

pin (Figure 18). This arrangement can be expected to minimize the frictional effects at the crack tip and was shown [41] to induce an error of less than 2% in the measured value of G_{IIc} due to crack surface friction, for AS1/3501–6 composites.

An analysis of the crack stability [44], however, showed that crack growth is unstable under fixed-grip conditions for $a/L < 0.7$, indicating that the specimen length should be chosen such that $a/L \geq 0.7$ to produce stable crack growth.

Precracking. The initial crack is usually introduced into the laminate by placing a thin ($\simeq 1$ mil) nonadhesive insert between the two midplane plies before curing, as shown in Figure 3. The crack tip at the end of the insert will be blunt, however, and will therefore result in higher measurements of G_{IIc} [42,46,47] than would be obtained from mode II delaminations that occur naturally from high interlaminar shear stresses. To produce a sharper crack tip, the ENF specimen is given either a static three-point bending load [42] or a low-amplitude cyclic load [47] sufficient to cause a small extension of the original embedded delamination. The natural crack extension caused by this "precracking" procedure will have a sharper crack tip and will therefore usually result in lower and more consistent measurements of G_{IIc}.

End-Loaded Split Test

A second, less frequently used test for pure mode II loading is the end-loaded split (ELS) specimen, which is similar to the double cantilever beam specimen but has different

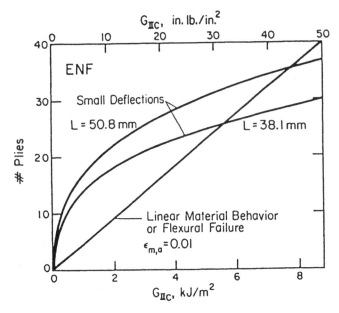

Figure 23 Minimum ENF specimen thickness required to maintain linear elastic behavior. (Adapted from Ref. 44.)

loading and boundary conditions, as shown in Figure 24. In Ref. 49 the elastic compliance is derived as

$$C(a) = \frac{L^3 + 3a^3}{2E_{11}Bh^3} \tag{26}$$

the mode II energy release rate is therefore given from Eq. (5) as

$$G_{II} = \frac{9P^2a^2}{4E_{11}B^2h^3} \tag{27}$$

Data Reduction Procedures

The compliance and energy release rate are similar to those of the ENF specimen, so similar data reduction procedures are used for calculating G_{IIc} from the test data.

Beam analysis method. The mode II energy release rate in Eq. (27) can be written as

$$G_{II} = \frac{9P^2a^2}{2B(L^3 + 3a^3)}C(a)$$

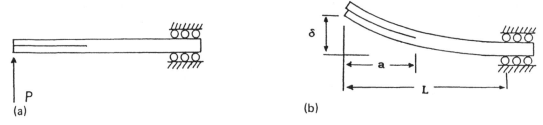

Figure 24 End-loaded split test specimen: (a) test specimen geometry; (b) deformed shape.

where $C(a) = \delta/P$ is the measured compliance. The fracture toughness is determined by calculating G_{II} at the critical load, P_c, that causes crack extension.

Compliance calibration method.   The compliance of the ELS specimen, given in Eq. (26), can be written as

$$C(a) = \frac{3}{2EBh^3}a^3 + \frac{L^3}{2EBh^3}$$
$$= ma^3 + \text{constant} \tag{28}$$

and the parameter m can be measured from a curve fit of the compliance versus a^3, as shown in Figure 20 for the ENF specimen. Equations (5) and (28) can be combined to give the mode II fracture toughness:

$$G_{IIc} = \frac{3P_c^2 a^2}{2B}m$$

where P_c is the critical load that causes a delamination of length a to propagate. Tests show [40] that there is more scatter in G_{IIc} measurements using the ELS specimen than there is for the ENF specimen because the crack length is harder to measure accurately.

Mixed-Mode Loading

Delamination can originate in mode I due to transverse tensile stress, in mode II due to in-plane shear stress, in mode III due to out-of-plane shear stress, or in any combination of modes I, II, and III due to a general, combined stress state at the ply interface. In actual structural applications, composite materials are subjected to a combined stress state. It is therefore necessary to evaluate the fracture resistance of composites under mixed-mode loading conditions, and to develop an appropriate *mixed-mode failure criterion* that describes the "failure envelope" for a given material under mixed-mode loading.

This section is divided into two parts. In the first part, two tests that produce mixed mode I–mode II loading conditions are presented and discussed. In the second part, an overview of mixed-mode failure criteria which are used to quantify interlaminar fracture resistance under combined mode I–mode II loading is presented.

Cracked-Lap Shear Test

The cracked-lap shear (CLS) specimen was originally developed for testing adhesive bonds between metals [7] and was first used to evaluate mixed-mode fracture toughness of composites in [50]. Progressive debonding of composite adhesive joints during fatigue loading is evaluated using CLS specimens in [32,37,51]

A typical composite CLS specimen is shown in Figure 25. The mode I component of the loading is due to the bending moment induced at the crack tip by the slightly eccentric load path. The compliance of the CLS specimen is given [50] by

$$C(a) = \frac{1}{EBh_1}\left[2L + a\left(\frac{h_1}{h_2} - 1\right)\right] \tag{29}$$

where h_1 and h_2 are the thickness of the two sections of the test specimen, as shown in Figure 25, $E = E_{11}$ is the stiffness in the fiber direction, and the other variables have the same definitions as used previously. From Eq. (5), the total energy release rate is given as

$$G_{\text{total}} = G_I + G_{II} = \frac{P^2}{2EB^2}\frac{h_1 - h_2}{h_1 h_2} \tag{30}$$

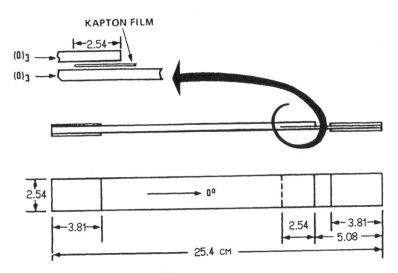

Figure 25 Cracked-lap shear test specimen. (Adapted from Ref. 50.)

The critical *total* energy release rate at the onset of crack extension is therefore determined by calculating G_{total} from Eq. (30) at the critical load P_c just before crack extension occurs. Some additional analysis is required to determine the relative amounts that G_I and G_{II} contribute to the total energy release rate. One means of accomplishing this is with finite element analysis. For the CLS specimen, a geometrically nonlinear finite element analysis is required to determine the proportions of mode I and mode II loading at the delamination crack tip [52] due to the out-of-plane displacements and rotations that result from the eccentric load path. Using a geometrically nonlinear finite element analysis with a local crack-closure method [53] to calculate the energy released during crack extension, it was shown in Ref. 50 for a unidirectional graphite–epoxy CLS specimen with a midplane delamination ($h_1 = 2h_2$) that the loading was 23.5% mode I; that is,

$$\frac{G_I}{G_{total}} = 0.235$$

and that this ratio is independent of crack length. An explicit solution for G_I in terms of the bending moment at the crack tip is given in Ref. 7. These calculations are supported by the results of tests that showed the loading to be approximately 20% mode I [42] for unidirectionally reinforced CLS specimens with midplane delaminations. The ratio of G_I to G_{II} can be varied by changing the relative thickness, h_1 and h_2, of the two sections [7,52].

Data Reduction Procedures: CLS Specimen

Two approaches that have been used to calculate the mixed-mode fracture toughness from CLS test data are described below.

Compliance calibration method. Equations (29) and (30) indicate that the compliance of the CLS specimen varies linearly with crack length, and that the critical load P_c that causes delamination is independent of crack length. This was verified by test measurements [37,50] and finite element analysis [52]. The slope calculated from a linear curve fit of

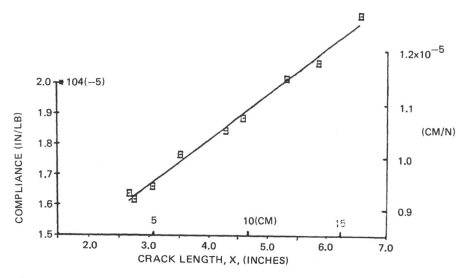

Figure 26 Compliance calibration curve for CLS specimen. (Adapted from Ref. 50.)

the compliance versus crack length can be used to calculate the total critical energy release rate using Eq. (5):

$$G_c^{\text{total}} = \frac{P_c^2}{2B} \frac{dC}{da}$$

where the derivative is taken as the slope of the data in Figure 26, and the critical load, P_c, is determined from the same load versus displacement data, as shown in Figure 27.

Strain gauge method.   Strain gauge measurements from a CLS specimen can be used to determine mixed-mode fracture toughness [42] in the following manner. The load P produces axial strains ϵ_1 and ϵ_2 in the two sections of the specimen such that

$$P = E_1 A_1 \epsilon_1 = E_2 A_2 \epsilon_2$$

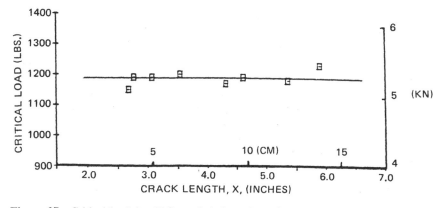

Figure 27 Critical load for CLS test is independent of crack length. (Adapted from Ref. 50.)

where E_i is the longitudinal modulus (E_{11}) in each of the sections, and $A_i = Bh_i$ are the cross-sectional areas. Therefore,

$$E_i Bh_i = \frac{dP}{d\epsilon_i} \qquad (i = 1 \text{ and } 2)$$

and Eq. (30) can be written

$$G_I + G_{II} = \frac{P^2}{2B}\left(\frac{d\epsilon_2}{dP} - \frac{d\epsilon_1}{dP}\right)$$

where the derivatives are determined from plots of strain versus load for each of the two sections.

Design Considerations: CLS Specimen

The CLS specimen can fail either in tension or by delamination. The specimen must therefore be designed such that the tensile failure load is higher than the load required to cause delamination. Following Ref. 54, we have, from Eq. (30),

$$P_c = B\sqrt{2Eh^*G_c} \tag{31}$$

where P_c is the critical load required to cause delamination, and

$$h^* = \frac{h_1 h_2}{h_1 - h_2}$$

To ensure that the load required for delamination is lower than that for tensile failure, the condition

$$P_c Bh_2 < S_{11} \tag{32}$$

must be satisfied, where S_{11} is the longitudinal tensile strength of the laminate. Substituting (31) into (32) gives the design requirement that the thickness of the CLS specimen be determined by

$$h_2\left(1 - \frac{h_2}{h_1}\right) > \frac{2EG_c}{(S_{11})^2}$$

to avoid tensile failure.

Edge-Delamination Tension Test

Edge delaminations are likely to initiate in a composite laminate under uniaxial tension due to the high interlaminar stresses near the free edge [55,56]. In Ref. 57, free-edge delamination specimens were designed by choosing the laminate ply orientations such that the Poisson's ratio mismatch between plies was maximized, making it susceptible to free-edge delamination. An approximate stress analysis of a $[\pm\theta_1,\theta_2]_{sym}$ HTS/ERLA 2256 graphite–epoxy laminate under uniaxial tension was used to show that values of $\theta_1 = 25°$ and $\theta_2 = 90°$ would maximize the interlaminar normal stress at the midplane near the free edge of the laminate, therefore initiating a pure mode I edge delamination. A mixed-mode delamination could initiate at the $-25/90$ interface [58], however, depending on the values of mode I and mixed-mode fracture toughness.

A similar approach was used in designing the edge-delamination tension (EDT) specimen [59–61] to estimate the mixed-mode fracture toughness for a T300/5208 composite. Interlaminar tensile stresses in an 11-ply $[(\pm30)_2,90,\overline{90}]_{sym}$ laminate cause free-edge de-

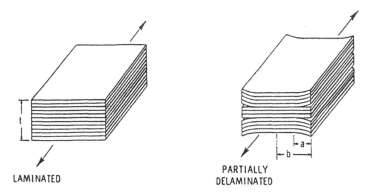

Figure 28 Edge-delamination test specimen before and after onset of delamination. (Adapted from Ref. 59.)

laminations to initiate under uniaxial tension at the $-30/90$ interfaces, as shown in Figure 28. As the load is progressively increased and the delaminations grow inward from the edges, radiographic examinations are used to measure the delaminated area. As shown in Figure 29, the measured delaminations are approximated by rectangles of equivalent area, so the stiffness after delamination can be estimated by

$$E = E_{\text{lam}} + (E^* - E_{\text{lam}})\frac{a}{b} \qquad (33)$$

where the rectangular delaminations have uniform length a and the width of the laminate is $2b$. The delaminations are assumed to grow symmetrically at both $-30/90$ interfaces. The delaminated ply groups in Figure 28 are also assumed to act independently in supporting the applied load, so that

$$E^* = \frac{8E_{\pm 30} + 3E_{90}}{11}$$

is the laminate stiffness in the loading direction for the fully delaminated case ($a = b$), and E_{lam} is the stiffness before delamination.

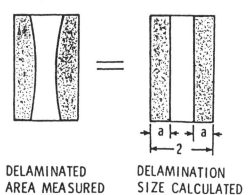

DELAMINATED
AREA MEASURED

DELAMINATION
SIZE CALCULATED

Figure 29 Idealization of edge delamination shape for calculation of equivalent crack length. (Adapted from Ref. 59.)

The energy release rate is determined by using Hooke's law in Eq. (10) and assuming that the work term vanishes due to fixed grip conditions. The total energy release rate is then approximated by

$$G_I + G_{II} = -V\frac{\epsilon^2}{2}\frac{dE}{dA}$$

where $V = 2blt$ is the volume of the test specimen, ϵ is the nominal strain in the loading direction, and the derivative represents the stiffness loss per unit crack area, obtained by differentiating Eq. (33). This gives

$$G_I + G_{II} = \frac{\epsilon^2 t}{2}(E_{lam} - E^*) \tag{34}$$

so the mixed-mode fracture toughness is determined using Eq. (34) and the critical strain level ϵ_c at which crack initiation occurs. This is indicated by the onset of nonlinear load–displacement behavior [61]. Equation (34) indicates that the energy release rate is independent of crack length. Finite element analysis [61] showed that the delamination is 57% mode I; that is,

$$\frac{G_I}{G_I + G_{II}} = 0.57$$

for the EDT specimen with this particular material and layup, although mode I percentages ranging from 22 to 90% have been achieved for T300/5208 EDT specimens [62] by varying the ply layup.

Examination of failed EDT specimens show that the multiple delaminations do not grow symmetrically; nor do they grow in a self-similar manner. Considerable intraply cracking occurs in the 90° plies as the cracks wander between the two $-30/90$ interfaces. A modified free-edge delamination test was therefore devised [63] to remedy these problems. In this case, nonadhesive inserts are used to initiate pure mode I delaminations along the midplane at the free edges of a $[\pm30/\mp30/90_2]_{sym}$ graphite–epoxy laminate. The inserts serve to promote self-similar crack growth between the two 90° plies and therefore eliminate any wandering of the crack that may otherwise occur. Thermal residual stresses are also shown to increase the calculated mode I fracture toughness [63,64] by approximately 15%, and should therefore be accounted for in the data reduction process. Measured values of G_{Ic} are considerably lower than those measured in double cantilever beam tests, however, for reasons that are unclear. Mixed-Mode toughness measurements for a variety of different composites are given in Table 3 on page 45.

Mixed-Mode Fracture Criteria

In this section we present several different criteria that have been used to predict the onset of delamination under mixed-mode loading conditions. One approach is to assume that crack growth occurs when the total energy release rate reaches a critical value:

$$G_I + G_{II} = G_c$$

This approach was used to predict the progressive interlaminar fracture at three different ply interfaces of a unidirectional glass–epoxy material [65,66] under a monotonically increasing tensile load, as shown in Figure 30. The analysis was performed by incorporating the failure criterion above into a finite element program and using singular elements to model the stresses near the crack tip.

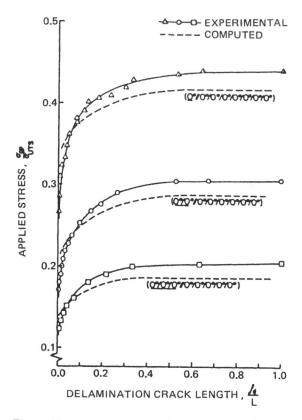

Figure 30 Mixed-mode delamination growth calculated with total energy release rate criterion. (Adapted from Ref. 66.)

A general mixed-mode failure criterion should be able to account for the observed differences in interlaminar fracture toughness with fracture mode, however. In brittle matrix composites, G_{IIc} can be as much as 10 times greater than G_{Ic} [28,39]. This difference is accounted for by including the values of G_{Ic} and G_{IIc} explicitly in the failure criterion. The modified form

$$\left[\frac{G_I}{G_{Ic}}\right]^n + \left[\frac{G_{II}}{G_{IIc}}\right]^n = 1 \tag{35}$$

was therefore proposed [32,42] as a general criterion for interlaminar fracture under an arbitrary mixed-mode loading ratio, G_I/G_{II}.

In Ref. 39, test data for three different epoxy–matrix composites over a range of mixed-mode loading ratios was used to determine an appropriate value for the exponent in Eq. (35). The data in Figure 31 were taken from mixed-mode fracture tests of three different types of composites. The matrix materials have widely different ductilities, and therefore the test results illustrate the wide range of fracture toughness that different epoxy–matrix composites can display.

Fracture toughness measurements are shown in Figure 31a for a brittle (Narmco 5208) epoxy-matrix composite, in Figure 31b for a more ductile (Hexcel Hx205) epoxy matrix composite with an extended polymer chain, and in Figure 31c for a composite with an

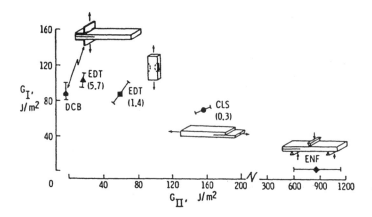

(a)

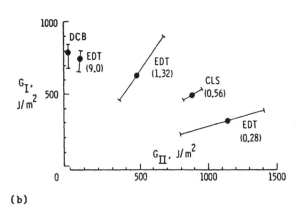

(b)

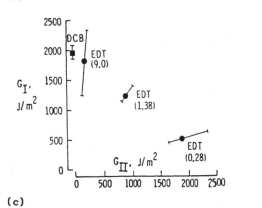

(c)

Figure 31 Interlaminar fracture toughness depends on crack propagation mode and matrix toughness: (a) fracture toughness of 5208 epoxy matrix composites; (b) fracture toughness of Hx205 epoxy matrix composites; (c) fracture toughness of F-185 epoxy matrix composites. (Adapted from Ref. 39.)

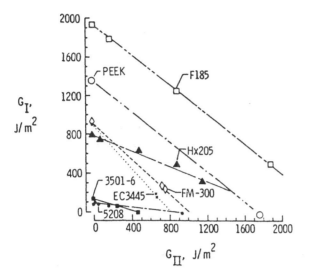

Figure 32 Fracture toughness calculated with linear mixed-mode failure criterion. (Adapted from Ref. 39.)

extended chain epoxy matrix (Hexcel F-185), modified with rubber additives for increased ductility and toughness. In all cases, a linear curve fit can be used to approximate the test data, as shown in Figure 32. This would suggest that an exponent of $n = 1$ can be used in the mixed-mode failure criterion, Eq. (35), regardless of material. This same failure criterion (with exponent $n = 1$) was also shown in Refs. 42 and 67 to accurately predict the initiation of delamination crack extension in AS1/3501–6 fracture specimens under a variety of mixed-mode loading ratios.

The results of a separate series of fracture tests [28] with a wide variety of different material systems are shown in Figure 33. In this case, the "failure envelopes" cannot all be represented by linear approximations, however; indicating that the exponent n in Eq. (35) may in fact be a variable, dependent on the particular fiber–matrix system.

RATE-DEPENDENT BEHAVIOR

Most of the research on rate-dependent fracture in composites has considered mode I loading, using the double cantilever beam test. The results presented and discussed in this section are therefore for mode I loading only, except where otherwise noted. Although the variation of fracture toughness with loading rate is most appropriately expressed in terms of crack propagation velocity, the most frequently used measure of loading rate is the speed at which the opening displacement is imposed on the DCB test specimen, as determined from the cross-head displacement rate on the test machine. Therefore, the opening displacement rate is used here as a common measure of loading rate, to compare the results of different tests.

Brittle-Matrix Composites

Most research in rate-dependent fracture behavior of composites has been on brittle-matrix materials. The earliest work in this area is reported in Ref. 10. E-glass/epoxy DCB specimens were tested over the range of opening displacement rates $5 \times 10^{-3} \leq \dot{\delta} \leq 5$

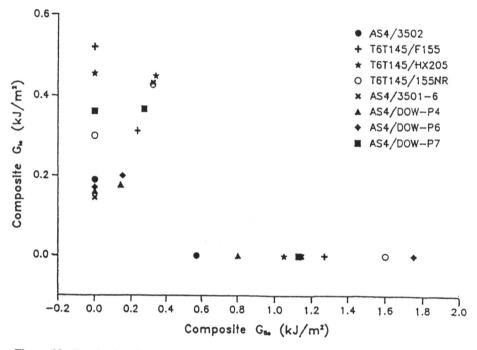

Figure 33 Interlaminar fracture toughness test data. (Adapted from Ref. 28.)

mm/s. It was shown that G_{Ic} almost doubled over this range of loading rates, with produced crack propagation speeds ($\dot{a}$) up to 1 mm/s. Calculating crack propagation speed from the cross-head displacement rate, they found the data to fit the trend

$$G_{Ic} = K\dot{a}^n \tag{36}$$

with the constants $K = 1288$ J/m^2 and $n = 0.1$, as shown in Figure 34. Results presented in Ref. 11 for AS4/3501–6 graphite–epoxy over a similar range of loading rates show no significant variation in G_{Ic} from the static fracture toughness. Fracture resistance increases at higher loading rates for the same material, however. A 28% increase in G_{Ic} was measured for AS4/3501–6 composites [68] for opening displacement rates in the range $0.009 \leq \dot{\delta} \leq 8.5$ mm/s, which produced crack propagation velocities up to 51 mm/s. The data follow a similar trend to that shown in Figure 34 and equation (36), with the parameters $K = 210$ J/m^2 and $n = 0.035$ determined from the curve fit. Similarly, a 25% increase in G_{Ic} was measured in tests of C6000/PMR-15 composites [22] over approximately the same range of loading rates.

Crack propagation speeds up to 26 m/s were obtained [19] in AS4/3501–6 composites using a double cantilevered beam specimen tapered in the height direction to have slightly decreasing compliance with crack length, to eliminate the intermittent crack-arrest (''slip-stick'') phenomenon. The test data, shown in Figure 35, were represented by a third-order curve fit of the form

$$\log G_{Ic} = \sum_{n=0}^{N} A_n (\log \dot{a})^n$$

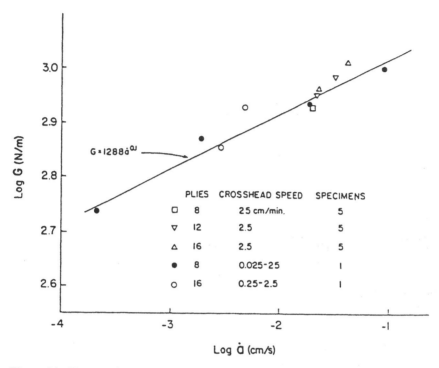

Figure 34 Energy release rate varies with crack speed for E-glass/epoxy DCB test. (Adapted from Ref. 10.)

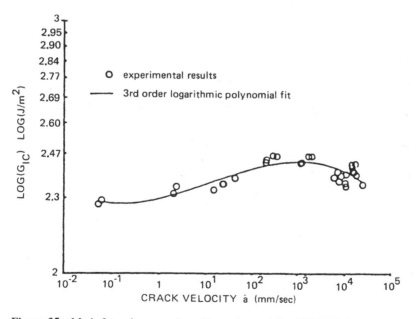

Figure 35 Mode I toughness varies with crack speed for AS4/3501-6 composite. (Adapted from Ref. 19.)

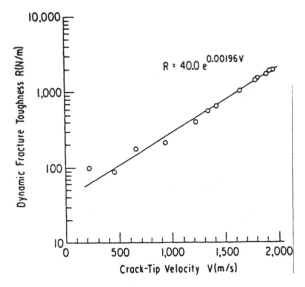

Figure 36 Mode I toughness for AS4/3501-6 at high crack speeds. (Adapted from Ref. 69.)

The fracture toughness increases with crack propagation velocity until it reaches its maximum value at $\dot{a} \simeq 1$ m/s, and decreases thereafter. The maximum G_{Ic} is approximately 46% higher than the fracture toughness measured under static loading conditions.

At very high loading rates, the fracture toughness of 90° AS/3501–6 laminates was shown [69] to increase exponentially with crack speed [70]:

$$G_{Ic} = C_1 \exp(C_2 \dot{a})$$

as shown in Figure 36.

Toughened-Matrix Composites

Relatively little research has been performed on the effects of loading rate on fracture in toughened-matrix composites, but the available data suggest that the mode I fracture toughness of the composite is determined to a large extent by the viscoelastic nature of the matrix.

Viscoelastic behavior of a neat elastomer-toughened epoxy causes the variation of mode I fracture energy with temperature and loading rate shown in Figure 37. At high loading rates, the matrix behaves in a brittle manner, and the fracture energy decreases to that of the unmodified, brittle epoxy because the crack tip deformation zone has less time to develop [11], so the material cannot redistribute the high crack tip stresses prior to fracture.

Composites with a toughened matrix also exhibit rate-dependent fracture behavior, although not to the extent observed in the neat matrix, because of the constraint on matrix deformation imposed by the fibers. Hexcel F-185 is an epoxy resin with CTBN rubber additives to increase ductility and toughness. In Ref. 8, the mode I fracture toughness of a T300/F-185 composite was measured at different loading rates by varying the displacement rate in DCB specimens over the range $0.0085 \le \dot{\delta} \le 8.5$ mm/s. Maximum crack propagation velocities were estimated at 21 mm/s from strain gauge measurements. The mode I fracture toughness decreased by 20% over this range of crack velocities, probably

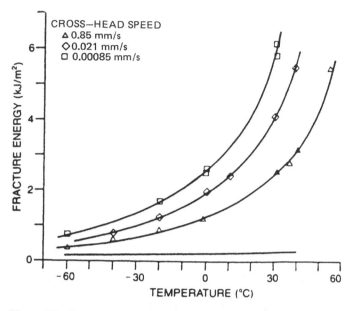

Figure 37 Fracture energy of a toughened epoxy varies with temperature and loading rate. (Adapted from Ref. 71.)

the result of a progressively decreasing size of the crack tip deformation zone, causing the composite to exhibit a more brittle fracture behavior at higher loading rates, as illustrated in Figure 38. The data followed a trend described by Eq. (36), where the values $n = 0.027$ and $K = 1.63$ kJ/m^2 were determined from a curve fit of the test data, shown in Figure 39.

At lower loading rates, no rate sensitivity in G_{Ic} is apparent. Three different graphite

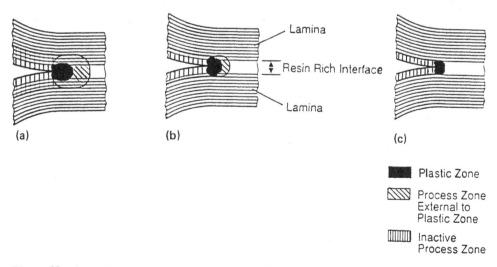

Figure 38 Crack-tip process zone development at (a) low loading rates, (b) intermediate loading rates, and (c) high loading rates. (Adapted from Ref. 11.)

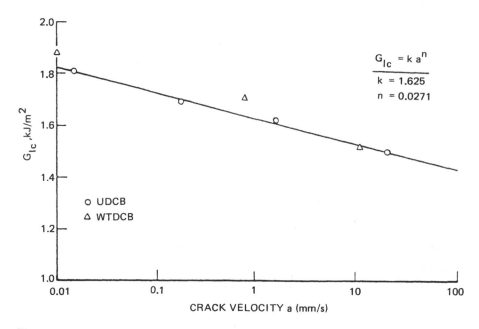

Figure 39 T300/F-185 fracture toughness decreases with loading rate. (Adapted from Ref. 8.)

fabric–epoxy matrix composites with varying amounts of elastomer additives were tested over a range of loading rates [72], by varying the displacement rate in DCB tests within the range $5 \times 10^{-4} < \delta < 1$ mm/s. Test results showed no significant variation in G_{Ic}.

Mode II Loading

Dynamic fracture toughness measurements under mode II loading are particularly difficult to make, because of the difficulty in measuring crack length and load under dynamic conditions in mode II specimens. A combined experimental–numerical approach is used in Refs. 73 and 74 to estimate G_{IIc} for T300/934 graphite–epoxy composites using cantilevered beam specimens. Finite element analysis [73] has verified that a symmetric laminate with a through-the-width delamination embedded along the midplane, as shown in Figure 40, would produce nearly pure mode II deformation at the crack tip under transverse impact loading.

A series of identical $[0/90]_{5s}$ cross-ply test specimens are impacted over a range of velocities, and the postimpact crack lengths are measured using ultrasonic C-scans. The critical level of impact energy required to cause a small extension of the initial embedded delamination is determined from a plot of postimpact delamination crack length, as shown in Figure 40. A finite element analysis of the test specimen impacted at the critical energy [74] is then used to calculate the time-dependent energy release rate, as shown in Figure 41. Since the analysis corresponds to the critical case in which the impact energy is just that required to cause a small extension of the initial crack, the maximum energy release rate should be equal to the fracture toughness, G_{IIc}, for the material. This was verified in Ref. 74 by showing that G_{IIc} determined in this manner was a material property, independent of delamination crack length.

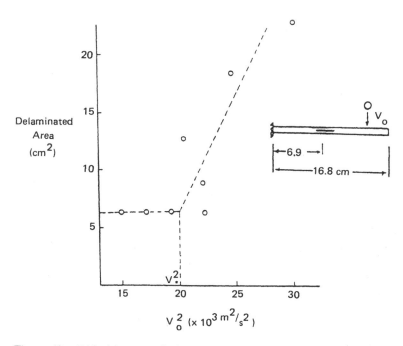

Figure 40 Critical impact velocity (V_*) is determined from postimpact measurements of delamination length. (Adapted from Ref. 73.)

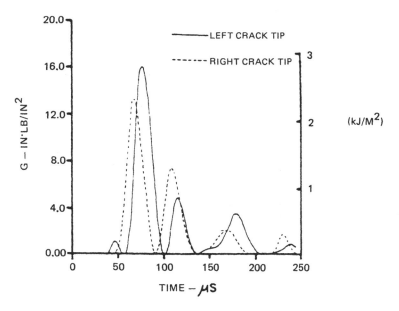

Figure 41 G_{IIc} is calculated from finite element analysis of impact test. (Adapted from Ref. 74.)

DESIGN OF TOUGH COMPOSITES

A variety of approaches have been used to design composite laminates for high interlaminar fracture resistance. Several of the most frequently used methods will be discussed here.

Matrix Properties

The fracture toughness of the neat resin is the most significant variable affecting the interlaminar fracture toughness of the composite. The primary factors that determine the fracture toughness of a polymer are its ductility and the extent of cross-linking in the polymer chain [1,34,75]. Matrix toughness increases with ductility and decreases with the amount of cross-linking. For example, Hercules 3502 is a highly cross-linked, brittle epoxy with mode I fracture toughness $G_{Ic} \simeq 70$ J/m^2 (Table 1), while Hexcel Hx205 has lower cross-link density [76] and a fracture toughness of 230 to 340 J/m^2. Resin ductility also increases fracture toughness. F-185 epoxy is the same as Hx205 except for the addition of 13.7% (by weight) carboxy-terminated butadiene acrylonitride (CTBN) rubber particles to increase ductility [17,39]. The data in Table 1 for the mode I fracture toughness of toughened F-185 epoxy have been measured at 5 to 8 kJ/m^2. Thorough discussions of the particle sizes and matrix properties required to achieve the optimum toughening by the addition of rubber particles to the neat resin is given in Refs. 34, 39 and 77. In Refs. 39 and 51 it is shown that toughening an epoxy with rubber additives does not increase G_{IIc} nearly as much as G_{Ic}. This is evident from the trend observed in Figure 31 for the three material systems with progressively increasing toughness, and is the result of the lack of matrix dilatation that is required near the crack tip under mode II loading [28,39].

Composite interlaminar fracture toughness varies in a complex way with neat resin toughness. For brittle polymers ($G_{Ic} \leq 200$ J/m^2), the fracture toughness of the composite is usually two to three times greater than the neat resin G_{Ic}. This difference has been attributed [30] to the additional energy-absorbing mechanisms of fiber pullout and fiber breakage that can occur in the composite. For the tougher polymers ($G_{Ic} > 200$ J/m^2), however, the size of the crack tip plastic zone in the composite is limited by the constraining effect of the surrounding fibers [28,30,77] as shown in Figure 42. This limits the ability of the matrix to redistribute the high stresses near the crack tip and causes the composite to fracture in a more brittle manner than the neat matrix. In Figure 43, the mode I interlaminar fracture toughness of a variety of different composites is plotted as a function of neat resin toughness. The trend is bilinear, changing slope near 200 J/m^2, which is approximately the point at which the crack tip plastic zone in the matrix is equal to the average fiber spacing [28,30] in the composite.

The large increases in neat resin toughness described above are therefore not fully transferred to the composite, due to the constraint imposed by the fibers on the size of the crack tip plastic zone. For example, the data from Ref. 36 in Table 1 indicate that the addition of 9% CTBN elastomer to the neat MY750 epoxy increased the toughness by a factor of 10 for the matrix, but only by a factor of 2 for the composite. In addition, although significant toughening of the matrix material can be achieved by modifying the matrix with elastomer additives such as CTBN, the elastomer also lowers the glass transition temperature of the matrix and therefore decreases the compressive strength of the composite in hot/wet environments [76].

Interply Layer

Interply layers are a means of toughening the composite without the large decrease in hot/wet compression strength that usually occurs when the composite is toughened using

Table 1   Mode I Interlaminar Fracture Toughness Measurements

Material	Test	Delamination interface	Fiber volume (%)	G_{Ic} (J/m^2)	Ref.
985					
T300/985	DCB	0/0	56	128	12
914					
T300/914	DCB	0/0	55	185	45
3502			0	69	26
3502			0	70	34
AS1/3502	DCB	0/0		155	26
AS4/3502	DCB	0/0		190	28
AS4/3502	DCB	0/0	76	225	26
5208	WTDCB		0	76	17
5208			0	80	30
T300/5208	DCB	0/0	56	84	22
T300/5208	DCB	+45/−45	72	84	31
T300/5208	DCB	0/45	72	86	31
T300/5208	DCB	0/0	68	88	5
T300/5208	DCB		65	100	30
T300/5208	DCB	0/0		103	32
Dow P4			0	80	28
AS4/Dow P4	DCB	0/0		160	28
3501-6			0	70	28
3501-6			0	95	30
AS1/3501-6	DCB	0/0	62	110	24
AS4/3501-6	DCB			144	28
AS4/3501-6	DCB		67	175	30
2220-3			0	95	33
AS4/2220-3	DCB	0/0	61	160	24
AS4/2220-3	DCB			221	16
AS6/2220-3	DCB	0/0	57	238	22
AS4/2220-3	DCB	0/0		250	33
Dow P6			0	150	28
AS4/Dow P6	DCB			160	28
3100					
T300/3100	DCB	0/0		170	37
DGEBA					
Glass/DGEBA	DCB	0/0	60	264	45
5245c					
AS6/5245c	DCB	0/0	57	287	22
PMR-15					
C6000/PMR-15	DCB	0/0	66	294	22
F-155 NR[a]			0	167	34
AS4/F-155 NR	DCB	0/0	54	335	28
Hx205			0	230	35
Hx205	WTDCB		0	270	17
Hx205	WTDCB		0	340	28

Table 1   Continued

Material	Test	Delamination interface	Fiber volume (%)	G_{Ic} (J/m²)	Ref.
T300/Hx205	DCB		58	380	30
C6000/Hx205	DCB	0/0		455	28
T300[b]/Hx205	WTDCB	[b]	61	600	17
C6000/Hx205	DCB		56	790	30
Glass[c]/Hx205	WTDCB	[c]	60	1000	17
BP907			0	325	28
T300/BP907	DCB	0/0	56	292	12
T300/BP907	DCB	0/45	62	333	31
T300/BP907	DCB	0/0		380	28
T300/BP907	DCB	0/0	62	382	31
T300/BP907	DCB	0/90	62	390	31
T300/BP907	DCB	+45/−45	62	403	31
MY750[d]	WTDCB		0	330	36
MY750[e]	WTDCB		0	1400	36
MY750[f]	WTDCB		0	2200	36
MY750[g]	WTDCB		0	3200	36
AS4/MY750[d]	WTDCB	0/0	60	280	36
AS4/MY750[e]	WTDCB	0/0	60	370	36
AS4/MY750[f]	WTDCB	0/0	60	360	36
AS4/MY750[g]	WTDCB	0/0	60	490	36
F-185 NR[h]			0	460	34
AS4/F-185 NR	DCB	0/0	58	455	34
R6376					
IM-6/R-6376	DCB	0/0		473	37
F-155			0	730	34
C6000/F-155	DCB	0/0	51	495	24
AS4/F-155	DCB	0/0	69	520	24
C6000/F-155	DCB	0/0		600	26
AS4/F-155	DCB	0/0	60	1015	34
Hx206			0	2200	35
T300/Hx206	DCB	0/0	55	830	30
C6000/Hx206	DCB	0/0		1200	28
C6000/Hx206	DCB	0/0	49	1550	30
Hx210			0	2800	28
C6000/Hx210	DCB			1800	28
F-263					
T300/F-263	DCB	0/0	56	119	12
T300[b]/F-263	WTDCB	[b]	64	360	17
P-1700[i]			0	2500	30
T300/P-1700	DCB	0/0	52	1200	30
P-1700	WTDCB		0	3200	38
PEEK[i]					
AS4/PEEK	DCB	0/0	58	1147	22
AS4/PEEK	DCB	0/0		1205	37
AS4/PEEK	DCB	0/0	66	1330	24

Table 1   Continued

Material	Test	Delamination interface	Fiber volume (%)	G_{Ic} (J/m^2)	Ref.
AS4/PEEK	DCB	0/0	61	1460	45
AS4/PEEK	DCB	0/0		1751	11
Ultem-1000[i]			0	3300	30
T300/Ultem	DCB	0/0	65	935	30
T300/Ultem	DCB	0/0	60	1060	30
F-185	WTDCB		0	5100	17
F-185			0	5830	35
F-185			0	6400	28
F-185			0	8000	34
T300/F-185	DCB	0/0	59	1960	30
AS4/F-185	DCB	0/0	57	2200	28
C6000/F-185	DCB		57	2250	30
C6000/F-185	DCB			2700	26
Glass[c]/F-185	WTDCB	[c]	60	4400	17
T300[b]/F-185		[b]	58	4600	17
Lexan			0	8100	28
AS4/Lexan	DCB	0/0		1600	28

[a] F-155 epoxy without CTBN additive.
[b] 13 × 12 weave.
[c] 7781 S-glass, 8HS weave.
[d] Unmodified epoxy.
[e] MY750 + 3.2% CTBN rubber.
[f] MY750 + 6.2% CTBN rubber.
[g] MY750 + 9.0% CTBN rubber.
[h] F-185 epoxy without CTBN additive.
[i] Thermoplastic.

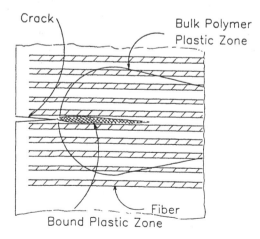

Figure 42   Fibers restrict development of crack tip plastic zone. (Adapted from Ref. 77.)

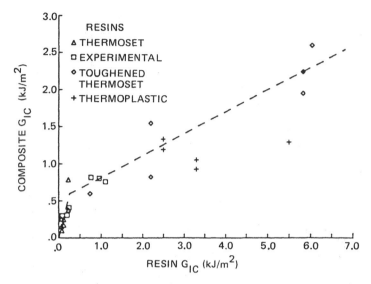

Figure 43 Strain energy release rates for steady interlaminar crack growth versus neat resin toughness. (Adapted from Ref. 30.)

matrix additives. Delaminations form at ply interfaces because the thin, resin-rich layer between plies cannot undergo much shear deformation, due to the brittleness of the matrix. Toughening can therefore be achieved by the addition of a discrete layer of a second, tougher resin between plies of the laminate, where high interlaminar shear stresses occur. This toughening approach was first used in the CYCOM HST-7 material [78,79] and was shown in Ref. 80 to double G_{IIc} in some graphite–epoxy systems, while 20 to 50% increases in G_{Ic} were measured [81]. In comparison, a sixfold increase in G_{Ic} and a fourfold increase in G_{IIc} were obtained for AS4/3502 graphite–epoxy [82] using an FM-300 adhesive interply layer, and an order-or-magnitude increase in G_{Ic} was measured [83] using a toughened AF-163 adhesive interply layer along the midplane of AS1/3502 DCB specimens. The interply layer also reduces the amount of delamination that occurs due to transverse impact loading. Impact damage usually initiates as transverse matrix cracks, which cause delaminations to form when they reach a ply interface. Tough interply layers were shown [78,79] to arrest impact-induced matrix cracks, thereby preventing delaminations from forming along the ply interfaces and increasing the compression strength after impact.

Similarly, tough adhesive strips of finite width can be embedded selectively, at delamination-prone locations, to arrest propagating delaminations. In Ref. 84, finite-width strips of American Cyanamid FM-1000 adhesive, placed as indicated in Figure 44, were shown to arrest edge delaminations in AS4/3501-6 laminates, resulting in an increase in static tensile strength and an extension of fatigue life, as shown in Figure 45.

Fiber–Matrix Bond

In brittle composites, an increase in interlaminar fracture toughness can be obtained by increasing the strength of the bond between fiber and matrix. In situ observations of delamination using a scanning electron microscope [85] revealed that mode I delaminations in brittle-matrix composites grew primarily by the progressive failure of the fiber–matrix interface region. In the tougher composites, crack propagation occurred primarily by

Table 2   Mode II Interlaminar Fracture Toughness Measurements

Material	Test	Delamination interface	Fiber volume (%)	G_{IIc} (J/m^2)	Ref.
3501-6					
AS1/3501-6	ENF	0/90		370	41
AS1/3501-6	ENF	0/45		438	41
AS1/3501-6	ENF	0/0		444	41
AS1/3501-6	ENF	0/0	62	605	24
AS4/3501-6	ENF			1150	28
914					
T300/914	ENF	0/0	55	518	45
3100					
T300/3100	ENF	0/0		548	37
3502					
AS4/3502	ELS	0/0		543	40
AS4/3502		0/0		570	28
AS4/3502	ENF	0/0		615	40
F-263					
T300/F-263	ELS	0/0	56	594	22
R-6376					
IM-6/R-6376	ENF	0/0		650	37
985					
T300/985	ELS	0/0	56	697	22
5208					
T300/5208	ELS	0/0	56	716	22
T300/5208	ENF	0/0		865	47
2220-3					
AS4/2220-3	ENF	0/0	61	750	24
AS6/2220-3	ELS	0/0	57	968	22
Dow P4					
AS4/Dow P4				800	28
F-155					
C6000/F-155	ENF	0/0	51	900	24
AS4/F-155				1500	28
5245c					
AS6/5245c	ELS	0/0	57	977	22
F-185 NR[a]					
AS4/F-185 NR				1050	28
BP907					
T300/BP907	ELS	0/0	56	1423	22
T300/BP907	ENF	0/0		2627	47
F-155 NR[b]					
AS4/F-155 NR				1660	28
Lexan					
AS4/Lexan				1700	28

Table 2 Continued

Material	Test	Delamination interface	Fiber volume (%)	G_{IIc} (J/m²)	Ref.
DGEBA					
Glass/DGEBA	ENF	0/0	60	1715	45
Glass/DGEBA	ELS	0/0	60	2110	45
Dow P6					
AS4/Dow P6				1750	28
PEEK[c]					
AS4/PEEK	ENF	0/0	61	1109	45
AS4/PEEK	ENF	0/0		1502	37
AS4/PEEK		0/0		1700	11
AS4/PEEK	ENF	0/0	66	1765	24
AS4/PEEK	ELS	0/0	61	1780	45
AS4/PEEK	ELS	0/0	58	2425	22
F-185					
AS4/F-185	ELS	0/0		2265	40
AS4/F-185	ENF	0/0		2354	40

[a] F-185 epoxy without CTBN additive.
[b] F-155 epoxy without CTBN additive.
[c] Thermoplastic.

fracture through the matrix, with little interfacial failure. In the latter case, the better interfacial bond results in the toughness of the resin being more fully utilized in the composite. A similar dependence of interlaminar fracture toughness with fiber–matrix bond strength was reported [41] for mode II loading.

The strength and/or toughness of the fiber–matrix bond can be increased by applying a polymer coating to the fiber surface [86,87]. A 50% increase in G_{Ic} was reported [88]

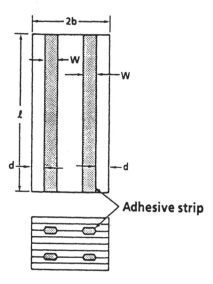

Figure 44 Location of adhesive interply layer in AS4/3501-6 laminate. (Adapted from Ref. 84.)

Table 3   Mixed-Mode Interlaminar Fracture Toughness Measurements

Material	Test	Delamination interface	$\dfrac{G_I}{G_{II}}$	G_{total}^c (J/m^2)	Ref.
5208					
T300/5208	EDT	0/0	1.33	84	59
T300/5208	CLS	45/45	0.29	460	32
Dow P4					
AS4/Dow P4		0/0	1.33	160	28
Dow P6					
AS4/Dow P6		0/0	1.33	160	28
3100					
T300/3100	CLS	0/0		170	37
3501-6					
AS4/3501-6		0/0	1.33	175	28
DGEBA					
Glass/DGEBA	a	0/0	1.14	264	45
Hx205					
AS4/Hx205		0/0	1.33	380	28
C6000/Hx205	EDT	0/0	1.33	790	59
R-6376					
IM-6/R-6376	CLS	0/0		473	37
F-155					
AS4/F-155		0/0	1.33	495	28
PEEK^b					
AS4/PEEK	CLS	0/0		1147	37
F-185					
C6000/F-185	EDT	0/0	1.38	5830	59

^{a} Test described in Ref. 45.
^{b} Thermoplastic.

for AS4/MDA (methylene dianiline epoxy) composites when a thin, tough copolymer layer was applied to the fibers using an electropolymerization process [89], as shown in Figure 46. However, this was accompanied by a large decrease in G_{IIc}, which was attributed to the failure of the matrix–interface bond. Coatings of toughened-epoxy adhesives applied to AS4 graphite fibers were shown [90] to increase the mode I fracture toughness of AS4/976 composites from their baseline value of 88 J/m^2. Coating the fibers with an elastomer-modified epoxy (AF-163-2 from 3M) increased toughness to 300 to 500 J/m^2, depending on the fiber volume fraction of the laminate. The fiber-coating method of toughening may adversely affect the matrix-dominated properties of the laminate, however. If a significant amount of low-modulus resin is added to the laminate, the compression strength can be decreased significantly.

Layup

Ply orientations and fiber architecture can affect delamination resistance through several different physical mechanisms. In Ref. 57, fabrication-induced residual stresses were

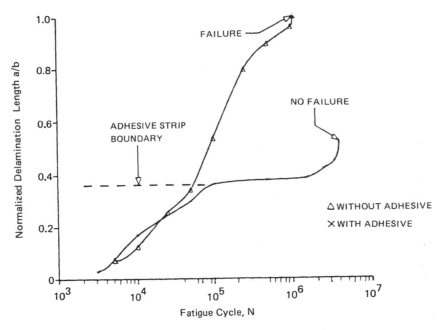

Figure 45 Adhesive interply layer arrests delamination growth due to fatigue loading. (Adapted from Ref. 84.)

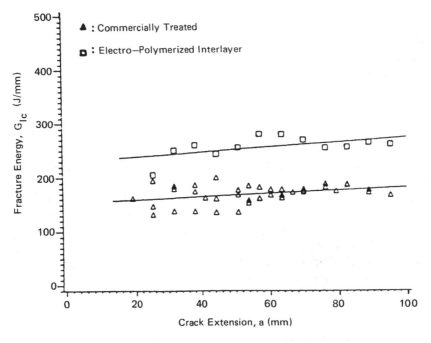

Figure 46 Increase in G_{Ic} due to electropolymerized interlayer on fiber surface. (Adapted from Ref. 88.)

shown to play an important role. An approximate analysis of the interlaminar stresses near the free edge of a laminate was used to design two types of laminates. A $[(\pm 25)_2/90]_{\text{sym}}$ laminate was shown to have maximum tensile residual stress at the midplane near the free edge, and is therefore prone to delaminate under uniaxial tension at that location. In contrast, a $[90/(\pm 25)_2]_{\text{sym}}$ laminate has compressive residual stresses at all interfaces and hence is more delamination resistant. Tests showed that edge delaminations initiated in the former laminate at approximately 50% of the ultimate failure load, while no delamination was visible prior to tensile failure in the latter specimen.

The data from Ref. 31 in Table 1 indicate no significant difference in G_{Ic} measured at 0/0, 0/45, and +45/−45 interfaces in DCB specimens of brittle T300/5208 or tougher T300/BP907 graphite–epoxy materials. This suggests that fracture toughness is a *material* property, independent of delamination interface. In Ref. 17, however, a *woven* glass or graphite fiber reinforcement is shown to increase resistance to delamination by a factor of 2 to 3, compared to unidirectional reinforcement. The additional fracture resistance is probably due to the irregular path the delamination must take to separate the plies.

The ply orientations in a laminate also effect the ability of fibers in neighboring plies to *nest* together [23,34], which in turn affects the thickness of the resin-rich *interply layer*

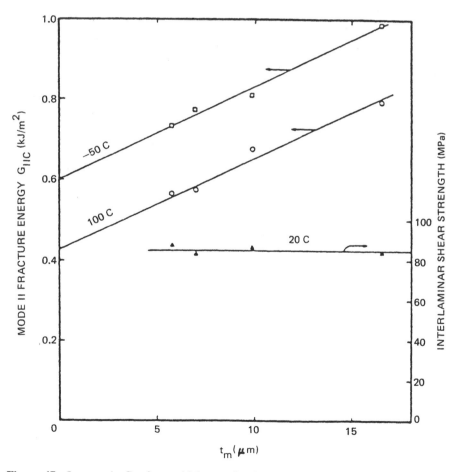

Figure 47 Increase in G_{IIc} due to thickness of resin-rich interply layer. (Adapted from Ref. 24.)

in the cured laminate by changing the amount of resin that bleeds away from the interlaminar region during the cure cycle. The thickness of the interply layer can therefore be controlled somewhat by varying the number of 90° "bleeder plies" [24] in the laminate. Using this approach, it is shown for AS/3501-6 laminates that G_{IIc} increases linearly with thickness of the interply layer over the range $5 \leq t_m \leq 15$ μm, and that the interlaminar shear strength was unaffected by the increase in interply layer thickness. A difference in G_{IIc} of $\simeq 50\%$ was measured, depending on the interply layer thickness, as shown in Figure 47. The effective mode I fracture toughness decreases with interply thickness, however, because of the reduced fiber volume fraction within the interply layer, which decreases the amount of fiber bridging [23] that occurs under mode I loading.

REFERENCES

1. J. G. Williams, *Fracture Mechanics of Polymers*, Ellis Horwood, Chichester, West Sussex, England, 1984.
2. D. Broek, *Fundamentals of Fracture Mechanics*, Martinus Nijhoff, Dordrecht, The Netherlands, 1982.
3. E. J. Ripling, S. Mostovoy, and R. L. Patrick, *ASTM STP-360*, p. 5 (1964).
4. J. M. Whitney, C. E. Browning, and W. Hoogsteden, *J. Reinf. Plast. Compos.*, 1: 297 (1982).
5. D. J. Wilkins, J. R. Eisenmann, R. A. Camin, W. S. Margolis, and R. A. Benson, *ASTM STP-775*, p. 168 (1982).
6. J. M. Whitney, C. E. Browning, and A. Mair, *ASTM STP-546*, p. 30 (1974).
7. T. R. Brussat, S. T. Chiu, and S. Mostovoy, *AFML TR-77-163* (1978).
8. I. M. Daniel, I. Shareef, and A. A. Aliyu, *ASTM STP-937*, p. 260 (1987).
9. J. M. Whitney, *Compos. Sci. Technol.*, 23: 201 (1985).
10. D. F. Devitt, R. A. Schapery, and W. L. Bradley, *J. Compos. Mater.* 14: 270 (1980).
11. J. W. Gillespie, Jr., L. A. Carlsson, and A. J. Smiley, *Compos. Sci. Technol.* 28: 1 (1987).
12. P. E. Keary, L. B. Ilcewicz, C. Shaar, and J. Trostle, *J. Compos. Mater.*, 19: 154 (1985).
13. J. P. Berry, *J. Appl. Phys.*, 34:62 (1963).
14. F. X. deCharentenay, J. M. Harry, Y. J. Prel, and M. L. Benzeggagh, *ASTM STP-836*, p. 84 (1984).
15. J. R. Rice, *J. Appl. Mech. Trans. ASME*, 379 (1968).
16. M. Ashizawa, *6th DOD/NASA Conference on Fibrous Composites in Structural Design*, 1983.
17. W. D. Bascom, J. L. Bitner, R. J. Moulton, and A. R. Siebert, *Composites*, 11: 9 (1980).
18. S. Mostovoy, P. B. Crosley, and E. J. Ripling, *J. Mater.* 2: 661 (1967).
19. G. Yaniv and I. M. Daniel, *ASTM STP-972*, p. 241 (1988).
20. D. L. Hunston and W. D. Bascom, *Compos. Tech. Rev.*, 5: 118 (1983).
21. D. J. Nicholls and J. P. Gallagher, *J. Reinf. Plast. Compos.*, 2: 2 (1983).
22. L. B. Ilcewicz, P. E. Keary, and J. Trostle, *Polym. Eng. Sci.*, 28: 592 (1988).
23. W. S. Johnson and P. D. Mangalgiri, *NASA TM-87716* (1986).
24. A. J. Russell, *Polym. Compos.*, 8: 342 (1987).
25. K. B. Su, *Proc. 5th International Conference on Composite Materials*, 1985, p. 995.
26. W. L. Bradley and R. N. Cohen, *ASTM STP-876*, p. 389 (1985).
27. D. C. Phillips and G. M. Wells, *J. Mater. Sci. Lett.* 1: 321 (1982).
28. W. L. Bradley, *Key Eng. Mater.* 37: 161 (1989).
29. J. W. Gillespie, Jr., L. A. Carlsson, R. B. Pipes, R. Rothschilds, B. Trethewey, and A. Smiley, *NASA CR-178066* (1986).
30. D. L. Hunston, R. J. Moulton, N. J. Johnston, and W. D. Bascom, *ASTM STP-937*, p. 74 (1987).
31. H. Chai, *Composites*, 15: 277 (1984).
32. R. L. Ramkumar and J. D. Whitcomb, *ASTM STP-876*, p. 315 (1985).

33. W. D. Bascom, G. W. Bullman, D. L. Hunston, and R. M. Jensen, *Proc. 29th SAMPE Symposium*, 1984.
34. W. M. Jordan and W. L. Bradley, *J. Mater. Sci. Lett.*, 7: 1362 (1988).
35. R. Y. Ting and R. L. Cottington, *J. App. Poly. Sci.*, 25: 1815 (1980).
36. J. M. Scott and D. C. Phillips, *J. Mater. Sci.*, 10: 551 (1975).
37. S. Mall, W. S. Johnson, and R. A. Everett, *Adhesive Joints*, Plenum Press, New York, 1984.
38. R. D. Gates and N. J. Mills, *Eng. Fract. Mech.*, 6: 93 (1974).
39. W. S. Johnson and P. D. Mangalgiri, *ASTM STP-937*, p. 295 (1987).
40. C. R. Corleto and W. L. Bradley, *ASTM STP-1012*, p. 201 (1989).
41. A. J. Russell, *DREP Mater. Rep. 82-0*, Victoria, British Columbia, 1982.
42. A. J. Russell and K. N. Street, *ASTM STP-876*; p. 349 (1985).
43. S. Mall and N. K. Kochhar, *J. Compos. Technol. Res.*, 8: 54 (1986).
44. L. A. Carlsson, J. W. Gillespie, and R. B. Pipes, *J. Compos. Mater.*, 20: 594 (1986).
45. Y. J. Prel, P. Davies, W. L. Benzeggagh, and F. X. deCharentenay, *ASTM STP-1012*, p. 251 (1989).
46. L. A. Carlsson, J. W. Gillespie, and B. R. Trethewey, *J. Reinf. Plast. Compos.* 5: 170 (1986).
47. T. K. O'Brien, G. B. Murri, and S. A. Salpekar, *ASTM STP-1012*, p. 222 (1989).
48. J. W. Gillespie, Jr., L. A. Carlsson, and R. B. Pipes, *Compos. Sci. Technol.*, 27: 177 (1986).
49. A. J. Russell and K. N. Street, *ASTM STP-937*, p. 275 (1987).
50. D. J. Wilkins, *NAV-GD-0037*, Naval Air Systems Command (1981).
51. S. Mall and W. S. Johnson, *ASTM STP-893*, p. 322 (1986).
52. B. Dattaguru, R. A. Everett, J. D. Whitcomb, and W. S. Johnson, *J. Eng. Mater. Technol.*, 106: 59 (1984).
53. E. F. Rybicki and M. F. Kanninen, *Eng. Fract. Mech.*, 9: 931 (1977).
54. P. D. Mangalgiri and W. S. Johnson, *ASTM D-30.02.02 Meeting*, Dallas, 1984.
55. A. H. Puppo and H. A. Evenson, *J. Compos. Mater.* 4: 204 (1970).
56. R. B. Pipes and N. J. Pagano, *J. Compos. Mater.*, 4: 538 (1970).
57. N. J. Pagano and R. B. Pipes, *Int. J. Mech. Sci.*, 15: 679 (1973).
58. A. S. Wang, *Key Eng. Mater.*, 37: 1 (1989).
59. T. K. O'Brien, *ASTM STP-775*, p. 140 (1982).
60. T. K. O'Brien, N. J. Johnston, D. H. Morris, and R. A. Simonds, *SAMPE J.*, Summer 1982, pp. 8–15.
61. T. K. O'Brien, N. J. Johnston, I. S. Raju, D. H. Morris, and R. A. Simonds, *ASTM STP-937*, p. 199 (1987).
62. T. K. O'Brien, *ASTM STP-836*, p. 125 (1984).
63. J. M. Whitney and M. K. Knight, *ASTM STP-876*, p. 298 (1985).
64. G. E. Law, *ASTM STP-836*, p. 143 (1984).
65. S. S. Wang and J. F. Mandell, *NASA CR-135248* (1977).
66. S. S. Wang, *ASTM STP-674*, p. 642 (1979).
67. R. A. Jurf and R. B. Pipes, *J. Compos. Mater.*, 16: 386 (1982).
68. A. A. Aliyu and I. M. Daniel, *ASTM STP-876*, p. 336 (1985).
69. J. H. Williams, S. S. Lee, and P. N. Kousiounelos, *Eng. Fract. Mech.*, 14: 427 (1981).
70. F. X. deCharentenay and M. L. Benzeggagh, in *Advances in Composite Materials*, Proc. ICCM-3, Vol. 1, 1980, p. 186.
71. J. L. Bitner, J. L. Rushford, W. S. Rose, D. L. Hunston, and C. K. Riew, *J. Adhes.* 13: 3 (1981).
72. A. G. Miller, P. E. Hertzberg, and V. W. Rantala, *Proc. 12th SAMPE Technical Conference*, 1980, p. 279.
73. J. E. Grady and C. T. Sun, *ASTM STP-907*, p. 5 (1986).
74. C. T. Sun and J. E. Grady, *Compos. Sci. Technol.*, 31: 55 (1988).
75. T. D. Chang and J. O. Brittain, *Polym. Eng. Sci.*, 22: 1228 (1982).
76. A. J. Klein, *Adv. Mater. Process.*, Oct. 1985, p. 43.

77. A. F. Yee, *ASTM STP-937*, p. 383 (1987).
78. R. B. Krieger, Jr., in *Progress in Advanced Materials and Processes*, G. Bartelds (ed.), Elsevier, Amsterdam, 1985, p. 189.
79. J. E. Masters. J. L. Courter, and R. E. Evans, *Proc. 31st SAMPE Symposium*, 1986, p. 844.
80. J. E. Masters, *Key Eng. Mater.*, 37: 317 (1989).
81. K. R. Hirschbuehler, *ASTM STP-937*, p. 61 (1987).
82. O. Ishai, H. Rosenthal, N. Sela, and E. Drukker, *Composites*, 19: 49 (1988).
83. C. E. Browning and H. S. Schwartz, *ASTM STP-893*, p. 256 (1986).
84. W. S. Chan, *ASTM STP-907*, p. 176 (1986).
85. M. F. Hibbs, M. K. Tse, and W. L. Bradley, *ASTM STP-937*, p. 115 (1987).
86. L. J. Broutman and B. D. Agarwal, *Polym. Eng. Sci.*, 14: 581 (1974).
87. L. E. Nielsen and D. G. Peiffer, *J. Appl. Polym. Sci.*, 23: 2253 (1979).
88. R. Joseph, J. P. Bell, and H. Rhee, *Polym. Eng. Sci.*, 28: 605 (1988).
89. J. P. Bell, J. Chang, H. Rhee, and R. Joseph, *J. Compos.*, *Mater.*, 8: 46 (1987).
90. H. S. Schwartz and J. T. Hartness, *ASTM STP-937*, p. 150 (1987).

2

Damage, Failure, and Fracture of Resin-Based Fibrous Composites: Mechanisms, Theories, and Criteria

S. Y. Zhang
Institute of Mechanics
Chinese Academy of Sciences
Beijing, China

INTRODUCTION

Fiber-reinforced plastics (FRPs) have evolved over the past 40 years. Initially, glass-fiber-reinforced plastics (GRPs) were fabricated into less-demanding applications and lower-load-bearing components. Then the material was exploited as a convenient substitute for traditional materials. More recently, the advent of more advanced fibers, such as boron, graphite, and Kevlar fibers, had led to the possibility of FRP constructions that are structurally and economically effective for high-grade load-carrying structures. Because the materials have the advantages of high specific stiffness and high specific strengths, they have been developing with astonishing speed.

The adoption of advanced fibrous composites by aerospace industry triggered a massive research effort. Obviously, the ability to design a structure composed of FRP must be matched by the ability to fabricate the component as well as by the knowledge of damage, failure, and fracture behavior in its service time. The use of fiber-reinforced composites for high-load-bearing applications raises completely new considerations for quality control, reliability, and safety assurance.

Our knowledge of damage, failure, and fracture of fiber composites is far less adequate than that for traditional metallic materials. Our experience of their mechanical behavior is limited. Our understanding of the micromechanisms of damage and its evolution under loading is imperfect.

As a consequence of their nature and methods of fabrication, there is a high probability that manufacturing flaws will be induced into the fabricated components. Furthermore, because composites consist of two or more materials with high disparate characteristics, it is impossible to avoid creating additional defects when putting into service. The service damages include cracks in matrix, debonding at the interface between the fiber and matrix, and breakage of fibers.

Our uncertainty about how these damages grow and accumulate is reflected in current design procedures. For example, this uncertainty results in very large safety factors for critical structures. Thus the weight and strength advantages of the materials are wasted.

To ensure a reliable application of these components, we must be able to identify these damages and to evaluate their significance. Moreover, we must be able to predict failure and fracture of the structures. In this respect, a great many research results have been published in the past two decades. Based on the existing results available, in this chapter we review the mechanisms theories, and criteria of composite failure and fracture. Because short fiber composites exhibit more involved phenomena, major emphasis will be focused on continuous fiber-reinforced composites, and on the popular composite systems of graphite-, glass-, and Kevlar-fiber-reinforced epoxy and polyester resins.

The chapter consists of three major sections. First is a brief account of damage in fiber composite materials. The defects produced in the manufacturing process and the damage created in service or by environmental conditions are summarized. These defects and damages are believed to be the origin leading to fracture or failure. To identify this damage, many nondestructive testing (NDT) methods have been developed. Several NDT techniques for detecting inherent defects or monitoring damage process are discussed briefly.

Next, the failure mechanisms of FRP are discussed. Damage phenomena, multiple fracture behavior, failure strength, and failure theories of unidirectional fibrous composites are introduced. Then the damage process, strength prediction, and postfailure theories of cross-ply laminates are described. Finally, the failure behavior of angle-ply laminates is discussed briefly.

At the end of the chapter we present the mechanics of composite fracture, and outline features of composite fracture in comparison with metallic materials. Microfracture mechanisms such as energy-absorbing models (e.g., interfacial debonding and fiber pullout) are reviewed. Then the macromechanics of composite fracture (i.e., phenomenological fracture mechanics) are introduced. Influential models, theories, and criteria on the onset of fracture and cracking direction are reviewed. Finally, the test methods for evaluating the damage tolerance and notch sensitivity of composite materials are reviewed briefly.

MANUFACTURING DEFECTS, SERVICE DAMAGE, AND NDT METHODS

In this section, various flaws and defects induced in the manufacturing process and the damages caused by load and environment conditions are summarized. Then a number of nondestructive testing methods are outlined.

Flaws in Fabricated Composites

Defects associated with resin

1. Wrong composition (e.g., for a given environment)
2. Wrong state of cure
3. Resin-rich zone
4. Voids or porosities in resin
5. Imperfect wetting of fibers (voids at interface)

Defects associated with fibers

1. Wrong fiber volume fraction V_f
2. Irregular arrangement of the fibers
3. Broken fibers
4. Nonstraight fibers: curved, whorled, or buckled fibers
5. Poorly aligned fibers

Defects associated with lamination

1. Poor arrangement of plies or wrong stack constructions
2. Translaminar cracking from thermal stress (residual stress)
3. Delamination regions in the fabricated composite laminates
4. Distortion or buckling due to residual stress

Defects associated with design

1. Machining defects (e.g., at drilled holes)
2. Adhesive joint defects (e.g., sharp change in cross section or debonding)
3. Design defects (e.g., cuts or openings where fibers are broken)

Typical Damage

Typical service and environmental damage is listed in Tables 1 and 2.

Nondestructive Testing

There is a strong demand to develop procedures that may be applied to detect damage in FRP, but it proved to be very difficult to establish cheap, reliable, and effective procedures. A number of NDT procedures are found to be suitable for evaluating composites.

Ultrasonic Methods

Ultrasonics may be used for detecting flaws or damages in FRP using a pulse-echo or attenuation measurement techniques [1–5]. The most commonly used technique is C-scan,

Table 1 Typical Service Damage

Damage	Cause
Cuts or scratches	Mishandling
Abrasion	Rain or grit erosion
Delaminations or disbonds	Impact or overload
Local buckling or kinking	Compression damages
Dents	Impact, walk-in areas, or hit by runway stones
Edge damages	Mishandling of doors and removable parts
Penetrations	Battle damage, service mishandling

using a calibrated attenuation method. The principle is based on the fact that delamination or porosity increase the level of attenuation. In the common application, a number of attenuation levels, corresponding to predetermined porosity or flaw levels, are set and the apparatus produces a figure with tones graded from black to white through three to eight gray levels, black corresponding to minimum attenuation and white the opposite.

The C-scan techniques [6] may also be used for evaluating delamination and the extent of adhesive bedonding in a suitable case. It is very sensitive to crack normal to the ultrasonic beam direction, but is not useful for detecting fine cracks parallel to the beam. The output may be processed by a computer, but the inherent drawback is the need to accurately calibrate the equipment against reliable standards.

X-Radiography

X-radiography has been improved to detect damages in fiber composites, but is not easily applied in the field [7–9]. A good deal of information about composite quality can be obtained by x-ray inspection [7]. One specialized technique is to use a radiopaque penetrant to reveal cracks or open flaws that intersect a free surface. This method is obviously limited to the detection of cracks that intersect a free surface.

In GRP, there is sufficient difference in x-ray attenuation between the glass fibers and the matrix to distinguish the individual fibers or bundles. However, this can only be usefully applied to thin sections to indicate the fiber orientation distribution [10].

Thermography

The principle of the method is that the rate of heat conduction through a composite sheet depends on the soundness of the material. Regions of delamination and porosity conduct less heat. When a uniform heat flux is applied to a plate, any anomalous variations in

Table 2 Typical Environmental Damage

Damage	Cause
Surface oxidation	Overheating, lightning strike, or battle damage (laser irradiation)
Delaminations or disbonds	Moisture expansion, thermal spike, freeze or thaw stresses
Core corrosion	Moisture penetration into honeycomb
Surface swelling	Use of undesirable solvents (e.g., paint stripper)

the material resulting from temperature distribution are indications of structural flaws in the material. Thus one side of a composite plate is uniformly irradiated with infrared rays and the temperature of the other side is monitored with a thermographic camera or by means of other indicators. Boron–aluminum [11], glass–epoxy [12], and carbon–epoxy [13] have been tested with thermal imaging technique. Williams [14] has described how visualization of these temperature distributions can be achieved by means of temperature-sensitive color change in cholesteric liquid crystals. An alternative application of this technique is to monitor a specimen continuously under the test of monotonically or cyclic loading.

Acoustic Emission

The acoustic emission (AE) testing method is not a nondestructive test in the sense that AE comes from occurrence of damage. But it is very useful for testing composites. In principle, any failure event will result in a transient energy release, and it is this "noise" that is monitored in acoustic emission analysis. The more correct term for this technique is *stress wave emission analysis*. It can be used for monitoring destructive tests (e.g., tension, bending, and fatigue) and for monitoring structures or components in service and in a proof test. There are a number of microfailure or damage mechanisms that give rise to stress waves [15]:

1. Fiber breakage [16,17]
2. Debonding of interface between fiber and matrix [18]
3. Matrix cracking or crazing
4. Transverse cracking and longitudinal splitting in cross-ply laminates
5. Delamination and individual ply failure in multilayer laminates

Each of these events will result in an energy pulse that may be characterized by its energy or frequency. Many experiments have been done and good correlation has been obtained between acoustic emission and detectable damages. However, at best the AE technique must be considered as a qualitative method.

A more sophisticated approach is to present an amplitude or energy spectrum of the emission. It is possible to relate different levels in the energy spectrum to individual micromechanisms [19]. For example, debonding produces low-energy emission, and fiber fracture gives rise to high-energy emission. However, it is difficult to separate the various events due to attenuation between the site of the event and the detection transducer and the fact that several different failure processes are occurring simultaneously. The use of acoustic emission methods to define acceptance criteria for composite structures is now becoming familiar to manufacturers [20].

FAILURE BEHAVIOR OF COMPOSITE MATERIALS

Failure behavior is pertinent to materials in which there are no holes, notches, or other discontinuities. Advanced fiber-reinforced plastics, components for high-technology applications are usually manufactured from unidirectional fiber laminae called *prepreg*. Prepreg laminae are stacked in the orientation sequence that will achieve the properties specified by the designer. An understanding of the failure behavior of unidirectional lamina is the basis of better laminate design.

Table 3   Typical Values of Elastic Constants, $v_t = 0.5^a$

Material	$E_{11}^T = E_{11}^C$ (GPa)	E_{22} (GPa)	G_{12} (GPa)	v_{12}
Glass polyester	35–40	8–12	3.5–5.5	0.26
Graphite epoxy	190–240	5–8	3–6	0.26
Kevlar 49 epoxy	65–75	4–5	2–3	0.35

[a] *Source*: Ref. 21. Superscript T, tensile; C, compression.

Failure Behavior of Unidirectional Composites

Each lamina is highly anisotropic and has a high stiffness (Young's modulus) in the fiber direction E_{11}; off the fiber axis, the stiffness decreases sharply to a minimum at 90°, E_{22}. The in-plane Poisson ratios also differ; v_{12} is much greater than v_{21}. Similarly, the longitudinal coefficient of thermal expansion is much smaller than the transverse coefficient. Some typical values of elastic constants of unidirectional laminae are summarized in Table 3. The highly anisotropic stiffness properties may exert a strong influence on the failure behavior of unidirectional laminae.

Strength of Unidirectional Laminae

A unidirectional lamina can fail in many different modes, depending on the external loading conditions. For the purpose of design, it is probably sufficient to know the fracture strengths associated with these fracture modes for different fiber–resin systems with a range of fiber volume fractions. The strengths of three commonly used composite systems are summarized in Table 4.

Longitudinal tensile stress can be related to the stresses in fibers and matrix using the following formula:

$$\sigma_{lc} = \sigma_f V_f + \sigma_m(1 - V_f) \tag{1}$$

where the subscript c refers to "composite."

The failure behavior depends on the relative strain to failure of matrix and fiber. There are two possible situations: $\epsilon_f^* < \epsilon_m^*$ and $\epsilon_f^* > \epsilon_m^*$, where ϵ_f^* and ϵ_m^* are strains to failure of fiber and matrix, respectively. For each of the two situations, the failure sequence is further dependent on the fiber volume fraction, V_f. This is depicted in Table 5 and in Figures 1 to 3. In a glass fiber–polyester resin lamina with $\epsilon_m^* = 0.02$ and $\epsilon_f^* = 0.025$, the response belongs to the first situation, whereas for a type I graphite–epoxy system, $\epsilon_f^* = 0.005$ and $\epsilon_m^* = 0.02$, the analysis for the second situation applies.

Table 4   Typical Values of Strength Properties of unidirectional Laminae^a

Material	σ_{1T}^* (MPa)	σ_{1C}^* (MPa)	σ_{2T}^* (MPa)	σ_{2C}^* (MPa)	τ_{12}^* (MPa)
Glass polyester	650–750	600–900	20–25	90–120	45–60
Graphite epoxy	850–1100	700–900	35–40	130–190	60–75
Kevlar 49 epoxy	1100–1250	240–290	20–30	110–140	40–60

[a] *Source*: Ref. 21. Superscript *, critical value; $V_f = 0.5$.

Table 5 **Failure Strength and Failure Mode for Different Relative Strain to Failure of Fiber and Matrix and Different V_f**

	V_f	$V_f < V_f'$	$V_f > V_f'$
$\epsilon_f^* > \epsilon_m^*$	Strength	$\sigma_1^* = \sigma_f' V_f + \sigma_m^*(1 - V_f)$	$\sigma_1^* = \sigma_f^* V_f$
	Failure mode	Single fracture	Multiple fracture
	V_f'	$V_f' = \sigma_m^*/(\sigma_f^* - \sigma_f' - \sigma_m^*)$, σ_f' is fiber stress when resin fractures	
	V_f	$V_f < V_f'$	$V_f > V_f'$
$\epsilon_f^* < \epsilon_m^*$	Strength	$\sigma_1^* = \sigma_m^* V_m$	$\sigma_1^* = \sigma_m'(1 - V_f) + \sigma_f^* V_f$
	Failure mode	Multiple fracture	Single fracture
	V_f'	$V_f' = (\sigma_m^* - \sigma_m')/(\sigma_f^* + \sigma_m^* - \sigma_m')$, σ_m' is stress borne by matrix when fiber fractures	

Multiple Failure in Unidirectional Laminae

Figure 1 illustrates the situation of $\epsilon_f^* > \epsilon_m^*$. The failure of the matrix is prior to the failure of the fibers. If fiber volume fraction V_f is smaller than a certain critical value, V_f' the matrix failure, represents total failure of the composite; multiple fractures are not observed. When $V_f > V_f'$, multiple resin cracking occurs, as described by Aveston and Kelly [22]. Multiple cracking is illustrated in Figure 2c. Multiple matrix cracking by no means implies complete unloading of the resin and does not result in final failure of the composite.

As the strain on the lamina increases, further resin cracking occurs. The final spacing of the cracks depends on the ratio E_f/G_m, the strength of the bond, and the difference in the failure strains of resin and fiber. After the first crack is formed, we can consider either of two limiting conditions to apply. In the first of these, matrix and fibers are considered to be ''unbonded'' in the sense that there is no connection between the elastic displacement in the two components, and provided that a limiting shear stress τ is exceeded, the fibers may be drawn through the matrix. The second limiting case is where the matrix remains bonded soundly to the fibers after it has cracked and remains elastic. For both cases, the fundamental equation governing load transfer between the two components is given as [22].

$$\frac{dF}{dy} = \frac{2V_f \tau}{r} \tag{2}$$

where dF is load transfer from fiber to matrix in distance dy, τ is the shear stress acting at the interface, and r the fiber radius.

For the unbonded case, τ is assumed to be a constant equal to frictional stress τ'. Setting $F = \sigma_m^* V_m$, we obtained from Eq. (2)

$$X' = \frac{V_m}{V_f} \frac{\sigma_m^* r}{2\tau'} \tag{3}$$

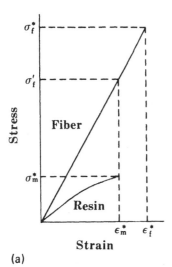

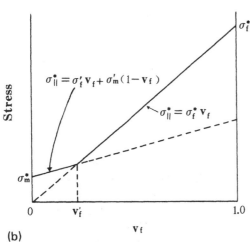

Figure 1 (a) Stress–strain curves of fiber and resin for $\epsilon_f^* > \epsilon_m^*$; (b) variation of fracture strength of unidirectional laminae with V_f for $\epsilon_f^* > \epsilon_m^*$. (From Ref. 21.)

Continued extension of the lamina will lead to the matrix being traversed by a set of parallel cracks spaced between x' and $2x'$ apart. The mean crack spacing is closer to x'.

For the bonded case, after the first crack has occurred in the matrix, an additional stress is placed on the fibers,

$$\Delta\sigma = \frac{\sigma_a}{V_f} - \epsilon_m^* E_f \qquad (4)$$

where σ_a is the applied stress. This additional stress has its maximum value $\Delta\sigma_0$ at the plane of matrix crack and decays with distance from the crack surface:

$$\Delta\sigma = \Delta\sigma_0 \exp(-\sqrt{\Phi}\, y) \qquad (5)$$

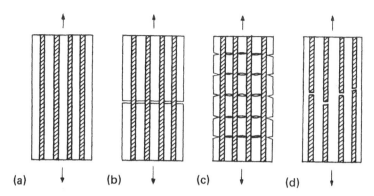

Figure 2 Schematic representation of longitudinal tensile failure of a unidirectional lamina; (a) before fracture; (b) resin fracture before fiber fracture $\epsilon_f^* > \epsilon_m^*$; (c) multiple resin cracking; (d) fiber fracture before resin fracture.

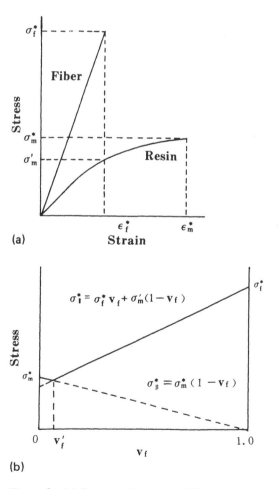

Figure 3 (a) Stress–strain curves of fiber and resin for $\epsilon_m^* > \epsilon_f^*$; (b) variation of fracture strength of unidirectional laminae with V_f for $\epsilon_m^* > \epsilon_f^*$.

where

$$\Phi^{1/2} = \left(\frac{2G_m E_c}{E_f E_m V_m}\right)^{1/2} \frac{1}{r[\ln(R/r)]^{1/2}} \tag{6}$$

where R is the radial distance from the center of the fiber at which the displacement in the matrix is equal to the average displacement in the matrix. The shear stress at the interface between fibers and matrix is given by

$$\tau = \frac{r}{2}\Delta\sigma_0 \sqrt{\Phi} \exp(-\sqrt{\Phi}\, y) \tag{7}$$

In this case, shear stress decays rapidly with distance y from the crack surface. From Eqs. (2) and (7) we obtained

$$F = V_f \Delta\sigma_0[1 - \exp(-\sqrt{\Phi}\, l)] \tag{8}$$

By setting $F = \sigma_m^* V_m$, we obtained from Eq. (8),

$$l = -\frac{1}{\sqrt{\Phi}}\ln\left(1 - \frac{\sigma_m^* V_m}{\Delta\sigma_0 V_f}\right), \qquad \Delta\sigma_0 \geq \frac{\sigma_m^* V_m}{V_f} \tag{9}$$

When $\Delta\sigma_0$ is larger than a few times $\sigma_m^* V_m/V_f$, Eq. (9) reduces to

$$l = \frac{1}{\sqrt{\Phi}}\frac{\sigma_m^* V_m}{\Delta\sigma_0 V_f} \tag{10}$$

We can prove that $\Delta\sigma_0/2 \sqrt{\Phi}\, r$ is equal to maximum shear stress in the interface.

The situation of $\epsilon_f^* < \epsilon_m^*$ is illustrated in Figure 3. There are again two different failure sequences depending on V_f. Clearly, when V_f is large the load transferred to the matrix when fiber fracture occurs is very large and cannot be supported by the resin. So failure of total composite lamina occurs. When fiber fracture occurs in low-V_f lamina, the extra load on the matrix is not sufficient to fracture the resin and multiple fiber fracture occurs. This is depicted in Figure 3.

After breakage of the fiber for the $\epsilon_f^* < \epsilon_m^*$ case, the fiber ends play an important role in the failure behavior. Figure 4 illustrates the deformation around a discontinuous fiber embodied in a matrix subjected to a tensile load parallel to the fibers. According to Cox's shear-lag analysis, the tensile stress along the fiber is given by

$$\sigma_f = E_f \epsilon_m \frac{1 - \cosh \beta(\frac{1}{2}l - x)}{\cosh \frac{1}{2}\beta l} \tag{11}$$

$$\beta = \left[\frac{2G_m}{E_f r^2 \ln(R/r)}\right]^{1/2} \tag{12}$$

where $2R$ is the interfiber spacing. The shear stress at the interface is given by

$$\tau = E_f \epsilon_m \left[\frac{G_m}{2E_f \ln(R/r)}\right]^{1/2} \frac{\sinh(\frac{1}{2}l - x)}{\cosh \frac{1}{2}\beta l} \tag{13}$$

The shear stress given by Eq. (13) may cause interfacial debonding from the fiber ends. Then the frictional force will still operate and allow the transfer of stress to the fiber. Assuming that the frictional force is constant (see Figure 5), we have

$$\pi r^2 \, d\sigma = 2\pi r\tau \, dx \quad or \quad \frac{d\sigma}{dx} = \frac{2\tau}{r} \tag{14}$$

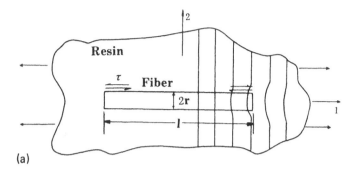

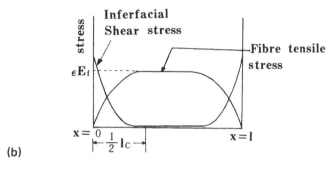

Figure 4 (a) Diagrammatic representation of deformation around a discontinuous fiber embedded in a matrix subjected to a tensile load parallel to the fiber; (b) variation along the fiber of tensile stress in the fiber and shear stress at the interface according to Cox. L_c is the critical value of the fiber length for maximum stress.

where τ is the interfacial frictional stress, and

$$l_c = \frac{\sigma_f^* r}{\tau} \tag{15}$$

the critical fiber length, represents the longest fiber length that can be extracted from the sheath of matrix. Thus before the fiber is pulled out from matrix, as the strain on the lamina increases the tensile stress on the fiber specified by Eq. (11) is increased. This results in further breakages in fibers. This fiber multiple fracture in epoxy- or polycarbonate-based composites can be clearly observed, since these two resins are transparent.

Orientation Dependence of Strength of Unidirectional Laminae

In this section the prediction of failure strength is considered with reference mainly to off-axis uniaxial tensile test. Three failure modes can be envisaged in a lamina tested in uniaxial tension at an angle θ to the fiber direction. These are longitudinal tensile, intra-laminar shear, and transverse tensile failures. The orientation dependence of failure strength for off-axis tests on unidirectional lamina of glass fiber–polyester resin composite is shown in Figure 6. It can be seen that at $0° \leq \theta \leq 4°$, longitudinal tensile failure will occur; shear failure will occur at $4° \leq \theta \leq 24°$, and transverse tensile failure occurs at $\theta > 24°$.

The strength of unidirectional lamina of type I carbon fiber epoxy resin ($V_f = 0.5$) under off-axial tension exhibits similar strong orientation dependence (see Figure 7).

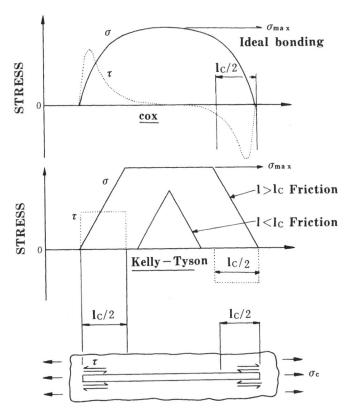

Figure 5 Stress transfer to a fiber by shear from the matrix according to the models proposed by Cox and by Kelly and Tyson.

Experiments showed that longitudinal tensile failure occurs close to $\theta = 0°$, intralaminar shear failure occurs at $45° \leq \theta \leq 20°$, and transverse tensile failure occurs at $45° \leq \theta \leq 95°$. In the range $20° \geq \theta \geq 45°$, a mixed mode can be observed, in which the interaction between different failure modes should be taken into account.

Failure Theories for Unidirectional Fibrous Composites

It should be pointed out that the prediction of failure strength in simple isotropic materials under multiaxial loading raises many difficulties, so it is not surprising that in highly anisotropic composites the problems are more complex. A vast number of experiments have been carried out. Many hypotheses and theories have been developed to predict failure strength under combined loading conditions. These have been reviewed by many authors. More than 25 theories are covered in Ref. 23.

All existing failure theories are phenomenological design criteria. The afore-described mechanistic analysis can be used to predict the composite strength properties from the strength properties of the component materials. The unidirectional strengths of lamina (i.e., the strength in fiber direction X, the transverse strength Y, and the shear strength S) can be measured experimentally by simple tests.

Lamina failure criteria are either in an independent failure mode, where both onset and mode of failure can be predicted, or in an interaction mode, where only the onset of

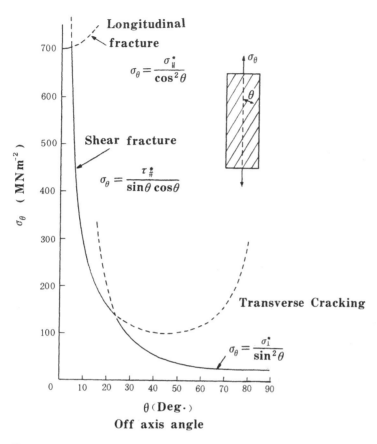

Figure 6 Orientation dependence of fracture strength for off-axis tests on unidirectional laminae predicted by maximum stress theory. (For glass fiber–polyester resin, $V_f = 0.5$.) (From Ref. 21.)

failure can be predicted. In the following, selected significant theories are summarized in tables. In Table 6, the criteria belong to the category of limiting theory; the criteria in Tables 7 and 8 are interaction failure criteria and tensor polynomial criteria, respectively. Our intention here is not to provide a complete survey, but to facilitate assessing the commonly used criteria for design purposes.

It should be mentioned that the failure envelopes are presented in six-stress-component space $(\sigma_x, \sigma_y, \sigma_z, \tau_{yz}, \tau_{xz}, \tau_{xy})$ or for the plane stress case, in three-stress-component space $(\sigma_x, \sigma_y, \tau_{xy})$.

Comparing the criteria listed in Tables 6 to 8, the following comments can be made:

1. Among the criteria given in the three tables, four failure theories are used most commonly by designers and researchers: maximum stress, maximum strain, Azzi–Tsai (also named as Tsai–Hill), and Tsai–Wu theories. For comparison, the failure envelopes of the four theories plus Hoffman theory are shown in Figure 8. In Figure 9, six theories are compared.

2. In Table 6, only three limit failure theories are listed, but there are several other theories extended from maximum stress theory [45,46]. In the maximum strain theory,

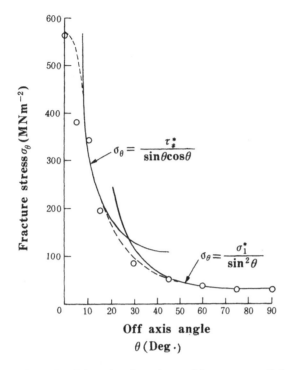

$$\sigma_\theta = \frac{\tau_{\#}^*}{\sin\theta\cos\theta}$$

$$\sigma_\theta = \frac{\sigma_1^*}{\sin^2\theta}$$

Figure 7 Orientation dependence of fracture strength for off-axis tests on unidirectional laminae. Data are for type I carbon fiber–epoxy resin, $V_f = 0.5$.

Table 6 Limit Failure Criteria Theories

Theory	Equation	Refs.	Notes
Maximum stress		24	Failure mode is predicted.
$\sigma_1 = X_t$ or X_c			
$\sigma_2 = Y_t$ or Y_c	(16)		
$\tau_{12} = S$			
Stowell–Lin		25	Strengths of fiber and matrix are taken into account
$\sigma_1 = X_f$			
$\sigma_2 = Y_m$	(17)		
$\tau_{12} = S_m$			
Maximum strain		24,26	Failure mode is determined. γ_u is ultimate shear strain.
$\epsilon_1 = \epsilon_{1t}$ or ϵ_{1c}			
$\epsilon_2 = \epsilon_{2t}$ or ϵ_{2c}	(18)		
$\gamma_{12} = \gamma_u$			

Table 7 Interaction Failure Theories

Theory	Equation	Refs.	Notes
Hill $$\frac{\sigma_1^2}{X^2} + \frac{\sigma_2^2}{Y^2} + \frac{\sigma_3^2}{Z^2} - \left(\frac{1}{X^2} + \frac{1}{Y^2} - \frac{1}{Z^2}\right)\sigma_1\sigma_2$$ $$- \left(\frac{1}{X^2} + \frac{1}{Z^2} - \frac{1}{Y^2}\right)\sigma_1\sigma_3$$ $$- \left(\frac{1}{Y^2} + \frac{1}{Z^2} - \frac{1}{X^2}\right)\sigma_2\sigma_3$$ $$+ \left(\frac{\tau_{12}^2}{S_{12}^2} + \frac{\tau_{23}^2}{S_{23}^2} + \frac{\tau_{13}^2}{S_{13}^2}\right) = 1$$	(19)	27	Three-dimensional yielding criterion for anisotropic materials; it does not consider different strengths for tensile and compressive modes.
$$\frac{\sigma_1^2}{X^2} + \frac{\sigma_2^2}{Y^2} - \left(\frac{1}{X^2} + \frac{1}{Y^2} - \frac{1}{Z^2}\right)\sigma_1\sigma_2$$ $$+ \frac{\tau_{12}^2}{S_{12}^2} = 1$$	(20)		For plane stress case.
Azzi–Tsai $$\frac{\sigma_1^2}{X^2} - \frac{\sigma_1\sigma_2}{X^2} + \frac{\sigma_2^2}{Y^2} + \frac{\tau_{12}^2}{S^2} = 1$$	(21)	28,29	Degenerated from Hill criterion for unidirectional lamina.
Marin $$(\sigma_1 - a)^2 + (\sigma_2 - b)^2 + (\sigma_3 - c)^2$$ $$+ q[(\sigma_1 - a)(\sigma_2 - b) + (\sigma_2 - b)$$ $$(\sigma_3 - c) + (\sigma_3 - c)(\sigma_1 - a)] = \sigma^2$$	(22)	30	Difference in tensile and compressive strengths is considered; $\sigma_1,\sigma_2,\sigma_3$ are principal stresses; a,b,c,q, and σ and K_1,K_2,K_3,K_4 can be found using different tests.
$$\sigma_1^2 + K_1\sigma_1\sigma_2 + \sigma_2^2 + K_2\sigma_1$$ $$+ K_3\sigma_2 = K_4$$	(23)		
Franklin $$K_1\sigma_1^2 + K_2\sigma_1\sigma_2 + K_3\sigma_2^2 + K_4\sigma_1$$ $$+ K_5\sigma_2 + K_6\tau_{12}^2 = 1$$	(24)	31	K_2 is determined from biaxial test; other K's from simple tests.
Norris–Mckinnon $$\frac{\sigma_1^2}{X^2} + \frac{\sigma_2^2}{Y^2} + \frac{\tau_{12}^2}{S^2} = 1$$	(25)	32	Degenerated from Azzi–Tsai.
Norris $$\frac{\sigma_1^2}{X^2} - \frac{\sigma_1\sigma_2}{XY} + \frac{\sigma_2^2}{Y^2} + \frac{\tau_{12}^2}{S_{12}^2} = 1$$ $$\frac{\sigma_2^2}{Y^2} - \frac{\sigma_2\sigma_3}{YZ} + \frac{\sigma_3^2}{Z^2} + \frac{\tau_{23}^2}{S_{23}^2} = 1$$ $$\frac{\sigma_3^2}{Z^2} - \frac{\sigma_3\sigma_1}{ZX} + \frac{\sigma_1^2}{X^2} + \frac{\tau_{13}^2}{S_{13}^2} = 1$$	(26)	33	More general than Azzi–Tsai theory.
Fischer $$\frac{\sigma_1^2}{X^2} + \frac{\sigma_2^2}{Y^2} + \frac{\tau_{12}^2}{S^2}$$ $$- \frac{E_1(1 + \nu_{21}) + E_2(1 + \nu_{12})}{2\sqrt{E_1E_2(1 + \nu_{12})(1 + \nu_{21})}} \frac{\sigma_1\sigma_2}{XY} = 1$$	(27)	34	Similar to Norris but includes elastic constants.

Table 7 **Continued**

Theory	Equation	Refs.	Notes
Griffith–Baldwin $$U_d = \frac{\sigma_1^2}{3}\left(S_{11} - \frac{S_{12} + S_{13}}{2}\right)$$ $$+ \frac{\sigma_2^2}{3}\left(S_{22} - \frac{S_{12} + S_{23}}{2}\right)$$ $$+ \frac{\sigma_1\sigma_2}{3}\left(2S_{12} - \frac{S_{11} + S_{22} + S_{13} + S_{23}}{2}\right)$$ $$+ S_{66}\tau_{12}^2 \qquad (28)$$		35	U_d is distortional energy. $[S_{ij}]$ is compliance matrix. When U_d equals to critical value failure occurs.
Chamis $$\frac{\sigma_1^2}{X^2} + \frac{\sigma_2^2}{Y^2} + \frac{\tau_{12}^2}{S^2} - KK'\frac{\sigma_1\sigma_2}{XY} = 1 \qquad (29)$$		36,37	$$K = \frac{[(1 + 4\nu_{12} - \nu_{13})}{[E_1E_2(2 + \nu_{12} + \nu_{13})}$$ $$\frac{E_2 + (1 - \nu_{23})E_1]}{(2 + \nu_{21} + \nu_{23})]^{1/2}}$$ K' is determined from biaxial test data.
Hoffman $$K_1(\sigma_2 - \sigma_3)^2 + K_2(\sigma_3 - \sigma_1)^2$$ $$+ K_3(\sigma_1 - \sigma_2)^2 + K_4\sigma_1 + K_5\sigma_2$$ $$+ K_6\sigma_3 + K_7\tau_{23}^2 + K_8\tau_{13}^2$$ $$+ K_9\tau_{12}^2 = 1 \qquad (30)$$ $$\frac{\sigma_1^2 - \sigma_1\sigma_2}{X_tX_c} + \frac{\sigma_2^2}{Y_tY_c} + \frac{X_c - X_t}{X_cX_t}\sigma_1$$ $$+ \frac{Y_t - Y_c}{Y_tY_c}\sigma_2 + \frac{\tau_{12}^2}{S^2} = 1 \qquad (31)$$		38	Similar to Hill, but accounts for different strengths in tension and compression; (31) is degenerated case for plane stress problem.
Zhang–Tsai $$\frac{\sigma_1^2}{X^2} + \frac{\sigma_2^2}{Y} - \frac{\nu_{12}}{E_{11}}\left(\frac{E_{11}}{X_2} + \frac{E_{22}}{Y_2}\right)\sigma_1\sigma_2$$ $$+ \frac{\tau_{12}^2}{S^2} = 1 \qquad (32)$$		39	Equation (32) is derived from strain energy density ratio (SEDR); also termed the SEDR criterion.

although there is no interaction between strain components, there does exist some interactions between stress components. In principal stress space, the failure envelope of maximum strain criterion is a parallelogram.

3. In the category of interaction failure theories, more than 20 theories could have been listed. Most of them are derived from Von Mises–Hencky maximum distortional energy theory by including anisotropy of strength properties. Some theories consider the difference in tensile and compressive strength properties. In Azzi–Tsai theory, for tensile stresses σ_1 and σ_2 (they are positive), X and Y are tensile strengths, whereas for compressive stresses (σ_1 and σ_2 are negative in value), X and Y are compressive strengths.

4. For the stress tensor polynomial theories, the most general theory is given by Eq. (33). From this equation, several degenerated forms are developed. Among them the

Table 8 Stress Tensor Polynomial Failure Theories

Theory	Equation	Refs.	Notes
Gol' denblat–Kopnov $(F_{ij}\sigma_{ij})^a + (F_{ijkl}\sigma_{ij}\sigma_{kl})^\beta$ $+ (F_{ijklmn}\sigma_{ij}\sigma_{kl}\sigma_{mn})^\gamma$ $+ \cdots = 1$	 (33)	40	It is the most general equation for criteria of this kind $i,j,k,l,m,n = 1,2,3$
Malmeister $F_{ij}\sigma_{ij} + F_{ijkl}\sigma_{ij}\sigma_{kl}$ $+ F_{ijklmn}\sigma_{ij}\sigma_{kl}\sigma_{mn} + \cdots = 1$	 (34)	41	$i,j,k,l,m,n = 1,2,3$ $F_{ij}F_{ijkl} \cdots$ are strength tensors. It is a special case of Gol' denblat–Kopnov.
Tsai–Wu $F_i\sigma_i + F_{ij}\sigma_i\sigma_j = 1$ $F_1\sigma_1 + F_2\sigma_2 + F_{11}\sigma_1^2$ $+ F_{22}\sigma_2^2 + 2F_{12}\sigma_1\sigma_2$ $+ F_{66}\tau_{12}^2 = 1$	 (35) (36)	42–44	$i,j, = 1,2,\cdots,6$ F_i, F_{ij} are strength tensors of second and fourth orders. Equation (36) is for two-dimensional problems.

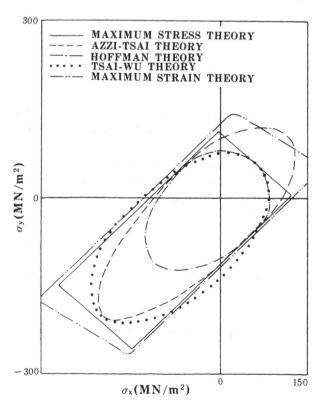

Figure 8 Comparison of the envelopes of the most commonly used failure theories. (From Ref. 23.)

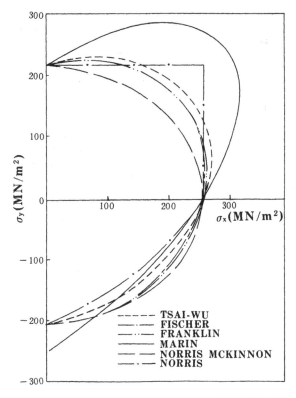

Figure 9 Comparison of the envelopes of six failure theories. (From Ref. 47.)

Tsai–Wu theory (36) is used most in practice. As a matter of fact, nearly all of the stress tensor polynomial theories have been used in their special cases, in which they reduce to the Tsai–Wu theory.

5. Equation (32) is derived from the strain energy density ratio (SEDR) criterion [39]. The strain energy density ratio, S_R, is calculated by using the stress and strain ratios in the following way. The stress ratios are defined as the stress components divided by the corresponding strengths:

$$\overline{\sigma_1} = \frac{\sigma_1}{X}, \qquad \overline{\sigma_2} = \frac{\sigma_2}{Y}, \qquad \overline{\tau_{12}} = \frac{\tau_{12}}{S}, \tag{37}$$

and the strain ratios are defined as the strain components divided by their critical values:

$$\overline{\epsilon_1} = \frac{\epsilon_1}{\epsilon_{1c}}, \qquad \overline{\epsilon_2} = \frac{\epsilon_2}{\epsilon_{2c}}, \qquad \overline{\gamma_{12}} = \frac{\gamma_{12}}{\gamma_{12c}} \tag{38}$$

Using the stress and strain ratios, the strain energy density ratio S_R can be calculated as

$$S_R = \tfrac{1}{2}(\overline{\sigma_1}\,\overline{\epsilon_1} + \overline{\sigma_2}\,\overline{\epsilon_2} + \overline{\tau_{12}}\,\overline{\gamma_{12}}) \tag{39}$$

The failure theory of SEDR postulates that failure occurs when twice S_R is equal to or larger than unity:

$$2S \geq 1 \tag{40}$$

Substituting Eqs. (37) to (39) into (40), the expression of Eq. (32) was obtained.

Failure Behavior of Cross-Ply Laminates

Cross-ply laminates can consist of different numbers of differently oriented laminae. For simplicity, our discussion will be centered on three-layer cross-ply laminates as illustrated in Figure 10. The thickness of the two outer layers is the same, so that the laminate is symmetrical about its midplane.

Before analyzing its failure behavior, some influential factors should be mentioned.

1. The thermal expansion mismatch results in different contraction between laminae on cooling to ambient temperature subsequent to elevated temperature curing or postcuring operations. This may cause buckling of unsymmetrical laminates. In a balanced symmetrical laminate, significant internal stress can be developed even though there is no obvious distortion. The residual strain between the adjacent 0° and 90° plies may cause transverse cracks in the 90° ply without external stress being applied, but generally they reduce the threshold strain at which these cracks are formed during 0° axial extension.

2. The effect of the Poisson ratio mismatch is that a tensile stress is induced in the transverse direction of 0° plies by interaction with the adjacent 90° ply. This may result in longitudinal splitting of the 0° plies.

3. In addition to these in-plane effects, out-plane stresses may be induced. These stresses, called *interlaminar normal stress* and *transverse shear stress*, may lead to delamination of the laminate. The extent of these interactions is influenced by the thickness of individual laminae. In the present section we describe the influence of laminae thicknesses on the development of damage under monotonic loading.

First-Ply Failure Strength and Multiple-Failure Phenomenon

A schematic section of a [0°/90°/0°] cross-ply laminate is as shown in Figure 10. The two outside 0° plies are of equal thickness *b* and the central 90° ply is of thickness *2d*; the

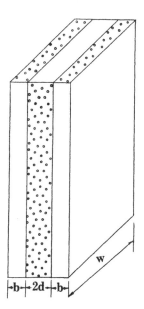

Figure 10 Cross-ply laminate.

width of the specimen is w. When the laminate is extended in tension, the initial failure is usually in the central 90° ply, which cracks in the direction normal to the applied load and parallel to the fibers of that layer. In general, the failure sequence in both GRP and CFRP laminates follows a similar pattern. The first cracking in 90° ply can be predicted by using the following formula:

$$\frac{\sigma_2^{90}}{\sigma_{c1}} = \frac{\frac{1}{2}E_2(E_1 + E_2) - v_{12}^2E_2^2}{\frac{1}{4}(E_1 + E_2)^2 - v_{12}^2E_2^2} \tag{41}$$

where the subscript c is for laminate, 1 is for longitudinal direction, 2 for transverse direction, and the superscript 90 is for 90° ply. When $\sigma_2^{90} = \sigma_2^*$, transverse cracking in 90° ply occurs.

This transverse cracking produces a knee point on the load–strain curve of the GRP, but this is not apparent in the CFRP. Typical load–strain curves and acoustic emission traces are shown in Figure 11. On further extension, more and more cracks are formed, with the crack spacing getting smaller and smaller until a saturated state is reached where the entire gauge portion of the specimen is filled with a regular array of cracks of minimum spacing. The strain at which the first crack occurs increases as the thickness ($2d$) of the middle 90° layer is reduced; at the same time the minimum crack spacing tends to be smaller. When the thickness of 90° ply is smaller than a limit value, no cracking in the middle 90° ply occurs (i.e., the transverse cracking is totally suppressed). Figure 12 illustrates the schematical representation of the laminate with regularly spaced cracks.

Clear study of the load–extension curve for GRP laminate reveals that the knee point (at around 0.55% strain) is associated with transverse cracking and the onset of acoustic emission. Before the onset of acoustic emission and transverse cracking, a visual whitening effect usually occurs. But for CFRP, no whitening effect can be observed. Although the

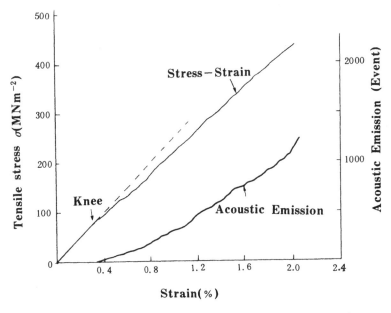

Figure 11 Typical stress–strain curve and acoustic emission output of a cross-ply laminate tested in uniaxial tension.

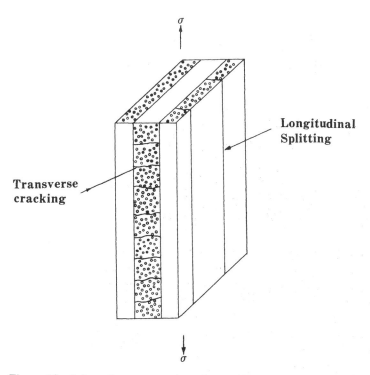

Figure 12 Schematic representation of regularly spaced cracks in transverse lamina and isolated cracks in longitudinal laminae of a cross-ply laminate.

first cracking coincides with the knee point, gradual cracking and multiple cracking require increasing of load. In Figure 13a–c, photographs of three different glass fiber–polyester resin cross-ply laminates strained in tension to 1.6% are given, showing that different crack spacing are due to different transverse-ply thickness. Figure 13d illustrates the polished section of a symmetric eight-layer cross-ply carbon fiber–epoxy resin laminate, showing similar transverse cracking phenomenon. Figure 14 shows longitudinal splitting in glass fiber-epoxy resin cross-ply laminate.

The shear-lag theory can be used to predict the redistribution of stress which occurs after a transverse crack has formed. When the layers are elastically bonded, there is an additional stress $\Delta\sigma$ on the longitudinal layers (see Figure 15). The additional stress $\Delta\sigma$ has a maximum value $\Delta\sigma_0$ in the plane of the transverse crack and decays along the longitudinal laminae as

$$\Delta\sigma = \Delta\sigma_0 \exp(-\Phi^{1/2} y) \tag{42}$$

where

$$\Phi = \frac{E_c G_{12}}{E_{11} E_{22}} \frac{b+d}{bd^2} \tag{43}$$

E_c is the initial composite modulus and can be approximated by

$$E_c = E_{11} \frac{b}{b+d} + E_{22} \frac{d}{b+d} \tag{44}$$

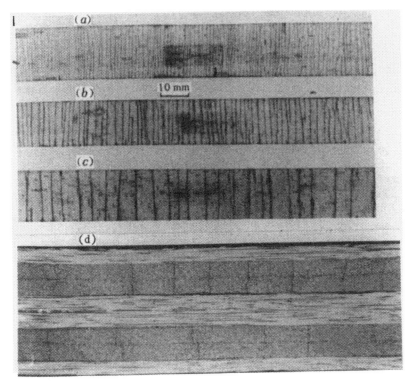

Figure 13 Transmitted-light photographs of three glass fiber–polyester resin cross-ply laminates, showing different spacings due to different transverse ply thickness: (a) 0.75 mm; (b) 1.5 mm; (c) 2.6 mm. (d) Section of a carbon fiber–epoxy resin (0°,90°) laminate showing regularly spaced cracks in transverse layers. (From Refs. 48 and 21.)

G_{12} is the shear modulus. The corresponding load F transferred back into the transverse layer at a distance y is given by

$$F = 2bw \, \Delta\sigma_0[1 - \exp(-\Phi^{1/2}y)] \tag{45}$$

If the first crack formed at a stress on the laminate $\sigma_c = \epsilon_2^* E_c$, which corresponds to a stress on the lamina σ_2^*, it is reasonable to assume that the next crack occurs when this stress is reached elsewhere in the transverse lamina. When σ_c is increased further, the next series of cracks forms midway between the earlier formed cracks, and the process continues until the crack spacing reaches a limiting value. Figure 16 shows an example of the change in average crack spacing with applied stress. The stepped curve represents the predicted spacing of the transverse cracks when the first crack forms in the center of the specimen. The upper and lower continuous curves indicate the predicted range of crack spacing for specimens of any length with an arbitrary first-crack position.

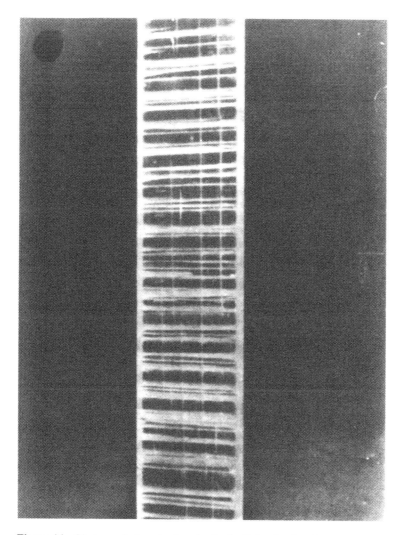

Figure 14 Photograph showing longitudinal splitting in glass fiber–epoxy resin cross-ply laminate. (From Ref. 49.)

At high stress levels (say, near the final failure stress) the slope of the stress–strain curve is approximately equal to

$$\frac{d\sigma}{d\epsilon} = E_{11} \frac{b}{b + d} \tag{46}$$

and the final failure strength is determined by

$$\sigma_c^* = \frac{\sigma_f^* b}{b + d} \tag{47}$$

Longitudinal Splitting

Longitudinal splitting was observed to occur in the 0° laminae of the cross-ply laminates at strains intermediate between the transverse cracking strain for the 90° lamina and the

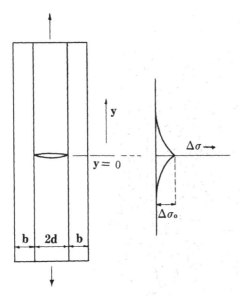

Figure 15 Transverse crack in a cross-ply laminate illustrating additional stress $\Delta\sigma$ on longitudinal layers.

final failure strain. The strain to initiate splitting increases as the thickness of the longitudinal plies is decreased. In the case of thicker longitudinal ply the splits form instantaneously, while for thinner plies the cracks propagate slowly as the strain is increased. The isolated cracks in longitudinal laminae were shown in Figure 12. Detailed quantitative analysis on longitudinal splitting in cross-ply laminate with consideration of thermal and Possion effects is given in Ref. 49. Figure 14 shows longitudinal-ply splitting in a [0°/90°/0°] GFRP cross-ply specimen.

Failure Behavior of Angle-Ply Laminates

It is common practice to make up the required laminate from a series of laminae stacked with the fibers aligned at 0°, 90°, and ± 45° to a reference direction. The 0° fibers are designed to carry the principal tensile or compressive loads, and 90° and ±45° laminae are added to bear any transverse or shear stresses, respectively. A sequence incorporating equal numbers of 0°, 90° and ±45° laminae yields a quasi-isotropic laminate.

To describe the failure mechanism of angle-ply laminates, it will suffice to deal with the simplest angle-ply laminates, whose layup construction is of [+θ°/−θ°] or in its particular case, [−45°/+45°]. For [±45] laminates, the main mechanical properties of the five layups tested in uniaxial extension are summarized in Table 9. Generally, the failure strains of [±45] specimens are much higher than that of [0°/90°] laminates and unidirectional laminates. It can be seen that the modulus at failure is much lower than that at low strains. This reflects the nonlinearity of the resin and the progressive cracking of the laminates. The load–strain curves for all the specimens of [±45] construction are complete nonlinear, as shown in Figure 17.

The failure mechanisms of these specimens follow roughly the same pattern at a low strain stage. The first failure event is the formation of edge cracks, which form as a

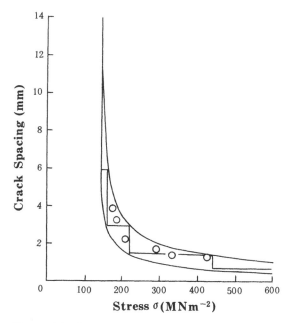

Figure 16 Comparison of experimental results with theoretical curves of crack spacings as a function of applied stress for a 190-mm-long glass fiber–epoxy resin cross-ply laminate with a transverse ply thickness of 1.2 mm. (From Ref. 50.)

consequence of the out-of-plane stresses that exist at the free edges (see Ref. 51). At high strains these cracks propagate along the fiber direction and eventually cause delamination and ultimate failure.

When angle-ply laminates are tested under uniaxial tensile stress, they exhibit a complicated failure process, in which interlaminar shear effects and intralaminar shear effects are always coupled together. The sequence of various failure events is dependent on the layup angle θ. Figure 18 illustrates the variation of the initial failure stress σ_c of the laminate with the layup angle θ. The data were obtained for a glass fiber–polyester resin with $V_f = 0.5$, using $\sigma_1^* = 700$ MPa, $\sigma_2^* = 22$ MPa, and $\tau_{12}^* = 50$ MPa. The most

Table 9 Mechanical Properties of $\pm 45°$ CFRP Laminates

Laminate	$[\pm 45]s$	$[+45/\\-45_2/+45]s$	$[+45_2/\\-45_2]s$	$[(+45/\\-45)_2]s$	$[+45_3/\\-45_3]s$
Low strain secant modulus (GPa)	17.3	18.2	19	17	14
Secant modulus at failure (GPa)	7.5	10.6	11.9	9.1	9.4
Failure stress (MPa)	126	152	135	125	89
Failure strain	0.017	0.016	0.0117	0.014	0.01
Acoustic emission threshold strain	0.008	0.006	0.007	0.01	0.007

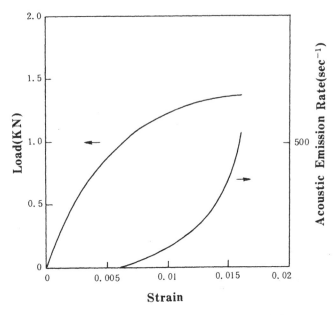

Figure 17 Load–strain curve and load-acoustic emission rate curve for a $[+45/-45_2/+45]_s$ specimen of glass fiber–epoxy resin.

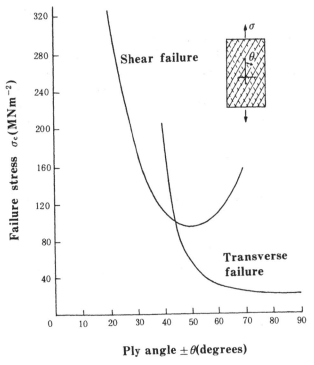

Figure 18 Failure stress predicted for glass fiber–polyester resin ($V_f = 0.5$). (From Ref. 21.)

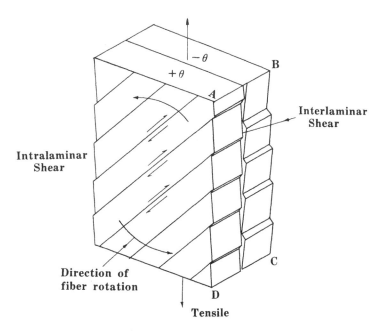

Figure 19 Schematic view of intralaminar and interlaminar shear associated with a tensile test on an angle-ply laminate.

likely initial failure mode can be predicted from these curves. It is noted that in Figure 18, the failure mode is predicted by the maximum stress failure criterion.

Angle-ply laminates undergoing uniaxial tension test show notable edge effects that stem primarily from the *scissors effect*, that is, the tendency of the individual laminae to rotate so that the fibers lie closer to the tensile axis. This means that the adjacent laminae move in opposite directions and shear displacement must occur between the laminae (interlaminar). Similarly, shear must take place within each lamina so that rotation of the fiber can occur (intralaminar). These effects are represented schematically in Figure 19. The edge effects play an important role on the tensile properties of the laminates. Figure 20 illustrates the effect of the specimen width on the tensile strength of glass fiber–polyester resin angle-ply laminate $V_f = 0.5$. The edge effect is more pronounced at low values of θ; say, for $\theta = 50°$, the transverse cracking will dominate. Experimentally, it is found that as the width decreases the applied stress at the onset of transverse cracking increases, and that below a critical width, transverse cracking is suppressed. These phenomena can be seen in Figure 21.

Initial Failure, Postfailure Behavior, and Direct Failure Prediction of Multilayer Laminates

Initial Failure Analysis

The primary difficulty lies in defining what is meant by strength since the complete failure of laminates is usually preceded by fracture of the individual laminae. In practice the relevant strength depends on the particular application of the composite material. The method for predicting the final failure strength of a laminate is as follows:

1. For a given laminate construction and applied loading conditions, stresses in individual laminae are calculated at progressively increasing load.

2. The laminae stresses are compared with the predicted failure stresses of the failure criteria described previously (Tables 6 to 8).

3. When the applied load is sufficiently large for the failure criterion to be satisfied for one of the laminae, it is assumed that failure of this lamina and all other laminae oriented in the same way occurs, and the load supported by these laminae is transferred to laminae oriented differently.

4. The stresses on the remaining laminae are recalculated for further increase in load until the failure criterion is satisfied for other laminae. At this stage, final failure of the laminate may occur.

So the failure strength is categorized into two types of failures: *first-ply failure* and *complete failure* of the laminate.

The first-ply failure strength of cross-ply laminates and some simple constructed angle-ply laminates can be predicted through the aforementioned procedures. In the following section, postfailure (post-first-ply failure) behavior is described.

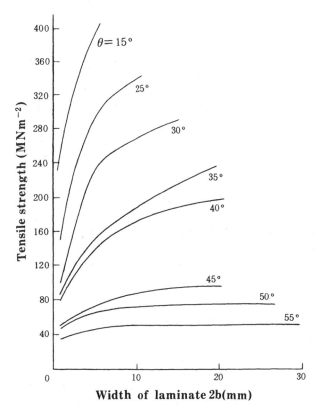

Figure 20 Effect of test specimen width on tensile fracture strength of glass fiber–polyester resin angle-ply laminate ($V_f = 0.5$). (From Ref. 21.)

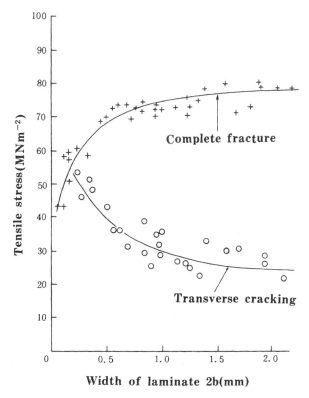

Figure 21 Effect of test specimen width on tensile stress for transverse cracking and complete fracture in angle-ply laminates (θ = 50°) for glass fiber–polyester resin ($V_f \approx 0.5$). (From Ref. 21.)

Postfailure Behavior of Laminates

As mentioned previously, the failure process consists of two stages: first-ply failure and final failure. First-ply failure in a laminate does not mean that the entire laminate lose its load-bearing capacity. Many researchers have investigated the postfailure behavior of laminated composites.

1. *Hahn–Tsai method.* This theory [52] considers that any failed lamina will support the load it was carrying when failure occurred, until the total laminate fails. This theory was applied to a cross-ply laminate under uniaxial extension. The laminate after the failure of 90° lamina will have a modulus equal to the modulus of 0° layers subjected to a correction for area reduction. This behavior is represented by Figure 22.

2. *Petit–Waddoups method.* In this theory [53] the failed lamina unloads gradually. Mathematically, this can be described by giving the tangent modulus a relatively high negative value as shown in Figure 23.

3. *Chiu method.* This method [54] is illustrated by using Figure 24, which indicates that when the transverse layer fails it unloads instantaneously in the direction in which the failure occurs.

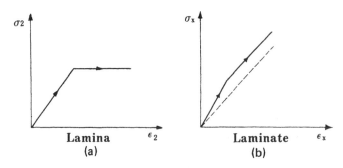

Figure 22 Postfailure model in Hahn–Tsai method.

4. *Sandhu method.* Sandhu's method [55] assumes that when the contribution to the degradation is dominantly transverse or shear, the lamina is still only capable of sustaining longitudinal load.

5. *Brown method.* This method [56] is similar to that of Sandhu's but is more general. In addition to Sandhu's assumption, Brown added another assumption: If the contribution to degradation is dominantly longitudinal, the lamina is assumed to be capable of carrying transverse load only.

6. *Puck–Forster–Knappe method.* This method [57] suggests that at resin cracking in a lamina, the transverse and shear elastic constants do not vanish immediately but sink gradually toward zero with increasing strain. The overexertion factor

$$U = \left[\left(\frac{\sigma_1}{X_m} \right)^2 + \left(\frac{\tau_{12}}{S} \right)^2 + \left(\frac{\sigma_2}{Y} \right)^2 \right]^{1/2} - 1 \tag{48}$$

is introduced, from which the correction factor η is obtained for each step. Figure 25 shows schematically the relation between the correction factor, η, and the overexertion factor, U.

7. *Hull–Legg–Spencer method.* Hull et al. [58] used Puck's failure criterion [59] to predict the onset to nonlinearity. Beyond the first-ply failure strength, "netting" analysis is employed to predict the total failure. In the netting analysis, the resin is assumed not to transfer any load.

8. *Nahas method.* In this approach [60] the failed layer unloads gradually in the direction of the failure following the exponential function, as shown in Figure 26.

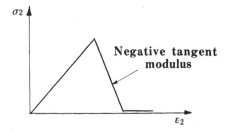

Figure 23 Postfailure model in Petit–Waddoups method.

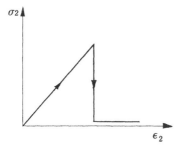

Figure 24 Postfailure model in Chiu method.

9. *Zhang method.* Zhang et al. [61] analyzed the damage process in the lap-shear specimen of chopped-strand-mat glass-fiber-reinforced polyester GRP. FEM was used in the calculations. The following four conditions were used to predict the damage process.

If $\tau \geq \tau_c$: $G_{12} = 0$

If $\sigma_2 \geq \sigma_{2c}$: $E_{22} = \nu_{21} = 0$ (49)

If $\sigma_1 \geq \sigma_{1c}$: $E_{11} = \nu_{12} = 0$

$$\text{If } \Phi = \sqrt{\left(\frac{\sigma_1}{\sigma_{1c}}\right)^2 + \left(\frac{\sigma_2}{\sigma_{2c}}\right)^2 - \frac{\sigma_1\sigma_2}{\sigma_{1c}\sigma_{2c}} + \left(\frac{\tau}{\tau_c}\right)^2} \geq 1: \quad E_{11} = E_{22} = G_{12} = 0$$

Direct Laminate Failure Theories

Although the failure behavior is very complicated in nature and depends on the layup constructions and properties of each lamina, many attempts have been made to predict the final failure of laminates using direct failure theories.

1. Wu and Schenblein [62] consider that the tensor polynomial failure criterion is also applicable to laminates. In this case the principal directions of strength are not necessarily orthogonal; hence the failure surface may not be ellipsoidal, and higher-order terms need to be included as in the following equation:

$$F_i\sigma_i + F_{ij}\sigma_i\sigma_j + F_{ijk}\sigma_i\sigma_j\sigma_k = 1 \tag{50}$$

All the strength parameters of noninteraction terms can be determined from uniaxial tests, but the interaction terms must be determined by biaxial tests under prescribed optimum stress ratios. Wu and Schenblein suggest a "hybrid" analysis, which is a combination

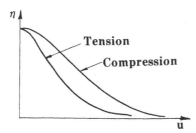

Figure 25 Correction factor for Puck–Forster–Knappe analysis.

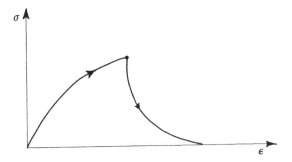

Figure 26 Postfailure model of Nahas.

of lamina failure and laminate failure analysis. In this way, better agreement was obtained between the experimental and predicted failure envelope values.

2.   Soden et al. [63] proposed that the following equations be applied directly to analyze the final failure of laminates:

$$\left(\frac{\sigma_x}{X}\right)^2 - \gamma_1 \frac{\sigma_x \sigma_y}{Y^2} + \gamma_1 \left(\frac{\sigma_y}{Y}\right)^2 = 1$$
$$\gamma_2 \left(\frac{\sigma_x}{X}\right)^2 - \gamma_2 \frac{\sigma_x \sigma_y}{Y^2} + \left(\frac{\sigma_y}{Y}\right)^2 = 1 \tag{51}$$

where the two constants γ_1 and γ_2 must be determined experimentally under biaxial test conditions. Their values may be quite different from one stress quadrant to another.

3. Guess and Gerstle [64] used two failure criteria directly to a laminate. The first is the maximum stress criterion; the second criterion is as given by the following equation:

$$\left(\frac{\sigma_x}{X}\right)^2 - \frac{\sigma_x}{X}\frac{\sigma_y}{Y} + \left(\frac{\sigma_y}{Y}\right)^2 = 1 \tag{52}$$

which is similar to that of the Norris lamina criterion (26), except for the absence of shear loading.

Because they lack general applicability, these theories and others are little used by other workers.

MECHANICS OF COMPOSITE FRACTURE

Main Features of Composite Fracture

The fracture phenomena of fibrous composite materials are complex in nature, owing to the complexity and diversity of composite systems and layup constructions. Unlike metallic materials, a composite (or a composite laminate) containing a hole or a notch fractures in a complicated way. The crack path is irregular and may be neither parallel to the original notch nor perpendicular to the applied stress direction. The direction of crack propagation in each ply of the laminate is essentially dependent on the orientation of fibers. In general, plies containing fibers that are at the same angle to the applied stress direction fail by cracking parallel to those fibers. To do so, these off-axis plies must delaminate first, as illustrated schematically in Figure 27. Thus only in rare cases, say,

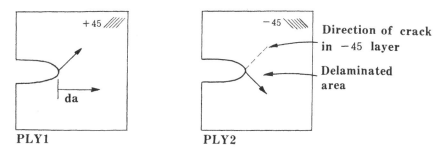

Figure 27 Schematic diagram of delamination between adjacent plies of [± 45°] angle-ply laminate.

mode I crack parallel to fibers, will the crack propagate self-similarly (or named colinearly). Besides, at the crack tip a large amount of damage will occur; for this reason, the well-established linear fracture mechanics for traditional metallic materials cannot be used without making modifications. A new theoretical system of fracture mechanics must be developed for predicting and analyzing the fracture behavior of composites.

Comparing with metallic materials, the fracture of composite materials possesses the following features:

1. The crack in composites frequently propagates non-self-similarly, and most composites or composite laminates fracture in mixed-mode fashion.

2. The materials manifest multiple fracture behavior, as described earlier.

3. A large damage zone at the crack tip will form prior to extension of that crack. In the damage zone, various damage mechanisms are intermingled.

4. Two different regimes may be identified in composite materials. When defects are small, general damage such as yielding in metals occurs as dispersed small cracks in the matrix, fiber breakages, or interface debonding; but when a defect is large, it may grow as a single crack, leading to final fracture with little damage or deformation in the material remote from the crack tip. Generally, there is no clear-cut boundary dividing the two regimes. For many kinds of materials, in the earlier stage of fracture it falls in the general damage fracture regime, while in the later stages larger damage areas (or cracks) will grow faster than others; eventually, the largest crack leads to final fracture. The two fracture modes are shown schematically in Figure 28.

5. Composite materials are strongly anisotropic, not only in their stiffness and strength properties, but also in their fracture behavior. For example, unidirectional composites possess different resistances to crack propagation in different directions. Cracks tend to extend along the fibers rather than normal to them. Kelly [65] has reported that for a typical glass-fiber-reinforced epoxy, the critical strain energy release rate G_{1c} for an opening mode crack running parallel to the fibers may be as small as 10^{-1} J/m^2, whereas the corresponding G_{1c} for a crack running normal to the fibers may be as large as 10^5 J m^{-2}.

Recently, Zhang [66] and Maiti [67] studied the mode I fracture toughness of two cases of a crack parallel to the fibers and a crack initially normal to the fibers. The results indicate that the fracture toughness for the first case, K_{Ic}^L is much smaller than that for the second case, K_{Ic}^T, although in the latter case the crack also propagates along the fiber

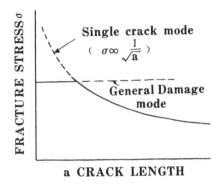

Figure 28 Schematic showing the stress/crack length dependence of single-crack failure mode and general damage failure mode.

direction. For a glass–epoxy composite the ratio of K_{I}^{T}/K_{Ic}^{L} is 3 to 4.5, and for a graphite–epoxy lamina K_{I}^{T}/K_{Ic}^{L} is 4 to 5.

Micromechanics of Composite Fracture

When a composite laminate fractures from preexisting flaws or notches, a series of energy-absorbing events can occur in the region surrounding the notch tip and in the crack wake: matrix cracking, fiber debonding from matrix, fiber breaking at weak point, and pullout of the broken fibers from the matrix sheath. The fracture process close to a crack tip in a unidirectional composite is shown schematically in Figure 29, where the fiber fracture strain is greater than the matrix fracture strain. According to Kelly [68], only a small

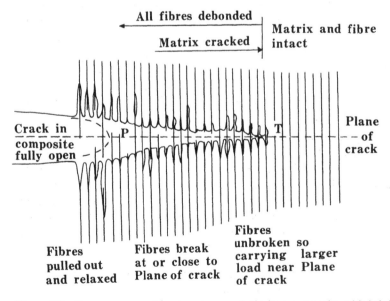

Figure 29 Processes occurring close to a crack tip in a composite with brittle fibers of fracture strain greater than that of the matrix.

amount of energy can be absorbed in resin cracking and fiber breaking, since the two processes are essentially brittle. The main energy-absorbing mechanisms are the interfacial debonding between the fiber and matrix and the fiber pullout. In the following section, these two fracture mechanisms are described.

Interfacial Debonding

The debonding energy that is absorbed during interfacial debonding can be given as [68]

$$W_D = \frac{1}{6}\pi r^2 \frac{\sigma_f^*}{E_f} x\sigma_f^* \tag{53}$$

where x is the debonding length, σ_f^* the ultimate strength of the fiber, and r the radius of fibers.

A more sophisticated investigation of the process of debonding–fracture–pullout of a fiber from the matrix has been carried out [69]. The energy-absorbing mechanisms can be illustrated using Figures 30 and 31. Figure 30 shows the stress distribution along the fiber after it debonds from matrix on the two sides of a matrix crack. It can be seen that the stress transfer from the fiber to the matrix is through friction on the interface, and the maximum stress point lies in the crack surface of the matrix. It is clear that it is possible for the fiber to fracture at another point rather than at the maximum stress point, since the weakest point in the fiber may not coincide with the matrix crack plane.

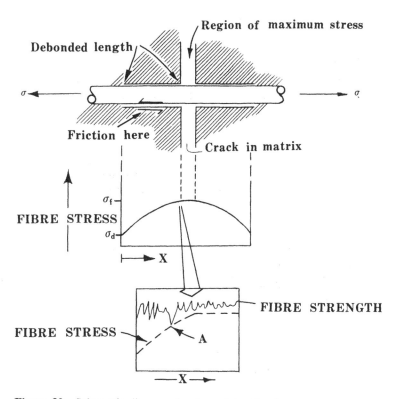

Figure 30 Schematic diagram showing origin of pullout.

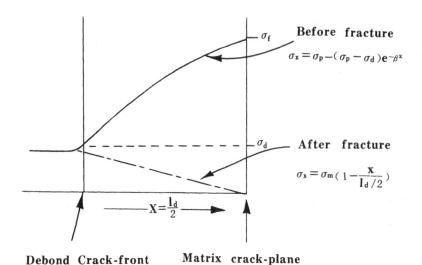

Figure 31 Schematic diagram of fiber stress distribution before and after fracture.

The schematic diagram in Figure 31 illustrates the fiber stress distribution before and after its fracture. Before fracture the fiber stress is given as

$$\sigma(x) = \sigma_p - (\sigma_p - \sigma_d)e^{-\beta x} \tag{54}$$

where

$$\sigma_p = \frac{\epsilon_0 E_f}{\nu_f}, \qquad \beta = \frac{2\mu\nu_f E_m}{E_f r_f(1 + \nu_m)} \tag{55}$$

ϵ_0 is the mismatch strain of Poisson's contraction between fiber and matrix, and μ is the coefficient of friction between fiber and matrix. σ_d is the debonding stress and can be obtained from [70]

$$\sigma_d = \left(\frac{4E_f G_{2c}}{r_f}\right)^{1/2} \tag{56}$$

G_{2c} is the critical strain energy release rate for interfacial shear cracking.

After fracture the fiber stress can be approximated by a straight line:

$$\sigma(x) = \sigma_m\left(1 - \frac{x}{l_d/2}\right) \tag{57}$$

where $l_d/2$ is half the debonded length. The fracture toughness equivalent to the release of stored elastic strain energy when a debonded fiber breaks suddenly can be calculated by integrating the two stress curves, which yields

$$G_f = \frac{V_f}{E_f}\left[\frac{\sigma_p^2 l_d}{2} - \frac{(\sigma_p - \sigma_d)^2(e^{-\beta l_d} - 1)}{2\beta} + \frac{2\sigma_p(\sigma_p - \sigma_\alpha)(e^{-\beta l_d/2} - 1)}{\beta} - \frac{\sigma_m^2 l_d}{6}\right] \tag{58}$$

Fiber Pullout Energy

The work to pullout a single fiber over distance l is

$$W = \int_o^l \pi r_f^2 \sigma(\chi)\,dx \tag{59}$$

where $\sigma(\chi)$ is fiber stress and has the expression [71]

$$\sigma(\chi) = \sigma_p(1 - e^{-\beta x}) \tag{60}$$

σ_p and β are as given in Eq. (55).

The fiber pullout energy can also be calculated using the following equations according to Kelly [65]:

$$W_e = \frac{V_f \sigma_f l_c^2}{12l} \qquad (l > l_c) \tag{61}$$

$$W_e = \frac{V_f \sigma_f l^2}{12 l_c} \qquad (l < l_c) \tag{62}$$

where l_c, the ineffective length of the fiber, can be determined using Eq. (15). The work of fracture due to pullout against fiber length is sketched in Figure 32.

Macromechanics of Composite Fracture

As mentioned earlier, macroscopically fibrous composites may fracture in two different modes: the general damage mode and the single-crack propagation mode.

General Damage Mechanics

Fibrous composites frequently contain manufacturing defects and weak fibers that break at low applied loads. General damage mechanics are concerned with the formation and aggregation of such dispersed defects, leading to failure. When a fiber has broken, the axial load formerly carried by that fiber is transferred to other fibers.

Rosen [72], Daniels [73], and Gucer and Gurland [74] modeled this process and assumed the load to be shared equally among all other fibers. Known as the *equal-load-sharing rule*, this takes no account of the stress concentration.

An alternative model, known as the *local-load-sharing rule*, was developed by Hedgepeth [75], Hedgepeth and Van Dyke [76], Scop and Argon [77], and Zweben [78,79]. In this model, the load on the fiber adjacent to broken fibers increases by a factor K_i above the average fiber stress in the composite. In a two-dimensional infinite large plate this factor is given by

$$K_i = \frac{4 \cdot 6 \cdot 8 \cdots (2i + 2)}{3 \cdot 5 \cdot 7 \cdots (2i + 1)} \tag{63}$$

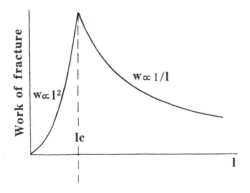

Figure 32 Work of fracture due to pullout against fiber length for a composite with discontinuous fibers. (From Ref. 65.)

where i refers to the number of broken fibers. Zweben's work assumes that when two neighboring fibers break, the composite plate fails (i.e., $i = 2$). This model was modified by Scop and Argon [77], who concluded that the failure was governed by the fracture of two fibers adjacent to a broken fiber (i.e., when $i = 3$). The predicted strength is given by a simplified form:

$$\sigma_u = \frac{\sigma_0}{K_3} \tag{64}$$

where σ_0 is the composite strength following the *rule of mixture*, $K_3 = 1.92$ for an elastic matrix, and $K_3 = 1.36$ for an elastic–plastic matrix.

Single-Crack Propagation Mechanics

The problem of whether linear elastic fracture mechanics (LEFM) established for metal materials is applicable to fiber composites has long been investigated by many researchers. Sih [80] has pointed out that when the crack surface is smooth and straight and the crack propagates self-similarly, the fracture problem can probably be analyzed by LEFM; otherwise, the results of LEFM may not be correct. Because in most cases the crack surface is neither smooth nor self-similar, the theories and criteria of LEFM must be modified to be suitable for a composite fracture.

In the case that LEFM is applicable to composite materials, it is assumed that [81] the following formulas from anisotropic fracture mechanics can be used to predict stress distribution in the vicinity of crack shown in Figure 33:

$$
\begin{aligned}
\sigma_x &= \frac{K_I}{\sqrt{2\pi r}} \operatorname{Re}\left[\frac{\mu_1 \mu_2}{\mu_1 - \mu_2} \left(\frac{\mu_2}{\psi_2} - \frac{\mu_1}{\psi_1} \right) \right] + \frac{K_{II}}{\sqrt{2\pi r}} \operatorname{Re}\left[\frac{1}{\mu_1 - \mu_2} \left(\frac{\mu_2^2}{\psi_2} - \frac{\mu_1^2}{\psi_1} \right) \right] \\
\sigma_y &= \frac{K_I}{\sqrt{2\pi r}} \operatorname{Re}\left[\frac{1}{\mu_1 - \mu_2} \left(\frac{\mu_1}{\psi_2} - \frac{\mu_2}{\psi_1} \right) \right] + \frac{K_{II}}{\sqrt{2\pi r}} \operatorname{Re}\left[\frac{1}{\mu_1 - \mu_2} \left(\frac{1}{\psi_2} - \frac{1}{\psi_1} \right) \right] \\
\sigma_x &= \frac{K_I}{\sqrt{2\pi r}} \operatorname{Re}\left[\frac{\mu_1 \mu_2}{\mu_1 - \mu_2} \left(\frac{1}{\psi_1} - \frac{1}{\psi_2} \right) \right] + \frac{K_{II}}{\sqrt{2\pi r}} \operatorname{Re}\left[\frac{1}{\mu_1 - \mu_2} \left(\frac{\mu_1}{\psi_1} - \frac{\mu_2}{\psi_2} \right) \right]
\end{aligned}
\tag{65}
$$

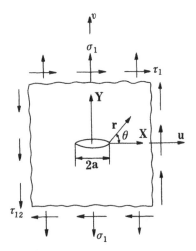

Figure 33 Anisotropic plate containing a central crack.

The displacement field in the vicinity of a crack can be calculated from

$$u = K_I \sqrt{\frac{2r}{\pi}} \, \text{Re} \left[\frac{1}{\mu_1 - \mu_2} (\mu_1 p_2 \psi_2 - \mu_2 p_1 \psi_1) \right] + K_{II} \sqrt{\frac{2r}{\pi}} \, \text{Re} \left[\frac{1}{\mu_1 - \mu_2} (p_2 \psi_2 - p_1 \psi_1) \right] \quad (66)$$

$$v = K_I \sqrt{\frac{2r}{\pi}} \, \text{Re} \left[\frac{1}{\mu_1 - \mu_2} (\mu_1 q_2 \psi_2 - \mu_2 q_1 \psi_1) \right]$$

$$+ K_{II} \sqrt{\frac{2r}{\pi}} \, \text{Re} \left[\frac{1}{\mu_1 - \mu_2} (q_2 \psi_2 - q_1 \psi_1) \right] \quad (67)$$

where

$$\psi_1 = (\cos \theta + \mu_1 \sin \theta)^{1/2} \tag{68}$$

$$\psi_2 = (\cos \theta + \mu_2 \sin \theta)^{1/2} \tag{69}$$

where Re means taking the real part of the complex number; μ_1 and μ_2 (or their conjugated complex numbers $\bar{\mu}_1$ and $\bar{\mu}_2$) are two different complex roots of the following equation:

$$a_{11}\mu^4 - 2a_{16}\mu^3 + (2a_{12} + a_{66})\mu^2 - 2a_{26}\mu + a_{22} = 0 \tag{70}$$

a_{ij} are the coefficients of the compliance matrix of the material.

$$p_1 = a_{11}\mu_1^2 + a_{12} - a_{16}\mu_1$$
$$p_2 = a_{11}\mu_2^2 + a_{12} - a_{16}\mu_2 \tag{71}$$

and

$$q_1 = a_{12}\mu_1 + \frac{a_{22}}{\mu_1} - a_{26}$$

$$q_2 = a_{12}\mu_2 + \frac{a_{22}}{\mu_2} - a_{26} \tag{72}$$

Using Eqs. (66) to (72), in Refs. 82 to 84, the mixed-mode fracture toughness of the chopped strand mate glass-fiber-reinforced polyester resin undergoing lap-shear testing is calculated.

Fracture Criteria for Composite Materials

In the course of developing fracture mechanics for composite materials, many theories and models have been proposed. Some important fracture criteria are summarized in Table 10.

Some comments and explanations regarding Table 10 follows.

1. Equation (73) is the well-known formula for fracture toughness of isotropic materials, where Y is dependent on the geometry (e.g., relative length of crack) of the specimen. But for anisotropic composite materials, Y depends not only on the geometry of the specimen but also on anisotropic elastic properties. By employing a finite element method and K calibration approach, it was demonstrated in Ref. 100 that the relative difference of Y values between isotropic and anisotropic materials can be as large as 20%.

2. Equations (74) and (75) are two double-parameter criteria. It is obvious that better fitting to experimental data can be obtained by using double-parameter criteria than by using single-parameter criteria. So Eq. (74) has been found to have more experimental evidence.

3. Equations (76) and (77) indicate that the stress singularity order at a crack tip in composites can be much less than -0.5, which is the stress singularity order at a crack

Table 10 Fracture Criteria for Composite Materials

Criterion	Equation		Refs.	Notes
LEFM K_c criterion	$K_c = Y\sigma_c^\infty \sqrt{\pi a}$	(73)	85	Single-parameter criterion a: Crack length Y: Geometry parameter of specimen
W.E.K. criterion intense energy zone model	$K_{IC} = \sigma_c^\infty \sqrt{\pi(c + a)}$	(74)	86	Two-parameter criterion C: Dimension of intense energy zone
Plastic zone correction model	$K_I = \sigma^\infty \left[W \tan\left(\dfrac{\pi a_e}{W}\right)\right]^{1/2}$ $a_e = a + r_y$ $r_y = \dfrac{1}{2\pi}\left(\dfrac{K_I}{\sigma_{ys}}\right)^2$	(75)	87	W: Width of specimen r_y: Size of plastic zone σ_{ys}: Yield stress
Mar and Lin and Caprino models	$\sigma_c = H_c(2a_c)^{-m}$ $\sigma_c = K_{Ic}(\pi L_0)^{-m}$	(76) (77)	88–91	$H_c K_{IC}$: Generalized stress intensity factor $\left(\dfrac{kN}{C_m^2 C_m^m}\right)$ L_0: Damage size m: Equals 0.3 to 0.33
G_C criterion, compliance calibration method	$G_I = \dfrac{P^2}{2B}\dfrac{d\lambda}{da}$ $\lambda = \dfrac{\Delta}{P}$ compliance	(78)	92	B: Thickness of specimen Δ: Displacement of loading point
J_C criterion	$J_1 = -\dfrac{1}{B}\dfrac{\partial U}{\partial a}\Big|_{\Delta=const}$	(79)	93	U: Strain energy stored in specimen

W. N. criteria

Point stress criterion

For a hole: $\dfrac{\sigma_N}{\sigma_0} = \dfrac{2}{2 + \xi_1^2 + 3\xi_1^4}$ (80)   94–96

$\xi_1 = \dfrac{R}{R + d_0}$

For a crack: $K_Q = \sigma_0\sqrt{\pi a(1 - \xi_3^2)}$ (81)

$\dfrac{\sigma_N}{\sigma_0} = \sqrt{1 - \xi_3^2}$ (82)

$\xi_3 = \dfrac{a}{a + d_0}$

- σ_N: Notched strength
- σ_0: Strength of material without hole
- d_0: Characteristic dimension
- R: Radius of hole
- a: Crack length

Average stress criterion

For a hole: $\dfrac{\sigma_N}{\sigma_0} = \dfrac{2(1 - \xi_2)}{2 - \xi_2^2 - 4\xi_2^4}$ (83)

$\xi_2 = \dfrac{R}{R + C_0}$

For a crack: $K_Q = \sqrt{\pi C_0 \xi_4}$ (84)

$\dfrac{\sigma_N}{\sigma_0} = \sqrt{\dfrac{C_0 \xi_4}{a}}$ (85)

$\xi = \dfrac{a}{2a + C_0}$

- C_0: Characteristic dimension for average stress criterion

S criterion

$\dfrac{\partial s}{\partial \theta} = 0$, when $\theta = \theta_C$ (86)   80

$S(K_I, K_I) = S_C$, if $\theta = \theta_c$ (87)

- S: Strain energy density factor
- θ_C: Crack growth direction

Dugdale model and CTOD criterion

$\dfrac{Y\sigma_f}{\sigma_0} = \dfrac{2}{\pi}\cos^{-1}\left[\exp\left(-\dfrac{\pi E(CTOD)_C}{8\sigma_0 a}\right)\right]$ (88)   97–99

$CTOD = \dfrac{4\sigma}{E}(a + d_0)\sqrt{1 - \left(\dfrac{a}{a + d_0}\right)^2}$ (89)

- CTOD: Displacement of crack tip (Fig. 34)
- σ_0: Material strength
- d_0: Plastic zone

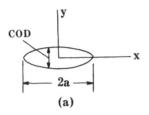

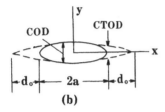

Figure 34 Relation of COD and CTOD (for Eqs. (88) and (89), in Table 10).

tip in metallic materials. The lower order of stress singularity implies that a blunting effect occurs in the fracture of composite materials.

4. The stress field in the vicinity of a circular hole of radius, R in an orthotropic laminate under uniaxial loading σ_y^∞ can be approximated by the relationship [101]

$$\sigma_y(x,0) = \frac{\sigma_y^\infty}{2}\left\{2 + \left(\frac{R}{x}\right)^2 + 3\left(\frac{R}{x}\right)^4 - (K_T^\infty - 3)\left[5\left(\frac{R}{x}\right)^6 - 7\left(\frac{R}{x}\right)^8\right]\right\} \tag{90}$$

where

$$K_T^\infty = 1 + \sqrt{2\left(\sqrt{\frac{E_y}{E_x}} - \nu_{yx} + \frac{E_y}{2G_{xy}}\right)} \tag{91}$$

For the isotropic case, $K_T^\infty = 3$;

$$\sigma_y(x,0) = \frac{\sigma_y^\infty}{2}\left[2 + \left(\frac{R}{x}\right)^2 + 3\left(\frac{R}{x}\right)^4\right], \qquad x \geq R \tag{92}$$

In the point stress criterion, failure is assumed to occur when the stress σ_y at a fixed distance d_0 ahead of the notch becomes equal to the unnotched tensile strength σ_0 of the material. For a circular hole, failure occurs when

$$\sigma_y(x,0)|_{x=R+d_0} = \sigma_0 \tag{93}$$

Using Eq. (93) in conjunction with Eq. (90) yields

$$\sigma_N^\infty = \frac{2\sigma_0}{2 + \zeta_2^2 + 3\zeta_2^4 - (K_T^\infty - 3)(5\zeta_2^6 - 7\zeta_2^8)} \tag{94}$$

It can be seen that Eq. (80) in Table 10 applies only for isotropic materials.

In the average stress criterion, failure is assumed to occur when the average stress value of σ_y over a fixed distance a_0 ahead of the notch first reaches the unnotched tensile strength of the material, that is, for the circular hole when

$$\frac{1}{a_0}\int_R^{R+a_0} \sigma_y(x,0)\,dx = \sigma_0 \tag{95}$$

Using Eq. (95) in conjunction with Eq. (90) yields

$$\sigma_N^\infty = \frac{2\sigma_0(1 - \zeta_4)}{2 - \zeta_4^2 - \zeta_4^4 + (K_T^\infty - 3)(\zeta_4^6 - \zeta_4^8)} \tag{96}$$

So in Table 10, Eq. (83) is again for isotropic materials.

5. As is well known for isotropic materials, the critical stress intensity factor K_C can be related with a critical energy release rate G_C or fracture surface energy γ. Analogous to the expression for isotropic materials, for composites we can give the following equations:

$$\gamma = \frac{G_C}{2} = \frac{K_C^2}{2E} \tag{97}$$

where

$$\frac{1}{E} = \left(\frac{b_{11}b_{22}}{2}\right)^{1/2} \left[\left(\frac{b_{22}}{b_{11}}\right)^{1/2} + \frac{2b_{12} + b_{66}}{2b_{11}}\right]^{1/2} \tag{98}$$

and

$$b_{11} = \frac{1}{E_1}\left(1 - \frac{E_1}{E_2}v_{12}^2\right), \qquad b_{22} = \frac{1}{E_2}(1 - v_{23}^2)$$

$$b_{12} = \frac{-v_{12}}{E_1}(1 + v_{23}), \qquad b_{66} = \frac{1}{G_{16}}$$

Prediction of Crack Growth Direction in Composites

In Table 10, only a small number of models and theories of macromechanics of composite fracture are summarized. In fact, a vast amount of work has been carried out dealing with fracture prediction of composites. They cannot be included here because of the limited space in this chapter.

It should be pointed out that the theories in Table 10 are concerned primarily with the prediction of onset of fracture, but not with the crack growth direction. As mentioned earlier, cracks in composites hardly propagate self-similarly. Therefore, prediction of cracking direction has the same significance with the prediction of fracture strength. In the following, some research work along these lines will be reviewed.

Wu Approach

Wu's approach [102] assumes that the intersection of the failure surface F of the material and the stress vector S defines the direction of fracture. The strength vector surface F is determined by employing the polynomial failure criterion:

$$F_1\sigma_1 + F_2\sigma_2 + F_{11}\sigma_1^2 + 2F_{12}\sigma_1\sigma_2 + F_{22}\sigma_2^2 + F_{66}\sigma_6^2 \leq 1 \tag{36}$$

The stress vector S is calculated at a radial distance r_o from the crack tip; r_o is assumed to be a dimension characteristic of a finite volume within which damage process occurs (Figure 35). Figure 36 shows an example of determining the fracture direction in unidirectional glass fiber–epoxy composite. In addition, Gregory and Herakovich [103] stipulate that the maximum value of the left-hand side of Eq. (36) predicts cracking direction in unidirectional composites.

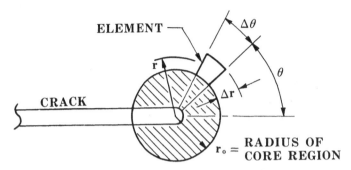

Figure 35 Crack model showing an element outside a core region around the crack tip.

Normal Stress Ratio Criterion

The normal stress ratio (NSR) criterion, proposed by Gregory and Herakovich [103] and Buezek and Herakovich [104], assumes that the cracking direction is determined by the ratio of normal stress acting on a radial plane $\sigma_{\theta\theta}$ to the related strength $T_{\theta\theta}$. Cracking will take place in the direction in which the ratio at a given distance r_o from the crack tip, $\overline{R}(r_o,\theta)$ is of maximum value. Here $\overline{R}(r_o,\theta)$ is defined as (see Figure 37)

$$\overline{R}(r_o,\theta) = \frac{\sigma_{\theta\theta}}{T_{\theta\theta}}\Big|_{r=r_o} \tag{99}$$

where $T_{\theta\theta}$ is the tensile strength, that is, the critical value of the stress acting on the radial plane and defined as

$$T_{\theta\theta} = X \sin^2\beta + Y \cos^2\beta \tag{100}$$

and $\sigma_{\theta\theta}$ is the normal stress, which can be calculated from

$$\sigma_{\theta\theta} = \frac{\sigma_x + \sigma_y}{2} + \frac{\sigma_x - \sigma_y}{2}\cos 2\theta + \tau_{xy}\sin 2\theta \tag{101}$$

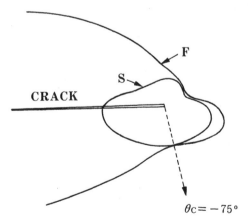

Figure 36 Stress vector and strength vector contours for a crack in a unidirectional glass fiber–epoxy resin composite under combined loading ($\tau^\infty/\sigma^\infty = 1$) with core radius $r_0 = 2$ mm.

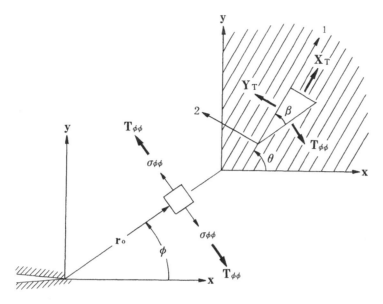

Figure 37 Normal stress ratio parameters.

where β is the angle between the radial direction and the first principal direction of the material and X and Y are tensile strengths in the two principal directions. Thus this criterion stipulates that

$$\overline{R}(r_o,\theta) = \overline{R}_{\max}(r_o,\theta) \qquad \text{when} \quad \theta = \theta_c \tag{102}$$

Strain Energy Density Ratio Criterion [105]

The strain energy density rate (SEDR) criterion was described earlier, where it was used as a failure criterion [see Eq. (32)]. Here its function of predicting fracture direction will be explained, although SEDR has the same expression.

This criterion postulates that cracking will occur in the radial direction along which the local SEDR value, S_R, reaches its minimum value, that is,

$$\frac{\partial S_R}{\partial \theta} = 0, \qquad \frac{\partial^2 S_R}{\partial \theta^2} > 0 \qquad \text{when} \quad \theta = \theta_c \tag{103}$$

If there is more than one minimum value around the point concerned, the cracking direction coincides with the maximum value; thus

$$S_R = (S_R)_{\min}^{\max} \qquad \text{when} \quad \theta = \theta_c \tag{104}$$

Extended Tsai–Hill Criterion

The Tsai–Hill criterion is usually used as a failure criterion [see Eq. (21)]. In Refs. [105 and 106] it has been extended to predict cracking direction. At first, the Tsai–Hill distortional energy is defined as

$$S_{\text{TH}} = \left(\frac{\sigma_x}{X}\right)^2 + \left(\frac{\sigma_2}{Y}\right)^2 - \frac{\sigma_1\sigma_2}{X^2} + \left(\frac{\tau_{12}}{S}\right)^2 \tag{105}$$

In analogy to SEDR criterion, the extended Tsai–Hill criterion can be expressed as

$$\frac{\partial S_{TH}}{\partial \theta} = 0, \qquad \frac{\partial^2 S_{TH}}{\partial \theta^2} > 0 \qquad \text{when} \quad \theta = \theta_c \tag{106}$$

or

$$S_{TH} = (S_{TH})^{\max}_{\min} \qquad \text{when} \quad \theta = \theta_c \tag{107}$$

Extended Norris Criterion [105,106]

The Norris distortional energy is defined as

$$S_N = \left(\frac{\sigma_1}{X}\right)^2 + \left(\frac{\sigma_2}{Y}\right)^2 - \frac{\sigma_1 \sigma_2}{XY} + \left(\frac{\tau_{12}}{S}\right)^2 \tag{108}$$

The extended Norris criterion is represented as

$$S_N = (S_N)^{\max}_{\min} \qquad \text{when} \quad \theta = \theta_c \tag{109}$$

Damage Tolerance and Notch Sensitivity of Composites

Good damage tolerance is one of the requirements of high-performance fiber-reinforced composite materials. As mentioned earlier, various damage and defects may develop during manufacturing process and in service time. Cracks may initiate and extend from free edges, holes, and areas of other defects. Once initiated they may propagate rapidly, leading to a loss of structural integrity and strength in components. Composites are particularly sensitive to this due to the anisotropy and heterogeneity that are present. Cracks will propagate in certain directions more readily than others. Fiber breakage normally requires high energy input; however, the propagation of cracks within the matrix and interface normally requires much lower energy. Therefore, composites are sensitive to crack propagation between plies (delamination) and within plies parallel to the fibers (splitting). The need to understand the damage tolerance and notch sensitivity has led to the development of a number of tests, which consist of comparative methods and intrinsic toughness measurements.

In Table 10, some intrinsic fracture toughness criteria have been reviewed. For the comparative toughness testing methods, there are tension tests in the presence of holes [107] and compression tests after low-energy impact [108] and through penetration impact [109]. Comparative tests are useful in comparing materials and in defining load or strain levels that can be used in service. However, the experimental results are geometry dependent and do not give a design parameter that can be used over a wide range of service conditions. Neither do they give any understanding of the reasons for the behavior observed. For this reason, a fracture mechanics approach (intrinsic toughness) is more useful, in which a critical value of the stress intensity factor (K_c) or more commonly, the strain energy release rate (G_c), is obtained.

Intrinsic Fracture Toughness Measurement

Two testing methods were used in assessing crack sensitivity in aromatic polymer composites [108,109] the notched flexure test and double cantilever beam (DCB) test. Figure 38a illustrates schematically the surface-notched three-point flexure specimen, and Figure 38b is the DCB specimen. For the two specimen configurations, K_c and G_c are defined as:

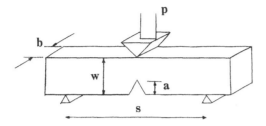

(a)

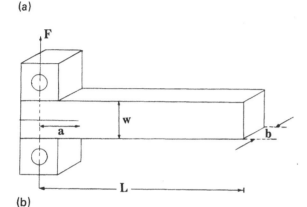

(b)

Figure 38 Fracture mechanics test specimens: (a) surface-notched three-point flexure specimen; (b) double cantilever beam specimen.

$$K_c = \sigma\infty Y \sqrt{a} \tag{110}$$

$$G_c = \frac{\Delta I}{\Delta A} \tag{111}$$

Equation (110) is actually the same as Eq. (73), where the geometry parameter Y is different for DCB specimen and the three-point flexure specimen. In Eq. (111), ΔI is an energy required to propagate a crack and create a new surface of area of ΔA. A simple derivation indicates that in actuality, Eq. (111) is equivalent to Eq. (78) in Table 10. The use of Eq. (78) needs to measure the change rate of compliance of the specimen.

Another approach to the determination of G_c is use of the following formula:

$$G_c = \frac{I_R}{bw\varphi} \tag{112}$$

where I_R is input energy (area under the load–deflection curve), b and w are specimen width and depth, and φ is a geometry factor defined in Refs. 108 and 110.

Comparative Toughness Evaluation

Open-hole tension test. The open-hole tension test determines the notched tensile strength or reduction in the tensile strength of a composite caused by a through-thickness crack. It has been shown by Morris and Hahn [111] that tensile coupons having different center defects, but with the same projected length perpendicular to the direction of the tensile load, fail at similar tensile loads. Hence the results from specimens with circular

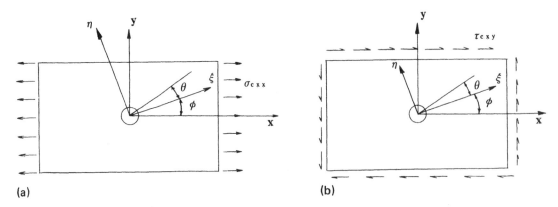

Figure 39 Stress concentrations depend on composite moduli (for Eqs. (113) and (114)].

holes used in these tests are relevant to determine the effect of other through-thickness defects. In Ref. [112], Chamis presents the equations for predicting the hoop stress concentration in the periphery of a circular hole in an anisotropic finite plate caused by in-plane loads (Figure 39):

$$\frac{\sigma_{c\theta\theta}}{\sigma_{cxx}} = \frac{E_{c\theta\theta}}{E_{cxx}}\left\{ R_o[(R_o + R_{D1})\sin^2\varphi]\cos^2\theta + [(1 + R_{D1})\cos^2\varphi]\sin^2\varphi \right.$$
$$\left. - \frac{R_{D1}}{4}(1 + R_o + R_{D1})\sin 2\varphi\sin 2\theta \right\} \quad (113)$$

$$\frac{\sigma_{c\theta\theta}}{\sigma_{cxy}} = \frac{E_{c\theta\theta}}{2E_{cxx}}(1 + R_o + R_{D1})\left\{ -R_{D1}\cos 2\varphi\sin 2\theta \right.$$
$$\left. + [(1 + R_o)\cos 2\theta + R_o - 1]\sin 2\varphi \right\} \quad (114)$$

where

$$R_o = \sqrt{\frac{E_{cxx}}{E_{cyy}}}, \qquad R_{D1} = \left[2\left(\frac{E_{cxx}}{E_{cyy}} - v_{cxy}\right) + \frac{E_{cxx}}{E_{cxy}} \right]^{1/2}$$

Compression test after low-level impact. This method has been advised by Byers [113] and Boeing [114]. Low-level impact can cause delamination of composite panels, which significantly reduces their residual compressive strength and strain to failure. Measurement of compression after impact involves tests on much large specimens. Panels of approximate dimensions (150 × 100 mm) are edge supported and impacted at the intercept of the diagonals. The specimens are then loaded into a compression jig that incorporates anti-buckling guides. Details of specimen preparation and test procedures are provided in Refs. 107 and 108.

Instrumented falling weight impact. This method involves the fracture of square panel specimens (75 × 75 mm). These are supported on a 50-mm-diameter ring and impacted with a cylinder of hemispherical nose diameter 12.5 mm. A test speed of 5 m s^{-1} is employed where sufficient energy is always used to fracture the specimen. The use of a force transducer attached to the impactor nose enables force–deflection curves to be determined during impact. Integration of the force–deflection curve provides a measure of energy to initiate cracking and total energy to fracture. Details of this technique are published in Refs. 109 and 115.

CONCLUDING REMARKS

The intent of the present chapter is to summarize the current understanding of damage, failure, and fracture of resin-based fiber-reinforced composite materials. Emphasis is placed on the mechanisms, theories, and criteria of failure and fracture. These topics are scientifically and technologically important to material scientists and design engineers.

It is recognized that failure and fracture are two of the most difficult aspects in mechanics of composite materials. While the failure behavior of materials is documented in the literature (e.g., Ref. 21), the fracture mechanics of FRP is still in an immature stage. It is not unusual that a fracture theory yields contradictory results in comparison with experiments of different composite systems or/and different stack constructions. Many new theories and hypotheses have emerged in recent years. Development of the fracture mechanics of composite materials is an arduous task that still confronts us.

ACKNOWLEDGMENTS

Appreciation is expressed to L. W. Tsai, who helped to prepare the text on mechanics of composite fracture. The present paper was supported by the Chinese National Science Foundation.

NOTATION

E	Young's modulus
F_i, F_{ij}	Strength parameters in Eq. (36)
G	Shear modulus; strain energy release rate
K	Stress intensity factor
R	Radius of a circular hole
S	Shear strength; strain energy density factor
V	Volume fraction of components in composite materials
W	Work done in interface debonding or to pull out a fiber
X, Y, Z	Strengths in the three principal directions of materials
a_{ij}	Compliance matrix of the material in Eq. (70)
b_{ij}	Compliance matrix of the material in Eq. (98)
l_c	Critical length of fibers defined in Eq. (15)
r	Radius of fibers
x	Transverse crack spacing in matrix
x, y	Coordinate directions
β	Parameter defined in Eq. (12)
γ	Shear strain
ϵ	Normal strain
θ	Off-axis angle of loading direction to fiber direction; argument in polar coordinate
ν	Poisson ratio
σ	Normal stress
$\Delta\sigma$	Additional stress

σ_a	Applied stress
τ	Shear stress

Subscript

f	Fiber
m	Matrix
c	Composites; compressive; critical
D, d	Debonding
e	For pullout in Eqs. (61) and (62)
t	Tensile
1, 2	Two principal directions of materials
I, II	For mode I or mode II fracture

REFERENCES

1. R. Truel, C. Elbaum, and B. B. Chick, *Ultrasonic Methods in Solid State Physics*, Academic Press, New York, 1969.
2. M. F. Marham, *Proc. AGARD Conference 63, Composite Materials*, Paper 4, 1970.
3. D. E. W. Stone, discussion in *Proc. Conference on Carbon Fibers: Their Composites and Applications*, Plastics Institute, London, 1974.
4. R. Prakash and C. N. Owston, *Composites*, 8: 100–102 (1971).
5. D. T. Hayford, E. G. Hennecke, and W. W. Stinchcombe, *J. Compos. Mater.*, 11: 429–444 (1977).
6. J. B. Sturgeon, *Br. J. Non-Destr. Test.*, Nov. 1978, pp. 303–309.
7. S. Torp, O. Forli, and J. Malmo, *Proc. 32nd Annual Technical Conference of SPI*, Paper 9-A, 1977.
8. F. Polato, P. Parrini, and G. Gianotti, *Proc. ICCM-3*, Vol. 2, Pergamon Press, Oxford, 1980, pp. 1050–1058.
9. W. N. Reynolds, *Proc. 24th Annual Technical Conference of SPI*, Paper 14-B, 1969.
10. M. W. Darlington and P. L. McGinley, *J. Mater. Sci.*, 10: 906 (1975).
11. W. W. Stinchcombe, K. L. Reifsneider, L. A. Marcus, and R. S. Williams, in *Fatigue of Composite Materials*, ASTM STP-596, ASTM, Philadelphia, 1975, pp. 115–129.
12. P. V. McLaughlin, E. V. McAssey, and R. C. Dietrich, *Non-Destr. Test Int.*, 13: 56–62 (1980).
13. C. J. Pye and R. D. Adams, *J. Phys. D. Appl. Phys.*, 14: 927–941 (1981).
14. J. H. Williams, S. H. Mansouri, and S. S. Lee, *Thermal Non-destructive Testing of Fiberglass Laminates Using Liquid Crystals*, Report of Composite Materials and Non-destructive Evaluation Laboratory, MIT, Cambridge, Mass., July 1979.
15. F. J. Guild, M. G. Phillips, and B. Harris, *Non-Destr. Test. Int.*, 13: 209–218 (1980).
16. M. Puwa, A. R. Bunsell, and B. Harris, *J. Mater. Sci.*, 10: 2060–2070 (1975).
17. D. O. Harris and B. L. Dunegan, in *Testing for Prediction of Materials Performance in Structures and Components*, ASTM, STP-515, ASTM, Philadelphia, 1972, pp. 158–170.
18. J. Becht, H. J. Schwalbe, and J. Eisenblaetter, *Composites*, 7: 245 (1976).
19. J. T. Ryder and J. R. Wadin, *Acoustic Emission Monitoring of a Quasi-isotropic Graphite/ Epoxy Laminate Under Fatigue Loading*, report from Dunegan Endevco, Inc. (1979).
20. T. J. Fowler, *Proc. ASCE Convention*, San Francisco, Preprint 3092 (1977).
21. D. Hull, *An Introduction to Composite Materials*, Cambridge University Press, Cambridge, 1981. pp. 128–130.
22. J. Aveston and A. Kelly, *J. Mater. Sci.*, 8: 252–262 (1973).
23. M. N. Nahas, *J. Compos. Technol. Res.*, 8(4): 138–153 (1986).

24. R. M. Jones, *Mechanics of Composite Materials*, McGraw-Hill, New York, 1967.
25. E. Z. Stowell and T. S. Lin, *J. Mech. Phys. Solids*, 9: 242–260 (1961).
26. L. W. Hu, *J. Franklin Inst.*, 265: 187–204 (1958).
27. R. Hill, *The Mathematical Theory of Plasticity*, Oxford University Press, Oxford, 1950.
28. V. D. Azzi and S. W. Tsai, *Exp. Mech.* 5: 283–288 (1965).
29. S. W. Tsai and V. Azzi, *AIAA J.*, 4: 296–301 (1966).
30. J. Marin, *J. Aeronaut. Sci.*, 24: 265–268 (1957).
31. H. G. Franklin, *Fiber Sci. Technol.*, 1: 137–150 (1968).
32. C. B. Norris and P. F. McKinnon, *Report 1328*, Forest Products Laboratory, 1946.
33. C. B. Norris, *Report 1816*, Forest Products Laboratory, 1962.
34. L. Fischer, *J. Eng. Ind.*, 89: 399–402 (1967).
35. J. E. Griffith and W. M. Baldwin, *Proc. First Southern Conference on Theoretical and Applied Mechanics*, 1962, pp. 410–420.
36. C. C. Chamis, *Composite Materials*: *Testing and Design*, ASTM STP-460, ASTM, Philadelphia, 1969, pp. 336–351.
37. C. C. Chamis, *US/USSR Seminar on Fracture of Composite Materials*, Paper 30, 1978, pp. 329–348.
38. O. Hoffman, *J. Compos. Mater.*, 1: 200–206 (1967).
39. Zhang Shuang-Yin and Tsai Liang-Wu, *Chin. Sci. Bull.* 34(14): 1173–1175 (1989).
40. I. I. Gol'denblat and V. A. Kopnov, *Polym. Mech.* (translation from *Mekh. Polim.*), 1: 54–59 (1966).
41. A. Malmeister, *Polym. Mech.* 2: 324–331 (1967).
42. S. W. Tsai and E. M. Wu, *J. Compos. Mater.*, 5: 58–80 (1971).
43. E. M. Wu, ''Phenomenological anisotropic failure criterion,'' in *Composite Materials*, L. J. Broutman and R. H. Krock (eds.), Vol. 2, Academic Press, New York, 1974.
44. E. M. Wu, *J. Compos. Mater.*, 6: 472–489 (1972).
45. A. Kelly and G. J. Davies, *Metall. Rev.*, 10: 1–77 (1965).
46. W. Prager, *J. Appl. Mech.*, 36: 542–544 (1969).
47. M. J. Owen and D. J. Rice, *Composites*, 12: 13–25 (1981).
48. K. W. Garrett and J. E. Bailey, *J. Mater. Sci.*, 12: 2189–2194 (1977).
49. J. E. Bailey, P. T. Curtis, and A. Parvizi, *Proc. R. Soc. London Ser. A.*, 366: 599–623 (1979).
50. A. Parvizi and J. E. Bailey, *J. Mater. Sci.*, 13: 2131–2136 (1978).
51. R. B. Pipes and N. J. Pagano, *J. Compos. Mater.*, 4: 538–548 (1970).
52. H. T. Hahn and S. W. Tsai, *J. Compos. Mater.*, 8: 280–305 (1974).
53. P. H. Petit and M. E. Waddoups, *J. Compos. Mater.*, 3: 2–19 (1969).
54. K. D. Chiu, *J. Compos. Mater.*, 3: 578–582 (1969).
55. R. S. Sandhu, *J. Aircraft*, 13: 104–111 (Feb. 1974).
56. G. E. Brown, ''Progressive failure of advanced composite laminates using the finite element method,'' M.Sc. thesis, University of Utah, Salt Lake City, Utah, 1976.
57. R. Forster and W. Knappe, *Kunststoffe*, 60: 1053–1059 (1970).
58. D. Hull, M. J. Legg, and B. Spencer, *Composites*, 9: 17–24 (1978).
59. A. Puck and W. Schneider, *Plast. Polym.*, 37: 33–43 (1969).
60. M. W. Nahas, *Compos. Struct.*, 6: 283–294 (1986).
61. S. Zhang and C. M. Leech, *Eng. Fract. Mech.*, 23(3): 521–535 (1986).
62. E. M. Wu and J. K. Schenblein, *Composite Materials*: *Testing and Design*, ASTM STP-546, ASTM, Philadelphia, 1974, pp. 188–206.
63. P. D. Soden, D. Leadbetter, P. R. Griggs, and G. C. Eckold, *Composites*, 9: 247–250 (1978).
64. T. R. Guess and F. P. Gerstle, *J. Compos. Mater.*, 11: 146–163 (1977).
65. A. Kelly, *Strong Solids*, 2nd ed., Clarendon Press, Oxford, (1973), pp. 207–208.
66. S. Y. Zhang, *Int. J. Fract.*, 35: R57–R60 (1987).

67. S. K. Maiti, *Int. J*. Fract., 32: R29–R32 (1986).

68. A. Kelly, *Proc. R. Soc*. London Ser. *A*, 319: 95–116 (1970).

69. P. W. R. Beaumont, "The mechanics of composite fracture and fatigue," in *Fiber Composites: Developing, Manufacturing, Testing and Maintaining Composite Structures and Products*, Edited by T. H. Mao and P. W. R. Beaumont. The Chinese Academy of Sciences and the Royal Society Present, Beijing, 1986.

70. J. D. Outwater and M. C. Murphy, *24th Annual Technical Conference of Composite Section, SPI*, Paper 11c, 1969.

71. J. K. Wells, Ph.D. dissertation, Cambridge University, Cambridge, 1982.

72. B. W. Rosen, *AIAA J.*, 2: 1985 (1964).

73. H. E. Daniels, *Proc. R. Soc.*, 183 A: 405 (1945).

74. D. E. Gucer and J. Gurland, *J. Mech. Phys. Solids*, 10: 365 (1962).

75. J. M. Hedgepeth, *NASA TN-D882*, Langley Research Center, Hampton, Va., 1961.

76. J. M. Hedgepeth and P. Van Dyke, *J. Compos. Mater.*, 1: 294 (1967).

77. R. M. Scop and A. S. Argon, *J. Compos.* Mater., 3: 30 (1969).

78. C. Zweben, *AIAA J.*, 6: 2325 (1968).

79. C. Zweben, *Eng. Fract. Mech.*, 6: 1 (1974).

80. G. C. Sih, *Proc. First USA/USSR Symposium on Fracture of Composite Materials*, Riga, Russia, 1978, pp. 111–129.

81. S. Parhiger, L. W. Zachary, and C. T. Sun, *Proc. 2nd USA/USSR Symposium on Fracture of Composite Materials*, Lehigh University, Bethlehem, Pa., (1982).

82. Zhang Shuang-Yin and C. M. Leech, *Appl. Math. Mech.* (Engl. ed.), 7(8): 741–753 (1986).

83. Zhang Shuang-Yin, *Proc. International Symposium on Composite Materials and Structures*, Beijing, China, 1986, pp. 598–603.

84. S. Y. Zhang, P. D. Soden, and P. M. Soden, *Composites*, 17(2): 100–110 (1986).

85. E. M. Wu, "Fracture mechanics of anisotropic plates," in *Composite Materials Workshop*, Technomic, Lancaster, Pa., 1968, p. 20.

86. E. M. Waddoups, J. R. Eisenmann, and B. E. Kamiski, *J. Compos. Mater.*, 5: 446–454 (1971).

87. M. J. Owen and P. T. Bishop, *J. Compos. Mater.*, 7: 146 (1973).

88. J. W. Mar and K. Y. Lin, *J. Aircraft*, 14: 703–704 (1977).

89. J. M. Mar and K. Y. Lin, *J. Compos. Mater.*, 11: 405–421 (1977).

90. G. Caprino, *J. Compos. Mater.*, 18: 508–518 (1984).

91. G. Caprino, *J. Mater. Sci.*, 18: 2269–2273 (1983).

92. K. S. Han and J. Kantsky, *J. Compos. Mater.*, 15: 371–388 (1981).

93. P. E. Keary, L. B. Ilcewicz, S. Shaar, and J. Trostle, *J. Compos. Mater.*, 19: 154–177 (1985).

94. J. M. Whitney and R. J. Nuismer, *J. Compos. Mater.*, 8: 253–265 (1974).

95. R. J. Nuismer and J. M. Whitney, *ASTM STP-593*, pp. 117–142 (1975).

96. J. M. Whitney and R. Y. Kim, paper presented at the *ASTM 4th National Conference on Composite Materials: Testing and Design*, Valley Forge, Pa., May 3–4, 1976.

97. C. E. Harris and D. H. Morris, *Exp. Mech.*, 25(2): 193–199 (1985).

98. S. W. Tsai and H. T. Hahn, in *ASTM ASME AMD-13*, pp. 73–96 (1975).

99. J. Awerbuch and H. T. Hahn, *Exp. Mech.*, 20: 334–344 (1980).

100. J. F. Mandell, F. J. McGarry, S. S. Wang and J. Im, *J. Compos. Mater.*, 8: 106–116 (1974).

101. H. J. Konish and J. M. Whitney, *J. Compos. Mater.*, 9(2): 157–166 (1975).

102. E. M. Wu, in *Composite Materials*, Vol. 5, *Fracture and Fatigue*, L. J. Broutman (ed.), Academic Press, New York, 1974, pp. 191–247.

103. M. A. Gregory and C. T. Herakovich, *J. Compos. Mater.*, 20(1): 67–85 (1986).

104. M. B. Buczek and C. T. Herakovich, in *Mechanics of Composite Materials*, AMD-Vol. 58, G. J. Dvorak (ed.), ASME, New York, 1983, pp. 75–82.

105. L. W. Tsai and S. Y. Zhang, *Compos. Sci. Technol.*, 31(2): 97–100 (1988).

106. S. Y. Zhang and L. W. Tsai, *Int. J. Fract.*, 40: R101–R104 (1989).
107. D. R. Carlile and D. C. Leach, *15th SAMPE Technical Conference*, Cincinnati, Ohio, Oct. 4–6, 1983, pp. 82–93.
108. D. C. Leach and D. R. Moore, *Compos. Sci. Technol.*, 23: 131–161 (1985).
109. R. A. Crick, D. C. Leach, and D. R. Moore, *31st SAMPE Symposium and Exhibition*, Las Vegas, Nev., Apr. 7–10, 1986.
110. E. Plati and J. G. Williams, *Polym. Eng. Sci.*, 15(6): 470–477 (1975).
111. D. H. Morris and H. T. Hahn, *J. Compos. Mater.*, 11(2): 124–138 (1977).
112. C. C. Chamis, *J. Compos. Technol. Res.*, 11(1): 3–14 (1989).
113. B. A. Byers, *NASA CR-159293* (Aug. 1980).
114. *Boeing Specification Support Standard BSS 7260* (1980).
115. C. J. Hooley, P. A. Gutteridge, D. R. Moore, S. Turner, and M. J. Williams, *Kunststoffe*, 72: 9 (1982).

3

Fracture Properties and Failure Mechanisms of Pure and Toughened Epoxy Resins

It-Meng Low
Applied Physics Department
Curtin University of Technology
Perth, Australia

Yiu-Wing Mai
Centre for Advanced Materials Technology
University of Sydney
Sydney, Australia

INTRODUCTION

Advanced composites based on epoxy resins are emerging as high-performance materials for many applications in the aerospace, defense, automotive, shipping, electrical, electronic, medical, and sporting goods industries. In short, they are gradually replacing the traditional aluminum and other structural metals. When the epoxy resins are suitably toughened or modified they possess a balance of desirable engineering properties, which include fracture toughness, tensile and flexural strengths, stiffness, and temperature resistance. Toughened epoxy resins are especially suitable for use as adhesives, tooling compounds, moldings, potting and encapsulating materials, and as matrices for fiber-reinforced composites. These toughened epoxy materials can be tailormade in composition and microstructure to give the properties required. Essentially, the microstructures of these materials are modified by dispersing a second phase in the pure epoxy matrix, which may be either rubbery or rigid in nature, or both. Indeed, the underylying concept for toughened epoxy resins parallels that for toughened polymers such as high-impact polystyrene (HIPS), acrylonitrile–butadiene–styrene (ABS) and styrene–acrylonitrile (SAN). All these specialty polymers are designed to impart a high toughness to the otherwise brittle matrices.

Complex mechanical and fracture properties have been observed for toughened epoxy resins, and significant research activities have been and are being directed toward uncovering the origins of toughening in these materials. In this chapter we present a detailed account of the various intrinsic (microstructural) and extrinsic (mechanical, environmental, etc.) factors controlling the fracture properties, deformation, and failure processes in toughened epoxy resins. The microstructure–property relationship and the toughening mechanisms for various toughened epoxy systems are highlighted and discussed.

EPOXY SYSTEMS AND PROPERTIES

Pure Epoxy Resins

Epoxy resins are characterized by having in the molecule a highly reactive oxirane ring of triangular configuration consisting of an oxygen atom bonded to two adjoining and bonded carbon atoms [1]. They are usually made by the reaction of epichlorohydrin with phenol compounds, but epoxidation is also done by the oxidation of a carbon-to-carbon double bond with an organic peracid such as peracetic acid. The latter types of resins usually have higher heat resistance. When mixed with a cross-linking agent (hardener), epoxy resins polymerize to a thermosetting solid. Many types of curing agents are used with epoxy resins, such as polyamines, polyamides, polysulfide, and melamine formaldehyde resins. Acid anhydrides as curing agents produce low toxicity, ease of handling, and improved flow. The commerical types of epoxy resins used are aliphatic epoxy resin, bisphenol A–epichlorohydrin resin, brominated epoxy resin, cycloaliphatic epoxy resin, epoxy novolac resin, flexibilizing epoxy resin, high-functionality epoxy resins, and the high-molecular-weight linearized epoxy resin.

The properties of cured epoxy resins depend on the type of epoxy, curing agent, and the curing process employed [2,3]. In general, those hardeners that require heat for a cure produce a higher chemical linkage and offer superior performance. Random substitution of hardeners should not be made unless chemical compatibility with the base resin is first established and mixing ratios properly determined. For good curing results, proportioning must be precise and mixing thorough.

For advanced engineering applications such as in the aerospace industry, only resins that can withstand high temperatures (ca. 177°C) are acceptable. In the new generation of high-speed civil transport aircraft, temperatures in the range 200 to 250°C are required. High-temperature epoxy resins include polyglycidyl ethers of novoclacs (Union Carbide ERR 0100), triglycidyl *p*-aminophenol (Union Carbide ERLA 0510), and tetraglycidyl diaminodiphenyl methane (TGDDM, Ciba-Geigy MY 720). Suitable curing agents are dicyaniamide (DICY), BF_3MEA, diamino diphenyl sulfone (DDS, Ciba-Geigy Eporal), methyl dianiline (MDA), *meta*-phenylenediamine (MPDA), and piperidine. The chemical compositions of these epoxies and curing agents have been studied elsewhere [4] and are given here in Appendix A.

The important properties attributed to epoxy resins are low cure shrinkage, outstanding adhesion to both metallic and nonmetallic surfaces, low moisture absorption (0.01 to 0.2% in 24 h), high strength in reinforced laminate form, excellent dielectric properties, and superb chemical or corrosion resistance. In addition, ozone, sunlight and weathering resistance are excellent. Resistance to oxidizing and nonoxidizing acids is also exceptional. The main drawback of epoxy resins is their poor fracture toughness, where the typical fracture energy (G_{Ic}) is about 80 to 300 J/m^2 either in bulk [6–9] or in mode I delamination in a composite [10–13] or as an adhesive [14–16]. Various methods have been developed to improve the fracture resistance of epoxy resins, and these are described later in the chapter.

Toughened Epoxy Resins

Rigid or rubbery particles, or both, may be uniformly dispersed in the matrix of an epoxy resin to modify its physical and mechanical properties. The mechanical behavior of these modified epoxies is a direct result of the complex interplay of the properties of the constituent phases: resin, filler, and interfacial regions. The principal relevant parameters are the volume fraction of filler, the particle size, the filler aspect ratio, the modulus and strength of the filler, the resin–filler adhesion, and the fracture toughness of the matrix [17].

Particulate-Filled Epoxies

Most rigid particulates are of an inorganic nature and vary from mica, silica, zirconium silicate, hydrated alumina, and iron powder to cork and hollow glass beads [18–29]. They may serve to decrease costs, improve heat transfer, impart thixotropism, and induce electrical conductivity in an otherwise electrical insulator [30]. In addition, the specific filler and percentage used has a controlling effect on such properties as the coefficient of thermal expansion, specific gravity, fracture resistance, strength, and stiffness. Generally, the last three properties of particulate-filled epoxy resins are higher than those of pure epoxy [25,26]. The change in these mechanical properties depends on the type of particulate used; its size, volume fraction, and surface treatment; and the strain rate and test environment. Various studies on the fracture toughness (K_{Ic}) of particulate-filled epoxies as a function of the particle volume fraction and surface treatment during fast [18,23–25,29] and slow [19] crack propagation show that K_{Ic} of these composites is generally increased. Increases in Young's modulus (E) and yield stress (σ_y) have also been reported (see Table 1).

Moloney et al. [17] have studied the influences of filler volume fraction, particle size, aspect ratio, modulus and strength of filler, resin–filler adhesion, and toughness of the matrix on the stiffness, strength, and toughness of particulate-filled epoxy resins (see Table 2). They highlighted several important points:

1. Increasing filler volume fraction increases the composite modulus and toughness.
2. Increasing filler particle size decreases the composite strength.
3. Increasing filler strength and modulus increases the composite modulus, strength, and fracture toughness (K_{Ic}).

Table 1 Influence of Fiber Particle Size and Volume Fraction on
E, σ_y, and K_{Ic} of an Epoxy–Silica Composite

Volume Fraction, V_p (%)	Mean Particle Size, d (μm)	K_{Ic} (MPa $\sqrt{m}$)	E (GPa)	σ_y (MPa)
40	300	1.76 ± 0.01		
	160	1.74 ± 0.02		
	100	1.87 ± 0.01		
	60	1.83 ± 0.02		
0		0.60	3.5	100
20		1.30	6.1	108
30		1.62	7.7	121
40		1.87	9.8	133
50		2.21	12.5	150
40 (silane treated)		1.90		

Source: Ref. 25.

4. K_{Ic} of the composite is insensitive to the type of filler and its particle size.

5. Improved particle–matrix adhesion does not improve K_{Ic} appreciably.

6. Good bonding is essential for good strength.

7. High strength and high fracture energy (G_{Ic}) are mutually exclusive.

Although considerable information is available on the mechanical behavior of particulate-filled polymers, very little is known on fatigue crack propagation in these materials. In a recent study, Gadkaree and Salee [20] reported that the fatigue crack growth rate for

Table 2 Parameters Determining the Modulus, Toughness, and Strength
of Particulate-Filled Epoxide Resins

Property	Effect on:		
	Modulus	K_c	Strength
Increase volume filler	Increase	Increase	~Constant at matrix value
Particle size	Constant	Constant	Decrease
Increase aspect ratio		Increase	Increase
Increase strength and modulus filler	Increase	Increase	Increase
Improved adhesion	Constant	Constant	Increase
Tougher matrix	Small decrease	Increase	Decrease

Source: Ref. 17.

a bisphenol A–terphalate/isophalate copolymer filled with an aluminosilicate (particle size ≈ 200 μm) decreases with increasing filler content. Interestingly, for low filler contents (i.e., up to about 10%), the fatigue crack propagation rate follows Paris power law [31], but not for high concentrations.

Rubber-Modified Epoxies

The toughening of thermoplastics by rubber has been in existence for over 40 years [32], but the addition of rubber to thermoset epoxy resins is only about 20 years old. The pioneering work of McGarry et al. [33–35] showed that improvements in toughness could be achieved by adding certain liquid rubbers to the epoxy formulations. Since then, a significant amount of information on toughened epoxies has been published [6–8,16,36–46]. The addition of rubber to a brittle resin modifies its various characteristics. For example, it causes a reduction in stiffness, lowers T_g, plasticizes the matrix, reduces the yield strength, and increases the linear thermal expansion coefficient; but it significantly increases the fracture resistance.

The properties of a rubber-toughened epoxy resin depend on the size and volume fraction of dispersed rubber, the strain rate, the temperature, and other environmental factors. Although a significant amount of information on the physical and mechanical behavior of toughened thermoplastic materials such as high-impact polystyrene (HIPS) and acrylonitrile–butadiene–styrene (ABS) is available [32], information on thermoset resins such as epoxies is rather limited. This is particularly so for high-temperature epoxies toughened with rubber. It is suspected that rubber toughening becomes ineffective at these high service temperatures.

Table 3 shows the variations of initial tensile modulus (E), tensile strength (σ_t), linear thermal coefficient of expansion (α), and glass transition temperature (T_g) with elastomer concentration for CTBN-toughened DGEBA (DER-332, Dow Chemical Co.) epoxide resin. Initially, E decreases to about 4.5% elastomer content and then remains constant up to about 15% CTBN before it declines sharply. Similarly, σ_t and T_g both decrease with increase in elastomer content. However, the opposite trend is true for α. These results are slightly different from the behaviour of two similar CTBN-toughened DGEBA (MY 750 and GY 250 Ciba-Geigy) epoxide resins studied by Scott and Phillips [47] and Lahiff [48], who have noticed only a marginal change in E and σ_t with rubber content. These properties are also affected by temperature and loading rates. σ_t and E are significantly reduced with increasing temperature [45,48]. The effect of strain rate on E is seen to be marginal, but σ_t increases with increasing strain rate [44,45,49,50].

Table 3 Mechanical Properties of a Rubber (CTBN)-Modified Epoxy (DGEBA)

CTBN Concentration (%)	Fracture Energy, G_{1c} (kJ/m²)	Tensile Strength, σ_{ty} (MPa)	Tensile Modulus, E (GPa)	Linear thermal Coefficient of Expansion, α (°C⁻¹ × 10⁻⁵)	Glass Transition Temperature, T_g (°C)
0	0.12	72	3.3	7.8	80
4.5	1.05	70	2.3	8.0	70
10	2.72	56	2.2	8.7	65
15	3.43	45	2.0	9.6	62
20	3.59	20	1.0	10.2	60
30	2.00	17	0.1	14.0	67

Source: Refs. 6 and 7.

The fracture energy (G_{Ic}) increases with increasing elastomer concentration up to about 16 to 17%, and then decreases (Figure 1 and Table 3). Such behavior is seen as a result of the change of the elastomer from a dispersion phase to a blend at high CTBN concentration [7]. These observations are in agreement with those reported by Yee and Pearson [49] and Lahiff [48]. In addition, Lahiff's data [48] for another CTBN-toughened epoxy (GY250) also support the trend for the toughness reported by Bascom et al. [7] (see Figure 1). While increasing temperature increases the fracture toughness [44,48], as shown in Figure 2, increasing the strain rate decreases the toughness [8,44].

The maximum fracture energy of rubber-modified epoxy is approximately 30 times that of the pure epoxy [7]. But the same increase is not obtained when the modified resin is used as an adhesive or as a matrix in fiber-reinforced composites [8,16,47] (see Table 4). The suppression of toughness in these examples has been explained [6,7,47] on the basis of the dependence of fracture energy on epoxy film thickness. The full toughening effect cannot be developed if the epoxy film is too thick, such as in adhesives and in composites [47]. Hunston et al. [51] recently showed that the toughness in rubber-toughened adhesives increases rather sharply with bond thickness to a maximum value and then decreases with further increase of bond thickness. They suggested that very thick bonds give toughness results similar to those of bulk specimens. When the bond thickness decreases, the increasing constraint causes an increase in the deformation zone length down the bond layer so that the toughness is increased. As the bond thickness continues to decrease, a point is reached such that the deformation zone volume is decreased so that the toughness decreases after passing though a maximum.

In the preceding paragraphs, it is shown that the addition of rubber to epoxy causes a significant increase in toughness. Such a trend is, however, not observed when modified epoxy specimens are subjected to cyclic stresses [36,53]. We return to this aspect later.

Hybrid Epoxy Systems

A simple and effective way to improve the fracture energy without loss of strength and elastic modulus is to incorporate inorganic fillers to rubber-toughened epoxies [54–58].

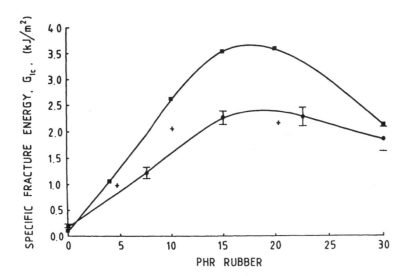

Figure 1 Variation of specific fracture energy. G_{Ic} with rubber content. +, Yee and Pearson [49]; ■, Bascom et al. [7]; and ●, Lahiff [48].

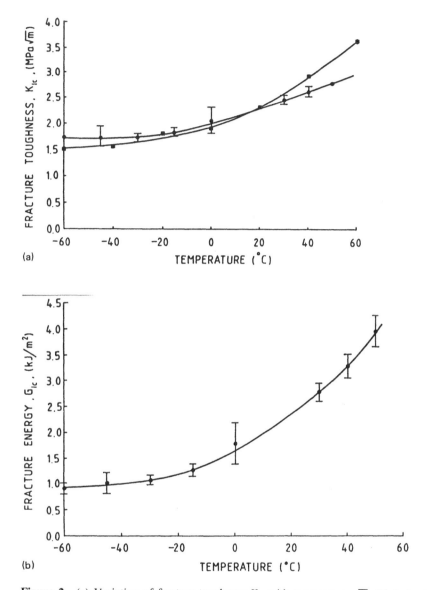

Figure 2 (a) Variation of fracture toughness K_{Ic} with temperature. ■, Kinloch et al. [44]; ●, Lahiff [48]. (b) Variation of fracture energy G_{Ic} with temperature (GY250/piperidine/CTBN ×1300). (From Ref. 48.)

Figure 3 shows that G_{Ic} for the hybrid glass-rubber-modified epoxy is much larger than the glass–unmodified epoxy for a given temperature and volume fraction of glass. If the interfacial adhesion between the glass particles and the matrix is increased by using a reactive silane bonding agent, the peak values of K_{Ic} (and G_{Ic}) for the hybrid composites can be further raised [56]. This synergistic effect of the presence of both dispersed rubbery and rigid fillers is recently confirmed for hybrid epoxies containing metastable zirconia particles [59], short alumina fibers [59–61], and glassy metal ribbons [62,63]. Figure 4 shows the substantial improvement in the fracture resistance of these hybrids as a result of the presence of both dispersed rubbery and rigid fillers. Large-aspect-ratio alumina

Table 4 Fracture Toughness for Bulk Resin, as Adhesive Film and as Matrix in Carbon-Fiber-Reinforced Composites

Rubber Content (%)	G_c (kJ/m^2)		
	Bulk Resin	Adhesive Film[a]	Composite
0.0	0.33	0.28	0.28
3.2[b]	1.4	0.33	0.37
6.2[b]	2.2	1.35	0.36
9.0[c]	3.2	1.50	0.49

Source: Ref. 47.
[a] Adhesive film thickness, ca. 200 μm.
[b] Resin MY 750 epoxy + CTBN with acrylonitrile content 24.2% + 6% piperidine.
[c] Resin MY 750 epoxy + CTBN with acrylonitrile content 28.2% + 6% piperidine.

fibers (40) and glassy metal ribbons (10) have a greater effect in toughening than do zirconia particulates. In addition, the former fillers appear to impart an increasing crack growth resistance (*R*-curve) for these hybrid composites [60,63].

FRACTURE IN EPOXY RESINS

Like most brittle solids, epoxy resins fracture at a stress of typically 40 to 100 MPa, which is well below the theoretical value of about 3 GPa. This shortfall in strengths suggests that flaws are inherently present in brittle materials and the dependence of fracture strength on flaw size forms the very basis of continuum fracture mechanics. This allows the fracture toughness of brittle materials to be measured using an appropriate test-piece geometry and a corresponding crack analysis.

Fracture Mechanics and Fracture Toughness

The behavior of fracture in a material may be analyzed by two interrelated approaches: (1) the energy approach and (2) the stress intensity factor approach. In the energy approach [64–66], fracture occurs when sufficient potential energy is released from the stress–strain fields by growth of the crack. The measurement of the energy required to extend a crack over a unit area in a suitable test geometry allows the fracture energy (G_{Ic}) to be computed such that

$$G_{Ic} = \frac{P_c^2}{2B} \frac{dC}{da} \tag{1}$$

where B is the width of crack front (thickness), C the compliance of the specimen, a the crack length, and P_c the load at crack propagation.

In the stress intensity factor approach [67], the stress field around a sharp crack is defined uniquely by a parameter called the stress intensity factor (K_I). Fracture initiation occurs when K_I at the crack tip approaches a critical value (K_{Ic}) known as the fracture toughness, given by

$$K_{Ic} = \sigma_c \sqrt{\pi a} \tag{2}$$

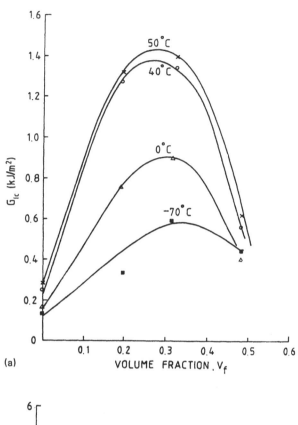

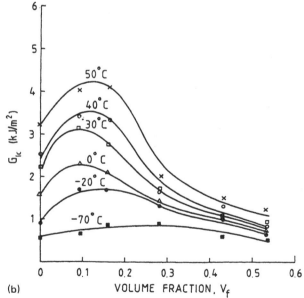

Figure 3 Fracture energy (G_{Ic}) against volume fraction (V_f) of glass particles for (a) an unmodified epoxy and (b) a rubber-modified epoxy.

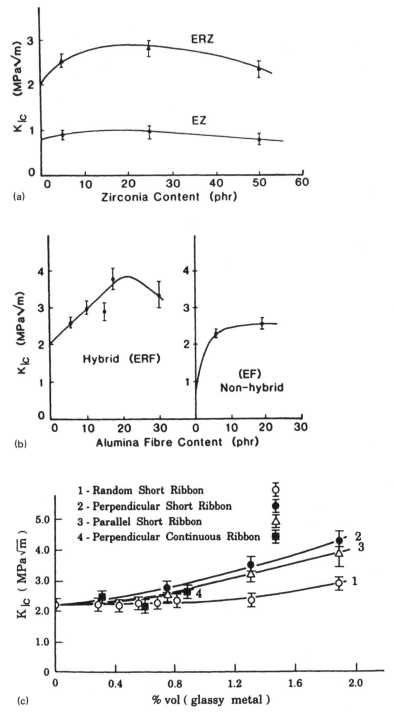

Figure 4 Fracture toughness K_{Ic} against (a) weight fraction of zirconia particles for a rubber-modified epoxy, (b) weight fraction of alumina fibers for a rubber-modified epoxy, and (c) weight fraction of glassy metal ribbons for a rubber-modified epoxy. [(a, b) From Ref. 59; (c) from Ref. 63.)

where σ_c is the critical stress normal to the plane of crack of length $2a$ in an infinite body. Equation (2) is valid only in a brittle solid where yielding near the crack tip is negligible (i.e., the plastic zone size is small compared to the crack length) and fracture can be treated by linear elastic fracture mechanics (LEFM). For cases where small plasticity occurs near the crack tip, the plastic zone can be replaced by a plasticity correction factor r_y [68] and

$$K_{Ic} = \sigma_c \sqrt{\pi(a + r_y)} \tag{3}$$

where

$$r_y \approx \frac{K_{Ic}^2}{2\pi\sigma_{ty}^2} \tag{4}$$

for plane stress and σ_{ty} is the tensile yield strength.

Substitution of Eq. (4) into Eq. (3) leads to

$$K_{Ic} \frac{\sigma_c \sqrt{\pi a}}{\sqrt{1 - \sigma_c^2/2\sigma_{ty}^2}} \tag{5}$$

Equation (5) is useful only if σ_c/σ_{ty} is small or the plastic deformations are confined to a small region at the crack tip. Under these conditions G_{Ic} and K_{Ic} are connected by the relationship $EG_{Ic} = K_{Ic}^2$. If yielding is significant, fracture is beyond the realm of LEFM and other methods (COD, R-curve, J-integral, etc.) [31,68] are more appropriate. The type of fracture (brittle or ductile) depends not only on the material but also on other factors, such as loading rate, temperature, environment and specimen size. [69]. Usually, materials at low temperatures and high loading rates fracture in a brittle manner with a low fracture toughness. As the temperature increases or the loading rate decreases, the toughness of the materials increases. Such effects of loading rate and temperature on toughness of a pure and a modified (rubber-toughened) epoxy are shown in Figure 5 [44]. Also, the factors that increase the toughness tend to reduce the yield stress and hence increase the plastic zone near the crack tip. This may invalidate the use of LEFM, especially for the toughened epoxy resins, which may behave in a ductile manner. To describe the effect of the crack tip yielding process on toughness in such cases, various models [6,7,40,70–72], such as those based on a critical plastic zone size (r_{cy}) or a critical opening displacement at the crack tip (δ_c) or the crack tip blunting model, have been proposed. A few of these models are discussed below.

Crack Tip Opening Displacement

Fracture occurs when the crack tip opening displacement (CTOD) exceeds a critical value δ_c. The critical CTOD is given by

$$\delta_c = \frac{G_{Ic}}{\lambda\sigma_{ty}} \tag{6}$$

where G_{Ic} is the critical mode I strain energy release rate and λ is a plastic zone correction factor. Although various values of λ have been published previously [68,73,74], Broek [68] suggests that λ equals to unity for both plane strain and plane stress conditions, that is,

$$G_{Ic} = \sigma_{ty}\delta_c \tag{7}$$

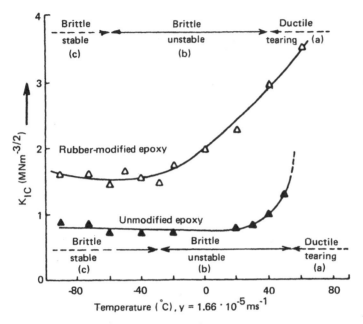

Figure 5 Stress intensity factor (K_{Ic}) at the onset of crack growth as a function of test temperature for unmodified and modified epoxies. (From Ref. 132.)

By measuring δ_c direct from experiments, it is possible to predict the fracture toughness. Bandyopadhyay and Silva [40] measured δ_c for a rubber-toughened epoxy using a scanning electron microscope (SEM) and obtained $\delta_c \approx 40 \pm 5$ μm. Using $\sigma_{ty} = 40$ MPa, they determined that $G_{Ic} = 1.6 \pm 2$ kJ/m², which is reasonable for such materials.

Plastic Zone Model

Bascom et al. [6,7] have used the plastic zone size in order to predict the toughness of rubber-toughened epoxies. For plane strain conditions, the critical plastic zone size (r_{cy}) at G_{Ic} is given by

$$r_{cy} = \frac{EG_{Ic}}{6\pi\sigma_{ty}^2} \tag{8}$$

Rearranging this equation and introducing the yield strain ($\epsilon_y = \sigma_{ty}/E$), we get

$$G_{Ic} = 6\pi\sigma_{ty}\epsilon_y r_{cy} \tag{9}$$

Thus, by measuring r_{cy} from fracture surfaces, G_{Ic} can be determined.

Both models above [i.e., Eqs. (7) and (9)] contain two independent parameters that need to be determined from experiments: σ_{ty} and δ_c for the CTOD model, and σ_{ty} and r_{cy} for the plastic zone model (provided that ϵ_y remains constant). This means that for each test condition, one must determine these parameters in order to compute the fracture toughness. Although the measurement of σ_{ty} is a simple matter by testing specimens under uniaxial tension, the accurate determination of r_{cy} and δ_c for the desired conditions may not always be possible. In fact, the yield stress, σ_{ty} in Eqs. (7) and (9) should correspond to the yield stress near the crack tip where high hydrostatic tensile stresses exist. These

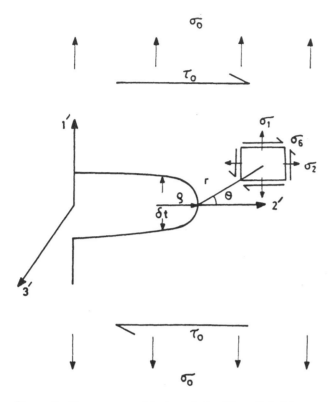

Figure 6 Stresses near a blunt crack tip. (From Ref. 4.)

may significantly lower the local yield stress. This effect is more pronounced in rubber-modified resins, as the cavitation of rubber particles lowers the yield stress [44,49]. Thus σ_{ty} based on the uniaxial tensile specimen may not be accurate enough to predict a reliable G_{Ic} using Eqs. (7) and (9).

Crack Tip Blunting Model

The stress distribution at a distance r ahead of a blunt crack tip of radius ρ at the axis of symmetry (Figure 6) for an applied tensile stress σ_0 (and in the absence of shear stress τ_0) is given by

$$\bar{\sigma}_1 = \frac{\sigma_0 \sqrt{\pi a}}{\sqrt{2\pi r}} \frac{1 + \rho/r}{(1 + \rho/2r)^{3/2}} \tag{10}$$

$$\bar{\sigma}_2 = \frac{\sigma_0 \sqrt{\pi a}}{\sqrt{2\pi r}} \frac{1}{(1 + \rho/2r)^{3/2}} \tag{11}$$

$$\bar{\sigma}_3 = \begin{cases} 0 & \text{for plane stress} \\ \nu(\bar{\sigma}_1 + \bar{\sigma}_2) & \text{for plane strain} \end{cases} \tag{12}$$

where ν is the Poisson ratio. According to Kinloch and Williams [70], fracture of a notched specimen occurs when a critical stress σ_{tc} is attained at a critical distance c ahead of the crack tip (i.e., $\bar{\sigma}_1 = \sigma_{tc}$ at $r = c$) and Eq. (10) becomes

$$\sigma_{tc} = \frac{\sigma_0 \sqrt{\pi a}}{\sqrt{2\pi c}} \frac{1 + 2\beta}{(1 + \beta)^{3/2}} \tag{13}$$

where $\beta = \rho/2c$. If we define

$$K_{Ic} = \sigma_0 \sqrt{\pi a} \tag{14a}$$

$$K_{Ics} = \sigma_{tc} \sqrt{2\pi c} \tag{14b}$$

Eq. (13) can be written as

$$\frac{K_{Ic}}{K_{Ics}} = \frac{(1 + \beta)^{3/2}}{1 + 2\beta} \tag{15}$$

K_{Ics} and K_{Ic} are interpreted as the critical stress intensity factors for sharp and blunt cracks, respectively. Equation (15) also gives the toughness enhancement K_{Ic} as a function of ρ with two adjustable parameters, σ_{tc} and c.

The experimental studies of Kinloch et al. [45] on epoxies have shown that these parameters (σ_{tc} and c) are independent of loading rate and temperature, but they depend on the elastomer concentration in a rubber-toughened epoxy. It is therefore possible to predict K_{Ic} for any given test conditions once σ_{tc} and c are established. The only parameter which depends on test conditions in Eq. (15) is β through the crack tip radius ρ, which can be expressed as a function of K_{Ic}, σ_{ty} and E, that is,

$$\rho = \frac{K_{Ic}^2}{E \sigma_{ty}} \tag{16}$$

where it has been assumed that the crack tip radius, ρ, at failure is equivalent to the crack tip opening displacement δ_c [45], as this gives better agreement with experimental results.

Thus K_{Ic} can be obtained as a function of $\sqrt{\rho}$ using Eq. (15) if σ_{tc} and c are already established. For example, Kinloch et al. [45] have determined that $c = 0.7$ μm, $\sigma_{tc} = 350$ MPa and $c = 10$μm, $\sigma_{tc} = 200$ MPa for the pure and modified epoxies, respectively. Using these values, they could fit their experiment data for various test conditions that affect the values of ρ as shown in Figure 7 [75].

A similar failure criterion for blunt notches in epoxies has been provided by Narisawa et al. [72]. They assumed that craze and fracture would initiate at the elastic–plastic boundary, a distance $r = d_y$ ahead of the notch tip, when the hydrostatic stress at this point becomes critical. The hydrostatic stress (S) at a point r ahead of a crack tip of radius ρ can be obtained from slip-line theory and Tresca's yield criterion. For a rigid perfectly plastic materials with Poisson's ratio of 0.5, it can be shown that [72]

$$S = \tau_y \left[1 + 2 \ln \left(1 + \frac{r}{\rho} \right) \right] \tag{17}$$

where τ_y is the shear yield stress.

Fracture occurs at $r = d_y$ when S attains a critical value S_c. Narisawa et al. have determined S_c and d_y for several epoxies at various loading rates. This failure criterion can be used to predict the crack growth behavior in a manner similar to the Kinloch and Williams [70] analysis. In terms of the principal stresses $\bar{\sigma}_i$ ($i = 1,2,3$) ahead of the crack tip, it can be shown that for plane strain using Eqs. (10) to (12),

$$S = \frac{1}{3}(\bar{\sigma}_1 + \bar{\sigma}_2 + \bar{\sigma}_3) \tag{18}$$

$$= \frac{\frac{2}{3}(1 + v) \sigma_0 \sqrt{\pi a}}{[2\pi d_y(1 + \rho/2d_y)]^{1/2}}$$

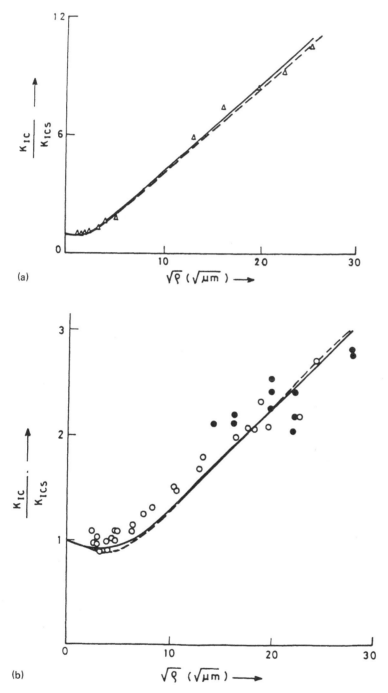

Figure 7 Variation of K_{Ic}/K_{Ics} with $\sqrt{\rho}$ for (a) pure epoxy and (b) rubber-modified epoxy. (From Ref. 4.)

At the onset of fracture, $S = S_c$, so that Eq. (18) becomes

$$\frac{K_{Ic}}{K_{Ics}} = \frac{\sigma_0\sqrt{\pi a}}{S_c\sqrt{2\pi d_y}} = \frac{1.5(1 + \rho/2d_y)^{1/2}}{1 + \nu}. \tag{19}$$

Equation (19) is similar to Eq. (15) in that S_c and d_y now become the two adjustable fracture parameters and K_{Ics} is equal to $S_c\sqrt{2\pi d_y}$. But unlike σ_{tc} and c, both S_c and d_y are strong functions of loading rate [72]. In this respect, Eq. (19) is not as useful a fracture criterion as Eq (15), which can be applied to a wide range of loading rates and test temperatures. It may, however, be noted that S_c and d_y have some physical significance, since S_c is the critical hydrostatic stress at a distance d_y from the notch tip when fracture initiates and d_y represents the distance from the notch tip to the elastic–plastic boundary. The physical interpretation of parameters σ_{tc} and c in Kinloch's et al. [45] model is not very clear. Although σ_{tc} could be interpreted as the constrained yield stress, c could not be seen as having any physical significance.

A further improvement to the Narisawa model is given by Garg and Mai [75], who considered both the complete stress state at the point of fracture initiation and the pressure dependence on fracture. They give the following fracture equation:

$$\frac{K_{Ic}}{K_{Ics}} = \sqrt{A^2 + B} - A \tag{20}$$

where

$$A = \frac{(k - 1)(1 + \nu)(1 + \beta)^{5/2}}{(1 + 2\beta)^2 - 2\beta - 4\nu(1 - \nu)(1 + \beta)^2} \tag{21a}$$

$$B = \frac{k(1 + \beta)^3}{(1 + 2\beta)^2 - 2\beta - 4\nu(1 - \nu)(1 + \beta)^2} \tag{21b}$$

Here $k = \sigma_{cc}/\sigma_{tc}$ (ratio of compressive to tensile strength). Equation (20) is given for plane strain, but for plane stress $\nu = 0$ in Eq. (21). Comparisons of the predicted K_{Ic} results for two epoxy polymers using Eq. (20) and that obtained by Kinloch and Williams [70] show good agreement between these two failure criteria for plane stress but somewhat poor agreement for plane strain.

Experimental Techniques

The fracture energy G_{Ic} and fracture toughness K_{Ic} can be determined by a variety of test geometries [32,76]. Figure 8 shows the common specimen geometries that have been employed to measure G_{Ic} and K_{Ic} of epoxy resins and other brittle polymers. It is worth noting that the measurements are highly dependent on the notching technique [77] and the temperature at which the notch is introduced [78]. A "natural" sharp crack must be used and this can be achieved by fatigue crack growth or by applying a controlled pressure to a razor blade placed at the bottom of the notch.

The most widely used specimen geometries for the study of crack initiation and propagation in epoxy resins have been the double-torsion (DT) [79–86] and compact-tension (CT) specimens [44,45,87–90]. Other configurations include the double cantilever beam (DCB) [91–94], tapered double cantilever beam (TDCB) [7,47,95–99], and three-point bend (TPB) specimens [49,50,72]. The DT and TDCB geometries are particularly useful in studying crack propagation because G_I or K_I is independent of crack length under a

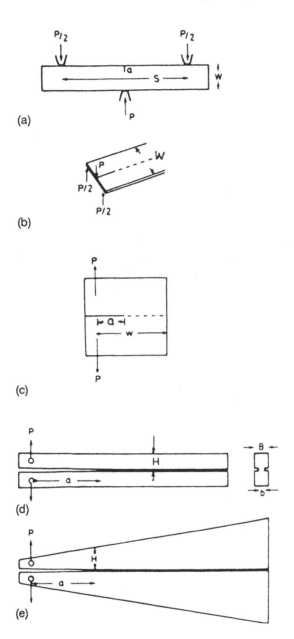

Figure 8 Test specimens for fracture toughness measurements: (a) three-point bending; (b) double torsion; (c) compact tension; (d) double cantilever beam; (e) tapered double cantilever beam.

constant applied load. In comparison with TPB specimens, CT, DT, DCB, and TDCB specimens have improved stability of crack propagation, which is essential for the accurate determination of fracture toughness values. However, TPB specimens are most suitable for the measurement of impact toughness. The fracture toughness equations for the various specimen geometries are given in Appendix B.

FAILURE MECHANISMS AND PROPERTIES

Failure Modes and Toughening Mechanisms

A detailed account of the microscopic and macroscopic aspects of fracture of epoxy resins has recently been reviewed by Bandyopadhyay [100]. Generally, epoxy resins fail at strains from a few percent to higher than 20%, depending on the type of resin, curing agent, and curing schedule. Certain types of epoxy resins can exhibit a maximum in the stress–strain curve, followed by strain softening up to the point of fracture [101]. Microscopically, fracture surfaces of pure epoxy resins can display (1) concurrent presence of a coarse initiation region, a smooth slow crack growth region and a rough fast crack growth region [102,103]; (2) fine craze fibrils [103–105]; (3) cleavage steps [103]; (4) shear bands [102–106]; (5) initiation/arrest lines [80]; and (6) parabolic crack fronts [72].

For toughened epoxy resins, the fracture surface can exhibit a series of complex features as a result of the interaction of the crack tip with dispersions of rubbery or rigid particles or both [59–63]. Possible failure processes that can contribute to the formation of these features are depicted in Figures 9 and 10 [4]. These include shear band formation near the rubber particles, fracture of rubber particles after cavitation, stretching, debonding

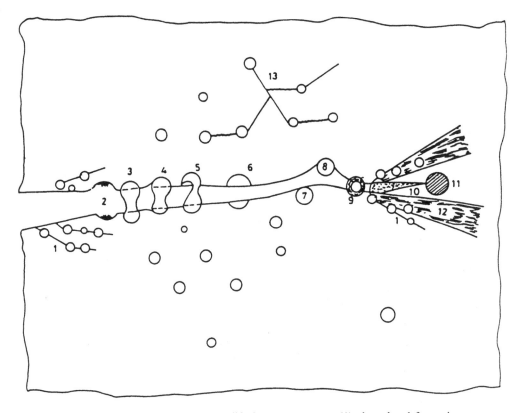

Figure 9 Crack failure mechanisms in modified epoxy systems: (1) shear band formation near rubber particles; (2) fracture of rubber particle after cavitation; (3) stretching; (4) debonding, and (5) tearing of rubber particles; (6) transparticle fracture; (7) debonding of hard particle; (8) crack deflection by hard particle; (9) voided/cavitated rubber particle; (10) crazing; (11) plastic zone at craze tip; (12) diffuse shear yielding; (13) shear band/craze interaction. (From Ref. 4.)

and tearing of rubber particles, trans-particle fracture, debonding of hard particles, crack deflection by hard particles, cavitated or voided rubber particle, crazing, plastic zone at craze tip, diffuse shear yielding, shear band/craze interaction, and pinning of crack front (Figure 10).

A few such failure modes may occur simultaneously in a toughened polymer, depending on the type of particles and the matrix. Each such mechanism contributes to the absorption of energy, and some of the relations to estimate such energies are given below.

1. *Stretching and tearing or debonding of rubber particle.* The increase in toughness ΔG_{Ic} contributed by the elastic energy stored in rubber during stretching, which is dissipated irreversibly when the particle fails either by debonding from the matrix or by tearing, is given by [39]

$$\Delta G_{Ic} = 4\gamma V_p \left(1 - \frac{6}{\lambda^2 + \lambda + 4} \right) \tag{22}$$

where λ is either the extension ratio at the time of debonding or rubber tearing, γ is either the energy per unit area of interface required to debond rubber from the matrix or the rubber tearing energy, and V_p is the volume fraction of rubber particles.

2. *Brittle fracture of a particle.* Assuming the particle to be elastic, the specific fracture energy (γ_p) of the particle of mean radius $\bar{r}$ is given by [107]

$$\gamma_p = \frac{2\bar{r}}{E} \mu^2 \sigma_f^2 \tag{23}$$

where μ is the stress concentration factor at the particle equator and σ_f is the fracture stress. The number of particles N of mean radius $\bar{r}$ per unit area of surface is [108]

$$N = \frac{3}{2} \frac{V_p}{\pi \bar{r}^2} \tag{24}$$

Thus the increase in toughness due to the fracture of particles is

$$\Delta G_{Ic} = 3\mu^2 \bar{r} E V_p \epsilon_f^2 \tag{25}$$

where $\epsilon_f (= \sigma_f/E)$ is the fracture strain of the particle. For an uniaxial stress field $\mu = 2$; hence

$$\Delta G_{Ic} = 12\bar{r}E\epsilon_f^2 V_p \tag{26}$$

3. *Crack deflection.* The crack is diverted by the particles, resulting in an increase of crack surface area. This causes an increase in fracture energy given by

$$\Delta G_{Ic} = \frac{3\gamma_m V_p}{2} \tag{27}$$

where γ_m is the specific fracture energy of the matrix. This relation is derived on the assumption that the increase in fracture surface area created by the deviation of the crack is equivalent to half the surface area of the particle (i.e., $2\pi\bar{r}^2$) minus the matrix area $\pi\bar{r}^2$ if the particle was not there. In addition, it has been shown that slender rods or fibers with high aspect ratios are more effective than disk-shaped particles or spheres in deflecting and twisting the propagating crack and hence in toughening [109,110].

4. *Crack pinning.* The hard filler particles create the obstructions to the propagation of the crack front and cause an increase in toughness by bowing out the crack front between

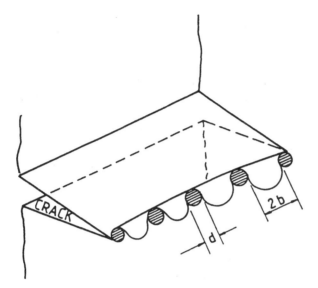

Figure 10 Crack-pinning mechanism. The bowed crack front is at the verge of breaking away from pinning. (From Ref. 18.)

the particles (Figure 10). Lange [18,22] has given a relation for the increase in fracture energy due to pinning as

$$\Delta G_{\text{Ic}} = \frac{T}{2b} \tag{28}$$

where T is the line energy of the crack front and $2b$ is the interparticle spacing. The interparticle spacing can be obtained from [18]

$$b = \frac{d(1 - V_p)}{3V_p} \tag{29}$$

where $d = 2\bar{r}$.

For a penny-shaped crack, Lange [111] showed that the line energy is

$$T = \frac{2\bar{r}}{E} \gamma_m \tag{30}$$

Thus

$$\Delta G_{\text{Ic}} = \frac{\bar{r}\gamma_m}{3b} \tag{31}$$

This equation predicts a linear relationship between the increase in fracture energy and the ratio $\bar{r}/b$. However, this is not observed in practice except for the case of a glass-filled alumina [112].

Consequently, Evans [29] modified the analysis by considering the increase in toughness to be given by the derivative of the strain energy increase with respect to the interparticle spacing. Moloney et al. [25] have since found good agreement of this modified theory with experimental results for alumina- and silica-filled epoxies at low volume fractions. However, at high volume fractions the agreement is not as good, probably because the "bow" crack front is semielliptical rather than semicircular, as assumed in the theory [25,113].

Stability of Cracking and Crack Growth Behavior

Generally, the crack growth behavior of both pure and toughened epoxies is the same. Three basic types of crack growth have been observed: (1) brittle stable crack growth in which cracking is continuous, (2) brittle unstable crack growth during which the crack proceeds in a stick-slip manner, and (3) ductile stable crack growth in which cracking is continuous but is dominated by gross plasticity in the specimen. Typical load–displacement records for crack growth behavior in a compact tension specimen geometry are given in Figure 11 and are termed, respectively, types C, B, and A by Kinloch et al. [44]. In between these three basic types, transitional crack growth behavior is also reported (see also Figure 11). Type A crack growth is only observed at high test temperatures, say, above 40°C, and under these conditions it is questionable whether valid K_{Ic} values can be obtained. For example, in a certain hybrid particulate epoxy composite, Kinloch et al. [56] obtained $K_{Ic} \approx 3.5$ MPa $\sqrt{m}$ and $\sigma_y < 50$ MPa. The required thickness must be larger than $2.5 (K_{Ic}/\sigma_y)^2$, which is about 12 mm and twice the actual specimen thickness. A possible mechanism of this type of ductile crack growth has also been suggested for rubber-toughened epoxies [114] in terms of the meniscus instability model of Taylor [115,116]. Although it is claimed that there is good agreement between the theoretical

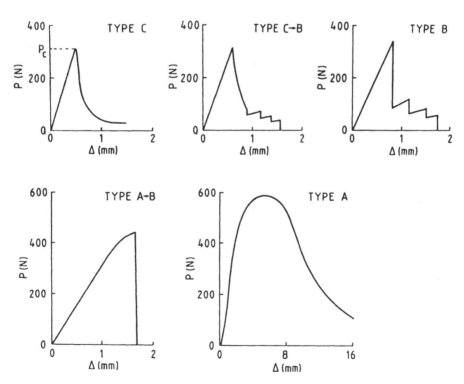

Figure 11 Load (P) versus deflection (Δ) curves for toughened epoxies associated with different types of crack growth in the compact tension geometry. Type C through to type A occurs with increasing temperature and decreasing test rate. (From Ref. 44.)

and experimental values of the critical wavelength of the fingerlike meniscus instabilities, doubts must be raised about the calculated plastic zone thickness based on apparently invalid K_{Ic} measurements. Using $K_{Ic} \approx 3.5$ MPa $\sqrt{m}$ and $\sigma_y \approx 324$ MPa [114], the required thickness for K_{Ic} is about 27 mm, which is again much larger than the actual thickness ($\sim$6 mm) used for the experiments. Under ductile fracture conditions the extent of the plastic zone thickness is geometry dependent, and neither K_{Ic} nor G_{Ic} can be used meaningfully to describe the fracture process. The specific essential work of fracture dissipated in the fracture zone can be obtained by the Cotterell–Mai experimental techniques [117–120]. Such essential fracture work is expected to be a material constant for a given sheet thickness [121]. The occurrence of stable continuous type C crack growth is promoted by low temperatures and high strain rates; conversely, type B stick-slip unstable crack growth is favored by high temperatures and low strain rates. The additions of rubber and other fillers can also change what is otherwise stable continuous crack growth in the pure epoxies to unstable stick-slip crack growth in the modified epoxies [44,56]. There is considerable debate as to the mechanics and mechanisms of stick-slip crack growth, and this concerns the very important problem of crack stability, which is discussed below.

The stability of cracks has been studied by a number of investigators [9,31,80,122–129] and it is shown that this depends on both the fracture toughness of the material and the specimen geometry used. For cracking to be stable the following mechanics equation must be satisfied [80,122,127]:

$$\frac{1}{G_c} \frac{dG_c}{dA} \geq g.s.f. \tag{32}$$

where G_c is the specific fracture resistance, A the crack area, and g.s.f. is the geometric stability factor, depending on the testing machine constraint. Thus

$$g.s.f. = \frac{d^2C/dA^2}{dC/dA} - \frac{2}{C} \frac{dC}{dA} \tag{33a}$$

for displacement control, and

$$g.s.f. = \frac{d^2C/dA^2}{dC/dA} \tag{33b}$$

for load control. Clearly, dG_c/dA is a material property and g.s.f. is a function of the specimen geometry. The more negative g.s.f. is, the more stable cracking is. Characteristic values of g.s.f. have already been tabulated for a variety of testpiece geometries in Refs. 31 and 123. So far, two explanations have been given for unstable stick-slip cracking in epoxies. Both require Eq. (32) to be violated (i.e., dG_c/dA is less than the g.s.f.). Why G_c changes with A probably explains the mechanism of crack instability. Mai and Atkins [9,31] first suggested that in terms of crack velocity $\dot{A}$ and crack acceleration $\ddot{A}$, Eq. (32) can be rewritten as

$$\frac{1}{G_c} \frac{dG_c}{d\dot{A}} \frac{\ddot{A}}{\dot{A}} > g.s.f. \tag{34}$$

If G_c decreases with crack velocity $\dot{A}$ sufficiently so that Eq. (34) is not satisfied, unstable cracking follows. Indeed, Mai and Atkins [9] as well as Andrews and Stevenson [129] have shown that for certain epoxies $dG_c/d\dot{A}$ is negative. Kinloch and Young [76] are uncertain whether this is the cause or the consequence of stick-slip crack propagation. Partly this is because Gledhill and Kinloch [98] could not promote stable cracking by improving the g.s.f. and partly because they obtained only positive $dG_c/d\dot{A}$ in the epoxies they studied. It must, however, be stressed that cracking can be made stable in otherwise unstable crack systems only if Eq. (32) is satisfied. If $dG_c/d\dot{A}$ is very negative, the g.s.f. may have to be very much improved with external stabilizers before any stable cracking can be obtained [9,31]. During stick-slip the crack velocity is not uniform, and only a mean crack velocity between upper and lower values can be estimated [126]. The corresponding mean fracture toughness is given by $\overline{G}_c = Q\sqrt{G_I G_A}$, where G_I and G_A are the initiation and arrest toughnesses and Q is a function of the specimen geometry [31]. The positive $dG_c/d\dot{A}$ results obtained by Gledhill and Kinloch refer only to the *initiation*, which must be higher for higher crosshead speeds of the testing machine. There are no measurements for K_{Ic} and $\dot{A}$ during the jump. Based on these arguments, it is felt that the negative $dG_c/d\dot{A}$ explanation for unstable stick-slip cracking in epoxies cannot be ruled out completely. In fact, this proposal has been supported by Leevers [126] and Maugis [128], who both assume that if G_c decreasing with $\dot{A}$ is a material property, stick-slip occurs when a mean crack velocity is imposed within this region. Both authors also discuss the effects of specimen geometry (i.e., g.s.f.) on controlling stick-slips. Why G_c should decrease with $\dot{A}$ is not clear. Because of the low crack velocity at which stick-slip can be observed in epoxies, Maugis [128] suggested that the mechanism may be one of internal friction [130] and not one of isothermal–adiabatic transition [131].

An alternative explanation for stick-slip crack propagation in epoxies is offered by Kinloch and Williams [70], who propose a crack tip blunting model. Indeed, direct observation of the crack tip deformation in a SEM by Bandyopadhyay et al. [41] shows that blunting occurs by localized shear yielding, and at fracture the shear yield strain is about 0.4. No crazing is observed. The mechanics of crack instability is because the crack tip blunts by localized plastic flow, and this relieves the effective stress intensity factor at the tip so that a higher applied stress intensity is required to cause crack initiation. Since the elastic energy stored in the blunt crack system is much larger than that absorbed when the sharp crack propagates, (i.e., $dG_c/dA < 0$), Eq. (32) may be violated unless the g.s.f. is negative enough to offset the negative dG_c/dA effect.

In principle it is possible to stabilize the crack provided that the excess energy can be absorbed elsewhere; but in practice it may be difficult to achieve this. The mechanism for the negative dG_c/dA is due to the varying degree of plastic blunting for different epoxies. Low-yield-stress epoxies have larger crack tip radii, and vice versa. Figure 7 shows the variation of the K_{Ic}/K_{Ics} ratio with the crack tip radius $\sqrt{\rho}$ for the unmodified DGEBA piperidine-cured epoxies and rubber-toughened epoxies. The theoretical curves for K_{Ic}/K_{Ics} are calculated using Eq. (15) [70,132,133]. Figure 12 shows the dependence of K_{Ic}/K_{Ics} on the tensile yield stress (σ_{ty}) of the pure, rubber-toughened, and hybrid particulate-toughened epoxies. The transition from stable cracking to unstable stick-slip crack propagation for the pure and rubber-toughened epoxies occurs at $\sigma_{ty} \leq 95$ MPa, indicating that there is a possible common deformation mechanism in these two materials. For the hybrid particulate composites this transition takes place at $\sigma_{ty} \leq 120$ MPa, which suggests that the addition of inorganic fillers to epoxies makes cracking more readily unstable [56]. The transition from type C to type B crack growth is characterized by a critical radius ρ_c of the blunt crack tip, which can be estimated from Eq. (16) using K_{Ic} at the transition and E and σ_{ty} from [132]. Thus we calculate $\rho_c = 2.9$, 12.1, and 9.1 μm, respectively, for the pure, rubber-modified, and hybrid particulate-toughened epoxies. Any crack tip radius bigger than ρ_c gives type A and B crack growth characteristics.

While the crack tip blunting model offers a simple mechanism for and an elegant

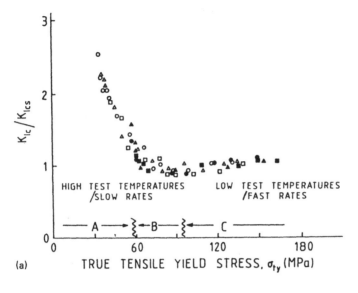

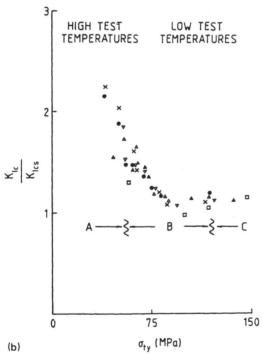

Figure 12 Relationship between K_{Ic}/K_{Ics} ratio and true tensile yield stress σ_{ty} showing the various types of crack growth: (a) pure and rubber modified epoxies; (b) hybrid particulate epoxy. (From Refs. 56, 132.)

analysis of crack instability in pure and toughened epoxies, it must be mentioned that Cherry and Thomson [92] have observed slow stable crack growth prior to unstable fracture in a certain epoxy system, and this result is in direct conflict with the crack tip blunting theory. It is also pertinent to ask whether the blunting model in which dG_{Ic}/dA or dK_{Ic}/dA is negative is any different from the negative $dG_{Ic}/d\dot{A}$ proposal. Figure 12 shows that in the instability region of type B crack growth, $dK_{Ic}/d\sigma_{ty} \approx 0$. It is common knowledge that in polymeric materials σ_{ty} increases with strain rate ($\dot{\epsilon}$). When crack tip blunting occurs, the crack tip is stationary and the strain rate in the crack tip zone ($\dot{\epsilon}_s$) is of the order of $\epsilon_y \dot{K}/K$ [31], where ϵ_y is the yield strain. When crack propagation eventually occurs, the strain rate for the moving crack is $\dot{\epsilon}_m$ ($= \dot{a}\pi E^2 \epsilon_y^3/K_{Ic}^3$) and is proportional to the crack tip velocity $\dot{a}$. Thus the σ_{ty}-axis in Figure 12 may be thought of as the $\dot{a}$-axis. Since $\dot{\epsilon}_m > \dot{\epsilon}_s$ there is an effective increase in σ_{ty} due to $\dot{a}$ when the blunt crack extends. At the same time, K_{Ic} decreases from its blunt tip value to K_{Ic} for a sharp crack, so that dK_{Ic}/dA (or $dK_{Ic}/d\sigma_{ty}$) and $dK_{Ic}/d\dot{A}$ are both negative and unstable cracking ensues by violating Eq. (32). It appears therefore that the crack tip blunting theory and the negative $dG_c/d\dot{A}$ model are qualitatively equivalent. Type C crack growth is stable because $dK_{Ic}/dA \approx 0$; however, the stable type A crack growth cannot be explained in terms of Eq. (32), for it applies only to linear elastic solids. Crack stability analysis in the presence of large plastic flow remote from the crack tip has been given by Mai and Atkins [31], Turner [134], and Paris et al. [135]. The remote plastic flow work accompanying crack growth acts as a sink for excess energy absorption, and this has the effect of stabilizing otherwise unstable fractures. As mentioned before, it must be cautioned whether valid K_{Ic} measurements can be made in this region.

Effect of Composition

The type of epoxy resin and curing agent used, the amount of curing agent, and the temperature and time of cure are some of the structural variables that dictate the micromechanisms of failure and the concomitant values of fracture toughness [76,132]. In addition, the thermomechanical properties of the cured resin will also be affected [136]. Table 5 shows values of the fracture energy G_{Ic} cured with various hardeners. Generally, the measured values of G_{Ic} and K_{Ic} are less than 500 J/m^2 and 1.0 MPa $\sqrt{m}$, respectively. These values, particularly G_{Ic}, are considerably higher than those for phenolics, polyesters, and other thermosetting polymers. Hancox [136] reported that the monomer content of a bisphenol A epoxide resin and the type of hardener (aromatic amine or anhydride) influence the thermomechanical properties of the cured resin. He observed a decrease in the room-temperature tensile strength and an increase in the glass transition temperature (T_g) of an amine-cured system as the monomer content is increased, and vice versa for the anhydride-cured system. Apparently, the amine induces the formation of an increased cross-link density, while the presence of anhydride results in decreased bonding because of hydroxyl groups leaving some hardener unreacted.

Considerable improvements in the fracture properties can be effected by dispersing second and/or third phases in the epoxy matrix to establish a multiphase microstructure. These dispersions may be rubbery [36–46] or rigid [18–229] in nature, or both [54–63]. Table 6 highlights the virtues of these dispersions in the multiphase epoxy systems without sacrificing other important properties, such as modulus, strength or glass transition temperature (T_g). Generally, the values of modulus (E) and fracture toughness (K_{Ic}) tend to increase with increasing volume fraction (V_f) of the rigid filler. The bending strength

Table 5 G_{Ic} Values Measured at Room Temperature for Diglycidyl Ether of Bisphenol A (DGEBA) Epoxy Resins Cured with Various Hardeners

Resin	Hardener (phr)[a]	G_{Ic} (J/m^2)
Epikote 828	10 DETA	172
	95 MNA + 0.5 BDMA	154
	27 DDM	340
	14.6 MPD	110
	4 DMP	180
CT200	13 PA	220
DER 332	5 PIP	121
	Various amounts of TEPA	52–227
	Various amounts of HHPA	158–262
MY 750	8.3 EDA	329
	12.2 TDA	489
	16.1 HDA	575
	11.5 DETA	130
	11.0 TETA	141
	15.0 TEPA	136

Source: Ref. 76.
[a] Hardeners [in parts per hundred of resin (phr)]: DETA, diethylenetriamine; TETA, triethylenetriamine; TEPA, tetraethylenetriamine; MNA, methylnadic anhydride; BDMA, benzyldimethylamine; DDM, diphenyldiaminomethane; MPD, *m*-phenylenediamine; PA, phthalic anhydride; EDA, ethylenediamine; TDA, tetramethylenediamine; HDA, hexamethylenediamine; PIP, piperidine; HHPA, hexahydrophthalic anhydride; DMP, tris(dimethylaminomethylphenol).

Table 6 Properties of Some Multiphase Epoxy Systems

Composition	K_{Ic} (MPa $\sqrt{m}$)	G_{Ic} (kj/m^2)	E (GPa)	σ_b^a (MPa)	T_g (°C)
Pure epoxy	0.8	0.23	2.8	100	100
Epoxy + glass (100/90)[b]	1.5	0.35	6.5	120	96
Epoxy + zirconia (100/25)	0.96		3.2		~100
Epoxy + alumina fiber (100/19)	2.55		3.62		~100
Epoxy + CTBN rubber (100/15)	2.18		2.58		96
Epoxy + rubber + glass (100/15/90)	3.04	1.54	6.0	80	96
Epoxy + rubber + zirconia (100/15/25)	2.84		2.75		96
Epoxy + rubber + Al$_2$O$_3$ fiber (100/15/19)	3.81		3.80		~100

Source: Refs. 30, 56, and 60–63.
[a] Flexural strength.
[b] Parts per hundred of resin by weight.

(σ_b) of the modified epoxy resins improved only with fillers that have high strength, high modulus, large aspect ratio and good adhesion [30].

In contrast to earlier studies [44,56], incorporation of rigid fillers into epoxy resins may stabilize fracture. For instance, the addition of 20 vol % or more silica or alumina changes the unstable crack propagation of an unfilled epoxy resin into a stable one [30]. This stability of crack propagation can be further enhanced by utilizing fillers with high aspect ratios [109]. In addition, the presence of rigid fillers can induce initiation of voids ahead of the crack tip when there is good adhesion at the filler–matrix interface [30]. For poorly bonded fillers, decohesion tends to occur at the interface ahead of the crack tip, resulting in crack tip blunting and unstable propagation [137,138].

The dispersion of rubbery particles in the epoxy matrix is a particularly elegant approach to achieve high toughness with minimal sacrifice in modulus, strength, and high-temperature performance [44]. The presence of these particles promotes stress whitening prior to fracture, which serves to facilitate the energy-dissipating deformations in the vicinity of the crack tip [45]. In general, the fracture energy (G_{Ic}) of rubber-toughened epoxies increases with the volume fraction (V_f) of rubber particles for up to 15%. At higher volume fractions, a blend of epoxy–rubber is formed instead. However, there is no unique relationship between toughness and the volume fraction of rubbery particles [89]. The relation between G_{Ic} and V_f depends greatly on the ability of the thermoset matrix to respond to the presence of the rubbery particles by undergoing shear yield deformations which increase the crack resistance of the polymer.

Shaw and Tod [167] have recently shown that the variations in curing conditions (temperature and time) for a given piperidine-cured CTBN rubber-modified epoxy have a significant effect on the fracture energy G_{Ic} in both bulk and adhesive joint form. Table 7 shows that G_{Ic} changes between 1.75 and 12.12 kJ/m^2 when the curing temperature varies between 120 and 160°C and the curing time from 2 to 6 h.

A combination of high strength, high stiffness, and excellent toughness can be achieved by dispersing both rubbery and rigid fillers in the epoxy matrix to produce a hybrid material [54–63] that possesses impact fracture toughness (K_{Ic}) and fracture energy (G_{Ic}) values of up to 4.0 MPa $\sqrt{m}$ and 6.0 kJ/m^2, respectively. In addition, there is an enhanced stable crack growth, as may be reflected by the large stress-whitening zone on the fracture

Table 7 Cure Temperature–Time Conditions on G_{Ic} of a Rubber-Toughened Epoxy

T (°C)	t (h)	G_{Ic} (kJ/m^2)
120	2	1.75
120	4	2.20
120	6	2.15
120	16	1.70
140	2	3.79
140	4	5.63
140	6	7.62
160	2	5.52
160	4	10.15
160	6	12.12

Source: Ref. 167.

surface. The presence of rubbery particles tends to offset the crack pinning/bowing effect of rigid fillers by promoting interface debonding and hence leading to prominent particle or fiber pullout [59–63] bridging the crack during fracture propagation. Consequently, this gives rise to an increasing crack growth resistance (R) curve [60–63]. The precise failure mechanisms in these hybrid materials are very complicated and depend on the volume fraction, size, and aspect ratio of fillers as well as the interface adhesion.

Effect of Temperature

The fracture behavior of pure and modified epoxy resins is strongly influenced by the variation of test temperatures. Figure 13 shows the crack propagation process for three different compositions of an epoxy resin cured under identical conditions but tested at different temperatures. Clearly, low temperatures tend to induce brittle stable (type C) propagation, while high temperatures are more favorable for brittle unstable (type B) propagation [139]. Ductile, stable (type A) crack propagation can occur only at higher

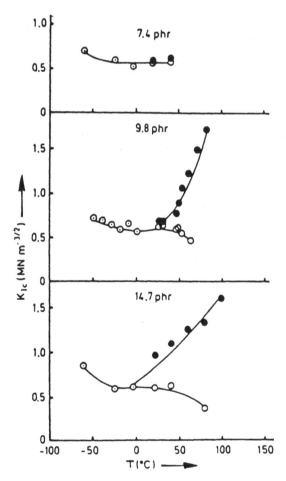

Figure 13 Variation of K_{Ic} with temperature for a DGEBA epoxy polymer cured with different-stated phr of TETA and tested at a cross-head speed of 0.5 mm/min. Brittle stable propagation: $\odot$, K_{Ic}; brittle unstable propagation: $\bullet$, K_{Ici}; $\bigcirc K_{Ica}$. (From Ref. 139.)

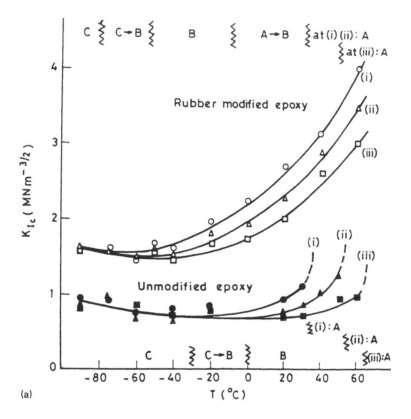

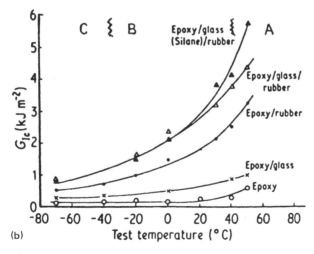

Figure 14 (a) Fracture toughness K_{Ic} as a function of test temperature and rate for pure and rubber-modified epoxy resins; (b) fracture energy G_{Ic} against test temperature for various epoxy systems. [(a) From Ref. 44; (b) from Ref. 56.)

test temperatures, approaching the glass transition temperature (T_g) of the polymer. This behavior of cracking is also typical of rigid particulate–epoxy resins and appears more pronounced for rubber-modified epoxies and hybrids (Figure 14).

The profound effect of temperature on the failure mechanisms and stability of crack growth in short-fiber-reinforced epoxies has recently been investigated by us [61]. Stress whitening became more pronounced as the test temperature was increased. In addition,

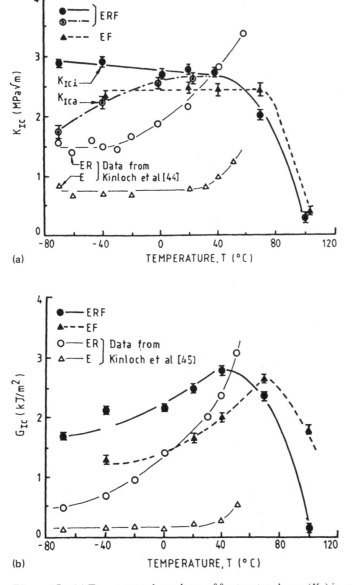

Figure 15 (a) Temperature dependence of fracture toughness (K_{Ic}) in epoxy–fiber (EF) and epoxy–rubber–fiber composites. (Data for pure (E) and rubber-modified (ER) epoxies are taken from Kinloch et al. [44].) (b) Temperature dependence of specific work of fracture (G_{Ic}) in EF and ERF composites. (Data for E and ER are taken from Kinloch et al. [56].) (From Ref. 61.)

the stability of crack growth increased as reflected by the difference between the values of K_{Ici} (initiation) and K_{Ica} (arrest), which diminished with rising temperatures (Figure 15a). An almost continuous stable crack growth was observed at high temperatures ($\geq 70°C$), but this was accompanied by some considerable plastic flow in the neighborhood of the fracture plane as the glass transition temperature ($T_g \approx 96°C$) was approached. Unstable crack growth prevailed in the epoxy–alumina fiber (EF) and epoxy–rubber–alumina fiber (ERF) composites at up to 40 and 70°C, respectively, and ductile tearing occurred in both cases at temperatures above 80°C. These fiber composites appear to maintain high values of K_{Ic} even at very low temperatures (i.e., $T < -40°C$), in contrast to those of pure (E) and rubber-modified (ER) epoxies. However, the fracture energy K_{Ic} of EF and ERF composites followed the same trend as the E and ER materials (i.e., G_{Ic} decreased with decreasing temperature for $T < 60°C$) (Figure 15b).

The fracture surfaces explicitly revealed the influence of temperature on the failure processes in the EF and ERF materials [62]. In particular, the extent of fiber–matrix debonding and fiber pullout increased rather dramatically with increasing temperatures. Beyond 70°C the fiber pullout lengths were the longest and the fiber surfaces appeared rather "clean," suggesting poor bonding occurred at elevated temperatures. Due to the addition of rubber particles, the hybrid ERF composite has a higher K_{Ic} or G_{Ic} than the EF composite at a given temperature (except at $T \geq 70°C$). This is because of toughening mechanisms involved in rubber tearing and cavitation and matrix plastic shear yielding, all of which increase with temperature.

Effect of Displacement/Strain Rate

The deformation and fracture processes of epoxy resins are both time and temperature dependent. Increasing the displacement rate has the same effect as decreasing the temperature, and vice versa. In essence, the crack growth behavior will change from brittle stable (type C) to brittle unstable (type B) and eventually to ductile tearing (type A) as the displacement rate decreases and/or test temperature increases.

In a study of an DGEBA epoxy resin cured with various amounts of the TETA, Young [139] has clearly shown the prominent effect of displacement rate on the measured value of K_{Ic} and the associated type of crack growth (Figure 16). For all compositions of the epoxy resin, increasing the displacement rate causes K_{Ici} (initiation) to fall, with K_{Ica} (arrest) remaining almost constant, so that for some compositions the amount of crack jumping decreases, and eventually there is a transition to continuous propagation (type C) at high displacement rates.

A similar phenomenon of cracking behavior is expected of rubber-modified epoxies. For a given test temperature, Kinloch et al. [45] observed that K_{Ic} decreased with increasing displacement rate. Similar reductions in both K_{Ic} and G_{Ic} have also been reported by Low et al [61] in their fiber-modified epoxies, where K_{Ic} decreases monotonically with increasing displacement rates (Figure 17). These observations are anticipated due to the time-dependent yield strength behavior of the composites at different displacement rates.

The effect of displacement/strain rate on failure mechanisms of polymers is most pronounced under impact loading conditions. Williams and co-workers [140,141] observed a very strong time (t) dependence on G_{Ic} ($\propto t^{-0.42}$) in a range of polymeric materials. A rate effect has also been observed in epoxy resins subjected to fatigue at different frequencies [52].

Low and Mai [142] have recently studied the rate and temperature effects on failure

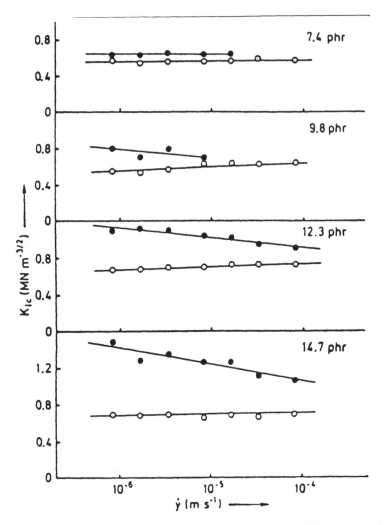

Figure 16 Variation of K_{Ic} with displacement rate $\dot{y}$ for a DGEBA epoxy resin cured with different stated phr of TETA and tested at 20°C. Symbols have same meaning as in Figure 13. (From Ref. 139.)

mechanisms in pure and modified epoxies. A substantial variation in fracture toughness, G_{Ic}, with rate was observed at both very low (down to 10^{-6} s^{-1}) and very high (up to 10^2 s^{-1}) strain rates (Figure 18). Under impact testing conditions, G_{Ic} for both pure and rubber-modified epoxies displayed peaks (Figure 19) at about 23 and $-80°C$, which appeared to correlate with the corresponding size of the crack tip plastic zone (Figure 20). Two separate crack tip blunting mechanisms were proposed to explain these rate- and temperature-dependent G_{Ic} results: thermal blunting due to crack tip adiabatic heating and plastic blunting associated with shear yield/flow processes. Thermal blunting was found to occur in the pure and rubber-modified epoxies under all impact testing conditions and temperatures above 0°C. For temperatures below $-20°C$ under impact conditions, the fracture toughness is primarily dependent on viscoelastic loss processes. At very slow

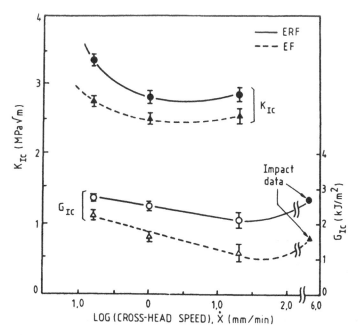

Figure 17 Variation of fracture toughness (K_{Ic}) and specific fracture work (G_{Ic}) with cross-head speed ($\dot{x}$) for epoxy–alumina fiber (EF) and epoxy–rubber–alumina fiber (ERF) composites. (From Ref. 61.)

strain rates ($<10^{-2}\,\mathrm{s}^{-1}$), plastic blunting was the predominant failure mechanism in these epoxy systems.

Strain rate also has a strong influence on the microstructural details on fracture surfaces of epoxy resins, particularly for those modified with second phase dispersions. Slow displacement rates tend to promote formation of a large stress-whitening zone (SWZ) in

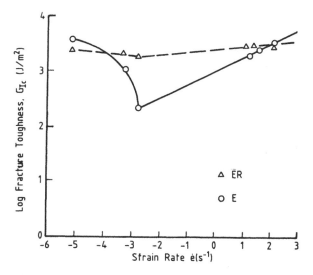

Figure 18 Variation of fracture toughness G_{Ic} with strain rate $\dot{e}$ for ($\bigcirc$) pure epoxy (E) and ($\triangle$) rubber-modified epoxy (ER). (From Ref. 142.)

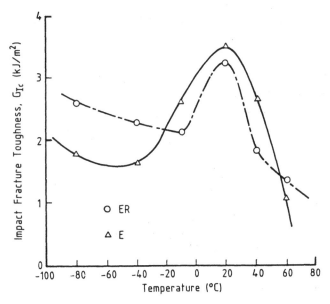

Figure 19 Temperature dependence of impact fracture toughness G_{Ic} in (Δ) pure and ($\circ$) rubber-modified epoxies. (From Ref. 142.)

rubber-modified epoxies with and without short alumina fibers [60]. In the former, fiber debonding, fiber bridging, and fiber pullout prevail [61]. These failure processes lead to a highly nonplanar fracture surface. At high rates, these microfailure processes are considerably suppressed, leading to a planar fracture surface with a greater frequency of fiber breakage and little fiber pullout. In rubber-toughened epoxy resins, the test conditions

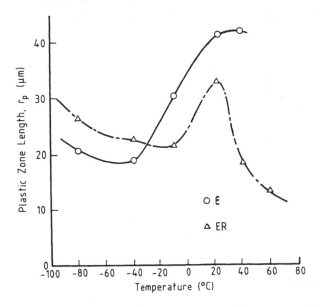

Figure 20 Variation of plastic zone size r_p with temperature for ($\circ$) pure and (Δ) rubber-modified epoxies under impact testing conditions. (From Ref. 142.)

dictate the degree of toughening [89]. At high rates or low test temperatures, the epoxy matrix cannot respond readily to the presence of the rubbery particles and so reduce the extent of shear-yield deformations in the vicinity of the crack tip. The reverse will be true for low rates or high test temperatures.

It is worth noting that the effects of rate and temperature described above are closely interrelated in view of the viscoelastic nature of epoxies. As such the time–temperature superposition principle in viscoelasticity is expected to apply to G_{Ic}. Hunston and co-workers have studied the interrelationship between rate and temperature with G_{Ic} for a range of epoxies [51,143]. By plotting the fracture energy (G_{Ic}) against time to failure (t_f) at various temperatures, they found that the curves could be shifted along the t_f-axis to obtain a master curve for G_{Ic}. An Arrhenius equation can be used to describe this shift factor (a_T) such that

$$a_T = \frac{\Delta E}{R}\left(\frac{1}{T} - \frac{1}{T_0}\right) \tag{35}$$

where R is the universal gas constant, T_0 the reference temperature at which the master curve is to be determined, and ΔE the activation energy of the particular failure process involved. Figure 21 shows a plot of a_T for various temperatures using the glass transition temperature (T_g) of the epoxy resin as the reference temperature (T_0). Though obtained from different test techniques a_T is independent of the compositions of epoxies. Using these shift factors and $T_0 = T_g$, the master curves for G_{Ic} of a pure epoxy and a rubber-modified epoxy are shown in Figure 22. These master G_{Ic} curves can be fitted to the following equation:

$$G_{Ic} = G_{Ics} + \beta t_f^m \exp\left[\frac{-m\,\Delta E}{R(1/T - 1/T_0)}\right] \tag{36}$$

Hunston and Bullman [143] have suggested that G_{Ics} measures the limiting toughness at low temperatures and high stress rates, β gives the magnitude of toughening and depends on the rubber content, and m assesses the time dependence of G_{Ic} and appears to be a function of rubber particle-size distribution. Equation (36) is thus not only useful for the prediction of G_{Ic} given any t_f and T, but also very usefully for comparing the fracture properties of epoxies of different compositions and predicting the degree of rubber toughening [89]. Indeed, Kinloch and Hunston [89] observed that the degree of toughening in rubber-modified epoxies did depend on the test conditions. At low values of t_f/a_T, when the yield stress of epoxy matrix is relatively large, the matrix undergoes only minimal shear yielding, and the toughness of the polymer is at its lowest, and vice versa at high values of t_f/a_T.

Effect of Residual Stresses

The differential thermal expansion is a very important parameter in any multiphase or composite system since it determines the residual stress–strain distributions after fabrication and can have overriding effects on the resultant mechanical and fracture properties [144,145]. Large stresses can be developed at the interface between the matrix and the dispersed phase due to thermal expansion mismatch. The magnitude of these stresses, both radial (σ_r) and tangential (σ_t), may be estimated from [146]

$$-\sigma_r = 2\sigma_t = \frac{\Delta\alpha\,\Delta T}{(1 + \nu\,/2E\,) + (1 - 2\nu\,/E\,)} \approx \Delta\alpha\,\Delta TE_m \tag{37}$$

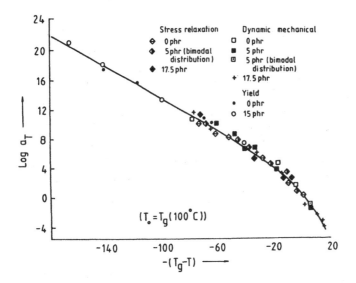

Figure 21 Shift factor a_T as a function of temperature needed for time–temperature superposition of G_{Ic} for rubber-modified epoxies. Rubber contents are indicated after each test method. (From Ref. 51.)

where $\Delta\alpha\,(= \alpha_m - \alpha_f)$, ΔT, ν, and E are the linear thermal expansion coefficient difference, temperature difference, Poisson's ratio, and elastic modulus, respectively. The subscripts refer to the matrix (m) and dispersed filler (f).

The nature of these stresses depends strictly on the thermal expansion characteristics

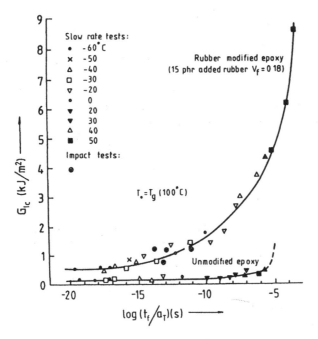

Figure 22 Fracture energy G_{Ic} versus reduced time to failure, t_f/a_T. (From Ref. 51.)

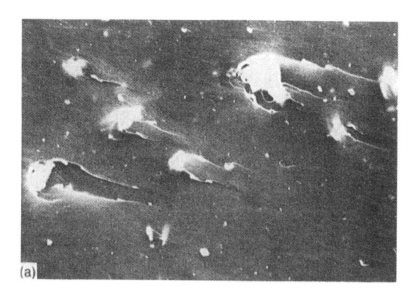

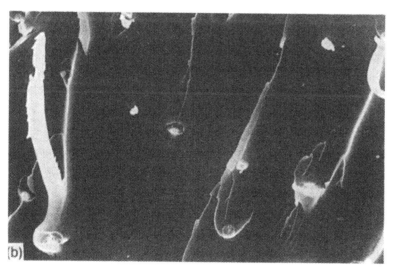

Figure 23 Fracture surfaces of (a) epoxy–ZrO$_2$ (EZ) system showing crack pinning and (b) epoxy–alumina fiber (EF) system showing crack pinning and fiber fracture.

of the two phases at the interface. When $\alpha_f < \alpha_m$, the stresses at the interface are compressive in the radial directions in both phases, but in the circumferential directions they are tensile in the matrix and compressive in the filler. The compressive stress in the interface would attract a crack to propagate across the strongly bonded interface to cause transgranular fracture, and this gives a low fracture toughness. On the other hand, when $\alpha_f > \alpha_m$, the interface will be subjected to radial tensile stresses. Two effects are possible, depending on the strength of the filler–matrix interface. If the interface is strong enough, the fillers will endeavor to strain the matrix in compression in the axial direction; this will increase the overall strain required to initiate failure and hence will result in an improvement in the fracture strength and toughness. Conversely, if the interface strength is not strong, interfacial debonding will occur and fracture will invariably be intergranular. Again, a high fracture toughness is obtained due to a longer and more tortuous crack path.

In the epoxy systems modified with fillers of ZrO_2 (EZ) and Al_2O_3 (EF), $\Delta\alpha$ is positive ($\alpha_{ZrO_2} = 9.5 \times 10^{-6}°C^{-1}$; $\alpha_{Al_2O_3} = 8.5 \times 10^{-6}°C^{-1}$). It follows from Eq. (37) that compressive radial stresses of about 16.7 and 17.0 MPa, respectively, are induced at the filler–matrix interface on cooling from the curing temperature of 120°C. These stresses serve to enhance the intrinsic bond developed at the interface. Consequently, processes for premature debonding at the interface in the vicinity of an advancing crack tip are suppressed. The stress fields surround the ZrO_2 particles or Al_2O_3 fibers serve to cause the crack front to be pinned and bowed before breaking away to produce "tail ends," which are the characteristic features of crack pinning (Figure 23). The strong bonding at the interface also provides an efficient stress transfer from the matrix to the filler. In the EF system, the large aspect ratio of Al_2O_3 fibers (≈ 1000) allows a rapid buildup of tensile stresses, which fracture the fibers when the ultimate tensile strength of the fiber is exceeded. Fiber debonding and pullout are thus unlikely. The nearly spherical shape of the ZrO_2 particles in the EZ system prevents the buildup of stresses sufficient to exceed their fracture strength. The crack front hence either deflects or shears through the interface. This absence of premature interface debonding is believed to prevent the formation of stress whitening during cracking, resulting in an unstable and catastrophic fracture with planar fracture surface. The poor fracture toughness recorded by the EZ system suggests that the principal toughening mechanism arises from work dissipation associated with crack pinning by the ZrO_2 particles. Phase transformation of these metastable particles had not taken place because not enough stresses are built up to exceed the critical transformation stress of these metastable ZrO_2 particles [147]. The large elastic modulus mismatch ($\Delta E = 197$ GPa) would certainly not be helpful in this case (Figure 24). In tandem with the compressive radial stresses at the interface is the presence of tensile tangential stresses in the matrix which are conducive for flaw generation. The formation of microcracking in the matrix of the EF system (Figure 24a) results from the endeavor of the fibers in keeping the matrix in tension in the axial direction.

The residual stress–strain distributions become more complicated when both rubbery and rigid fillers are dispersed in the epoxy matrix. This complication arises because the rubbery phase has a larger thermal expansion coefficient than that of epoxy. The converse is true for the rigid phase. The following discussion on the hybrid epoxies is only approximate. In the rubber-modified epoxy system (ER), $\Delta\alpha$ is negative. Hence, from Eq. (37), tensile radial stresses of about 6 MPa are induced at the rubber–matrix interface. The chemical bonding between rubbery particles and matrix ensures a strong interface adhesion which can support the resulting tensile stresses. Consequently, the matrix is in compression in the axial direction and a larger average strain is required to initiate failure.

The addition of a third filler with low thermal expansion to the epoxy matrix complicates the mechanism above. The thermal expansion mismatch effect of the two fillers generates residual stresses that are opposite to each other. In the ERF system, tensile stresses of about 6 MPa are induced at the rubber–epoxy interface with the concomitant generation of compressive tangential stress at the surrounding matrix. On the other hand, compressive radial stresses of about 17 MPa are induced at the fiber-matrix interface in concert with the production of tensile tangential stresses in the matrix. However, due to the interference from the rubbery particles, the interface bonding between the fiber and the matrix is somewhat reduced by the presence of tensile radial stresses. Under this circumstance, the tendency for premature debonding at the interface is largely enhanced. This results in a

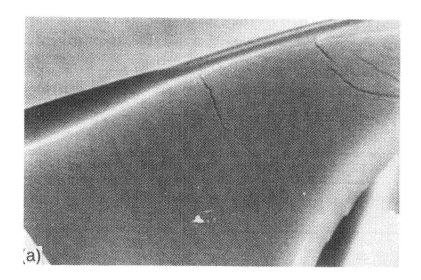

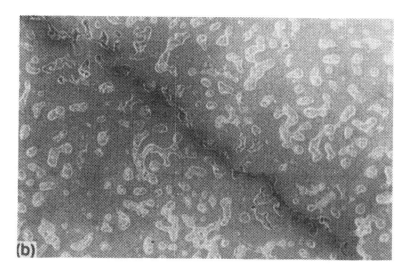

Figure 24 (a) Formation of microcracks in the epoxy matrix containing short Al_2O_3 fibers; (b) particle debonding in a polished surface of epoxy matrix containing metastable ZrO_2 particles.

poor stress transfer by shear and only a moderate level of stress may be built up in the fibers, which is insufficient to cause fiber fracture. Hence those fibers shorter than the critical transfer length (l_c) will be fully pulled out (Figure 25). The formation of an extensive SWZ in this system is due largely to the concurrent display of debonding at the fiber interface and the cavitation of rubber particles. These newly created free surfaces serve to alter the light refractive index at the SWZ [44]. In this case the crack follows a fiber avoidance path, resulting in a highly tortuous fracture surface. Avoidance of fibers by the advancing crack tip is accomplished by the initial tilting at the debonded interface and subsequent twisting of the crack front between fibers [109]. Invariably, these sequential processes of fiber debonding, crack bridging, and fiber pullout result in the stabilization of crack growth and a display of an *R*-curve effect [60].

Similar failure processes are anticipated for the ERZ materials. Again, the presence of tensile radial stresses at the rubber–matrix interface substantially reduces the bonding strength at the ZrO_2–matrix interface. Consequently, premature debonding at the ZrO_2–matrix interface is greatly enhanced, leading to a substantial loss in the crack pinning capability of these ZrO_2 particles (Figure 25). The enhanced stress whitening during cracking in this system arises largely from the combined effects of premature debonding at the ZrO_2–matrix interface and the cavitation of rubbery particles. The energy dissipated during this enhanced stress whitening is believed to be responsible for the pronounced nonlinearity in the load–displacement curve and the enhanced stabilization of crack growth.

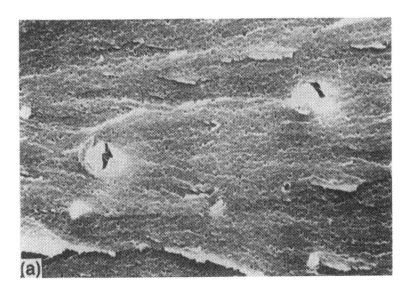

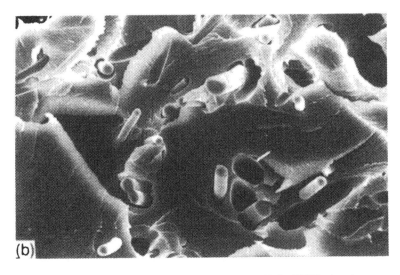

Figure 25 Fracture surfaces of (a) epoxy–rubber–ZrO_2 (ERZ) showing a much reduced crack-pinning effect and (b) epoxy–rubber–alumina fiber (ERF) system showing fiber debonding and pullout.

Effect of Cyclic Loading

Structural weakening of a material due to repetitive or cyclic loading at peak stresses much less than the yield is known as *fatigue*. The failure mechanism of polymer fatigue depends primarily on the magnitude of the cyclic stress range ($\Delta\sigma$). At high values of $\Delta\sigma$, hysteretic heating may lead to temperatures high enough to induce failure by softening or even melting [52,148]. In contrast, at low values of $\Delta\sigma$, failure may occur by the propagation of a single crack in a nominally brittle manner.

In the presence of flaws or cracks, fatigue failure is usually characterized by the Paris power law equation [31]:

$$\frac{da}{dN} = D \, \Delta K^m \tag{38}$$

where D and m are numerical constants, da/dN is the fatigue crack growth per cycle, and ΔK is the applied stress intensity factor range. Although the fatigue resistance of several polymeric matrices [53] have been improved by the addition of rubbery phases, such a beneficial effect is not always observed for rubber-toughened epoxies [52]. In a study on fatigue of CTBN-modified DGEBA (DER 311, Dow Chemical Co.), Manson and co-workers [36,52] showed that at low test frequencies (10 to 15 Hz) the FCP rates at a given value of ΔK were not affected by the presence of rubber. Stable crack growth could be observed at higher ΔK values as the rubber content was increased from 5 to 15 phr (parts per hundred of resin) (Figure 26a). However, the fatigue life was only marginally increased. At higher frequencies (20 and 50 Hz) the FCP rates for the rubber-modified epoxies (except that with 15-phr rubber) were essentially the same as at lower frequencies. The FCP rates of neat resins showed more sensitivity to cyclic frequency effects. The higher FCP rates at 50 Hz for the resins with 15-phr rubber must have been caused by some weakening mechanisms at this high frequency. Figure 26b shows these frequency effects on FCP rates of rubber-toughened epoxies more clearly. The effects of temperature ($-40°$ to 60°C) and mean stress ($R = 0.1$ to 0.5) are shown in Figures 27 and 28 for a similar rubber-toughened epoxy resin containing 15-phr rubber at 5 Hz. It appears that these effects, if any, are minimal [149,150]. Hwang et al. [151] have recently shown that the FCP resistance is linearly dependent on the square root of the apparent molecular weight between cross-links on the rubber-toughened epoxies, and it increases with the static fracture toughness K_{Ic} of the material.

TOUGHENING MECHANISMS

As mentioned earlier, the fracture energies of pure epoxy resins (≈ 100 to 300 J/m^2) can be readily improved several fold by incorporating a second phase, which may be either rubbery (up to 3 kJ/m^2) or rigid (≈ 1 kJ/m^2) in the matrix. This demonstrates that the presence of the dispersions allows various energy absorption processes to take place at the crack tip. These toughening mechanisms may be classified into two general categories. The first includes all such processes that occur along the crack plane, such as crack bridging, crack bowing/pinning, or fiber pullout, which directly reduce the crack tip stress intensity factor. In contrast, toughening mechanisms such as phase transformation, microcracking, plastic void growth, shear deformation, and so on, which dictate the toughness by means of events occurring in a finite-width fracture process zone, form the second category. Mechanisms of the latter type result in a toughness that typically scales with

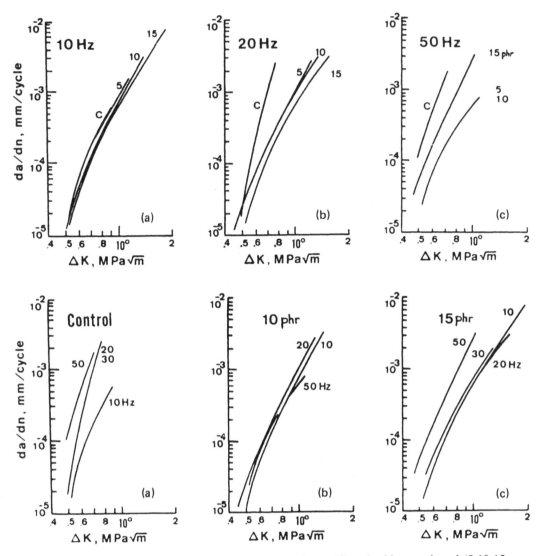

Figure 26 (a) Effect of rubber content on FCP rate of pure (C) and rubber-toughened (5,10,15 phr) epoxies; (b) effect of frequency at constant rubber content (5,10,15 phr) and on FCP rates of pure and toughened epoxies. (From Ref. 52.)

the width of the process zone [152]. A summary of the complex interactions between the crack tip and the dispersions in toughened epoxy systems was given in Figure 9.

In pure epoxy resins, the primary source of energy dissipation is plastic-shear yielding [31], although crazing [103–105] has also been reported. The latter is only plausible for particular cases, such as when straining very thin epoxy films [105] or for resins with an extremely low cross-link density [153]. Plastic shear yielding in the vicinity of the crack tip may promote blunting of a propagating sharp crack [30,82], resulting in a concomitant increase in fracture resistance. The energy dissipation process becomes very complex in multiphase epoxy resins by virtue of the complex interactions between the crack tip and the dispersions. The plausible toughening mechanisms in individual toughened epoxy systems are discussed below.

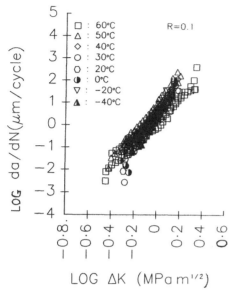

Figure 27 Effect of temperature on FCP rates of rubber-toughened epoxies at $R = 0.1$. (From Refs. 149 and 150.)

Particulate-Filled Epoxies

Possible toughening mechanisms suggested for particulate-filled epoxies include [18] (1) increase in fracture surface area due to the irregular path of the crack, (2) plastic deformation of matrix around the particles, and (3) crack pinning. The first mechanism may account for only a small increase in toughness but not the substantial increase observed in many of the particulate epoxy systems. The second mechanism always exists, as the particles may act as stress concentrations to produce shear bands, causing localized yielding.

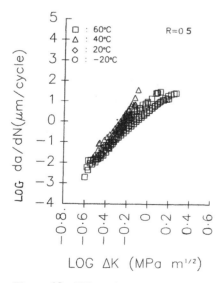

Figure 28 Effect of temperature on FCP rates of rubber-toughened epoxies at $R = 0.5$. (From Refs. 149 and 150.)

However, it is the third mechanism, crack pinning, that is the most significant source of toughening.

After studying the epoxide resins filled with particles as alumina and silica, Moloney et al. [25,26] concluded that the increase in toughness is the result of crack pinning. However, crack pinning does not apply to epoxies with weak filler particles, such as dolomite and aluminum hydroxide [25], and beyond a critical filler volume fraction (ca. 20%) the crack passes through these particles, leading to the trans-particle fracture. In glass-particle-filled epoxy resins Spanoudakis and Young [23] found that in addition to pinning, crack tip blunting caused toughening of the epoxy matrix. There is generally a competition between crack pinning and crack tip blunting, depending on the strength of resin–filler adhesion. Good interfacial adhesion tends to encourage crack pinning, while poor bonding leads to decohesion at the interface and results in crack tip blunting [137,138]. The filler aspect ratio is an important parameter in determining fracture mechanisms. For particles with high aspect ratios (e.g., short fibers), other energy dissipation mechanisms, such as fiber pullout, fiber deformation, fiber fracture, and crack deflection, prevail.

Rubber-Toughened Epoxies

Several theories have been proposed to explain the toughening mechanisms in rubber-modified epoxies [6–8,16,32–35,38–45,49,50,154–156]. Many of these are based on the mechanisms observed in rubber-toughened thermoplastics [32]. In such materials, the toughening is explained by deformation mechanisms involving crazing, shear yielding, and interaction of crazes with shear bands. The rubber particles have been treated as the sites for initiation of crazes and shear bands. Also, the particles may act as obstructions to the propagating crazes, thereby controlling their sizes.

McGarry and co-workers [33–35] first investigated the behavior of rubber-toughened epoxies. They explained that the toughening mechanism involved was similar to that of modified thermoplastics (i.e., by generation of crazes in the vicinity of rubber particles). Such a suggestion was thought to be supported by the presence of a stress-whitened zone near the crack tip. However, due to the high hydrostatic stress beneath the blunt crack tip, microcavitation of the rubber particles caused stress whitening to occur, which was misinterpreted as crazing. Yee and Pearson [49,50] recently confirmed that such stress whitening was indeed caused by the initiation and growth of voids in the rubber particles.

Bucknall [32] also suggested that crazing is a toughening mechanism in rubber-toughened epoxies. He observed that pure epoxy deformed by a shear mechanism (i.e., by shear banding), but rubber-toughened epoxy deformed by an additional crazing mechanism. However, Yee and Pearson [49,50] have expressed their doubts about the presence of crazing since volume dilatation can also be caused by voiding of the rubber particles. The chance of craze formation in fully cured resins, whether pure or modified, is considered to be negligible due to the very small chain lengths between chemical entanglements (cross-links), as the epoxies have high cross-link densities [157]. Even so, other authors [103,104,158,159] have felt strongly about the presence of craze formation in thermoset resins. Morgan and co-workers [103,104] have extensively studied micrographic features of various epoxies in different conditions. On fractured pure epoxy specimen surfaces, they have noticed the fibrillar and nodular structures and cavities. They have interpreted the formation of fibrillar and nodular structures to be caused by fracture of craze fibrils, and the formation of cavities by the void growth and coalescence through the center of craze. Kinloch [132] suggests that such features could have been microcracks and the nodular structure of the fracture surface of pure epoxies might have been caused by the nonuniform structure created by the curing agent [44].

Kunz-Douglass et al. [39] have proposed a different toughening mechanism of rubber stretching and tearing for rubber-toughened epoxies. They did not notice any significant shear yielding in pure and modified epoxies and ignored the contribution of shear yielding to toughness. Nor did they obtain any evidence of crazing. Based on such observations, Kunz et al. have proposed a simple model [Eq. (22)] to compute the fracture energy of rubber-toughened epoxies. However, as noted by Yee and Pearson [49,50], this quantitative theory can account for an increase in toughness by only a factor of 1 or 2, whereas the actual increase in toughness due to rubber particles is seen to be an order of magnitude greater (Table 3). Sayre et al. [160] acknowledged this point, but they explained that the discrepancy was due to the rubber tear energy γ used in Eq. (22) being considerably underestimated in quasi-static tests. In fracture experiments the tearing rate for the rubber particles would be three to four orders of magnitude larger, so that γ is at least 10 times bigger. If so, better agreement would be obtained between theory and experiment.

According to Kinloch et al. [44] there are also a few discrepancies between the Kunz et al. model and their own experimental observations. For example, the model does not explain the stress-whitening phenomenon, the high toughness at high temperatures and the transitions between various types of crack growth; it also predicts the wrong trend for the time–temperature dependence of G_{Ic}, and it ignores the very important contributions to toughness by rubber cavitation and shear yielding in the matrix. It is difficult to resolve these differences due to the sensitivity of the mechanical properties to curing hardeners and curing conditions [173]. However, the rubber stretching and tearing mode cannot be completely dismissed. Bandyopadhyay and co-workers [40,41] have clearly identified particle stretching and bridging behind the advancing crack tip in rubber-toughened epoxies from direct scanning electron microscopy. Toughening by crack interface bridging behind the propagating tip has been observed in cementitious materials [161–163] and in ceramics [164–166]. This gives rise to so-called R-curve behavior [127], in which the crack growth resistance as measured by either G_c or K_c increases with crack growth (Δa). It is expected that an R-curve should be obtained if rubber particle bridging is present behind the tip of the crack. The condition for a propagating crack is that at the crack tip the effective stress intensity factor K_e is equal to that of the pure matrix K_m. Thus

$$K_e = K_a + K_r(\Delta a) = K_m \tag{39}$$

where K_r is the stress intensity factor due to the rubber particles bridging over a distance Δa and its magnitude is negative, K_a is the applied stress intensity factor and at fracture it measured the critical value of the fracture resistance K_c. Therefore, the crack growth resistance of the modified epoxy is

$$K_c = K_m - K_r(\Delta a) \tag{40}$$

K_r depends on the properties of the rubber particles and K_c reaches a maximum value K_∞ when full bridging is established over a saturated distance Δa_s, and at this point the rubber particles at the original crack tip position just begin to tear. From experimental evidence given in Ref. 160 it seems that Δa_s is on the order of several millimeters and a R-curve should have been obtained. Unfortunately, no such curves have been published in the literature for rubber-modified epoxies. Kunz-Douglass et al. [39] have shown that the increased fracture toughness in terms of G is given by

$$G_{Ic} = G_m(1-V_p) + \Delta G_{Ic} \tag{41}$$

Since ΔG_{Ic} is given by Eq. (22) and corresponds to rubber tear, G_{Ic} calculated is therefore related to the maximum value K_∞ of the R-curve:

$$K_\infty^2 = EG_{Ic} = E[G_m(1 - V_p) + \Delta G_{Ic}] \tag{42}$$

Scott and Phillips [47] have also studied rubber-toughened epoxies and have discussed the possibilities of mechanisms such as tearing of rubber particles, localized crazing, and plastic zone effect by plasticization of the resin. However, no conclusive toughening mechanisms have been identified. Only in more recent extensive studies by Bascom and co-workers [6,16], Yee and Pearson [49,50], and Kinloch et al. [44,45] have more definite toughening mechanisms been verified. Bascom et al. have noticed in the fracture surface of specimens of toughened epoxies a stress-whitening zone in the slow crack propagation region. Such stress whitening was observed for neither pure epoxy nor for the modified epoxy in the region of fast crack growth or during an impact test. Detailed microscopic examination of the stress-whitening region showed the presence of small closely spaced holes (larger than the original particle size) in the matrix and yielding (seen as tear markings in epoxy) of the matrix resin around the particles. Such large holes are interpreted to be caused by the dilational deformation of the particles and the matrix. Also, the dilatation of rubber particles nucleate local shear yielding of the epoxy matrix, causing a significant crack tip deformation. Thus volume dilatation of rubber particles and the surrounding matrix as well as shear yielding of the epoxy are proposed to be two mechanisms responsible for the increase in fracture toughness of elastomer-modified epoxies.

Yee and Pearson [49,50] and Kinloch et al. [44,132] have also proposed similar mechanisms responsible for the increase in toughness of rubber-modified epoxies (i.e., the crack growth resistance in the rubber-modified epoxy arises from the large energy-dissipating deformations occurring in the vicinity of the crack tip). The deformation processes are (1) localized cavitation in the rubber, or at the particle–matrix interface caused by dilatation near the crack tip, and (2) plastic shear yielding in the epoxy matrix. Shear yielding is considered to be the major source of energy dissipation. Due to the interactions between the shear field ahead of the crack tip and the rubbery particles, shear yielding is much more important for the modified epoxies than the pure epoxies. The shear deformations near the crack tip produce blunting of the crack tip, leading to the reduction of local stress concentration, which consequently improves the fracture toughness for crack propagation.

While the rubber cavitation and matrix shear yielding toughening mechanisms described above are supported by experimental evidence, no quantitative expressions for these mechanisms have been given. Recently, Evans et al. [152] have proposed a toughening model for rubber-modified epoxies by considering both crack bridging [39] behind the advancing crack tip and rubber cavitation and shear band formation [44,45,49,50] in a process zone ahead of the crack tip. This quantitative mathematical model is based on the theories of transformation and microcrack toughening of ceramic materials [168,169]. In particular, the model indicates the synergistic effects of rubber stretching/tearing and plastic dilation for cavitation and void growth if these two mechanisms occur simultaneously. Quantitative comparison with this model has not been made to date since it requires careful experimental evaluation of various microstructural parameters associated with those different mechanisms. It is also noted that the model predicts the possibility of an R-curve for rubber-modified epoxies similar to Eq. (40). Δa_s is approximately $1.25K_\infty^2/\sigma_y^2$, which is about five times the process zone width [168]; that is, for a rubber-modified epoxy [132] with $K_\infty = 2.7$ MPa $\sqrt{m}$, $\sigma_y = 75$ MPa and $\Delta a_s = 2$ mm.

Hybrid Epoxy Resins

The highest values of fracture energy (G_{Ic}) and fracture toughness (K_{Ic}) are possessed by hybrid epoxies [54–63], containing both rubbery and rigid fillers. The toughening mechanisms in these hybrids are expected to come from the total sum of the individual contributions observed in the rubber-toughened and particulate-filled epoxy resins and any synergistic effects.

The rubber particles increase the fracture resistance by promoting energy absorption processes via cavitation in the rubber or at the particle–matrix interface, and initiation of multiple but localized plastic-shear yielding in the matrix [56]. The rigid fillers enhance the crack resistance by crack pinning and/or inducing additional matrix shear deformations. The former mechanism is expected to be most pronounced at low test temperatures or when there is strong interfacial adhesion between the particle and the matrix. Weak interfacial adhesion invariably impairs the efficiency of the crack-pinning mechanism if the fillers are debonded [170]. Rigid fillers may introduce local stress concentrations which serve to initiate shear deformations, particularly at high temperatures, when the yield stress of the matrix is relatively low [56]. In addition, they may also provide sites for further interfacial debonding because of a large thermal expansion mismatch between the matrix, rubber, and the particles [143]. Together with the localized cavitation of rubber, these failure mechanisms give rise to an extensive stress-whitening zone [171] on the fracture surface and a large fracture energy absorption. Garg and Mai [4] suggest crack pinning as the main toughening mechanism in hybrid particulate epoxies tested at low temperatures and propose that fracture can be described by the critical crack-opening displacement criterion [172]. At high temperatures, where plastic flow of matrix becomes more prominent, the crack tip blunting model [70] is more relevant for the fracture process.

Very complex energy dissipative processes have occurred in rubber-toughened epoxies containing short alumina fibers [60,61]. These hybrids display a range of failure mechanisms, which include rubber tearing and cavitation, fiber debonding, breakage, and pullout, fiber bridging behind the advancing crack tip, and enhanced plastic-shear de-

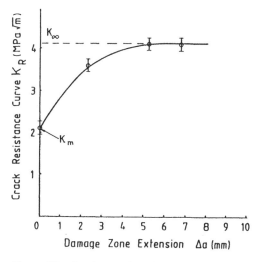

Figure 29 Crack growth resistance curve for an epoxy–rubber—alumina fiber hybrid. (From Ref. 60.)

formation of the matrix. In addition, strongly bonded fibers promote crack deflection processes [109] by tilting and twisting of the advancing crack front to result in a highly nonplanar fracture surface. This process of toughening has been observed in some ceramic materials [109,110]. A broken fiber can also act as a site for crack initiation. The development of a very large stress-whitened zone as a direct result of the above-mentioned failure processes gives rise to a pronounced crack-resistance curve, as shown in Figure 29 for the ERF material.

CONCLUDING REMARKS

There is an increasing trend in utilizing epoxy resins as structural adhesives and high-performance matrices for advanced composites. This trend arises because of the recent advances in the development of new toughened epoxy systems and is set to escalate as quantitative information on their fracture behavior is firmly established. More rigorous studies on the failure micromechanisms, fracture mechanics, and the structure–property relationships are still needed before new epoxy resins with optimum toughness and strength properties can be manufactured. This calls for a greater collaboration and concerted effort in areas of materials science and fracture mechanics from scientists and engineers of interdisciplinary fields.

Further research needs to be done in areas that are still poorly understood. These include the effects of loading rate on G_{Ic} under impact conditions, the fatigue crack propagation behavior, and the hygrothermal aging and creep effects on service performance. Since engineering materials are required to operate under various loading and environmental conditions during service, greater attention should be directed toward conducting fracture experiments under both static and cyclic loadings and at conditions covering the useful temperature and moisture range of the material.

Future research on toughened epoxies should be directed toward the hybrid systems, which have hitherto been most promising in terms of toughness and strength properties. In anticipation of greater applications in the aviation industry, such as the HSCT program, which requires epoxies with a T_g of not less than 200°C, much further work is required for high-temperature epoxy resins (e.g., TGMDA resin) and more effective toughening agents also need to be developed.

ACKNOWLEDGMENTS

The work reported in this chapter was supported by a research contract to Y.-W. Mai of the Centre for Advanced Materials Technology, University of Sydney, by the DSTO Materials Research Laboratories through the program coordinator, Dr. S. Bandyopadhyay. We wish to thank our past and present colleagues: B. Cotterell, M. Dunn, C. Engdahl, C. Foley, A. Garg, G. Hewitt, H.-S. Kim, J.-K. Kim, and H. Lahiff for their contributions in the form of either original data or useful discussions. I.-M. Low was postdoctoral fellow at Sydney University, supported by DSTO, when this work was done.

APPENDIX A: Structure of some epoxy resins and curing agents [4]

Epoxies

1. Polyglycidyl ethers of the novoclacs (Union Carbide ERR 0100)

2. Triglycidyl *p*-aminophenol (Union Carbide ERLA 0510)

3. Tetraglycidyl diaminodiphenyl methane (TG DDM, Ciba-Geigy MY 720):

4. Diglycidyl ether of bisphenol A (DGEBA)

Curing Agents

1. Dicyaniamide (DICY)

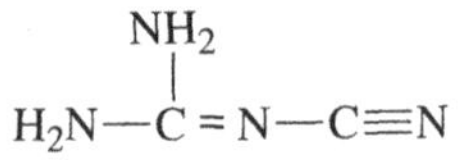

$$H_2N-\underset{\underset{NH_2}{|}}{C}=N-C\equiv N$$

2. Borontrifluoride–monoethylene amine (BF_3 MEA)

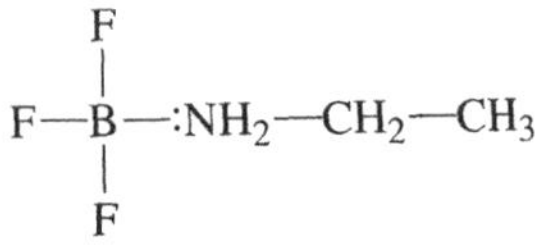

$$F-\underset{\underset{F}{|}}{\overset{\overset{F}{|}}{B}}-:NH_2-CH_2-CH_3$$

3. Diamine diphenyl sulfone (DDS, Ciba-Geigy Eporal)

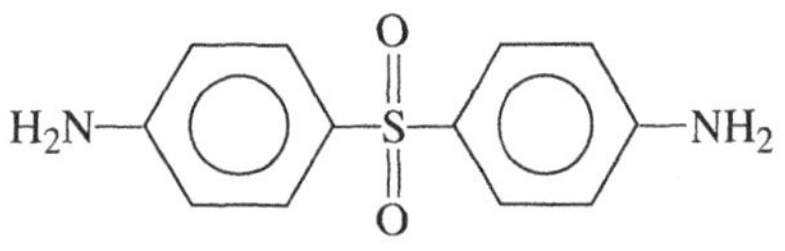

4. Methyl dianiline (MDA)

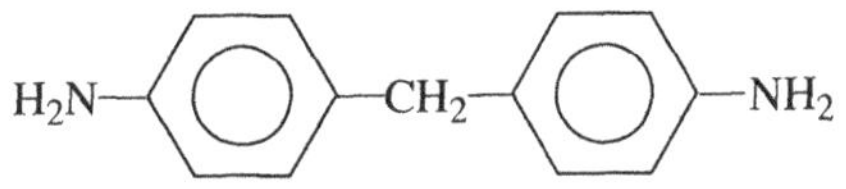

5. *meta*-Phenylenediamine (MPDA)

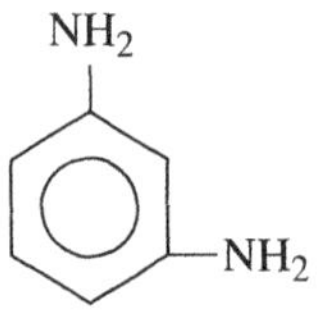

6. Diethylenetriamine (DETA)

$$H_2N-CH_2-CH_2-NH-CH_2-CH_2-NH_2$$

7. Piperidine

$(CH_2)_5NH$

Chemical Formula for Carboxyl-Terminated Butadiene–acrylonitrile (CTBN) Rubber:

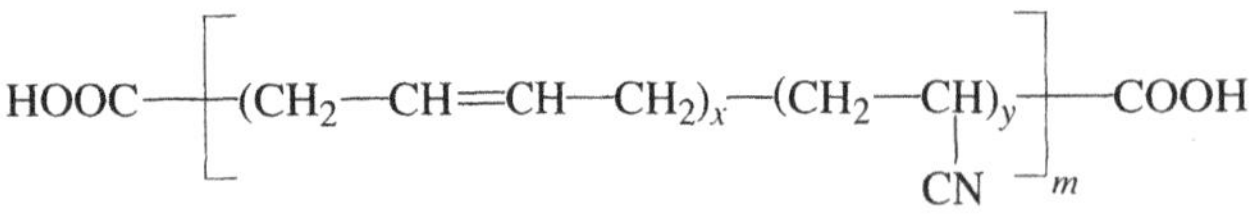

$$HOOC-\left[(CH_2-CH=CH-CH_2)_x-(CH_2-\underset{\underset{CN}{|}}{CH})_y\right]_m-COOH$$

where on average $x = 5$, $y = 1$, and $m = 10$, giving a molecular weight of 3320 g/mol.

APPENDIX B: Formulas/equations for computing values of K_{Ic} or G_{Ic} of various fracture mechanics specimens

Refer to Fig. 8 and Ref. 31. In the following equations, we use the following notation:

P_m, P_c = Maximum and critical loads at fracture

$\quad\quad a$ = Crack length

$\quad\quad W$ = Specimen width or depth

$\quad\quad W_m$ = Moment arm of DT specimens

$\quad\quad B$ = Specimen thickness

$\quad\quad b$ = Specimen thickness in the crack plane

$\quad\quad S$ = Span of notched bend test

$\quad\quad H$ = Height of cantilever beams

$\quad\quad \nu$ = Poisson ratio

$\quad\quad k$ = Geometry constant

Three-Point Bend (TPB)

$$K_{Ic} = \left(\frac{P_m S}{BW^{3/2}}\right) f\left(\frac{a}{W}\right) \tag{B1}$$

where

$$f\left(\frac{a}{W}\right) = \frac{3(a/W)^{1/2}[1.99 - (a/W)(1 - a/W)*(2.15 - 3.93a/W + 2.7a^2/W^2)]}{2(1 + 2a/W)(1 - a/W)^{3/2}} \tag{B2}$$

Double Torsion (DT)

$$K_{Ic} = P_c W_m [WB^3 b(1 - \nu)k]^{-1/2} \tag{B3}$$

Compact Tension (CT)

$$K_{Ic} = \frac{P_c f(a/W)}{B\sqrt{W}} \tag{B4}$$

where

$$f\left(\frac{a}{W}\right) = 29.6 \left(\frac{a}{W}\right)^{1/2} - 185.5 \left(\frac{a}{W}\right)^{3/2} + 655.7 \left(\frac{a}{W}\right)^{5/2} - 1017 \left(\frac{a}{W}\right)^{7/2} + 638.9 \left(\frac{a}{W}\right)^{9/2} \tag{B5}$$

Double Cantilever Beam (DCB)

$$K_{Ic} = \frac{3.46 P_c(a/H + 0.7)}{[Bb H(1 - \nu^2)]^{1/2}} \tag{B6}$$

Tapered Double Cantilever Beam (TDCB)

$$K_{Ic} = \frac{2P_c}{\sqrt{Bb}} \left(\frac{3a^2}{H^3} + \frac{1}{H} \right)^{1/2} \tag{B7}$$

REFERENCES

1. C. Brady and H. R. Clauser (eds.), *Materials Handbook*, 11th ed., McGraw-Hill, New York, 1977.
2. R. J. Morgan, "Structure–property relations of epoxies used as composite matrices," in *Advances in Polymer Science*, Vol. 72, *Epoxy Resins and Composites I*, K. Dusek (ed.), Springer-Verlag, New York, 1985, pp. 1–43.
3. C. A. May, *Resins for Aerospace*, ACS Symposium Series 285, American Chemical Society, Washington, D.C., 1985, pp. 557–580.
4. A. C. Garg and Y.-W. Mai, *Compos. Sci. Technol.*, 31: 179 (1988).
5. I. Mellan (ed.), *Corrosion-Resistant Materials Handbook*, 3rd ed., Noyes, Park Ridge, N. J., 1976.
6. W. D. Bascom and D. L. Hunston, "The fracture of epoxy and elastomer modified epoxy polymers," in *Adhesion*, Vol. 6, K. W. Allen (ed.), Applied Science, Barking, Essex, England, 1980.
7. W. D. Bascom, R. L. Cottington, R. L. Jones, and P. Peyser, *J. Appl. Polym. Sci.*, 19: 2545 (1975).
8. W. D. Bascom, R. Y. Ting, T. J. Moulton, C. K. Riew, and A. R. Siebert, *J. Mater. Sci.*, 16: 2657 (1981).
9. Y.-W. Mai and A. G. Atkins, *J. Mater. Sci.*, 10: 2000 (1975).
10. P. E. Keary, L. B. Ilcewicz, C. Shaor, and T. Trostle, *J. Compos. Mater.*, 19: 154 (1985).
11. J. M. Whitney, C. E. Browning, and W. Hoogsteden, *J. Reinf. Plast.* Compos., 1: 297 (1982).
12. T. K. O'Brian, *Interlaminar Fracture of Composites*, NASA TM-85768 (1984).
13. D. L. Hunston, *Compos. Technol. Rev.*, 6: 176 (1984).
14. S. Mostovoy and E. J. Ripling, *J. Appl. Polym. Sci.*, 15: 644 (1971).
15. Y.-W. Mai, *J. Adhes.*, 7: 141 (1975).
16. D. L. Hunston and W. D. Bascom, "Failure behavior of rubber toughened epoxies in bulk, adhesive and composite geometries," in *Rubber Modified Thermoset Resins*, ACS Advances in Chemistry Series 208, C. Riew and J. K. Gillham (eds.), American Chemical Society, Washington, D.C., 1984, pp. 83-99.
17. A. C. Moloney, H. H. Kausch, T. Kaiser, and H. R. Beer, *J. Mater. Sci.*, 22: 381 (1987).
18. F. F. Lange, "Fracture of brittle matrix particulate composites," in *Composite Materials*, Vol. 5, *Fracture and Fatigue*, L. J. Broutman (ed.), Academic Press, New York, 1974, pp. 2–44.
19. R. J. Young and P. W. R. Beaumont, *J. Mater. Sci.*, 10: 1343 (1975).
20. K. P. Gadkaree and G. Salee, *Polym. Compos.*, 4: 19 (1983).
21. R. Griffiths and D. G. Holloway, *J. Mater. Sci.*, 5: 302 (1970).
22. F. F. Lange and K. C. Radford, *J. Mater. Sci.*, 6: 1197 (1971).
23. J. Spanoudakis and R. J. Young, *J. Mater. Sci.*, 19: 473 (1984).
24. J. Spanoudakis and R. J. Young, *J. Mater. Sci.*, 19 (1984) 487.
25. A. C. Moloney, H. H. Kausch, and H. R. Steiger, *J. Mater.* Sci., 18: 208 (1983).
26. A. C. Moloney, H. H. Kausch, and H. R. Steiger, *J. Mater. Sci.*, 19: 1125 (1984).
27. A. C. Moloney, H. H. Kausch, and H. R. Steiger, "Interfacial properties of filled epoxy resins," in *Adhesive Joints*, K. L. Mittal (ed), Plenum, New York, 1984, pp. 883–904.
28. A. C. Moloney and H. H. Kausch, *J. Mater. Sci. Lett.*, 4: 289 (1985).
29. A. G. Evans, *Philos. Mag.*, 26: 1327 (1972).

30. H. R. Clauser (ed.), *Encyclopedia and Handbook of Materials, Parts and Finishes*, Technomic, Lancaster, Pa., 1976.

31. A. G. Atkins and Y.-W. Mai, *Elastic and Plastic Fracture: Metals, Polymers, Ceramics, Composites, Biological Materials*, Ellis Horwood/Wiley, Chichester, West Sussex, England, 1985, Chap. 7.

32. C. B. Bucknall, *Toughened Plastics*, Applied Science, Barking, Essex, England, 1977.

33. J. N. Sultan and F. J. McGarry, *Polym. Eng. Sci.*, 13: 29 (1973).

34. F. J. McGarry, *Proc. R. Soc. London Ser.* A, 319: 59 (1970).

35. J. N. Sultan, R. C. Laible, and F. J. McGarry, *J. Appl. Polym. Sci.*, 6: 627 (1971).

36. D. N. Shah, G. Attalla, J. A. Manson, G. M. Connelly, and R. W. Hertzberg, "Effect of monotonic and cyclic loading on some rubber modified epoxies," in *Rubber Modified Thermoset Resins, ACS Advances in Chemistry Series 208*, American Chemical Society, Washington, D.C. C. Riew and J. K. Gillham (eds.), 1984, pp. 117–135.

37. C. K. Riew, E. G. Rowe, and A. R. Siebert, in *Toughness and Brittleness of Plastics*, ACS Advances in Chemistry Series 154, D. Deanin and A. M. Crugnola (eds.), American Chemical Society, Washington, D.C., 1976, p. 326.

38. W. D. Bascom and D. L. Hunston, "Adhesive fracture behavior of CTBN-modified epoxy polymers," Paper 22, *Proc. International Conference on Toughening of Plastics*, July 4–6, 1976

39. S. Kunz-Douglass, P. W. R. Beaumont, and M. F. Ashby, *J. Mater. Sci.*, 15: 1109–1123 (1980).

40. S. Bandyopadhyay and V. M. Silva, "Crack propagation studies of a rubber-toughened epoxy resin in the SEM," *Proc. 6th International Conference on Fracture*, New Delhi, A. R. Valluri et al. (eds.), Pergamon Press, Elmsford, N.Y., 1984, pp. 2971–2978.

41. S. Bandyopadhyay, F. J. Pearce, and S. A. Mestan, "Crack tip micromechanisms and fracture properties of rubber toughened epoxy resins," Paper 18, *Proc. Churchill Conference on Deformation, Yield and Fracture of Polymers*, 1985.

42. S. Bandyopadhyay, *J. Mater. Sci. Lett.*, 3: 39 (1984).

43. R. S. Drake and A. R. Siebert, *Q. SAMPE*, 6: 11 (1975).

44. A. J. Kinloch, S. J. Shaw, D. A. Tod, and D. L. Hunston, *Polymer*, 24: 1341 (1983).

45. A. J. Kinloch, S. J. Shaw, D. A. Tod, and D. L. Hunston, *Polymer*, 24: 1355 (1983).

46. E. H. Rowe, A. R. Siebert, and R. S. Drake, *Mod. Plast.*, 49: 110 (1970).

47. J. M. Scott and D. C. Phillips, *J. Mater. Sci.*, 10: 551 (1975).

48. H. M. Lahiff, "Rubber-toughened epoxy resins," B. E. thesis, Department of Mechanical Engineering, University of Sydney, Sydney, Australia, 1986.

49. A. F. Yee and R. A. Pearson, *J. Mater. Sci.*, 21: 2462 (1986).

50. R. A. Pearson and A. F. Yee, *J. Mater. Sci.*, 21: 2475 (1986).

51. D. L. Hunston, A. J. Kinloch, S. F. Shaw, and S. S. Wang, "Characterization of the fracture behavior of adhesive joints," in *Adhesive Joints*, K. L. Mittal (ed.), Plenum, New York, 1984, pp. 789–807.

52. J. A. Manson, R. W. Hertzberg, G. M. Connelly, and J. Hwang, "Fatigue in rubber-modified epoxies and other polyblends," in *Multicomponent Polymer Materials*, ACS Advances in Chemistry Series 211, C. K. Riew and J. K. Gillham (eds.), American Chemical Society, Washington, D. C., 1986, pp. 291–311.

53. R. W. Hertzberg and J. A. Manson, *Fatigue of Engineering Plastics*, Academic Press, New York, 1980.

54. D. Maxwell, R. J. Young, and A. J. Kinloch, *J. Mater. Sci.* Lett., 3: 9 (1984).

55. A. J. Kinloch, D. Maxwell, and R. J. Young, *J. Mater. Sci. Lett.*, 4: 1276 (1985).

56. A. J. Kinloch, D. Maxwell, and R. J. Young, *J. Mater. Sci.*, 20: 4169 (1985).

57. R. J. Young, D. L. Maxwell, and A. J. Kinloch, *J. Mater. Sci.*, 21: 380 (1986).

58. M. F. Tse, *J. Appl. Polym. Sci.*, 30: 3625 (1985).

59. S. Bandyopadhyay, V. M. Silva, I. M. Low, and Y.-W. Mai, *Plast. Rubber Process. Appl.*, 10: 193 (1988).
60. I. M. Low, Y.-W. Mai, S. Bandyopadhyay, and V. M. Silva, *Mater. Forum*, 10: 241 (1987).
61. I. M. Low, Y.-W. Mai, and S. Bandyopadhyay, *Compos. Sci. Technol.*, (1991), In Press (1991).
62. I. M. Low, G. Hewitt, Y.-W. Mai, and C. Foley, *Proc. 11th Australasian Conference on the Mechanics of Structures and Materials*, University of Auckland, New Zealand, Aug. 1988, p. 43.
63. I. M. Low, G. Hewitt, Y.-W. Mai, and C. Foley, *Compos. Sci. Technol.*, 34: 267 (1989).
64. A. A. Griffith, *Philos. Trans. R. Soc.*, A221: 163 (1920).
65. E. Orowan, *Rep. Prog. Phys.*, 12: 185 (1948).
66. C. Gurney and J. Hunt, *Proc. R. Soc.*, A299: 508 (1967).
67. G. R. Irwin, in *Fracture: An Advanced Treatise*, Vol. 3, H. Liebowitz (ed.), Academic Press, New York, 1971, p. 13.
68. D. Broek, *Engineering Fracture Mechanics*, Sijthoff en Noordhoff, Alpen aan den Rijn, The Netherlands, 1978.
69. Y.-W. Mai, *Mater. Forum*, 11: 232 (1988).
70. A. J. Kinloch and J. G. Williams, *J. Mater. Sci.*, 15: 987 (1980).
71. S. Hashemi and J. G. Williams, *J. Mater. Sci.*, 20: 922 (1985).
72. I. Narisawa, T. Murayama, and H. Ogawa, *Polymer*, 23: 291 (1985).
73. J. R. Rice, *J. Appl. Mech.*, 35: 379 (1968).
74. R. M. McMeeking, *J. Mech. Phys. Solids*, 25: 357 (1977).
75. A. C. Garg and Y.-W. Mai, *Compos. Sci.* Technol. 31: 225 (1988).
76. A. J. Kinloch and R. J. Young, *Fracture Behavior of Polymers*, Applied Science, Barking, Essex, England, 1983.
77. G. P. Marshall, L. E. Culver, and J. G. Williams, *Int. J. Fract. Mech.*, 9: 295 (1973).
78. L. C. Cessna and S. S. Sternstein, *J. Polym. Sci.*, B3: 825 (1965).
79. S. Yamini and R. J. Young, *J. Mater. Sci.*, 15: 1814 (1980).
80. S. Yamini and R. J. Young, *Polymer*, 18: 1075 (1977).
81. D. C. Phillips, J. M. Scott, and M. Jones, *J. Mater. Sci.*, 13: 311 (1978).
82. S. Yamini and R. J. Young, *J. Mater. Sci.*, 15: 1823 (1980).
83. R. A. Gledhill, A. J. Kinloch, S. Yamini, and R. J. Young, *Polymer*, 19: 574 (1978).
84. R. J. Young and P. W. R. Beaumont, *J. Mater. Sci.*, 10: 1343 (1975).
85. S. Bandyopadhyay, S. A. Mestan, P. J. Pearce, and C. E. M. Morris, "Fracture characteristics of toughened epoxies," *Proc. Polymer 85 International Symposium on Characterisation of Polymers*, Melbourne, Australia, November 11–14, 1985, Polymer Division, Royal Australian Chemical Institute, 1985, pp. 140–142.
86. S. Yamini and R. J. Young, *J. Mater. Sci.*, 14: 1609 (1979).
87. A. J. Kinloch and S. J. Shaw, *J. Adhes.*, 12: 59 (1981).
88. R. Y. Ting and R. L. Cottington, *Polym. Bull*, 2: 211 (1980).
89. A. J. Kinloch and D. L. Hunston, *J. Mater. Sci. Lett.*, 6: 131 (1987).
90. J. A. Clarke, *Polym. Commun.*, 26: 113 (1985).
91. J. N. Sultan, R. C. Liable, and F. J. McGarry, *J. Appl. Polym. Sci. Appl. Polym. Symp.*, 16: 127 (1971).
92. B. W. Cherry and K. W. Thomson, *J. Mater. Sci.*, 16: 1913 (1981).
93. C. B. Bucknall and T. Yoshii, *Br. Polym. J.*, 10: 53 (1978).
94. K. Mizutani and T. Iwatsu, *J. Appl. Poly. Sci.*, 26: 3447 (1981).
95. S. Mostovoy, P. B. Crosley, and E. J. Ripling, *Proc. 6th Conference on Cracks and Fracture*, ASTM Spec. Tech. Publ. 601, ASTM, Philadelphia, 1976, p. 234.
96. J. S. Mijovic and J. A. Koutsky, *J. Appl. Polym. Sci.*, 23: 1037 (1979).
97. J. S. Mijovic and J. A. Koutsky, *Polymer*, 20: 1095 (1979).
98. R. A. Gledhill and A. J. Kinloch, *Polym. Eng. Sci.*, 19: 82 (1979).

99. S. Bandyopadhyay and C. E. M. Morris, *Micron*, 13: 269 (1982).
100. S. Bandyopadhyay, *Mater. Sci. Eng.*, 125: 157 (1990).
101. A. S. Wronski and T. V. Parry, *J. Mater. Sci.*, 17: 1047 (1982).
102. S. Bandyopadhyay, "Fracture mechanisms in structural epoxies," *Proc. 6th Polymer Technology Convention Australian Section of Plastics and Rubber Institute*, Canberra, Australia, Oct. 5–7, 1983.
103. R. J. Morgan, E. T. Mones, and W. J. Steele, *Polymer*, 23: 295 (1982).
104. R. J. Morgan and J. E. O'Neal, *Polym. Plast. Technol. Eng.*, 10: 49 (1978).
105. R. J. Morgan and J. E. O'Neal, *J. Mater. Sci.*, 10: 1966 (1977).
106. A. S. Wronski and M. Pich, *J. Mater. Sci.*, 12: 28 (1977).
107. J. Gurland, "Fracture of metal matrix particulate composites," in *Composite Materials*, Vol. 5, *Fracture and Fatigue*, L. J. Broutman (ed.), Academic Press, New York, 1975 pp. 45–91.
108. E. E. Underwood, in *Quantitative Microscopy*, R. T. Detoff and F. N. Rhines (eds.), McGraw-Hill, New York, (1968), p. 149.
109. K. T. Faber and A. G. Evans, *Acta Metall.*, 31: 565 (1983).
110. K. T. Faber and A. G. Evans, *Acta Metall.*, 31: 577 (1983).
111. F. F. Lange, *Philos. Mag.*, 22: 983 (1970).
112. F. F. Lange, *J. Am. Ceram. Soc.*, 54: 614 (1971).
113. D. J. Green, P. S. Nicholson, and J. D. Embery, *J. Mater. Sci.*, 14: 1657 (1979).
114. A. J. Kinloch, D. G. Gilbert, and S. J. Shaw, *J. Mater. Sci.*, 21: 1051 (1986).
115. G. I. Taylor, *Proc. R. Soc.*, A201: 192 (1950).
116. P. G. Saffman and G. I. Taylor, *Proc. R. Soc.*, A245: 312 (1958).
117. B. Cotterell and J. K. Reddell, *Int. J. Fract.*, 13: 267 (1977).
118. Y.-W. Mai and B. Cotterell, *Int. J. Fract.*, 32: 105 (1986).
119. Y.-W. Mai, B. Cotterell, G. Vigna, and R. Horlyck, *Polym. Eng. Sci.*, 27: 804 (1987).
120. Y.-W. Mai and B. Cotterell, *J. Mater. Sci.*, 15: 2296 (1980).
121. Y.-W. Mai and B. Cotterell, *Eng. Fract. Mech.*, 21: 123 (1986).
122. C. Gurney and Y.-W. Mai, *Eng. Fract. Mech.*, 4: 853 (1972).
123. Y.-W. Mai and A. G. Atkins, *J. Strain Anal.*, 15: 63 (1980).
124. H. T. Corten, in *Fracture: An Advanced Treatise*, Vol. 7, H. Liebowitz (ed.), Academic Press, New York, 1972, pp. 675–769.
125. Y.-W. Mai, A. G. Atkins, and R. W. Caddell, *Int. J. Fract.*, 11: 939 (1975).
126. P. S. Leevers, *Theor. Appl. Fract. Mech.*, 6: 45 (1986).
127. Y.-W. Mai and B. R. Lawn, *Annu. Rev. Mater. Sci.*, 16: 415 (1986).
128. D. Maugis, *J. Mater. Sci.*, 20: 3041 (1985).
129. E. H. Andrews and A. Stevenson, *J. Mater. Sci.*, 13: 1680 (1978).
130. F. A. Johnson and J. C. Radon, *Eng. Fract. Mech.*, 4: 555 (1972).
131. Y.-W. Mai and N. B. Leete, *J. Mater. Sci.*, 14: 2264 (1979).
132. A. J. Kinloch, in *Advances in Polymer Science*, Vol. 72, K. Dúsek (ed.), Springer-Verlag, Berlin, 1986, pp. 45–67.
133. J. G. Williams, *Fracture Mechanics of Polymers*, Ellis Horwood/Wiley, Chichester, West Sussex, England, 1984.
134. C. E. Turner, in *ASTM STP-677*, 614–628 (1979).
135. P. C. Paris, H. Tada, A. Zahoor, and H. Ernst, in *ASTM STP-688*, pp. 5–36 (1979).
136. N. L. Hancox, *J. Mater. Sci. Lett.*, 6: 337 (1987).
137. A. B. Owen, *J. Mater. Sci.*, 14: 2523 (1979).
138. K. B. Su and N. P. Suh, *Soc. Plast. Eng.*, 27: 46 (1981).
139. R. J. Young, in *Developments in Polymer Fracture*, E. H. Andrews (ed.), Applied Science, Barking, Essex, England, 1979, pp. 183–222.
140. M. W. Birch and J. G. Williams, *Int. J. Fract.*, 14: 69 (1978).
141. J. G. Williams and J. M. Hodgkinson, *Proc. R. Soc. London Ser. A*, 375: 231 (1981).

142. I. M. Low and Y.-W. Mai, *J. Mater. Sci.*, 24: 1634 (1989).
143. D. L. Hunston and W. G. Bullman, "Viscoelastic fracture behavior for different rubber-modified epoxy adhesive formulations," to be published.
144. I. M. Low, *J. Mater. Sci.*, 25: 2144 (1990).
145. I. W. Donald and D. P. W. McMillan, *J. Mater. Sci.*, 11: 949 (1976).
146. J. Selsing, *J. Am. Ceram. Soc.*, 44: 419 (1961).
147. I. M. Low, *J. Mater. Sci. Lett.*, 7: 241 (1988).
148. C. B. Bucknall and W. W. Stevens, in *Toughening of Plastics*, Plastics and Rubber Institute, London, 1978, p. 24.
149. H. S. Kim, C. Engdahl, M. Dunn, and Y.-W. Mai, "Effects of temperature and stress ratio in a toughened epoxy resin," presented at *AeroMat'90*, Long Beach, Calif., May 21–24, 1990.
150. M. Dunn, "Effects of temperature and stress ratio on fatigue crack propagation in toughened epoxy resins," B. E. thesis, University of Sydney, Sydney, Australia, 1988.
151. J.-F. Hwang, J. A. Manson, R. W. Hertzberg, G. A. Miller, and L. H. Sperling, *Polym. Eng. Sci.*, 29: 1479 (1989).
152. A. G. Evans, Z. B. Ahmad, D. G. Gilbert, and P. W. R. Beaumont, *Acta Metall.*, 34: 79 (1986).
153. A. Van Den Boogaart, in *Physical Basis of Yield in Glassy Polymers*, R. N. Haward (ed.), Institute of Physics, London, 1966.
154. A. J. Kinloch, *Met. Sci.*, 14: 305 (1982).
155. A. J. Kinloch, S. J. Shaw, and D. A. Tod, "Rubber-toughened polyimides," in *Rubber Modified Thermoset Resins*, ACS Advances in Chemistry Series 201; C. K. Riew and J. K. Gillham (eds.), American Chemical Society, Washington, D.C. 1984, pp. 101–115.
156. S. K. Kunz, J. A. Sayre and R. A. Asink, *Polymer*, 28: 1897 (1982).
157. A. M. Donald and E. J. Kramer, *J. Mater. Sci.*, 17: 1871 (1982).
158. J. Lilley and D. G. Holloway, *Philos. Mag.*, 28: 215 (1973).
159. S. H. Carr, "What we should be doing," in *Proc. Critical Review, Techniques for the Characterization of Composite Materials*, AMMRC MS 82–3, Army Materials and Mechanics Research Center, Watertown, Mass., 1983, p. 493.
160. J. A. Sayre, S. C. Kunz, and R. A. Assink, "Effect of rubber cross-link density and tear energy on the toughness of rubber-modified epoxies," in *Rubber Modified Thermoset Resins*, ACS Advances in Chemistry Series 208, C. K. Riew and J. K. Gillham (eds.), American Chemical Society, Washington, D.C., 1984, p. 215.
161. A. Hillerborg, M. Modeer, and P. E. Petersson, *Cem. Concr. Res.*, 6: 773 (1976).
162. M. Wecharatana and S. P. Shah, *Cem. Concr. Res.*, 10: 833 (1980).
163. B. Cotterell and Y.-W. Mai, *J. Mater. Sci.*, 22: 2734 (1987).
164. Y.-W. Mai and B. R. Lawn, *J. Am. Ceram. Soc.*, 70: 289 (1987).
165. P. L. Swanson, C. J. Fairbanks, B. R. Lawn, Y.-W. Mai, and B. J. Hockey, *J. Am. Ceram. Soc.*, 70: 279 (1987).
166. R. F. Cook, C. J. Fairbanks, B. R. Lawn and Y.-W. Mai, *J. Mater. Res.*, 2: 345 (1987).
167. S. J. Shaw and D. A. Tod, *J. Adhes.*, 28: 231 (1989).
168. R. M. McMeeking and A. G. Evans, *J. Am. Ceram. Soc.*, 65: 242 (1982).
169. A. G. Evans and K. T. Faber, *J. Am. Ceram. Soc.*, 67: 255 (1984).
170. D. G. Green, P. S. Nicholson, and J. D. Embrury, *J. Mater. Sci.*, 14: 1413 (1979).
171. I. M. Low and Y.-W. Mai, *Compos. Sci. Technol.*, 33: 191 (1988).
172. J. G. Williams, *Int. J. Fract. Mech.*, 8: 393 (1972).

4

Fretting Wear and Fretting Fatigue of Continuous Fiber-Reinforced Epoxy Resin Laminates

Olaf Jacobs*
Technical University
Hamburg, Germany

INTRODUCTION

Relative cyclic sliding motion of two surfaces in intimate contact constitutes a special wear process called *fretting* [1]. Quite often this type of surface damage is caused by cyclic straining due to a fatigue loading of one or both of the mating components. In this case, the surface damage induced by fretting (e.g., microcracking) may act as a nucleation

---

* Current affiliation: *Deutsche Airbus GmbH, Bremen, Germany*

point for an accelerated fatigue failure development leading to a reduced life of the whole structure. Since this interaction of fretting and fatigue (called *fretting fatigue*) influences a variety of engineering applications, such as multilayer leaf springs, bolted joints, and flanges, this problem has attracted numerous research activities in the case of metallic materials [2–4].

Continuous fiber reinforced polymer composites replace metals in an increasing number of engineering applications. High strength and stiffness combined with low specific weight predestinate these materials for application in lightweight constructions. This, of course, is largely exploited by the aircraft industry, but during the last decade, high-performance composites also found increasing interest in civil engineering applications where high accelerations occur (e.g., robot kinematics [5]). Here, saving of weight leads to lesser mechanical loadings of the machinery and enables conservation of energy. Another important advantage of high-performance (especially, carbon-fiber reinforced) polymer composites is their outstanding good fatigue performance. The endurance limit of these materials amounts to about 50 to 70% of the static strength [6]. This suggests the use of continuous fiber-reinforced polymer composites in fatigue-loaded systems (e.g., leaf springs).

The application in dynamically loaded systems may be accompanied by typical fretting wear and fretting fatigue situations. In contrast to the extensive research on sliding and abrasive wear of polymeric materials (e.g., Refs. 7 and 8), there exist only a few studies on the fretting wear of reinforced polymers [9] and in particular of CF/EP laminates [10–12]. As far as the author knows, fretting fatigue is investigated only by the group around Friedrich et al. [13–17], to which the author belongs.

In this chapter we describe the present state of the research on fretting fatigue of high-performance composites. In a first step, the fretting wear of several fiber–matrix combinations is studied. Special attention is focused on the effect of interface dynamics and the influence of loading conditions on the fretting wear of carbon-fiber-reinforced epoxy resin composites. Based on this knowledge, the fretting fatigue behavior of monolithic carbon fiber and aramid fiber composites, as well as of carbon–aramid hybrid composites, is compared with their performance under plain fatigue loading. The mechanisms of interaction between surface damage due to fretting and bulk fatigue are investigated, and a data reduction method is established which gives a quantitative measure for the degree of fretting fatigue damage.

TESTING METHODS

Friction and wear properties cannot simply be attributed to a material but depend on the entire tribological system [18]. The intrinsic properties of the material tested are only one system factor. Other important parameters are the material of the counterpart, the contact geometry, the interface conditions (lubrication, third-body), surface topography, loading parameters, and environmental conditions. Even small changes in parts of this system may lead to significant changes in the friction and wear performance of a material. Furthermore, a tribological system is usually time dependent [19]. The tribological process itself affects the surface and subsurface properties of the material and the counterpart material as well as their surface topographies. The debris produced by wear changes the interface conditions, and frictional heat may influence the material properties. A systematic procedure for the analysis of a tribological system is fixed in the German DIN standards [20].

The system dependence of wear properties makes it difficult to standardize testing methods in such a way that they deliver comparable data usable for the design of a real system. Accordingly, there are hundreds of different friction and wear testing devices described in the literature [21]. A schematic classification of the customary apparatus is given by Kragelski [22]. Some efforts toward more uniformity of friction and wear testing are currently in progress [23]. Nevertheless, commonly accepted ASTM standards exist only for some particular testing procedures [24]. Especially, there is a general lack of standardization of fretting wear and fretting fatigue testing. Just recently, ASTM Committee E-9, on fatigue, has started a project leading to the standardization of fretting fatigue tests [25].

The term *fretting* describes a special wear situation [1]:

1. Two solids perform a tangential oscillatory sliding motion against each other, resulting in cyclic stressing of the surfaces and subsurface regions with changing loading direction.

2. The slip amplitude is small in comparison with the dimensions of the apparent contact area (Figure 1). The mating surfaces are perpetually in close contact, thus causing entrapment of wear debris [26–28] and reduced heat transportation away from the contact zone as well as restricted access of the environment.

This implies that the contact geometry is of particular importance in fretting situations. Generally, two different types of contact geometry can be distinguished: (1) contacts, where one or both of the contacting surfaces possess a spherical shape with different curvature (sphere-on-flat, crossed cylinders), and (2) flat contacts. In flat contact geometries, the apparent (geometrical) contact area is much more extended and a typical fretting situation (mutual coverage of the mating surfaces) is given over a wider range of oscillation amplitudes. The contact pressure is macroscopically homogeneous distributed over the contact zone, and time independent. However, a careful adjustment is required to assure these well-defined contact conditions. Contacts of the first category are less sensitive to a misadjustment. The initial contact conditions are well defined, but the contact pressure is inhomogeneously distributed [29]; the contact area, and thus the contact pressure, vary with time, while the wear process changes the shape of the contacting surfaces.

Fretting Wear Testing

Fretting Test Device

For the present study, flat-on-flat contact geometry was chosen to assure a constant contact geometry and a defined contact pressure over the entire testing time. Figure 2a illustrates

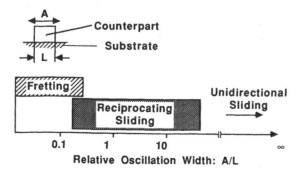

Figure 1 Schematic presentation of the fretting, reciprocating sliding, and unidirectional sliding regime.

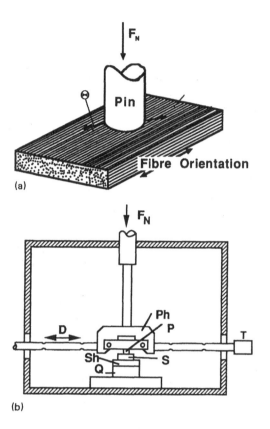

Figure 2   (a) Principle of the loading conditions and (b) schematic of the SRV testing system. S, specimen; Sh, specimen holder; P, fretting pin; Ph, pin holder; F_N, normal load applied by a spring mechanisms; Q, quartz force measuring unit; D, inductive drive; T, inductive displacement transducer; an air circulation heating enabled the adjustment of a defined environmental temperature, T_e.

the loading situation. Fretting pins with flat front surfaces were pressed onto the specimen and forced to an oscillatory sliding motion parallel to the fiber orientation.

The fretting test device was a SRV testing system, commercially available from Optimol Instruments. A schematic presentation of the system is given in Figure 2b. This apparatus allows it to adjust the loading parameters mutually independently in a wide range. A disadvantage of this system is that the course of the driving force is kept invariably sinusoidal instead of controlling the motion of the fretting pin. The superposition of the driving force with the friction force, whose particular course depends on the loading conditions, causes a waveform of the motion, which also changes with the experimental situation. This may sometime affect the wear of the sample, as pointed out by Heinz and Heinke [9].

Specimens, S, of the composite were fixed on a quartz force measurement unit, Q. The fretting pins, P, were pressed by a spring mechanism onto the specimen surface with a defined normal load, F_N. An inductive drive, D, produced the tangential oscillatory motion with the turning-point-to-turning-point oscillation width, A. An air circulation heating allowed the adjustment of the environmental temperature, T_e, in a range from room temperature up to 150°C. Friction caused shearing of the quartz element, whose piezoelectric response was used as a measure for the tangential force acting on the sample. The pin displacement could be measured by an inductive displacement transducer, T. With this setup, the course of the coefficient of friction for each load cycle, averaged from peak-to-peak as a function of the testing time, could be recorded.

The total displacement of the fretting pin includes elastic deformation due to the compliance of the entire apparatus, as well as real sliding. Continuous recording of the friction force versus pin displacement allowed a separation of the real sliding phase from the elastic deformation. The amplitude of real sliding, A, was adjusted and controlled during the fretting tests.

The standard loading parameters in the experiments reported here were

$F_N = 300$ N (contact pressure $p = 15.3$ MPa)

$A = 700$ μm

$v = 20$ Hz

$T_e = 30$ °C

Loading parameters deviating from these standard conditions but applied in some of the studies are mentioned specifically in the text or figure captions.

The mean surface temperature of the specimen in the contact zone is not directly accessible because the poor heat conductivity of the polymeric material creates a steep temperature gradient from the surface to the bulk. As a substitute, the mean surface temperature of the metallic counterpart was measured. A hole was drilled into the fretting pin 1 mm above the fretting contact and a Ni/Cr–Ni thermocouple was placed in this hole. With a second thermocouple positioned 5 mm higher, the temperature gradient within the pin was controlled to extrapolate the temperature measured in the vicinity of the surface to the interface temperature.

Wear Rates

A wear process, in general, can be subdivided into two distinct phases: (1) a running-in period of relatively fast but successively retarding material removal, and (2) a steady state [30]. During steady state, wear frequently proceeds linearly with time and applied normal load. This suggests the definition of a constant $\dot{w}_s$, conventionally called *specific wear rate* [3]. The inverse of the specific wear rate is commonly referred to as *wear resistance*:

$$\dot{w}_s = \frac{\Delta V}{\Delta L\, F_N}$$
$$= \frac{\Delta h}{pv\, \Delta t} \tag{1}$$

where

ΔV = volumetric mass loss
ΔL = sliding distance
Δh = linear material loss
v = sliding velocity
Δt = running time

For a given tribological system, the quantity $\dot{w}_s$ is a type of material constant independent on the loading conditions, which are represented by the product of contact pressure, p, and sliding velocity, v. However, the specific wear rate becomes a steadily increasing function [32] of the loading conditions if the pv product exceeds a critical value pv_{crit}.

For the present study, the mass loss Δm was measured with a laboratory balance (accuracy = 0.01 mg) after removal of loose wear particles from the fretted region by the aid of an air-blowing device. Reference specimens were weighed simultaneously to separate the weight changes due to sorption of moisture. From the mass loss Δm, the specific wear rate $\dot{w}_S$ was calculated according to Eq. (1) by replacing ΔV by $\Delta m/\rho$ and inserting $2A \, \Delta N$ for ΔL (ρ = density of the sample material and ΔN = number of load cycles).

Fretting Fatigue Testing

Design of a Fretting Fatigue Test Device

Fretting fatigue studies are usually performed to explore how an additional fretting load may affect the fatigue performance of a material under certain fatigue loading conditions. Fretting fatigue testing should be carried out in the same servohydraulic test machine, in which the plain fatigue experiments were also performed.

Derived from a fretting fatigue test device used by Gaul and Duquette [33], for this investigation a system was developed by Friedrich et al. [13] which can be attached to the grips of the servohydraulic test machine. The system consists of a positioning device (Figure 3) for cylindrical fretting pins and a fretting load frame (Figure 4). The positioning device was fixed to the upper grips. This device consisted mainly of a steel block (A) into which two mutual perpendicular slids had been machined, to which two plates (B) opposite to each other were fixed. The plates could be positioned with their central holes either parallel or perpendicular to the length axis of the gripping system. Each plate contained five central holes (C) through which two fretting pins could symmetrically act on the two opposite sides of the specimen. Thus, by changing the gauge length, five different slip amplitudes could be selected. Besides the holes for the fretting pins, the plates contained two parallel rows of holes for positioning the fretting load frame (D).

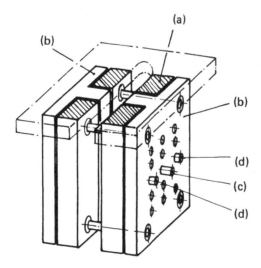

Figure 3 Positioning device for fretting pins. (a), steel block; (b), plates; (c), fretting pin; (d), pins for holding the fretting load frame in position.

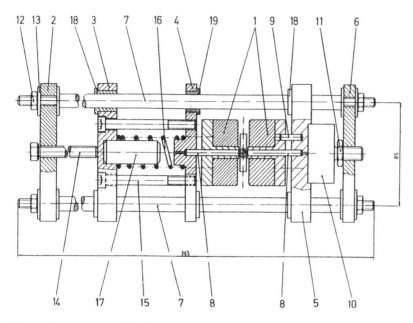

Figure 4 Assembly of the fretting load frame. 1, Cross section through the positioning device; 2 to 6, tension plates; 7, guide rod; 8, fretting pins; 9, positioning pins (D in Figure 2); 10, load cell; 16, coil spring; 17, stabilization bolt; 18 and 19, ball bearings.

The fretting load frame allows symmetrical loading, exact adjustment, and control of the fretting load. The heart of this system is a small load cell on one side and a screw-adjustable coil spring on the other side. Depending on the stiffness of the coil spring, the fretting load can self-reduce more or less quickly with material removal at the fretting contact points. As the load is permanently controlled by the load cell, readjustment can be made from time to time to maintain a constant contact pressure during the test.

Figure 5a is a photograph of the whole system. The upper grips were connected to the load cell of the servohydraulic test machine while the lower grips were attached to the actuator applying the fatigue load. Fretting pins (G) were pressed against the specimen (F). The cyclic straining of the laminate caused a relative motion between fretting pins and specimen.

Loading Conditions

For this investigation, a sinusoidal tension–tension fatigue load was applied with a frequency of 10 s^{-1} at an *R*-ratio of $R = \sigma_1/\sigma_u = 0.1$ (σ_u = upper and σ_1 = minimum fatigue stress in each consecutive load cycle). The relative slip, *A*, between the pins and the laminate can be estimated according to (see Figure 5b)

$$A = \frac{\Delta\sigma}{\sigma_T}\epsilon_T l_l \tag{2}$$

Where $\Delta\sigma$ is the stress amplitude, σ_T the ultimate tensile strength (= 850 MPa), ϵ_T the strain to failure (1.2%), and l_1 the distance of the fretting pins from the upper, fixed clamping grip.

For a stress amplitude of 700 MPa (upper fatigue stress σ_u = 780 MPa), Eq. (2) gives

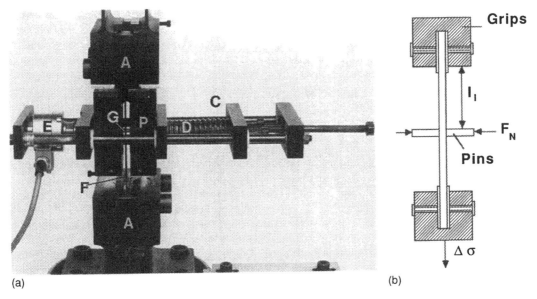

(a)   (b)

Figure 5   Assembly of the complete fretting fatigue device. A, upper and lower grips; P, positioning device; C, fretting fatigue load frame with coil spring (D) and load cell (E); F, test sample; G, fretting pins.

a slip of 650 μm. A measurement with a mechanical displacement transducer resulted in a value of about 700 μm, which is slightly higher than the calculated value because the laminate already performs a small strain motion within the clamping grips.

   In the fretting contact zone, the superposition of tensile, normal, and shear stresses builds up a very complex stress field. Detailed calculations on this stress field have been published by O'Conner [34] and Broszeit et al. [35], among others. Figure 6 schematically illustrates the loading situation of the specimen in the fretting contact, but of course, the

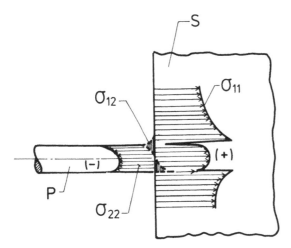

Figure 6   Principal drawing of the stress field in the fretting contact according to Ref. 35. σ_{11}, Tensile stress inside the composite specimen (S); σ_{22}, contact pressure between fretting pin (P) and specimen; σ_{12}, shear stress due to friction.

Figure 8 Optical micrograph of laminate CF2. The arrows point at the resin rich surface layers, left by the peel ply.

Hybrid composites (HC) were manufactured with the same stacking sequence as CF/EP laminate CF2 and the AF/EP laminate. The 0° plies at the surface were replaced by AF layers, thus forming a laminate of the type $[0°(AF)_2, 90°(CF)_2, 0°(CF)_2, 90°(CF)_2]_s$. The intention of this procedure was to combine the fatigue performance of the bulk CF/EP laminate with the wear properties of AF/EP under fretting against soft metals (aluminum).

For fatigue and fretting fatigue tests, the laminates were cut with a diamond saw to rectangular beam-shaped test pieces which were clamped to the servohydraulic test machine

Table 2 Dimensions of Fretting Fatigue Specimens

Sample	Length (mm)	Width (mm)	Thickness (mm)
CF1	280	6.3	6.2
CF2	300	8.3	2
CF3	280	6.3	6.3
AF	300	8.3	2.1
HC	300	8.3	2

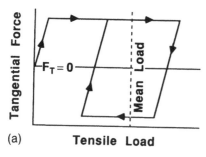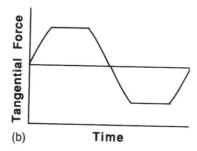

(a)   (b)

Figure 7 Schematic illustration of the tangential force fluctuations during a single load cycle: (a) tangential force versus applied tensile load, (b) tangential force versus time.

particular shape of the stress field changes during the load cycle. The fluctuations of the shear load during one load cycle were analyzed by Rooke and Edwards [36]. Phases of plain elastic deformation of the system specimen/counterpart/load frame alternate with phases of real sliding. During elastic deformation, the tangential force increases linearly with the applied tensile stress and remains on a constant level in the sliding phase (Figure 7a). The resulting friction force waveforms for sinusoidal tensile–tensile fatigue loading are illustrated in Figure 7b.

Sample Materials

Fretting Fatigue Specimens

The sample materials investigated consisted in epoxy matrix reinforced by continuous fibers. As reinforcements, carbon fibers and aramid fibers were employed. Table 1 lists the materials tested under fretting fatigue. Most of these laminates were manufactured using a peel ply. This is a layer of an open-weave material which separates during curing the laminate from the bleeder cloth, which picks up excessive matrix resin. After curing the peel ply is removed, but it leaves a resin-rich layer with the impression of the fabric on the surface of the sample (Figure 8). To study the effect of this surface structure, one set of CF/EP specimens of the CF2-type laminate was produced without peel plies, having plain surfaces.

Table 1 Sample Materials Tested Under Fretting Fatigue Conditions

	Matrix	Fiber	Stacking Sequence	Strength (MPa)	Modulus (GPa)
CF1	Ciba-Geigy 914 C	T 300	$[\pm 45°, 0°, \pm 45°_3, 90°, \pm 45°]_{2s}$	296	Unknown
CF2	BASF R 5212	T 300	$[0°_2, 90°_2]_{2s}$	850	76
CF3	Ciba-Geigy 914 C	T 300	$[0°_2, \pm 45°, 0°_2, \pm 45°, 90°]_s)_3$	799	70
AF	BASF R 5212	Kevlar 49	$[0°_2, 90°_2]_{2s}$	648	36
HC	BASF R 5212	K 49, T 300	$[0°(AF)_2, 90°(CF)_2, 0°(CF)_2, 90°(CF)_2]_s$	629	59

Table 3 Materials Employed in Fretting Wear Tests

Designation[a]	Fiber Volume Content (%)	Density ρ (g/cm^3)	Manufacturer
EP	0	1.17	BASF
GF/EP	61	2.06	BASF
CF/EP	59	1.62	BASF
AF/EP	62	1.36	BASF

[a] EP, epoxy resin R5212; CF, carbon fibers (T300); AF, Kevlar 49; GF, E-glass fibers.

by grips. Table 2 gives the dimensions of the fretting fatigue specimens. Aluminum end tabs enabled easy load transfer and protected the specimen surface against the action of the grips.

Fretting Wear Specimens

For fretting wear tests, simple unidirectional laminates produced were employed. The matrix was the R 5212 system from BASF, reinforced by continuous carbon (T 300), glass (E-glass), or aramid (Kelvar 49) fibers. The fiber content amounted to about 60% by volume. The materials tested under fretting wear are listed in Table 3. The laminates were manufactured without using peel plies so that they possessed smooth surfaces. These sample materials were cut to small test pieces of about 15 mm length and 8 mm width. The thickness of the laminates was about 2 mm for AF and GF composites and 6 mm for CF composites.

Fretting Pins

Cylindrical metal pins (diameter = 5 mm) with flat front surfaces were pressed against the two opposite-width sides of the specimen, thus serving as fretting pins. The front surfaces were ground and polished. The final surface roughness amounted to about 0.10 μm. The edges of the front surfaces were rounded off to avoid cutting of the laminate by sharp edges. Prior to starting the tests, the pins were cleaned with acetone. A low-carbon NiCr steel (Vickers hardness HV = 296), a titanium alloy (HV = 305), an aluminum alloy (HV = 135), and a brass (HV = 160) were chosen as pin materials.

INTERFACE DYNAMICS IN FRETTING OF POLYMER COMPOSITES

The Load Cycle

Figure 9 illustrates a typical course of the tangential force and the pin displacement during one load cycle. Similar tangential force versus displacement loops have been published by other authors [37,38]. The driving force obeyed a sinusoidal function. At the beginning of motion, there is a phase of static friction (*S*) during which the specimen, the counterpart, and their fixtures were elastically deformed. Above a special value for the oscillation width, the elastic deformation force exceeded the static friction force and the fretting pin started to slide across the specimen surface (*S-D*). Subsequently, the system changed over to real sliding (*D*), while the tangential force diminished to the value of dynamic friction. At the turning point of motion, the pin again adhered to the sample and the entire system performed elastic characteristic vibrations (*V*), which occurred only in the F_N–*t* diagram

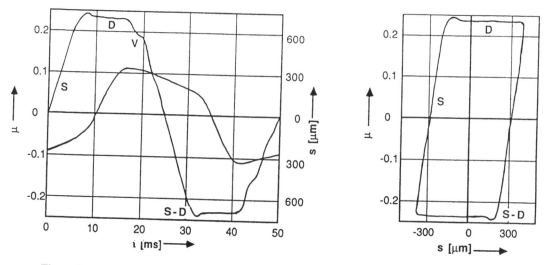

Figure 9 Course of tangential force during one single load cycle for fretting wear of CF/EP against steel: (a) friction coefficient, μ, and pin displacement, *s*, versus time, *t*; (b) friction coefficient as a function of pin displacement. *S*, phase of static friction; *S-D*, transition from static to dynamic friction; *D*, real sliding; *V*, elastic vibrations.

because in the F_N–s diagram they take place along the straight line of elastic deformation.

It should be mentioned that the friction hysteresis presented in Figure 9 is characteristic for hard-elastic materials. Thermoplastics with a softening point (T_G) in the region or below the test temperature give friction loops as shown in Figure 10. The static value of the coefficient of friction was lower than the dynamic value, which again depended on sliding speed. Thermoplastics with a high T_G, as, for example, PEEK and neat thermosets, produce friction–displacement curves with a shape between the extremes represented by Figures 9 and 10. At first sight, the curve appears entirely quadrangular with a clear transition from static to dynamic friction, but the adhesion peak is little pronounced and the dynamic coefficient of friction depends weakly on the sliding velocity.

Only the phase of real sliding (*D*) can contribute to the actual wear process. The elastic deformations (*S*) may cause at the most some subsurface fatigue. Figure 11 shows an EP specimen that was subjected for 30 h to fretting with an amplitude slightly below the transition to sliding. The main part of the fretting contact exhibited a smooth and undamaged appearance even after loading. Only around some impurities included in the interface and along the rim of the impression, where the contact pressure was lower, small scratches indicate that real sliding with small amplitude took place. Here the small slip produced a formation and extrusion of polymeric rolls and scratches, but the mass loss was too small to be measured gravimetrically. However, the surface cracks may act as notches when the material simultaneously is subjected to a tensile or fatigue load.

Effect of Fiber Reinforcements

The surface conditions of the two rubbing bodies and the interphase composition are a function of time. The particular physical processes contributing to the several phases of the wear process (running in and steady state) vary significantly with the materials considered.

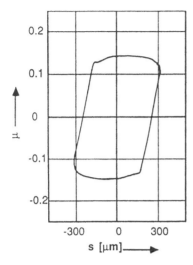

Figure 10 Course of the tangential force during a single load cycle for fretting wear of neat PA 6.6 (TG ≈ 50°C) against steel.

CF Composites

Figure 12a presents the peak-to-peak averaged friction coefficient of CF/EP fretted versus steel as a function of testing time. The initial value of the friction coefficient was on a rather low level near 0.1 (not resolved in Figure 12a because of the large time scale). The friction loop possessed a nearly quadrangular shape at the beginning of the fretting process, with no peaks at the transition points from static to dynamic friction (Figure 13a). This referred to a low contribution of adhesive mechanisms to the friction process. The reason is that initially the real contact area of the two rubbing bodies was very small, due to the microrough surface topography [29].

During the first 2000 load cycles, the asperities in the tribocontact were evened up and the real contact area grew. Therefore, not only the friction coefficient as averaged from

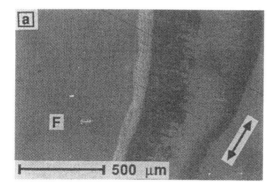
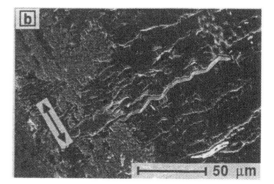

Figure 11 (a) Optical and (b) scanning electron micrograph of a neat EP specimen subjected to fretting with an amplitude slightly below transition to real sliding. The arrow assigns the motion direction, the letter F the fretting contact region.

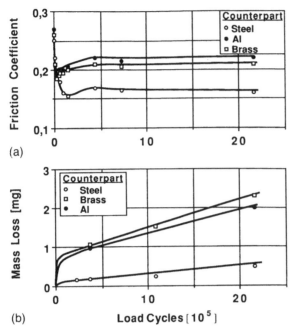

(a)

(b)

Figure 12 Fretting of CF/EP: (a) coefficient of friction as averaged from peak to peak; (b) mass loss versus time for different counterparts.

peak to peak drastically increased to a value near 0.25 (Figure 12a), but simultaneously, a pronounced adhesion peak developed in the friction loop at the transition from static to dynamic friction (Figure 13b). The latter indicated an increasing contribution of adhesive friction mechanisms. This mutual adoption of the rubbing surfaces to each other during the initial wear process was accompanied by fast material removal (Figure 12b).

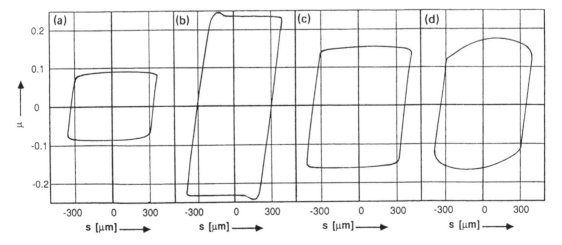

Figure 13 Fretting of CF/EP against steel. Development of the course of the tangential force during one single load cycle with time: (a) immediately after test start; (b) after 2 min (2400 load cycles); (c) after 2 h (144,000 load cycles); (d) after 25 h (1.8 million load cycles).

Once the wear process has begun, the wear debris was entrapped in the contact region [39] and led to gradual separation of the rubbing bodies [26]. Adhesive friction mechanisms became less effective. The friction coefficient decreased (Figure 12a) and the adhesion peak in the friction loop disappeared after some hundred thousand cycles (Figure 13c and d). Instead, the formation of a "third body" in the interfacial region caused a transition from solid body friction to a kind of dry lubricated friction. The tangential force no longer remained constant during the sliding period of the load cycle but varied with sliding speed (Figure 13d).

The average friction coefficient followed a basically similar course (Figure 12a) for fretting against aluminum and brass in comparison to fretting against steel pins. But in contrast to steel pins, aluminum and brass pins led to much smaller maxima at the beginning of the wear process and a minimum occurred already at about 8000 load cycles. This behavior corresponds to a faster transition from two-body to three-body friction, caused by the higher initial rate of material removal. After passing the minimum, the friction coefficient increased again significantly. The steady-state values of the friction coefficient amounted to 0.22 for aluminum pins and 0.21 for brass pins. These high values can be attributed to a stronger influence of abrasive mechanisms caused by scratching of fiber debris through the soft metals. Accordingly, the wear rates of the CF/EP specimens are higher for the softer aluminum and brass than for the hard steel counterparts. These observations are in general agreement with results on sliding wear of CF composites against several counterparts obtained by Giltrow and Lancaster [40].

The rate of material removal became constant when the coefficient of friction entered into the stationary state (Figure 12b). There are three possible reasons for the transition from running-in to steady-state wear: (1) changes in the surface quality of the composite (e.g., protective layers of back-transfer), (2) the formation of a wear-reducing transfer on the fretting pin, and (3) the development of the third body in the interface region. To check weather or not one of these possible mechanisms is dominant, (1) the mass loss of a fresh specimen fretted against an aluminum pin used in a previous fretting experiment and (2) the mass loss of an already fretted specimen against a fresh aluminum were measured. In both cases, the mass loss during the first 108,000 load cycles (i.e., including the running-in wear) was the same as for a fresh specimen worn against a fresh pin. This result indicated that neither the formation of the transfer on the pin nor the changes in surface morphology of the CF/EP specimen during running-in was mainly responsible for the transition to steady-state wear but rather, formation of the third body. The difference in mass loss was smaller for steady-state conditions than for the running-in period (Figure 14). The partial separation of the rubbing bodies by entrapped wear debris reduced the real contact area between pin and specimen and thus the influence of the counterpart material.

GF Composites

Figure 15a depicts that the maximum in the friction coefficient is much less pronounced for GF than for CF composites. Accordingly, the adhesion peak in the friction hysteresis (Figure 16b) appears only very weakly. The probable reason is that the brittle glass fibers get quite rapidly fractured, leading to the fast formation of a debris granulate between the sliding surfaces. This assumption was also supported by the high initial rate of material removal (Figure 15b). After the coefficient of friction passed a minimum, it increases again to a value even slightly above the peak at the beginning.

A new feature of the force–displacement loop arose after the running-in process: a

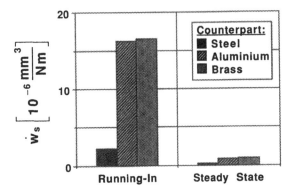

Figure 14 Fretting wear of CF/EP against different counterparts. Specific wear rate, $\dot{w}_s$, as averaged over the first hour of the test and under steady-state conditions.

sharp peak at the turning point of motion (Figure 16d). This peak originated in bumping of the fretting pin against the wall of the hole produced by fretting wear. The appearance of this peak required the existence of a sufficiently deep fretting pin impression. This was given for fretting wear of AF and GF composites, which were worn rather fast. For CF/EP this peak occurred only under particular conditions, which caused high wear rates (very long testing period, aluminum or brass pin, high contact load, high frequency).

The steady-state coefficient of friction for fretting against steel was about twice the value as for CF composites. This directs to a less lubricating effect of the GF debris in comparison with the GF debris, while the GF particles act much more abrasive than CF debris [41].

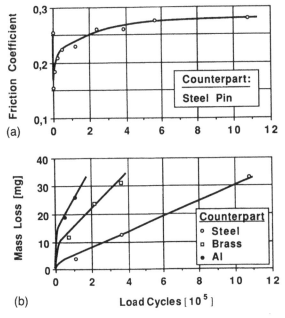

Figure 15 Fretting of GF/EP: (a) coefficient of friction as average from peak to peak (counterpart steel); (b) mass loss versus time for different pin materials.

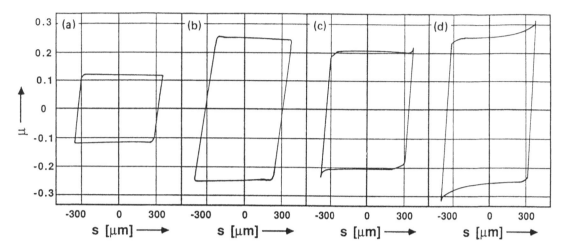

Figure 16 Fretting of GF/EP against steel. Development of the course of the tangential force during a single load cycle with time: (a) immediately after test start; (b) after 2 min (2400 load cycles); (c) after 2 h (144,000 load cycles); (d) after 25 h (1.8 million load cycles).

AF Composites

Aramid fibers consist of a highly oriented liquid crystal polymer. In contrast to glass and carbon fibers, AFs are rather soft and flexible. Under wear loading, AFs tend to fibrillate and smeer off, whereas no brittle fracture occurs. The wear debris does not build up as a granulated structure but forms as flat thin lumps [42]. This type of interlayer, of course,

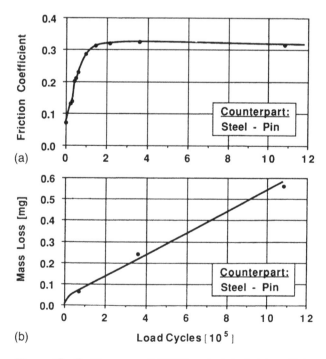

Figure 17 Fretting wear of AF/EP against steel: (a) coefficient of friction as averaged from peak to peak; (b) mass loss versus time.

is less mobile than the powder of GF and CF and may strongly adhere to one or both of the contacting surfaces. The debris patches also separate the rubbing bodies but cannot accommodate the relative motion. The tribological process thus remains as a solid body friction during the full testing time.

Figure 17a shows the average coefficient of friction as a function of time. The course did not exhibit any sharp peaks as found in case of CF and GF composites. Initially, the friction coefficient increased (this is slower than for the CF and GF specimens) and remained subsequently on a high steady-state level at $\mu \approx 0.32$ (this is in the region of the maximum values for CF and GF composites). Briscoe and Tabor also found this type of friction curve for multiple-pass sliding of polymers, which build up a lumpy transfer [43].

Figure 18 presents the resulting development of the friction hysteresis with time. At the beginning of the test, the force–displacement loop possessed a quadrangular shape (Figure 18a) as was also found for CF and GF composites. During the following period, there developed a strongly pronounced adhesion peak (Figure 18b), which again diminished a little while the wear process continued. But the adhesion peak remained much more distinct than for CF and GF composites. This directs to a dominance of adhesive friction mechanisms and supports the denial of a lubricating effect of the interface layer built up by the AF wear debris. Accordingly, the adhesion peak in the friction log was strongly pronounced and remained so over the total testing time. The small changes in friction mechanisms during fretting of AF composites were responsible for the relatively constant rate of material removal (Figure 17b).

Comprehensive Comments

Friction and wear (even if unlubricated) are not a two-body but a three-body problem [28]. This is especially important for fretting conditions, where small slip amplitudes favor the entrapment of debris [44]. In fretting under steady-state conditions, the bulk material properties become less important because the wear process is governed by the wear debris elimination rate [37]. In the preceding section, it was shown that different

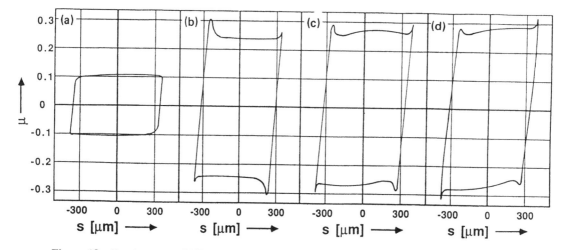

Figure 18 Fretting wear of AF/EP against steel. Development of the course of the tangential force during a single load cycle with time: (a) immediately after test start; (b) after 0.5 h (36,000 load cycles); (c) after 2 h (144,000 load cycles); (d) after 15 h (1.08 million load cycles).

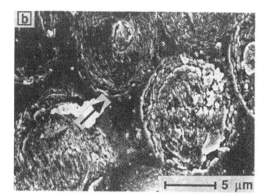

Figure 19 Scanning electron micrographs of CF/EP subjected to sliding wear against smooth steel. The fibers were oriented normally to the sliding plane. The arrows indicate the sliding direction. (a) Development of cracks in the fibers transverse to the sliding direction. In a more advanced state of damage development (b), the single graphite crystals are more or less separated. The crystallites are arranged tangentially at the surface and randomly in the interior as it is typical for PAN-based carbon fibers.

fiber reinforcements produced different third bodies, leading to different interface dynamics.

1. CF and GF composites led to the formation of debris in granulate form. Relative motion between the rubbing surfaces was accommodated by shear ''flow'' within the powder bed.

2. CF debris exhibited some lubricating effect [45]. For GF composites, this lubricating effect was superposed by the additional action of strong abrasive friction mechanisms. The micrographs in Figure 19 show that the carbon fibers could be decomposed into graphite crystals, which are frequently used as a solid lubricant. This is possible because the chemical bonds transverse to the fiber axis are weak van der Waals forces. In contrast, the glass fibers consist of a three-dimensional network of covalent bonds, which causes the formation of hard and sharp-edged particles after fracture.

3. On the other side, the fiber particles (especially of GF) may act as a third-body abrasive, thus superposing the beneficial lubrication effect.

4. AF composites build up an interface layer of relatively large flat lumps. The first bodies were also separated, but the sliding velocity is accommodated at the surfaces of the specimen and the counterpart instead of a lubricating interphase.

FRETTING WEAR STUDIES

Fretting Against Steel: Role of the Fiber–Matrix Combination

The histogram in Figure 20 compares the specific steady-state wear rates of the several EP composites exposed to fretting against steel.

Mild Loading Conditions

At mild loading conditions (frequency $\nu = 10$ Hz, oscillation width $A = 250$ μm), the lowest wear rates were found for the neat matrix resin. Any type of fiber reinforcement

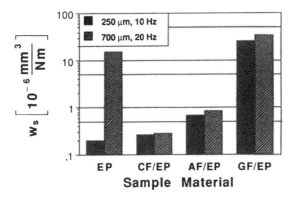

Figure 20 Specific steady-state wear rates for the materials tested for fretting against steel under two different loading conditions.

accelerated material removal. However, the specific wear rates of the CF composites were only slightly above those for neat EP. The highest wear rates were found for GF composites, the lowest wear rates for CF composites.

The difference between running-in and steady-state wear rates (Table 4) was very small for neat EP and AF/EP and significant for CF and even more for GF composites, which produced debris in form of loose granulate. The high lubricating efficiency and of the CF composites and their low abrasiveness caused a strong reduction in the wear rate during the transition from running-in to steady state. The strongly abrasive effect of the GF particles were responsible for a rather high level of the wear rate even in the steady state.

Severe Loading Conditions

At more severe loading conditions (v = 20 Hz, A = 700 μm), the neat polymer was worn very rapidly (Figure 20). Now a fiber reinforcement acted beneficially. The specific wear rates of the CF and AF composites are almost the same as under mild loading conditions. Only the composites reinforced by the strongly abrasive GF still exhibit higher wear rates than the pure EP.

The intense brown coloring of the EP surface after fretting is the result of thermal decomposition. Additionally, the friction heat promotes creeping of the polymer. The incorporation of fiber reinforcement improves the thermal conductivity of the polymeric sample, thus reducing heat concentration in the fretting contact, and enhances the load-

Table 4 Fretting Wear Against Steel

Material	Specific Wear Rates[a]		
	$\dot{w}_{s,i}$ (10^{-6} mm³/N·m)	$\dot{w}_{s,s}$ (10^{-6} mm³/N·m)	$\dot{w}_{s,i}/\dot{w}_{s,s}$
EP	12.3	11.0	1.1
GF/EP	53.9	34.4	1.6
CF/EP	2.23	0.28	8.0
AF/EP	1.11	0.85	1.3

[a] Averaged over the first hour of the test $\dot{w}_{s,i}$ and during steady-state $\dot{w}_{s,s}$ for all material tested.

carrying capacity and resistance to creep [46]. This explains why composites exhibit a higher wear resistance than neat EP at severe loading conditions.

Microscopic Examinations

The SEM micrograph in Figure 21 shows that the neat EP predominantly was worn by surface fatigue and adhesive mechanisms. The surface is covered by a dense system of cracks perpendicular to the sliding direction. Incorporated carbon fibers stop the development of these cracks. At mild loading conditions, the worn surfaces of CF and GF composites exhibit mainly polished wear-thinned fibers protruding from a smooth matrix. When the severity of the loading is increased, there occur more and more fiber fractures and finally, removal of extended chips of the sample material. GF are even more susceptible to brittle fracture than CF. The micrograph in Figure 22 summarizes the wear mechanisms that might occur in CF/EP composite materials. The worn surface exhibits polished fibers and matrix, broken single fibers (F), and more extended areas of delaminated chips (D). Particles of broken fibers may act as a third body abrasive. Accordingly, micrographs of the fretted CF composites exhibited many more scratches than did the neat EP specimen (Figure 23). The carbon fibers are less abrasive than glass fibers because they can easily be decomposed to small graphite crystals (Figure 19).

Aramid fibers are flexible and soft due to their polymeric nature and are therefore not very abrasive. The micrograph in Figure 24 shows that the worn surfaces of the AF composites exhibit no scratches. Typical flat lumps of worn material back-transferred to the specimen surface can be seen in Figure 25. The poor fiber–matrix bonding, which is characteristic of AF composites [47], causes extended fibrillation of the aramid fibers [48], mainly along the rim of the contact region (Figure 26).

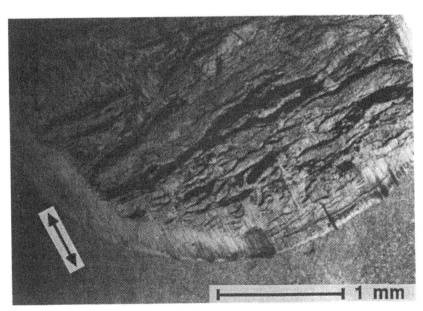

Figure 21 Optical micrograph of a neat EP specimen after fretting against steel (amplitude $A =$ 250 μm, frequency $\nu = 10$ Hz). The arrow indicates the sliding direction. Scratches are visible only along the rim of the impression; the main part of the contact zone exhibits a dense system of cracks.

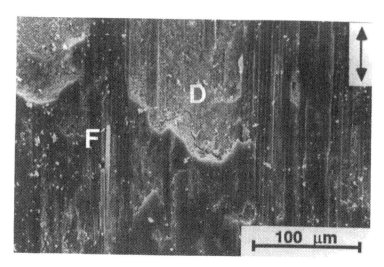

Figure 22 SEM micrograph of a CF/EP surface subjected to fretting wear against aluminum. $F_N = 400$ N ($p = 20$ MPa), peak-to-peak oscillation width $A = 700$ μm, 110,000 load cycles.

Effect of Counterpart Material

The specific wear rates of EP and AF/EP did not noticeably depend on the counterpart material. This is because the metallic counterparts are much harder than the samples. In contrast, the fretting wear of CF/EP and GF/EP was strongly affected by the counterpart. As Figure 27 illustrates, steel and titanium pins, which possessed almost the same hardness, led to equal specific wear rates. The soft aluminum pins produced a far more rapid material removal; the specific wear rate of CF/EP against brass (with an intermediate hardness) was nearly the same as for aluminum but slightly higher. Figure 12b depicts that this

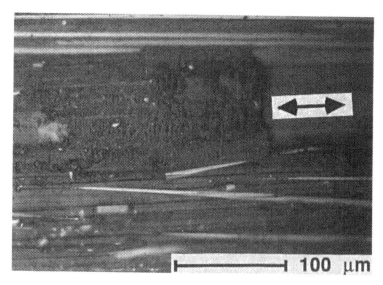

Figure 23 Light micrograph of a CF/EP specimen after exposure to fretting against steel. The arrow indicates the sliding direction. $A = 250$ μm, $v = 10$ Hz.

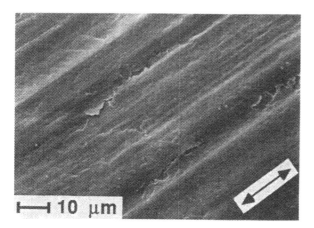

Figure 24 Scanning electron micrograph of a AF/EP specimen after fretting against steel. The arrow indicates the sliding direction.

influence of the counterpart is smaller in steady state than during running-in. A similar trend was found for GF composites, with the exception that the aluminum pin led to a higher wear rate than the brass pin. Furthermore, the absolute values of the wear rates were higher for GF/EP than for CF/EP and the differences between the several counterpart materials was more pronounced. Several reasons could have been responsible for these trends:

1. The hard carbon and glass fibers were able to scratch through the counterpart surfaces, thus producing metallic debris. This metallic debris, together with the fiber particles, were entrapped in the interface. The softer pin materials were more susceptible to abrasion and led to a higher concentration of metallic particles in the interface than the harder counterpart materials. However, when the hardness

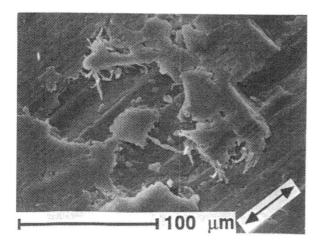

Figure 25 Scanning electron micrograph of a AF/EP specimen after fretting against steel. Flat patches of back-transferred material. The arrow indicates the sliding direction.

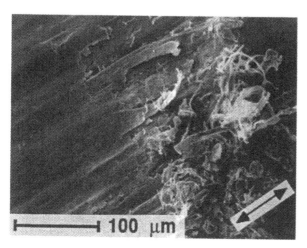

Figure 26 Scanning electron micrograph of a AF/EP specimen after fretting against steel. Fibrillation at the rim of the fretted region. The arrow indicates the sliding direction.

of the metallic particles is small compared with the fibers, their abrasive efficiency drops.

2. Fiber debris and metallic particles entrapped in the interface region roughened the counterpart surface, thus increasing the abrasiveness of the counterpart. The harder materials were more resistant against this roughening by the third body (Figure 28). On the other hand, the abrasiveness of the produced asperities increases with hardness.

3. Protruding edges of broken but still embedded fibers are pressed into the counterpart surface, and this is easier the softer the pins are. Subsequent relative motion causes cracking and removal of these fibers.

Conclusively, one can say: the softer the pin material, the more it can be attacked by the fibers while, simultaneously, the abrasive efficiency of the roughened counterfaces

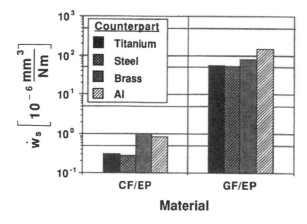

Figure 27 Specific wear rates of CF and GF composites for fretting against several counterparts. Standard loading conditions.

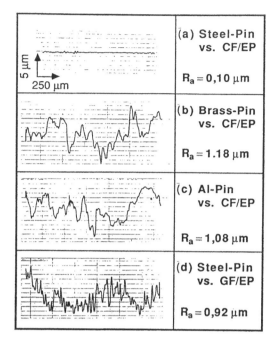

	(a) Steel-Pin vs. CF/EP $R_a = 0,10 \, \mu m$
	(b) Brass-Pin vs. CF/EP $R_a = 1.18 \, \mu m$
	(c) Al-Pin vs. CF/EP $R_a = 1,08 \, \mu m$
	(d) Steel-Pin vs. GF/EP $R_a = 0,92 \, \mu m$

Figure 28 Surface profiles of steel, brass, and aluminum pins after fretting against CF/EP and GF/EP, respectively. Standard loading conditions.

increases with increasing counterpart hardness. This may explain why the specific wear rate of CF/EP exhibited a maximum for the brass pin with intermediate hardness. Glass fibers can be softened by frictional heat [49]. The maximum of the specific wear rate may be shifted to lower counterpart hardness than for CF/EP, so that the GF/EP is more rapidly worn by aluminum than by brass.

Figure 29 shows some optical micrographs of steel and aluminum pins, respectively, which were fretted against CF/EP and of the according composite specimens. It can clearly be seen that the aluminum pin was much more attacked by the wear process while the surface of the steel pin still was rather smooth and reflective. The specimen that was worn against steel exhibited some scratches and some cracking of single fibers. The specimen fretted against aluminum contained many more broken fibers and there were numerous extended areas of fiber fracture.

Fretting of CF/EP Against Aluminum: Parameter Studies

Continuous fiber-reinforced polymer composites are usually employed in lightweight constructions. In these applications, of course, the probable counterpart in a fretting contact would be rather aluminum than steel (rivet joints in airplane structures). Therefore, the fretting wear of CF/EP against aluminum was studied in more detail. Another reason for this decision emerged from the fretting fatigue tests, which delivered, with the apparatus employed, more significant effects of aluminum counterparts than for steel counterparts.

Interface Temperature

In many tribilogical systems, the applicability of polymeric materials is restricted by their sensitivity to elevated temperatures. The mean temperature in the contact region of two

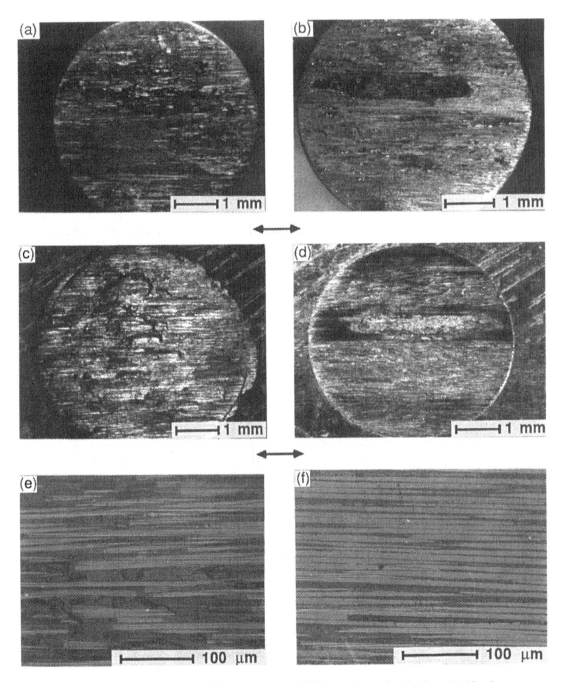

Figure 29 Optical micrographs of fretting pins and CF/EP specimen after fretting. (a) Aluminum pin after 5 h fretting against CF/EP at standard loading conditions; (c) the according CF/EP specimen; (e) detail magnification of (c). (b) Steel pin after 30 h fretting against CF/EP at standard loading conditions; (d) the according specimen; (f) detail magnification of (d).

rubbing bodies, T_a, is composed of the environmental temperature, T_e, and the temperature increase produced by friction, ΔT_F[29]:

$$T_a = T_e + \Delta T_F \tag{3}$$

The temperature increase starts from zero and finally approaches an equilibrium value which linearly depends on the frictional work dissipated per time and area unit:

$$\Delta T = CF_N\mu v \tag{4}$$

The constant, C, comprises (1) geometrical effects concerning the heat conductivity of the system, and (2) materials properties as the heat capacity and thermal conductivity of the sample material and the counterpart. For fretting conditions, the sliding velocity, v, can be replaced by the sliding speed averaged over one load cycle:

$$v_{mean} = 2Av \tag{5}$$

Figure 30 shows that T_a indeed increased linearly with the frequency (the friction coefficient proved essentially constant). It can be seen that the temperature at $v = 20$ Hz (standard testing frequency) was near 35°C. At this temperature, heat-induced distortion or decomposition in extended regions of the material is inprobable because the softening point of the epoxy matrix is in the region between 110 and 130°C [50]. However, it has to be considered that the heat is not produced uniformly over the entire apparent contact area but concentrated at single asperity contacts. These so-called flash temperatures may significantly exceed the average interface temperature [51,52]. Indeed, the neat EP specimens exhibited some light brown coloring after fretting under standard conditions.

Figure 31 presents a plot of the material removed during the first 72,000 load cycles versus the environmental temperature adjusted by the air circulation heating ($A = 300$ µm, $F_N = 300$ N, $v = 10$ Hz). Wear proceeded linearly with environmental temperature up to 150°C, and no unsteadiness occurred even beyond the softening temperature.

Loading Frequency

The effect of frequency variation on the wear process is twofold: (1) increasing the frequency causes softening of the material via an intensified heat production [Eq. (4)]; (2) enhancement of the frequency leads to higher strain rates and thus to a more brittle reaction of the material. Both effects may accelerate wear.

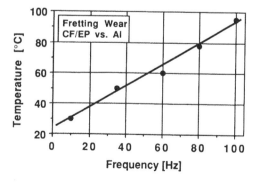

Figure 30 Fretting of CF/EP against aluminum. Mean interface temperature as a function of oscillation frequency. Environmental temperature $T_e = 22$°C (see intercept with temperature axis). $F_N = 300$ N, $A = 300$ µm.

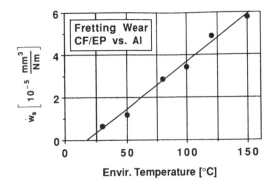

Figure 31 Fretting of CF/EP against aluminum. Specific wear rate as averaged over the first 72,000 load cycles, $\dot{w}_s$, versus environmental temperature, T_e.

Figure 32 shows the results of a test series with different frequencies. The duration of the tests varied inversely with frequency to keep the total sliding distance constant at a value of $L = 65$ m. Below $\nu = 30$ Hz there was no visible influence of the frequency on mass loss. Above 40 Hz the mass loss increased steadily with the frequency; beyond 80 Hz, the slope curve again diminished. The mean sliding speed according to the transition point ($\nu = 35$ Hz) amounts to $v_{mean} = 2.5 \times 10^{-2}$ m/s ($A = 350$ μm). From Figure 30 it can be deduced that the interface temperature at 50 Hz amounts to 60°C, and Figure 31 shows that no wear transition occurred at this temperature. However, the flash temperatures may significantly exceed this mean temperature, and in fact, the SEM micrographs presented in Figure 33 indicate considerable softening of the matrix at high frequencies. The transition frequency shifts to lower values when the environmental temperature is increased [12].

Fretting Amplitude

The influence of the oscillation width on wear was determined for two different frequencies: 10 Hz and 20 Hz (Figure 34). In both cases, the curves started from the coordinate origin and increased at the beginning proportionally with the slip amplitude. Above a critical value for the slip amplitude, the material removal proceeded much faster but still linearly

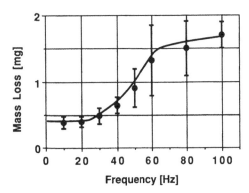

Figure 32 Fretting of CF/EP against aluminum. Mass loss as a function of oscillation frequency. $F_N = 300$ N, $A = 300$ μm.

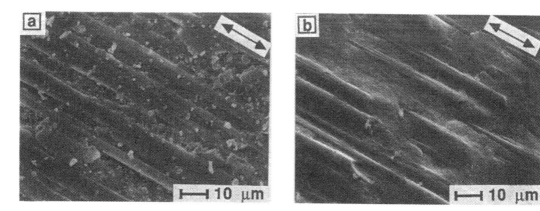

Figure 33 Scanning electron micrographs of CF/EP after fretting against aluminum. $T_e = 22°C$, $F_N = 300$ N. (a) $A = 200$ μm, $v = 10$ Hz; (b) $A = 350$ μm, $v = 100$ Hz.

with the amplitude. The critical value of the amplitude was for 10 Hz about twice the value (860 μm) as for 20 Hz (430 μm). Thus the average sliding speed at the critical point was the same for both frequencies ($v_{mean} = 1.7 \times 10^{-2}$ m/s). This value is considerably below the value found for the frequency variation. The reason is difficult to determine because the waveform of the motion and accordingly, the waveform of the sliding velocity changed with the frequency, and this may have affected the wear mechanisms [9].

Contact Pressure

The relatively low compressive strength of polymeric materials is an important limitation for their applicability in tribologically loaded systems. It leads to a low value for the limiting pv product, beyond which the material removal proceeds unacceptably fast. Figure 35 illustrates the pressure dependence of the specific wear rate as averaged over the first 108,000 load cycles. This test series was accomplished with fretting pins of 4 mm diameter because for the 5-mm pins, the friction force exceeded the capacity of the inductive drive

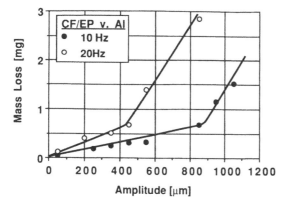

Figure 34 Fretting of CF/EP against aluminum. Mass loss versus oscillation width for two different frequencies. $F_N = 300$ N.

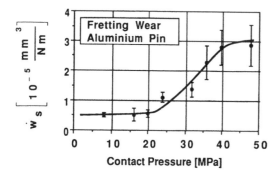

Figure 35 Fretting of CF/EP against aluminum. Specific wear rate as averaged over the first 108,000 load cycles as a function of contact pressure. $F_N = 300$ N, $A = 700$ μm.

at contact pressures above 25 MPa. Up to a contact pressure of $p \approx 20$ MPa, the specific wear rate remained constant; between 20 and 40 MPa, it increased by a factor of about 6. The flattening of the curve above 40 MPa may be an artifact because the inductive drive reaches the limits of its capacity and no constant slip amplitude could be assured.

Synopsis

The mass loss depends proportionally on the contact pressure, the slip amplitude, and the loading frequency until these parameters exceed special critical values, even for the initial phase of the wear process. Obviously, the product of contact pressure and mean sliding speed (pv_{mean}) is a wear-determining parameter analog to sliding wear conditions [32]. Furthermore, there exists a critical value of this product beyond which material removal proceeds accelerated. This critical value can be derived from experiments with varying contact pressure ($A = 700$ μm, $v = 10$ Hz, $p_{crit} = 20$ MPa) to amount to 0.28 MPa·m/s. Almost the same value ($pv_{mean,crit} = 0.26$ MPa·m/s) emerges from the amplitude dependence ($p = 15.3$ MPa, $v_{mean,crit} = 0.17$ MPa·m/s). In contrast, $pv_{mean,crit} = 0.38$ for the frequency series. The possible reason for this discrepancy are discussed above.

FRETTING FATIGUE OF CF/EP

Testing Results

Friedrich et al. pointed out that an additional fretting component can drastically reduce the fatigue life of a CF/EP laminate when the covering plies possess a 0° orientation [14]. Off-axis plies contribute little to the load-carrying capacity of a laminate containing 0° plies. Their damage due to fretting therefore affects the fatigue performance of these laminates insignificantly. Figure 36 shows a plot of the applied upper fatigue stress versus a logarithmic scale of the resulting fatigue life (Wöhler diagram or σ–log N curve) for laminate CF1. The fibers in the covering layers of this laminate were oriented in a 45° direction relative to the applied fatigue load. The plain fatigue curve and the fretting fatigue curve coincided.

The Wöhler diagrams fretting fatigue of the laminates CF2 and CF3 are presented in Figures 37 and 38, respectively. These laminates contain covering layers with 0° orientation. The plain fatigue curves can be approached by a linear function [6] according to Eq. (6):

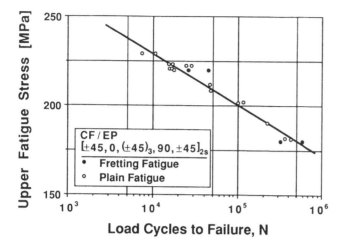

Figure 36 Fretting fatigue of laminate CF1. Applied upper fatigue stress versus resulting lifetime. Counterpart aluminum, contact pressure $p = 23$ MPa. (From Ref. 14.)

$$\sigma_F = \sigma_T(1 - m \log N) \tag{6}$$

where σ_T is the ultimate tensile strength and m is the experimental parameter. In the present case, $m \approx 0.02$ for CF2 and 0.06 for CF3, respectively. Obviously, CF3 reacts much more sensitively than CF2 to a change of the fatigue load.

Application of an additional fretting load leads to a deviation of the Wöhler curve from the simple logarithmic rule of Eq. (6). This deviation strongly depended on the particular loading conditions. The hard steel pins did not produce any significant fretting fatigue effect up to a contact pressure of 23 MPa ($F_N = 450$ N). When the fretting pins consisted of aluminum, the fretting fatigue life of the laminate depended strongly on the contact pressure. A similar effect was found for brass pins. This fretting fatigue behavior is consistent with the fretting wear behavior discussed above. However, Friedrich et al. reported an opposite trend in Ref. 14. The fatigue life was stronger reduced when the

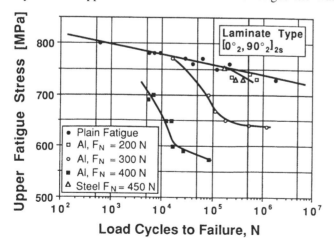

Figure 37 Upper fatigue load versus lifetime for plain fatigue and fretting fatigue of laminate CF2 at various contact loads, F_N. Aluminum pins are assigned by "Al," steel pins by "Steel."

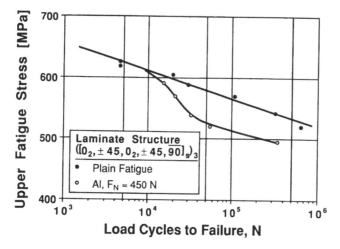

Figure 38 Upper fatigue load versus lifetime for plain fatigue and fretting fatigue (against aluminum pins) of laminate L2.

fretting pins consisted of the hard TiAl alloy than in case of the soft Al pins. This difference can be traced back to the different pin preparations. The pins used by Friedrich et al. possessed flutes due to turning on the front surfaces, whereas the front surfaces of the pins employed by the present author were ground and polished. The turning flutes were indented into the composite surface [15]—the more so the harder the pin material.

Damage Mechanisms

The development of fretting fatigue damage in homogeneous, isotropic materials can be described by means of fracture mechanics concepts [53,54]. In the area of the fretting contact, local stress peaks due to friction and normal forces initiate cracks. These cracks act as sharp notches and lead to a high concentration of the fatigue stress at their tips. As a result, the cracks propagate into the bulk and thus cause premature failure.

In continuous fiber-reinforced composites, the situation is considerably different. The simple application of fracture mechanics is not possible because cracks do not always propagate perpendicular to the main loading direction but advance preferentially parallel to the fibers [55]. Neither the initiation and accelerated growth of single cracks, as in homogeneous, isotropic materials, nor morphological changes characterize failure mechanisms in laminated composites under fatigue loading [56]. Instead, multiple matrix cracking along the fibers causes a reduced load-carrying capacity of the off-axis plies and accordingly, enhanced stresses in the 0° layers. Starting from intersections of matrix cracks and from edges, delamination between the differently oriented plies develop [57,58]. Subsequently, the stress in the off-axis plies is reduced, while the 0° plies have to carry an increasing part of the applied load. Final failure occurs when the stress in the 0° plies locally exceeds their strength [58], which may additionally be reduced by random cracking of 0° fibers [59]. Figure 39 schematically illustrates the stress situation in a 0° layer of a cross-ply laminate containing transverse cracks in the 90° plies.

Laminate CF3 contains a large number of 45° plies, which have a higher load-carrying capacity than the 90° layers in CF2. Therefore, it is reasonable that the residual strength after a given number of load cycles is more reduced in laminate CF3 than in CF2.

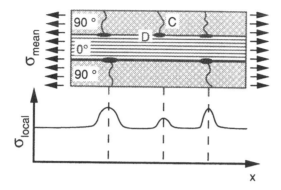

Figure 39 Schematic of the stress distribution in a 0° layer after cracking (C) and local delamination (D) of neighboring 90° plies.

The effect of an additional fretting component on the fatigue damage mechanisms of CF/EP laminates is described in detail in Ref. 15. Only some essential features will be reported here and connected to a closed view of the fretting fatigue damage development. The x-ray radiographs in Figure 40 visualize several damage states of laminate CF2. Figure 40a represents a specimen prior loading: the radiograph exhibits homogeneous darkening, indicating the absence of any cracks or delaminations. After plain fatigue loading (Figure 40b), the specimen contains transverse matrix cracks in the 90° layers together with some longitudinal cracks in the 0° plies. Figure 40c is taken from a specimen after fretting fatigue loading. The white spot at the fretting contact suggests that pressing the pins against the specimen hinders delamination. However, strong longitudinal cracks and delaminations starting from the fretted region grow along the 0°/90° interphase. The optical micrograph in Figure 41a depicts that this is due to peeling off of cracked bundles of 0° fibers.

The cracking of 0° layers could be caused by high equivalent stresses in the region of

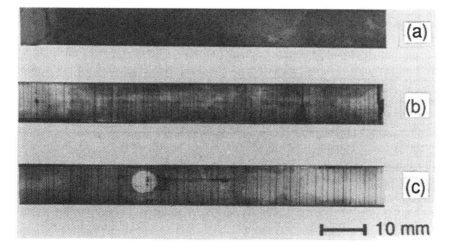

Figure 40 X-ray radiographs of specimens of laminate L1: (a) prior loading; (b) after exposing to plain fatigue; (c) after fretting fatigue.

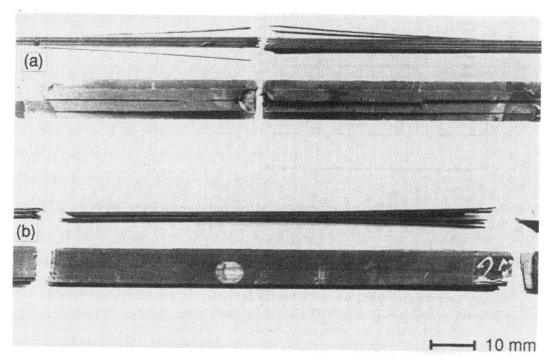

Figure 41 Optical micrographs of specimens (laminate L1) after fretting fatigue: (a) versus an aluminum pin (F_N = 400 N, 30,000 load cycles); (b) versus a steel pin (F_N = 450 N, 270,000 load cycles).

the fretting contact. But this explanation seems not to be very probable because the applied apparent contact pressures (<20 MPa) are very small in comparison to the fatigue stress level (>700 MPa). In fact, when a contact pressure of 22 MPa is applied via steel pins instead of aluminum pins, no cracking of the 0° plies was found (Figure 41b). Obviously, the effect of the normal load (in the regarded range), with which the pins are pressed against the specimens, on the equivalent stress is not the crucial parameter that controls the development of the fretting fatigue damage.

Another assumption considers the initiation and advance of *fretting fatigue* damage to be controlled by the *fretting wear* performance of the laminate. This explanation is supported by the fact that the fretting wear of CF/EP versus aluminum is more severe than against steel counterparts according to the more detrimental effect of the aluminum pins on the fretting fatigue performance.

Once the fibers or fiber bundles are broken, they can no longer support the load-carrying capacity of the laminate, although they are actually not yet worn. Therefore, it must be expected that the *fretting fatigue* damage proceeds considerably faster than plain *fretting wear* as measured gravimetrically. When the specimen shown in Figure 22 is exposed to a tensile fatigue load, shear stresses arise along the interface between the cracked and the undamaged fiber bundles, which are enhanced by the friction force. This leads to the observed splitting and peeling off of the cracked 0° layers. The SEM micrograph in Figure 42 shows cracks starting from regions of severe surface damage and advancing along the direction of fiber orientation. The final result of this process is the total splitting and delamination of the damaged 0° layers (Figure 41a) [60]. This delamination was also

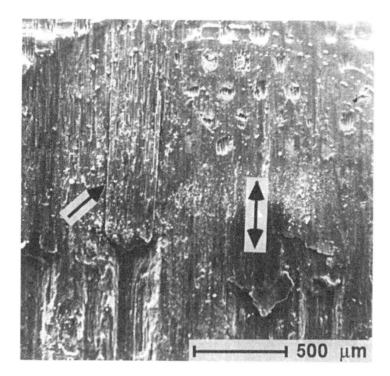

Figure 42 SEM micrograph of the fretting pin mark on a CF/EP specimen subjected to fretting fatigue against aluminum. The double arrow assigns the loading direction, the plain arrow points to logitudinal cracks starting from regions of severe surface damage.

observed when specimens predamaged by plain fretting were exposed to a quasi-static tensile stress. The following conclusions can be drawn:

1. An additional fretting component is able to reduce the fatigue life of a CF/EP laminate drastically if the fibers exposed to fretting possess a 0° orientation.

2. Fretting fatigue of continuous fiber-reinforced laminates cannot be treated in terms of fracture mechanics descriptions of crack initiation and growth. Notch effects in the classical meaning do not occur.

3. In the considered range of contact pressures, an additional fretting component influences the fatigue performance of a laminate primarily not via the enhancement of the equivalent stresses due to the contact stresses but instead by fretting-induced surface damage.

4. Fretting fatigue damage penetrates faster into the laminate than pure material removal due to plain fretting wear. Fiber bundles that are predamaged by fretting tend to break and peel off under fretting fatigue conditions.

Mathematical Model for Damage Development

The conclusions drawn from the observation of the fretting fatigue failure mechanisms suggest:

1. The absence of notch effects causes a relatively uniform distribution of tensile stresses across any cross section of the undamaged 0° plies. Therefore, the change of the load-carrying capacity, dF, should be proportional to the reduction of the cross section dQ of the load-bearing 0° layers in the fretting contact region between n and $N + dN$ load cycles:

$$dF = \sigma_u \, dQ \tag{7}$$

   with σ_u = upper fatigue load at which the specimen fails after N load cycles.

2. Since a notch effect with any accelerated crack growth can be denied, the damage development can be considered to proceed proportional to the time:

$$\frac{dQ}{dN} = \text{const} \tag{8}$$

Actually, the reduction of the cross section of the 0° layers in the fretted region leads to an enhanced fatigue stress in the remaining 0° fibers. The specimen finally fails when this stress exceeds the strength of the 0° plies. However, the actual cross section is not continuously monitored, and thus the *true* stress in the laminate between the fretting pins cannot be determined. But the consideration can be turned out as follows: The *apparent* upper fatigue stress σ_u, at which the specimen fails after N load cycles, diminishes proportionally with the cross section of the load-bearing 0° layers. Accordingly, σ_u can be calculated by replacing Eq. (6) by

$$\sigma_u = \sigma_T \left(1 + \frac{dQ}{Q_0 \, dN} N \right) (1 - m \log N) \tag{9}$$

with Q_0 = initial total cross section of the 0° plies. The first term in parentheses considers the reduction of the tensile strength by reducing Q_0 to Q. Inserting Eqs. (9) and (8) into Eq. (7) gives

$$d\sigma = \frac{dF}{Q_0} = \sigma_T \left(1 + \frac{dQ}{Q_0 dN} N \right) (1 - m \log N) \frac{dQ}{Q_0 dN} dN \tag{10}$$

Integration of Eq. (10) leads to

$$\int_{\sigma_F}^{\sigma_{FF}} d\sigma = \frac{\sigma_F}{Q_0} \frac{dQ}{dN} N + \frac{1}{2} \frac{\sigma_F}{Q_0^2} \left(\frac{dQ}{dN} N \right)^2 + m \log e \, \frac{\sigma_T}{Q_0} \frac{dQ}{dN} \left(1 + \frac{dQ}{4 Q_0 \, dN} N \right)$$

or defining a parameter $\Delta\sigma_{\text{rel}}$:

$$\Delta\sigma_{\text{rel}} = \frac{\sigma_F - \sigma_{FF}}{\sigma_F}$$

$$= \frac{dQ}{Q_0 dN} N + \frac{1}{2} \left(\frac{dQ}{Q_0 dN} N \right)^2 + m \log e \, \frac{\sigma_T}{\sigma_F} \frac{dQ}{Q_0 dN} \left(1 + \frac{dQ}{4 Q_0 dN} N \right) \tag{11}$$

where σ_F is the upper fatigue load at which the specimen fails after N load cycles and σ_{FF} is the upper fretting fatigue load at which the specimen fails after N cycles. The first summation term means that in first-order approximation the laminate behaves as if its cross section was reduced prior to fatigue loading. The quadratic correction term considers the successive nature of the progress of fretting fatigue damage. The last summation term is caused by the fact that the fibers removed at the beginning of the test possessed a somewhat higher load-carrying capacity than the fibers that had already experienced a

longer fatigue loading history. However, this last term is smaller than 0.003, while the first term is in the range of 0.1. Therefore, the last term will be neglected in further considerations. The physical reason for this is the high resistance of the carbon fibers against fatigue loading. Conclusively, one can write

$$\Delta\sigma_{rel} = \frac{\Delta Q}{Q_0} - 0.5 \left(\frac{\Delta Q}{Q}\right)^2 \tag{12}$$

The cross section reduction, ΔQ, accumulated during N load cycles is a positive quantity while $\frac{dQ}{dN} N$ is negative for decreasing cross section Q.

It should be emphasized that Eq. (12) does not enable the deduction of the fretting fatigue life of a laminate from simple material and loading parameters. But $\Delta\sigma_{rel}$ gives a quantitative measure for the degree of fretting fatigue damage and provides a tool for the investigation of fretting fatigue damage development.

Presuming the validity of Eq. (8), Eq. (12) suggests that, in first-order approximation, $\Delta\sigma_{rel}$ increases proportionally with the time. To check this assumption, Figure 43 presents a plot of $\Delta\sigma_{rel}$ versus the fretting fatigue life, N, for two different laminates. Initially, all curves increase proportionally with time but reach a constant value after a certain number of load cycles. Maximum fretting fatigue damage is expected when the covering 0° layers totally penetrated by the fretting pins. Further fretting damages only off-axis plies and thus influences the fatigue performance of the laminate insignificantly [14]. Figure 44

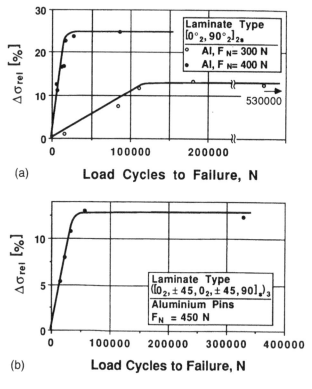

Figure 43 Relative fatigue strength reduction ($\Delta\sigma_{rel}$) as a function of fretting fatigue life for laminate CF2 (a) and CF3 (b) at several contact loads.

Figure 44 Optical micrograph of a specimen of laminate CF3, subjected to fretting fatigue against an aluminum pin. $F_N = 450$ N, failure after 345,900 load cycles.

presents an optical micrograph of a specimen of laminate CF3 which failed after 345,900 load cycles, after reaching the plateau region of the $\Delta\sigma_{rel}$ versus N curve. The covering 0° layers are totally penetrated and fretting already took place on the 45° plies. Based on these considerations, the maximum value for $\Delta\sigma_{rel}$ can be calculated from the maximum reduction of the cross section of the 0° layers (see Figure 45):

$$\Delta Q_{max} = \frac{d}{w} \frac{i_{Fmax}}{i_{tot}} \tag{13}$$

where

d = diameter of the pin (= 5 mm)

w = width of the specimen (= 6.3 for CF3 and 8.4 for CF2)

$i_{F,max}$ = maximum number of fretted 0° plies (= 4)

i_{tot} = total number of 0° plies (8 for CF2 and 24 for CF3)

Substituting ΔQ_{max} for ΔQ in Eq. (12) gives

$$\Delta\sigma_{rel,max} = \begin{cases} 0.25 & \text{for CF2} \\ 0.125 & \text{for CF3} \end{cases}$$

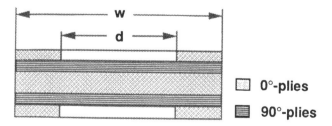

Figure 45 Schematic of the cross section of a specimen in the region of fretting contact.

These values are in good agreement with the experimental results (Figure 43). However, at lower contact pressures, the maximum fretting fatigue damage is considerably smaller than the calculated value. This effect is presented and discussed later in the chapter.

From the liner part of the $\Delta\sigma_{rel}$–N curve, the rate of damage development dQ/dN can be derived. For reasons of comparability with the fretting wear tests, this quantity will be transformed to a "specific pseudo-wear rate" $\dot{w}_s^*$, in analogy to the concept of the specific wear $\dot{w}_s$:

$$\dot{w}_s^* = \frac{\Delta V}{2F_N L} = \frac{\pi}{16} \frac{d \, \Delta Q}{F_N AN} \tag{14}$$

where ΔV is the volume of removed or cracked 0° fibers and L is the total sliding distance ($= 2\,AN$). The factor $\frac{1}{2}$ considers that two pins rub against the laminate simultaneously and only one-half of the accumulated damage of the laminate can be related to one pin. Resolving Eq. (12) to ΔQ and inserting this result into Eq. (14) gives

$$\dot{w}_s^* = \frac{\pi}{16} \frac{dQ_0}{F_N AN} (1 - \sqrt{1 - 2\Delta\sigma_{rel}}) \tag{15}$$

The quantity $\dot{w}_s^*$ measures how deep fretting fatigue damage penetrates the laminate and includes material removal due to pure fretting wear as well as cracking of predamaged fibers and fiber bundles. Therefore, $\dot{w}_s^*$ should be greater than $\dot{w}_s$.

Additionally, cracking and peeling off of fiber layers predamaged by fretting disturbs the confinement of wear debris in the interfacial layer, thus preventing the interface dynamics from reaching steady-state conditions. This means that the fretting fatigue behavior of CF/EP is controlled rather by the material loss during running-in than by the steady-state fretting wear rate. Figure 46 schematically illustrates the wear process in the fretting contact under fretting fatigue: When the material removal during running-in exceeds a special value, ply cracking and delamination occur at the point indicated by the first arrow. Thereby, a part of the third body entrapped in the contact zone may be thrown out. Further fretting takes place on fresh plies under conditions similar to running-in until repeated cracking of fiber bundles. The resulting mass loss versus time relationship represented by the solid line in Figure 46 is much steeper than the virtual wear curves for plain fretting.

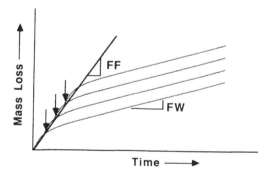

Figure 46 Effect of cracking and peeling-off on material removal. Cracking and peeling-off of predamaged fiber bundles takes place at the positions assigned by the arrows before the wear process enters steady state, FW. The resulting wear rate, FF, is much higher than FW.

Experimental Proof of the Model

To avoid confusion, it should be emphasized that the measured points along the $\Delta\sigma_{rel}$ versus N curve represent different specimens which were subjected to different fatigue stress amplitudes. The fact that these points follow a straight line means primarily that the fretting fatigue damage accumulated until final failure is proportional to the fretting fatigue life, regardless of the fatigue stress level. On the other hand, the presented model states that during the test of one particular specimen the effective cross section of the load-bearing 0° layers, when exposed to fretting, decreases proportionally with time. Therefore, an additional test was necessary to check this assumption of the model.

Several specimens of laminate CF2 were exposed to a fretting fatigue loading equivalent to one-half of the expected fretting fatigue life. The contact load was set to $F_N = 300$ N, the upper fatigue stress was $\sigma_u = 700$ MPa. Figure 37 shows that a lifetime of 79,000 load cycles can be expected under these conditions. After 45,000 load cycles, the test was interrupted according to a theoretical fatigue strength reduction of 4.5% (Figure 43). Subsequently, the specimens were continued under plain fatigue conditions ($\sigma_u = 740$ MPa) until final failure. Table 5 lists the resulting lifes and the according relative deviation from the σ versus log N curve for plain fatigue of the undamaged laminate ($\Delta\sigma_{rel}$). The measured mean value of $\Delta\sigma_{rel}$ amounts to 4.5%, which is in good agreement with the theoretical value, which was calculated under the assumption that the fretting fatigue damage proceeds proportional to time during a single test.

Influence of Loading Conditions

Slip Amplitude

Aluminum pins pressed onto laminate CF2 at three different positions ($l_1 = 40$, 65, and 90 mm). According to Eq. (2), each pin position correlates with a slip amplitude at a given stress amplitude. Figure 47 plots the calculated slip amplitude A for all three pin positions versus the upper fatigue stress σ_u. The applied upper fatigue stress varied between 650 and 700 MPa. The range of the respective realized slip amplitudes reaches from 360 µm ($\sigma_u = 640$ MPa, $l_1 = 40$ mm) to 890 µm ($\sigma_u = 700$ MPa, $l_1 = 90$ mm). The contact pressure was set to $F_N = 400$ N ($p \approx 20$ MPa). Table 6 lists the resulting lifetimes of the laminae CF2. It can be seen that under the conditions selected, no significant effect of the slip amplitude can be observed. At the applied high contact pressure, the slip amplitude is probably already above its critical value even for pin position $l_1 = 40$ mm.

Pressure and Counterpart Material

Figure 48 shows the specific pseudo-wear rate $\dot{w}_s^*$ calculated from the linear part of the $\Delta\sigma_{rel}$ versus N curve of laminate CF2 as a function of the contact pressure p. Obviously,

Table 5 Residual Fatigue Life and Relative Fatigue Strength Reduction of CF2 Laminates ($\sigma_u = 740$ MPa) Predamaged by 45,000 Load Cycles Under Fretting Fatigue ($F_n = 300$ N, $\sigma_u = 700$ MPa) Conditions

No.	Number to Fatigue Cycles to Failure	$\Delta\sigma_{rel}$ (%)	$\Delta\sigma_{rel,mean}$ (%)
1	7,320	5.1	
2	29,970	3.8	4.5
3	15,630	4.5	

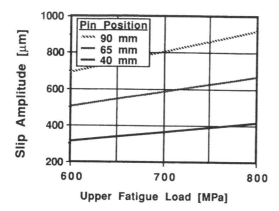

Figure 47 Slip amplitude versus applied upper fatigue load.

there existed a critical value (boundary value) where the propagation of damage development jumps from an insignificant to a very high level. The specific fretting wear rate $\dot{w}_s$, in principle, obeyed a similar behavior (Figure 32).

For aluminum pins, $\dot{w}_s^*$ is smaller than 8×10^{-6} mm³/N·m at a contact load of 10 MPa ($F_N = 200$ N) and thus has a similar magnitude as the specific fretting wear rate (5×10^{-6} mm³/N·m). At 15 MPa, still below the boundary value of the contact pressure, $\dot{w}_s^*$ is already considerably increased to a value of about 3×10^{-5} mm³/N·m, while the specific fretting wear rate, $\dot{w}_s$, remains constant below its boundary value.

Above the boundary value for the contact pressure, fretting wear and fretting fatigue both advanced accelerated. But under fretting fatigue conditions, this boundary pressure lies at clearly lower values ($p_{\text{crit}} \approx 17$ MPa) than under plain fretting ($p_{\text{crit}} \approx 25$ MPa for aluminum pins). Above the boundary value of the contact pressure, fretting fatigue damage proceeded much faster than pure material removal due to plain fretting wear:

$$\dot{w}_s^* \, (p \approx 20 \text{ MPa}) = 5 \times 10^{-4} \text{ mm}^3/\text{N·m}$$

$$\dot{w}_s \, (p \approx 45 \text{ MPa}) = 2.8 \times 10^{-5} \text{ mm}^3/\text{N·m}$$

(15)

A similar behavior of the specific pseudo-wear rate as a function of contact pressure can be found for the brass pins. However, the boundary value of the pressure is shifted to lower values if compared to aluminum. This correlates with the fretting wear of CF/EP, which was found to be slightly higher for brass than for aluminum counterparts. When steel pins were used as counterparts, no stepwise increase in the specific pseudo-

Table 6 Fretting Fatigue Life of Laminate CF2 (Number of Cycles to Failure) for Different Pin Positions I_1 and Fatigue Stress Levels σ_u[a]

σ_u	$I_1 = 40$ mm	$I_1 = 65$ mm	$I_1 = 90$ mm
650	16,600	15,000	18,000
	14,500	12,000	—
700	6,800	6,800	6,500
	5,400	6,400	5,800

[a] Counterpart: aluminum, $F_N = 400$ N.

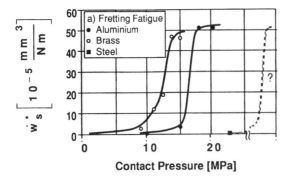

Figure 48 Specific pseudo-wear rate (a) and specific fretting wear rate (b) of laminate CF2 as a function of contact pressure for different counterpart materials.

wear rate was observed up to a contact pressure of 23 MPa (F_N = 450 N), according to the low fretting wear rate of CF/EP against steel pins. The boundary value of the pressure for steel pins probably lies at higher values, which could not be realized with the apparatus employed.

Another effect of the contact load can be seen in Figure 43a. At a contact pressure of p = 15 MPa (F_N = 300 N), the relative fatigue strength reduction, $\Delta\sigma_{rel}$, becomes constant after about 150,000 load cycles at a value of $\Delta\sigma_{rel}$ = 14% and does not reach the theoretical value of 25% as was found for p = 23 MPa (F_N = 450 N).

Several authors [61–63] reported that in metallic materials the influence of an additional fretting component on material fatigue is effective only during crack initiation and the early stage of crack growth. After a certain number of load cycles, the crack propagation rate under fretting fatigue approaches that under plain fatigue [64]. This can be explained as follows. The initial value of the stress concentration at the crack tip is enhanced by normal and shear stresses applied by the fretting pin. This stress concentration diminishes to that value which is also active under plain fatigue conditions while the crack propagates into the interior of the specimen [65]. This crack propagation behavior would be in agreement with the fact that $\Delta\sigma_{rel}$ remains constant after a certain number of load cycles.

However, the absence of notch effects in continuous fiber-reinforced layers with 0° orientation opposes the application of this fracture mechanics model to the CF/EP laminates investigated here. In fact, it was observed that specimens exposed to fatigue stresses above 640 MPa (contact load F_N = 300 N) exhibited a more pronounced surface damage after a given number of cycles than specimens that were fatigued at a lower stress level. Especially, no peeling off of fiber bundles occurred at fatigue stress levels below 640 MPa even after several hundred thousand load cycles (Figure 49a). These observations suggest that the falling of the fatigue stress level below a critical value ($\sigma_{u,crit}$ = 640 MPa) is responsible for the lowering of the propagation rate of fretting fatigue damage rather than exceeding a special number of load cycles. In this case, probably, tensile stresses in the fretting region and shear stresses between cracked and undamaged layers are not high enough to cause amplification of the surface damage due to fretting. As a consequence, the confinement of the developing "third body" is assured, and the fretting wear rate may reduce with time while the interface dynamics approach steady-state conditions. In this context it should be remembered that the several points along the $\Delta\sigma_{rel}$ curve (Figure 43) represent differently loaded specimens. The damage progress in the

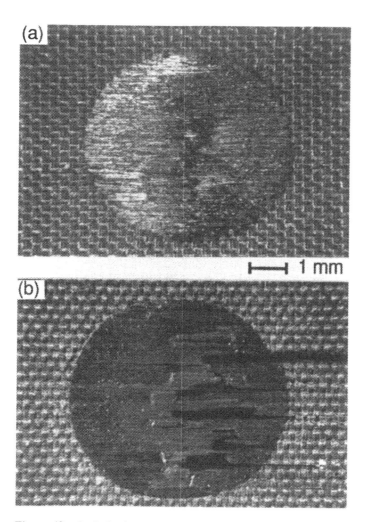

Figure 49 Optical micrographs of laminate CF2 after exposure to fretting fatigue. Specimen (a) was exposed to fretting fatigue against aluminum ($F_N = 300$ N) at an upper fatigue load of $\sigma_U = 640$ MPa for 200,000 load cycles. Afterward it was run to failure (b) without fretting at an upper fatigue load of 750 MPa.

specimens that failed in the plateau region may not have proceeded along the drawn line but any other course with a smaller slope that ends at the measured points.

According to the foregoing considerations on damage development, it can be assumed that the $\Delta\sigma_{rel}$ versus N curve for each individual specimen follows a straight line until final failure. Table 7 represents an experimental check of this concept. Some specimens were exposed to fretting fatigue at an upper fatigue stress of 640 MPa and a contact pressure of $p = 15$ MPa ($F_N = 300$ N) for 200,000 load cycles. According to Figure 37, the expected fretting fatigue life of these specimens amounts to about 500,000 load cycles. Figure 49a depicts that these specimens did not exhibit any peeling off after 200,000 load cycles, in contrast to specimens subjected to higher fatigue loads. Presuming a linear increase of $\Delta\sigma_{rel}$ until final failure after 500,000 load cycles at $\Delta\sigma_{rel} = 14\%$, a

Table 7 Relative Fatigue Strength Reduction of Laminate CF2 Due to Fretting Fatigue Predamage[a]

No.	Number of Cycles to Failure	$\Delta\sigma_{rel}$ (%)	$\Delta\sigma_{rel.mean}$ (%)
1	390	6.6	
2	1380	5.4	6.0
3	780	6.0	

relative fatigue strength reduction after 200,000 load cycles of about 5.5% would be expected (see Figure 43). Subsequent to fretting fatigue loading, the specimens were run under plain fatigue at $\sigma_u = 750$ MPa until final failure occurred. During this time, layers predamaged by fretting rapidly cracked and delaminated (Figure 49b). Table 7 lists the number of plain fatigue cycles to failure and the resulting deviation from the σ_u versus log N curves for the undamaged laminate. The mean value for this difference is $\Delta\sigma_{rel} = 6\%$. The coincidence of this result with the expected value confirms the assumptions that:

1. Fretting fatigue damage proceeds proportional with time in continuous fiber-reinforced laminates.

2. At low contact loads ($F_N \leq 300$ N), the fretting fatigue damage propagation rate is controlled by the upper fatigue stress, σ_u.

However, for high contact loads ($F_N \geq 400$ N) the propagation rate is independent of the fatigue stress level (see above). This can be explained by assuming the equivalent stress amplitude in the subsurface region instead of the fatigue stress level to be responsible for the amplification of fretting-induced surface damage.

Effect of Surface Structure

All experiments described above were performed with a flat-on-flat fretting contact geometry. Of course, it is very difficult to adjust these flat surfaces really parallel, and if the contacting surfaces include a small angle, the initial contact pressure distribution is inhomogeneous and can exhibit peaks, as illustrated in Figure 50. Therefore, the fretting pin locally penetrates very fast into the laminate at the beginning of the fretting fatigue

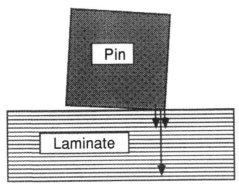

Figure 50 Schematic illustration of a contact stress peak due to small misadjustment of pin and specimen.

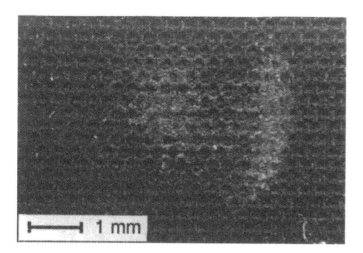

Figure 51 Surface of laminate CF2 (manufactured with peel ply) after exposure to fretting fatigue versus aluminum ($F_N = 300$ N) for about 100 load cycles. At the beginning the pin did not touch the surface over the whole contact area. While the pin penetrates the resin-rich surface layer, the real contact area increased.

loading. If the fibers are protected by a resin-rich surface layer left by peel plies (Figure 8), the real contact area is significantly increased and the contact stress peaks are reduced before the fretting pin touches the fibers (Figure 51).

Laminates that were produced without using peel plies do not possess the resin-rich surface layer. The contact pressure peaks due to a misadjustment of the fretting pins act directly on the fibers. As a result, very fast wear with subsequent cracking of fiber bundles can be expected at the beginning of the test. Figure 52 shows the resulting $\Delta\sigma_{rel}$ versus N curve. The curve in principle exhibits a similar shape as for the laminates with peel plies (Figure 43a). However, the intersection with the $\Delta\sigma_{rel}$ axis is shifted from 0 to about 6%. Subsequently, the curve again increases linearly with time and reaches a plateau. The maximum value of $\Delta\sigma_{rel}$ is also shifted by about 6% to higher values when compared to the laminates produced with peel ply. Thus the span of $\Delta\sigma_{rel}$ remains at about 14%.

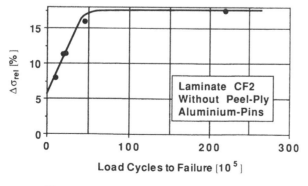

Figure 52 Relative fatigue strength reduction of laminate L1 (without peel ply) versus fretting fatigue life. Counterpart: aluminum, $F_N = 300$ N.

FRETTING FATIGUE OF HYBRID COMPOSITES

In previous sections the fretting fatigue performance of CF/EP was established to be controlled by the fretting wear process, as it is active during the early stages of the fretting loading. The rapid running-in wear of CF/EP when fretted against aluminum caused a drastic fatigue life reduction. AF/EP exhibited a higher wear resistance under fretting against aluminum, and the running-in period was much less pronounced with respect to the wear rate. Accordingly, the fatigue performance of AF laminates is expected to be significantly less susceptible to an additional fretting component applied by aluminum pins than CF composites. In practical applications carbon fibers may be more appropriate for several other reasons (low compression strength of AF [66], electrical conductivity of CF, poor fatigue performance of AF composites [67]). Therefore, a hybrid laminate was produced and tested with carbon fibers in the bulk and aramid fibers in the covering plies, which are subjected to the fretting loading.

Mechanical Properties

Table 1 lists the Young's modulus and the tensile strength of the laminates tested. The modulus of the hybrid, E_{HC}, was the same as the mean value of the moduli of the AF composite, E_{AF}, and the CF composite, E_{CF}, and could be calculated in terms of a rule-of-mixtures approach as Figure 53 depicts:

$$E_{HC} = E_{AF} V_{AF,0°} + E_{CF} V_{CF,0°} \tag{16}$$

where $V_{AF,0°}$ is the portion of 0° layers that contain AF and $V_{CF,0°}$ is the portion of 0° layers that contain CF. Only the 0° plies were considered in Eq. (16) because these plies determined the mechanical properties.

The tensile strength of the hybrid was lower than the fracture stresses of each of the constituents. On first sight, this effect seems astonishing, but it can be explained by the criteria of constant strain to failure [68]: As long as $V_{AF,0°}$ is below a special value, $V_{AF,0°}^*$, the main part of the load, is supported by the CF layers. Fracture of the CF plies under 0° orientation, which possess a smaller strain to failure than the AF, is followed

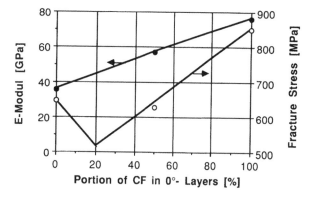

Figure 53 Young's modulus and tensile strength for the laminates AF (left side of the diagram), CF2 (right side), and for the hybrid (middle) represented as a function of the contribution of CF to the 0° layers. The lines emerge from the rule-of-mixtures approach (E modulus) and from the criteria of constant strain to failure.

immediately by final failure of the entire hybrid. The resulting tensile strength of the laminate is described by

$$\sigma_{B,H} = \sigma_{B,CF}V_{CF,0^\circ} + \sigma_{AF}^*(1 - V_{CF,0^\circ}) \qquad (17)$$

where σ_{AF}^* is the stress carried by the 0° AF plies at the moment of final failure ($\approx \epsilon_{CF}E_{AF}$). Above $V_{AF,0^\circ}^*$, the strength of the laminate is solely determined by the AF under 0° orientation:

$$\sigma_{B,H} = \sigma_{B,AF}V_{AF} = \sigma_{B,AF}(1 - V_{CF,0^\circ}) \qquad (18)$$

According to this approach, the solid line in Figure 53 is calculated from the data for the monolithic composites and it represented the measured strength of the hybrid composite well.

Fatigue and Fretting Fatigue

Figure 54 shows the Wöhler diagram for fatigue and fretting fatigue of the AF laminate. In fact, the fretting fatigue lives were near the lifetimes obtained under plan fatigue. But some further features are worth mentioning: (1) The scatter of the results was significant, and (2) the curve obtained for plain fatigue exhibited a rather odd shape. Up to about 10^5 load cycles, almost no effect of the fatigue loading on the residual strength was found. This high initial resistance of AF composites to fatigue was already reported by several other authors [67,69,70]. Beyond 10^5 load cycles, the curve suddenly became much steeper, thereby indicating a change in the damage mechanisms. According to Fernando et al. [69], this transition arises from degradation of fiber strength due to internal disruption or local abrasion of the fibers caused by fretting at the fiber–matrix interface.

The fatigue and fretting behavior of the hybrid laminate is represented by the Wöhler diagrams shown in Figure 55. The scatter of the results was suspiciously small in comparison with the monolithic AF composite and more typical for CF laminates. The entire curve again is composed of a more flat and a steep branch. The relative slope of the flat part [$m = 0.02$ according to Eq. (6)] was almost the same as for the neat CF/EP laminate, whereas in the steep part the curve goes parallel to the curve obtained for the AF composite. This means that during each phase the respective weakest component dominated the fatigue

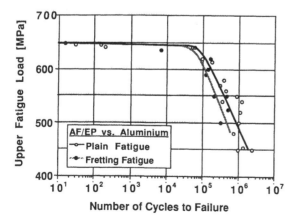

Figure 54 Applied upper fatigue stress versus lifetime resulting under plain fatigue and fretting fatigue of laminate AF. Counterpart: aluminum.

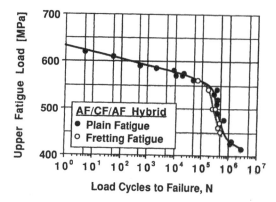

Figure 55 Applied upper fatigue stress versus lifetime resulting under plain fatigue and fretting fatigue of laminate HC. Counterpart: aluminum.

performance of the hybrid, thus leading to a kind of negative hybrid effect. The failure of the weaker component is followed instantaneously by the fracture of the whole laminate because each component carries to much load to be transferred to the other component. This appears to be contradictory to the results published by Fernando et al. [69], who stated the σ–log N curve for the hybrid to be calculable from the curves of the monolithic composites by means of a simple rule of mixtures.

The effect of an additional fretting load on the fatigue performance for the hybrid composite was even smaller than for the AF laminate. The Wöhler curves for fretting fatigue and plain fretting were nearly indistinguishable. Because of the higher stiffness of the CF component, the AF layers supported the minor part of the load applied. Therefore, fretting damage of the AF plies was less relevant for the total load-carrying capacity of the laminate than in the case of monolithic composites. Lesser straining of the AF at a given fatigue stress level, additionally, delayed cracking of fiber bundles predamaged by fretting.

A further beneficial influence of the hybridization on the susceptibility of the AF surface layers to fretting fatigue may arise from the high thermal conductivity of the CF component. Internal friction in the composite during repetitive loading creates a temperature increase which might affect the mechanical characteristics of the polymeric matrix or of the aramid

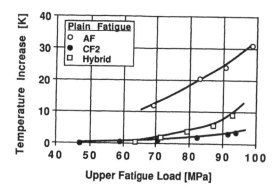

Figure 56 Mean temperature increase in different laminates caused by fatigue loading.

fibers. Figure 56 shows the temperature increase in different laminate types, measured with a thermocouple fixed on the specimen surface, as a function of applied upper fatigue load. Only a small temperature increase was obtained for the CF composite because the good thermal conductivity enabled transportation of heat from the laminate to the metallic clamping grips. AF are very poor thermal conductors, thus causing the accumulation of the produced heat in the specimen. The temperature increase of the hybrids was in the region of the temperature increase of the CF laminae. This may be important if a fretting loading additionally produces heat at the surface of the AF layers. Accordingly, partial brown coloring on surfaces of AF laminates exposed to fretting fatigue indicated the action of thermally induced processes (as degradation and oxidation), whereas this coloring appeared much weaker and less extended on the hybrid surfaces.

CONCLUSIONS

1. An additional fretting component may drastically reduce the fatigue life of an CF/EP laminate if the fibers subjected to fretting are under 0° orientation.

2. The occurrence and magnitude of a fretting influence on fatigue performance sensitively depends on the particular loading conditions (counterpart material, contact pressure). The use of hard steel counterparts prevents a fretting fatigue effect, at least up to a contact pressure of 23 MPa under conditions given in the experiments described previously. For softer counterparts (aluminum, brass), a fretting fatigue effect, which was very significant for CF/EP, occurred above a critical contact pressure. AF composites exhibited a higher resistance to fretting fatigue against aluminum than CF/EP.

3. As a quantitative measure for the degree of fretting fatigue damage, the relative fatigue strength reduction

$$\Delta\sigma_{rel} = \frac{\sigma_F - \sigma_{FF}}{\sigma_F}$$

was proposed. A fretting fatigue damage propagation rate

$$\dot{w}_s^* = \frac{\pi}{16} \frac{dQ_0}{F_N AN} (1 - \sqrt{1 - 2\Delta\sigma_{rel}})$$

could be introduced in analogy to the concept of specific wear rate. Based on this model, the fretting fatigue damage was found to penetrate proportional with time into the laminate. No notch effects were observed. The fretting fatigue performance of a laminate is controlled by the early stages of the wear process in the fretting contact.

4. The mechanisms of interaction between surface damage due to fretting and fatigue are different for different loading conditions. Figure 57 schematically distinguishes three regimes of fretting fatigue as they were found for fretting fatigue of CF/EP against aluminum. At low contact pressures (*regime I*), there exists no synergism between fretting and fatigue. The rate of damage propagation is small and determined by the resistance against plain fretting wear. The specimens live nearly as long as under plain fatigue. At higher pressures (*regime II*), the fretting fatigue damage proceeds considerably faster than pure material removal due to plain fretting wear. Fiber bundles exposed to fretting tend to crack and delaminate. However, if the fatigue stress descends below a certain value, the mutual amplification of fretting and fatigue damage becomes less effective because cracking and peeling off of fiber bundles decelerates. Above a critical contact pressure

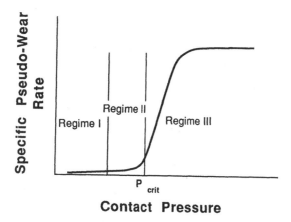

Figure 57 Schematic presentation of the course of specific pseudo-wear rate as a function of contact pressure.

(*regime III*), fretting fatigue damage proceeds about 15 times faster than pure material removal due to fretting. Fiber bundles that are predamaged by fretting rapidly crack and peel off, thus preventing the fretting wear process from entering steady state. The fretting fatigue damage proceeds until the fretting pins reach off-axis plies, which carry only a small part of the applied fatigue stress.

The particular value of the critical contact pressure in Figure 58 depends on the hardness of the counterpart material. For hard steel pins, the boundary pressure is beyond the sensitivity range of the employed fretting device apparatus; for soft aluminum counterparts it is near $p_{crit} = 17$ MPa. The lowest value ($p_{crit} = 12$ MPa) for the critical pressure was found for brass pins with intermediate hardness. The intermediate hardness acts most detrimental because these pins can easily be roughened by fiber debris, but the arising asperities are harder, and thus more abrasive, than in the case of softer counterparts.

5. Some comparative experiments showed that CF laminates manufactured without peel plies are more sensitive than the other CF laminates to fretting fatigue. This is especially true for the early stage of fretting fatigue loading. The resin-rich layer, caused by the use of peel plies during curing of the laminate, assures a more homogeneous contact pressure distribution until the fretting pins touch the fibers.

6. Hybridazation of AF and CF reinforcement caused a negative synergism concerning the fatigue properties, while the AF plies in the skin layers proved to be very resistant to fretting fatigue. Conclusively, the application of a very thin AF surface layer without any load-carrying function may improve the fretting fatigue performance of a CF composite against soft counterparts such as aluminum.

ACKNOWLEDGEMENTS

The author gratefully acknowledges the support by Professor K. Friedrich (University of Kaiserslautern) and Dr. K. Schulte (D.L.R. Köln), who launched the investigation program and accompanied the whole project with helpful discussions and comments. Thanks are also due to BASF for supplying the sample materials free of charge. The project was fully financed by the German Ministry for Research and Technology (BMFT 03 M 1022 A).

REFERENCES

1. R. B. Waterhouse, "Fretting," in *Treatise on Materials Science and Technology*, Vol. 13, *Wear*, D. Scott (ed.), Academic Press, New York, 1979.
2. R. B. Waterhouse (ed.), *Fretting Fatigue*, Applied Science, Barking, Essex, England, 1981.
3. *Wear*, 125 (1988), special issue on fretting wear and fretting fatigue.
4. T. C. Chivers and D. C. Gordelier, *Wear*, 96:153 (1984).
5. W. Hollstein and D. Pingel, "Leichte und steife Roboterkinematiken aus Carbonfaserver-stärktem Kunststoff," *Proc. Verbundwerk 88*, Wiesbaden, Germany, 1988.
6. K. L. Reifsnider, K. Schulte, and J. C. Duke, in *Long Term Fatigue Behavior of Composite Materials*, ASTM STP-813, T. K. O'Brien (ed.), ASTM, Philadelphia, 1983, p. 136.
7. K. Friedrich (ed.), *Friction and Wear of Polymer Composites*, Elsevier, Amsterdam, 1986.
8. H. Uetz and J. Wiedemeyer, *Tribologie der Polymere*, Carl Hanser Verlag, Munich, 1985.
9. R. Heinz and G. Heinke, "Die Vorgänge beim Schwingungsverschleiß in Abhängigkeit von Beanspruchung und Werkstoff," in *Tribologie—Reibung, Verschleiß, Schmierung (Documentation of the German Ministry for Science and Technology)*, Springer-Verlag, Berlin, 1981.
10. M. Godet and D. Play, "Third-body formation and elimination on carbon-fibre/epoxy composite," *Proc. First European Space Tribology Symposium*, Frascati, Apr., European Space Agency, Paris, 1975.
11. N. Ohmae, K. Kobayashi, and T. Tsukizoe, *Wear* 29: 345 (1974).
12. O. Jacobs, K. Friedrich, G. Marom, K. Schulte, and H. D. Wagner, *Wear*, 135: 207 (1990).
13. K. Friedrich, S. Kutter, and K. Schulte, *Compos. Sci. Technol.*, 30: 19 (1987).
14. K. Schulte, K. Friedrich, and S. Kutter, *Compos. Sci. Technol.*, 30: 203 (1987).
15. K. Schulte, K. Friedrich, and S. Kutter, *Compos. Sci. Technol.*, 33: 155 (1988).
16. O. Jacobs, K. Friedrich, and K. Schulte, *Wear*, 145: 167 (1991).
17. O. Jacobs, J. Schulte, and K. Friedrich, "Fretting fatigue of continuous fibre reinforced epoxy laminates," *Symposium on Standardization of Fretting Fatigue Testing Methods and Equipment*, San Antonio, Tex., Nov. 12–13, 1990; to be published as ASTM-STP.
18. H. Czichos, *Tribology: A Systems Approach to the Science and Technology of Friction, Lubrication and Wear*, Elsevier, Amsterdam, 1978.
19. H. Czichos, Systemanalyse und Physik tribologischer Vorgänge, *Schmiertech. Tribol.*, 22(6): 126 (1975).
20. *Wear: System Analysis of Wear Process, Classification of the Field of Wear*, German Standard DIN 50320, Beuth-Verlag, Berlin, 1979.
21. *Catalogue of Friction and Wear Devices*, American Society of Lubrication Engineering, Park Ridge, Ill., 1977.
22. I. W. Kragelski, *Reibung und Verschleib*, Carl Hanser Verlag, Munich, 1971.
23. A. W. Ruff, *Wear*, 134: 49 (1989).
24. J. K. Lancaster, "Abrasion and wear," in *Encyclopedia of Polymer Science and Engineering*, H. F. Mark, N. Bikales, C. G. Overberger, G. Menges, and J. I. Kroschwitz (eds.), Wiley, New York, 1985, p. 1.
25. *Symposium on Standardization of Fretting Fatigue Test Methods and Equipment*, San Antonio, Tex., Nov. 12–13, 1990; to be published as ASTM-STP (in press).
26. M. Godet, *Wear*, 100: 437 (1984).
27. Y. Berthier, L. Vincent, and M. Godet, *Wear*, 125: 25 (1988).
28. M. Godet, *Wear*, 136: 29 (1990).
29. F. P. Bowden and D. Tabor, *Friction and Lubrication of Solids*, Clarendon Press, Oxford, 1950.
30. J. Y. Lin and H. S. Chang, *J. Tribol.*, 111: 468 (1989).
31. D. C. Evens and J. K. Lancaster, "The wear of polymers," in *Treatise on Materials Science and Technology*, Vol. 13, *Wear*, D. Scott (ed.), Academic Press, New York, 1979, p. 85.
32. J. K. Lancaster, *Tribology*, 6: 219 (1973).
33. D. J. Gaul and D. J. Duquette, *Metall. Trans.*, 11A: 1555 (1980).

34. J. J. O'Conner, "The role of elastic stress analysis in the interpretation of fretting fatigue failures," in *Fretting Fatigue*, R. B. Waterhouse (ed.), Applied Science Publisher, London, 1981, p. 23.
35. E. Broszeit, K. H. Kloos, and B. O. Schweighöfer, *Werkstofftechnik*, 16: 187 (1985).
36. D. P. Rooke and P. R. Edwards, *Fatigue Fract. Eng. Mater. Struct.*, 116: 447 (1988).
37. Y. Berthier, C. Colombie, G. Lofficial, L. Vincent, and M. Godet, "First and third body effects: a source and sink problem," in *Mechanisms and Surface Distress*, D. Dowson, C. M. Taylor, M. Godet, and D. Berthe (eds.), Butterworth, London, 1986.
38. O. Vingsbo and S. Söderberg, *Wear*, 126: 131 (1988).
39. Y. Berthier, Ch. Colobié, L. Vincent, and M. Gobet, *J. Tribol.* 110: 517 (1988).
40. J. P. Giltrow and J. K. Lancaster, *Wear*, 16: 359 (1970).
41. J. K. Lancaster, *J. Phys. D Ser. 2*, 1: 549 (1968).
42. K. Friedrich and O. Jacobs, *Compos. Sci. Technol.* (1991), in press.
43. D. Tabor, "The wear of non-metallic materials: a brief review," in *The Wear of Non-metallic Materials*, D. Dowson, M. Godet, and C. M. Taylor (eds.), Mechanical Engineering Publications, London, 1976, p. 3.
44. R. E. Pendlebury, Formation, *Wear*, 125: 3 (1988).
45. T. Tsukizoe, and N. Ohmae, *Ind. Lubr. Tribol.* Jan./Feb. 1976, p. 19.
46. N. N., *LNP. Bull. 254–988*, LNP Plastics Nederland B.V., Raamsdonksveer, The Netherlands, 1990.
47. R. W. Lang, H. Stutz, M. Heym, and D. Nissen, *Angew. Makromol. Chem.*, 145/146: 267 (1986).
48. K. Friedrich, O. Jacobs, M. Cirino, and G. Marom, 'Hybrid effects on sliding wear of polymer composites," *Proc. International Conference on Tribology of Composite Materials*, Oak Ridge, Tenn., May 1–3, 1990.
49. K. Tanaka, "A review of recent studies on polymer friction and wear in Japan," *Proc. International Solid Lubrication Symposium*, Tokyo, 1975.
50. R. W. Land, H. Tesch, and G. H. Herrmann, "125°-curable epoxy matrix with improved toughness and hot/wet performance," in *New Generation Materials and Processes*, F. Saporiti, W. Merati, and L. Peroni (eds.), Milan, Italy, 1988.
51. J. F. Archard, *Wear*, 2: 438 (1959).
52. J. F. Archard, *Wear*, 128: 1 (1988).
53. P. R. Edwards, "The application of fracture mechanics to predicting fretting fatigue," in *Fretting Fatigue*, R. B. Waterhouse (ed.), Applied Science Publishers, Barking, Essex, England, 1981.
54. D. A. Hills, D. Nowell, and J. J. O'Conner, *Wear*, 125: 129 (1988).
55. J. Krey, K. Friedrich, and K.-H. Schwalbe, *J. Mater. Sci. Lett.*, 6: 851 (1987).
56. K. Schulte, "Damage developoment under cyclic loading," in *Proc. European Symposium on Damage Development and Failure Processes in Composite Materials*, I. Verpoest and M. Wevers (eds.), Leuven, Belgium, May 4–6, 1987, p. 39.
57. R. D. Jamison, K. Schulte, K. L. Reifsnider, and W. W. Stinchcomb, *Characterization and Analysis of Damage Mechanisms in Tension: Tension Fatigue of Graphite/Epoxy Laminates*, ASTM STP-836, ASTM, Philadelphia, 1984 p. 21.
58. T. K. O'Brien and M. Rigamonti, C. Zanotti, *Int. J. Fatigue*, 11: 379 (1989).
59. M. Bader, "Modelling fiber and composite failure," in *Proc. European Symposium on Damage Development and Failure Processes in Composite Materials*, I. Verpoest and M. Wevers (eds.), Leuven, Belgium, May 4–6, 1987, p. 8.
60. M. D. Thouless, H. C. Cao, and P. A. Mataga, *J. Mater. Sci.*, 24: 1406 (1989).
61. K. Endo and H. Goto, *Wear*, 38: 311 (1976).
62. J. A. Alic and A. L. Hawley, *Wear*, 56: 377 (1979).
63. K. Sato, *Wear*, 125: 163 (1988).
64. K. Sato, H Fujii, and S. Kodama, *Wear*, 107: 245 (1986).

65. D. W. Hoeppner, *Wear*, 43: 267 (1977).
66. J. H. Greenwood and P. G. Rose, *J. Mater*. Sci. 9: 1809 (1974).
67. R. Talreja, *Fatigue of Composite Materials*, Technomic, Basel, 1987.
68. T. W. Chou and A. Kelly, *Annu. Rev. Mater*. Sci. 10: 229 (1980).
69. G. Fernando, R. F. Dickson, T. Adam, H. Reiter, and B. Harris, *J. Mater. Sci.* 23: 3732 (1988).
70. G. Marom, H. Harel, S. Neumann, K. Friedrich, K. Schulte, and H. D. Wagner, *Composites*, 20: 537 (1989).

5

Basic Chemistry and Mechanical Properties of Polyurethane and Glass-Fiber-Reinforced Composites

Abdul Mateen
Institute of Chemical Engineering and Technology
University of the Punjab,
Lahore, Pakistan

Saadat Anwar Siddiqi and Ghazala Anwar
Centre for Solid State Physics
University of the Punjab,
Lahore, Pakistan

INTRODUCTION

In 1937, O. Bayer discovered that a reaction between aliphatic diisocyanate and aliphatic diols (glycols) went smoothly under reflux conditions to build a linear polymer of high molecular weight [1]. Polymers so formed, called polyurethanes, can be drawn into high-tenacity yarn (perlon) or used as thermoplastics for injection molding (Durethan). These initial discoveries laid the foundation for present polyurethane chemical technology.

A urethane is a reaction product of an alcohol and isocynate, but in many examples, the reaction occurs not only between isocyanate and other groups, such as COOH and NH_3. The linkage so formed is not urethane but occurs along with them in the final polymer. So the term *polyurethane* is basically one of convenience and is used broadly to cover the polymeric and other reaction products of polyisocyanates. Polyurethanes include those polymers that contain a significant number of urethane groups which are not necessarily repeated in a regular order. Polyurethanes can be prepared by a number of methods, the most widely used being the reaction of a polyfunctional hydroxyl compound, based on either a polyester or polyether, with a polyfunctional isocyanate, resulting in the generalized structure

$$HO-R'-OH + OCN-R-NCO-HO-R'-OH + OCN-R-NCO + \cdots \longrightarrow$$

$$\cdots O-R'-O-\underset{\underset{O}{\|}}{C}-NH-R-NH-\underset{\underset{O}{\|}}{C}-O-R'-O-\underset{\underset{O}{\|}}{C}-NH-R-NH-\underset{\underset{O}{\|}}{C}\cdots$$

The functionality of the hydroxyl-containing component as well as of the isocyanate can be increased to three or more to form branched or cross-linked polymers. With the large number of possible polyfunctional species that are available for the manufacture of polyurethane, almost any degree of elasticity and cross-linking can be attained [2]. For these reasons the polyurethanes are almost unique in that cross-linking, chain flexibility, and intermolecular forces can be varied widely and almost independently. It is therefore not unexpected that the polyurethanes include such products as fibers, flexible and rigid foams, coating, soft and hard elastomers, and highly cross-linked plastics. The topics discussed here are (1) the basic chemistry of polyurethanes (PU[4]), (2) the reaction injection molding (RIM) process, and (3) the physical properties of high-glass-transition-temperature (T_g) polyurethanes and glass-fiber-reinforced composites.

BASIC CHEMISTRY OF POLYURETHANES

The basic chemistry of polyurethanes is that 1 mol of a given di- or polyisocyanate reacted with 1 mol of di- or polyfunction hydrogen donar will polymerize into a solid mass. No by-products are given off unless water or carboxyl groups are present, in which case carbon dioxide is given off. Common isocyanate can be divided into two main classes.

Primary Reactions

Reactions of isocyanate with compounds containing reactive hydrogen to give additional products:

$$R-N{=}C{=}O + H_2O \longrightarrow R-NH-\underset{\underset{O}{\|}}{C}-OH \tag{1}$$

(unstable
carbanic acid)

$$R-NH-\overset{\overset{\text{O}}{\|}}{C}-OH \longrightarrow R-NH_2 + CO_2 \tag{2}$$
$$(amine)$$

$$R-N=C=O + R'-OH \rightleftharpoons R-NH-\overset{\overset{\text{O}}{\|}}{C}-OR' \tag{3}$$
$$(urethane)$$

$$R-N=C=O + H_2NR' \longrightarrow R-NH-\overset{\overset{\text{O}}{\|}}{C}-NHR' \tag{4}$$
$$(urea)$$

$$R-N=C=O + R'-COOH \longrightarrow RNH-\overset{\overset{\text{O}}{\|}}{C}-O-C-R' \tag{5}$$
$$(unstable\ mixed\ anhydrides)$$

$$RNH-\overset{\overset{\text{O}}{\|}}{C}-O-\overset{\overset{\text{O}}{\|}}{C}-R' \longrightarrow R-NH-\overset{\overset{\text{O}}{\|}}{C}-R' + CO_2 \tag{6}$$
$$(amide)$$

The addition of reactive hydrogen compounds, primary reactions, are the most useful in the formation of polyurethanes. Isocyanate reacts with hydroxyl compounds and amines to give urethanes and ureas, respectively. Similar reactions with water and carboxic acids produce carbanic acid and mixed anhydrides. In both of the cases the intermediate products break down, and in the case of water an amine is formed, which reacts further with more isocyanate to give urea. In the case of carboxic acid, the mixed anhydrides break down to form amide groups.

Secondary Reactions

In secondary reactions isocyanate further reacts with the groups containing active hydrogen that are present in all the products of primary reactions. Thus isocyanate reacts with urethanes to give allophanates and with ureas to give biurets.

$$R-N=C=O + R-NH-\overset{\overset{\text{}}{}}{C}-OR' \longrightarrow R-N-\overset{\overset{\text{O}}{\|}}{C}-OR' \tag{7}$$
$$\overset{\|}{O} \qquad\qquad\qquad O=\overset{\|}{C}-NH-R$$
$$(urethane) \qquad\qquad (allophanate)$$

$$R-N=C=O + R-NH-\overset{\overset{\text{}}{}}{C}-NH-R' \longrightarrow R-N-\overset{\overset{\text{O}}{\|}}{C}-NH-R' \tag{8}$$
$$\overset{\|}{O} \qquad\qquad\qquad O=\overset{\|}{C}-NH-R$$
$$(urea) \qquad\qquad\qquad (biuret)$$

$$R-N=C=O + R-NH-CO-R' \longrightarrow R-NH-CO-NR-CO-R' \tag{9}$$
$$(amide) \qquad\qquad\qquad (acyl\ urea)$$

The secondary reactions occur to a much lesser extent than the primary reactions. The importance of these reactions is that the formation of allophanates, in particular biurets, is responsible for cross-linking and branching, which have an important effect on the properties of the end product.

The urethane industry seldom uses a simple mole-to-mole chemistry in the formulation of urethane and must take into account every possible constituent in all of the intermediates. Impurities such as water, acids, chlorine, and all similar materials must be considered; their reaction must be calculated and the true equivalent weights of the intermediates must be used in the stoichiometry of the reaction.

Characteristics of Isocyanate Reactions

The primary reactions of isocyanates with active hydrogen compounds occur with remarkable ease, normally proceeding at ordinary temperatures with the evolution of heat. The rates of reaction differ quite considerably, depending on the reactant used. In an alcohol–isocyanate reaction that forms polyurethane, aromatic isocyanate are more reactive than the aliphatic isocyanates. Similarly, active hydrogen-containing compounds can differ considerably from alcohols in their reactivity. A scale of different rate of reaction with different types of group is as follows [3]: aliphatic NH_2 > aromatic NH_2 > primary OH > water > secondary OH > tertiary OH > phenolic OH > COOH and $R \cdot NH \cdot R'$ > $R \cdot CO \cdot NHR'$ > $R \cdot NH \cdot COOR'$.

This scale shows that the most isocyanate-reactive compounds are primary amines, followed by hydroxyl compounds and water. The primary isocyanate products, such as urea, amide, and urethane, are the least reactive.

Control of Reactions in Polyurethane Formation

Polyurethanes are a class of materials having a wide range of properties, and versatility is associated with a greater-than-normal complexity, which calls for a greater degree of control of chemical reactions to manufacture reproducible materials. During PU formation many reactions occur simultaneosly. The rate at which these reactions occur relative to one another has an important bearing on the properties of the final polymer. Discussed below are important variables that can control these reactions to a considerable extent.

Temperature

The general rule that elevated temperatures will increase the rate of reaction is also applicable in the formation of polyurethanes. At low temperatures, about 50°C, reactions of isocyanate with hydroxyl to give urethanes, with amine to give urea, and with water to give amine will proceed relatively easily, and a linear polyurethane, lacking any cross-links or branches, will result. Therefore, for the preparation of linear PU balanced with an acceptable rate of reaction, a low temperature is required. For the introduction of cross-links and branches in the polyurethane in the absence of catalyst, higher temperatures, between 50 and 150°C, are used to encourage the formation of isocyanate, biurate, and allophanate links through secondary reactions.

Catalyst

Most polyurethane reactions are too slow for commercial manufacturing. Therefore, catalysts are used to increase the rate of reaction. Catalysts used in PU formation not only act as accelerators for all reactions but in fact can be used to increase selectively—the rate of one reaction rather than another. Considerable information on the activity of catalyst has been published by various authors [4–7].

Inhibitors

Any acid or material that yields acids can act as inhibitors or retarders in the polyurethane reactions. In fact, these acidic materials do not act as true inhibitors but function by neutralizing the small portions of basic materials, which would otherwise catalyze iso-cyanate reactions to an undesirable extent during PU formation. Thus very small amounts of HCl gas, adipic acid, and various sulfonic acids are well-known retarders. Acid chlorides such as acetyl chloride, which are considered to give rise to carboxylic acid and HCl under the conditions of PU formation, especially in the presence of traces of water necessary to hydrolyze them, are also well known to the industry.

Types of Isocyanate and Active Hydrogen Components

Polymerized urethanes exhibit a wide variety of structures, depending on the type of isocyanates and active hydrogen components used in formulation. The aromatic isocyanates will give more urea groups than alliphatic isocyanates in the finished product. Excess NCO in the formation will react with atmospheric moisture to form urea and biuret groups. Aromtic diamines tend to yield a great many urea groups in the structure with some amide and aromatic rings.

Polyester components will give far more urethane groups, along with a few ester groups in some cases. The urethane linkage will be most prominent. Polyester components give a very large number of allophanates, with a few ether groups.

The nature of the polyurethane can also vary by varying the functionalities of the isocyanates and the active hydrogen components. Among the isocyanates the most prominent and important are the two high-volume product groups: toluene disocyanate (TDI) and diphenyl methane diisocyanate (MDI). TDI or TDI quasi-prepolymers, however, are not suitable for the RIM proces because of the lower reactivity and long mold residence time required for suitable demolding.

REACTION INJECTION MOLDING

Conventionally injection molding (IM) of (principally) thermoplastics and processing of sheet molding compounds (SMC[4]) are employed for the fabrication of polymer materials. RIM is a relatively new technology which has already found wide applications in the automobile, building, marine, furniture, and footwear industries because of its obvious advantages over other molding processes. To now, the process has been applied commercially to PU more or less exclusively. However, in principle, there are no serious problems in its application to the processing of other polymeric materials, such as vinyl monomers, epoxy resins, unsaturated polyester resins, or even nylon. An indication of this direction is the investigation of the processing of pseudo interpenetrating networks [8] and the processing of styrene monomers [9].

The basic principles of the RIM process were developed in Germany by Bayer during the period 1966–1969, and since then a lot of research work was done on the development of this technology, yet an all-purpose process is still in the making.

Basic Process Details

As is evident from the name itself, RIM is essentially a single operation process. The RIM process starts with two (more are possible) liquid streams, a polyol premix and an

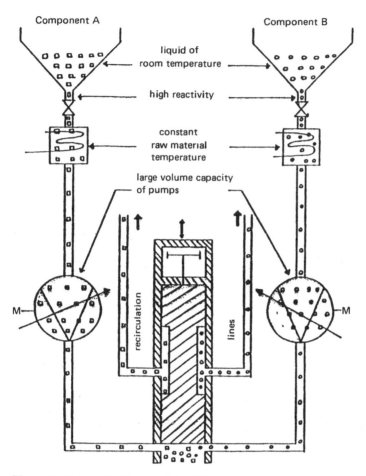

Figure 1 Schematic diagram of the RIM process.

isocyanate prepolymer. These streams are metered by a positive displacement to a mixhead and from there delivered to a mold, where it reacts and gels rapidly into a resilient elastomer or rigid plastic, depending on the formulation used. A schematic description of the process is given in Figure 1, and discussed briefly below.

Raw Material Storage

The raw materials for PU formation are polyol or polyol premix (the premix also includes ''minor'' ingredients, such as catalyst, cross-linkers or chain extenders, blowing agents, etc.) and isocyanate-based materials of low viscosity. These materials are stored in the molten state at a fixed temperature. Steel tanks are used for storage, and care should be taken to avoid contamination with moisture, which may react with polyol to produce considerable gas formation during the gelling reaction. On the isocyanate side, water can react to form a hard urea precipitate and reduce isocyanate contents, thereby affecting the properties of end products.

Metering

Since maintaining the proper stoichiometric ratios in the RIM formulations is extremely significant, the most important part of any RIM machine is the high-pressure metering unit, which meters precisely the required amounts of each component to the mixhead. This unit takes the highly reactive starting ingredients, polyol premix and isocyanate, and within a few seconds injects them into the mold without allowing the reaction to start. This delivery must be accomplished with highly accurate metering and shot ratios and must be controlled to levels between 1 and 2% in order to obtain consistent part properties [10]. The delivery must be accomplished at high throughput rates to provide high impingement velocity at the mixing point, thus producing turbulence and the required blending of ingredients.

The most important unit of the metering system is the pump. The pumps most commonly used are either axial piston or radial piston pumps. Whatever the type, to achieve adequate mixing, the pump must be capable of delivering the precursors to the mixing head at a pressure of 2000 to 3000 psi during injection.

Mixing

The second important element of modern RIM technology is the high-pressure, impingement-type, self-cleaning mixhead. Liquid streams enter the mixing chamber at high flow rates and their residence time is essentially very short and can vary, within limits, with different formulations. To achieve good mixing in the mixing chamber, the mixhead must perform the following functions [11]:

1. Developing high-velocity delivery to each stream to the mixing chamber
2. Effecting the delivery under precisely synchronized conditions
3. Developing turbulence in the mix chamber to mix the two liquid streams intimately
4. Cleaning the mix chamber in such a way that progressive polymer buildup does not occur and impede subsequent and repeated operation

The mixheads used on most advanced high-pressure RIM machines operate on the principle of impingement mixing. The component streams at high pressure (1000 to 3000 psi) are fed through very small holes in the opposite side of the mixing block. The streams impinge at an angle of about 180°, and the turbulence produced is responsible for intimate mixing.

The advent of the self-cleaning mixhead was a revolutionary step in the development of RIM technology, and presently all mixheads, irrespective of the manufacturer, work on the principle of operation for this type of device suggested by Suh and Malguarneva [12].

According to them, the momentum ratio of the fluid streams is the key and should be 1. That is, both liquid streams should be equal in their momentum at the time of collision. These workers also suggest that the mix chamber geometry is not a significant variable. Other industry sources suggest different important variables, such as stream pressure and flow rate.

All agree that the Reynolds number, $Re = pVD/\mu$ (where p is the density of the liquid, μ the dynamic viscosity, and D the mixhead orifice diameter), of both streams should be as high as possible. Values of 100 to 1000 have been proposed, but it should definitely be more than 50.

In certain demanding application and process areas, where 100% effective mixing cannot be achieved by available mixheads, additional mixing capability (after mixing) can be provided downstream from the mixhead [13]. Like conventional injection molding, there is a transition zone in RIM between the mixhead and the mold. The transition zone includes a sprue runner and gate system and an after mixer, if necessary. The function of the RIM transition zone is to convert the mixhead stream from a turbulent state to a laminar flow; turbulence can result, entrapping air which cannot be released from trapping the reaction mixture because of the increasing viscosity of the rapidly gelling liquid. A detailed account of the functions of the transition zone has been presented by Silverwood [14].

Molding

In this step, the mixed intermediates flow out and take the shape of the mold, gel, and polymerize to final part properties. To produce a uniform surface finish the molds are usually heated and maintained at constant temperature. The effect of mold temperature on part density for certain PU systems has been discussed by Thomson and co-workers [15].

Because of low molding pressure in the RIM process, there are a number of options open for the mold manufacturers with regard to the material of construction. Machined steel, machined aluminum, and formed nickel liners in aluminum or epoxy bases are being used successfully in productin with the RIM process where excellent surface quality is required.

In the RIM process a flawless construction of mold is of utmost importance because the RIM materials enter the mold as low-viscosity liquids, accurately reproducing the mold surface. Moreover, both mold halves, as well as all moving parts, must be closely fitted to avoid excess leakage and flash of the low-viscosity liquid mixtures. The close-fit requirement must be balanced, however, with the need to dispel all the air in the mold within a very short fill time.

The ultimate object of the RIM process is the completely automatic production of large parts within shortest possible overall cycle times. The latest RIM techniques allow moldings to be produced on a 60-second cycle. A key development in this regard permits the incorporation of inmold release agents into the feedstock. This reduces cycle times and handling problems by not requiring the spraying of the mold cavity surfaces with a release agent before each injection operation [16].

Advantages of the RIM Process

Lower Material Cost

The costs of the starting materials for RIM (liquids) have been shown to be less than those for IM (ready-formed) given that similar articles with similar chemical and mechanical properties are to be produced by the two processes [17]. In addition, advances in the design of a mixing head in RIM machinary have resulted in better end products and a reduction in the rejection rate, which is now favorably comparable with that for IM.

Low Capital Investment

In the RIM process the internal mold pressures are low (< 100 psi). Therefore, large parts can be molded on relatively low tonnage equipment, resulting in capital investment, cost, and floor space savings. Moreover, a major advantage of RIM in economic terms is in the use of multiple mixing units, molds, and clamps, all of which are driven from a single hydraulic power source. It is known that RIM equipment is three to four times cheaper than IM equipment for the same output.

Low Energy Consumption

In RIM and IM, materials are converted directly into finished articles, whereas with sheet molding compound (SMC) two stages are required. However, RIM uses lower temperatures and pressures since the process handles low-viscosity reactants directly rather than solid polymers, which have to be converted into a high-viscosity melt before molding. In addition, with respect to filled materials, RIM allows high loading of fillers to be introduced into the liquid reactants as a part of a one-stage process. In injection molding of filled polymers, an additional compounding stage, usually extrusion, is necessary prior to molding.

Rapid Production of Large and Complex Parts

The RIM process starts with liquid reactants. The reactants enter the mold at low viscosities, reproducing the exact surface of the mold. All this happens in a single process and very short cycle times compared to SMC, which is a two-stage process [18]. In addition, large parts of varying cross sections with inserts can be produced without the sink mark problems of IM.

Varying Chemistry

The RIM process is based on a versatile urethane chemistry which provides polymers of high physical properties over a wide range of stiffness. Material properties of polymers processed by RIM can be varied systematically and dynamically even during the formation of a single article by varying feed components and reactant ratios. Such versatality is not possible in either sheet molding compound (SMC) or injection molding (IM).

Reinforced Reaction Injection Molding Process

In the reinforced reaction injection molding (RRIM) process, the incorporation of reinforcing fillers to the liquid reactants create problems in the resultant slurries. Viscosities of slurries increase considerably (1000 to 6000 cP) at ambient temperature, depending on the fiber length and a particular blend of polyol and isocyanate. These slurries cannot possibly be processed by conventional machinary without a great deal of wear and abrasion. Settling of the slurries in the storage tank is another problem that can be dealt with, to some extent, by continuous low-speed agitation of the materials.

Substantial research and development work has been done and is being done in Europe and America. Difficulties and improvements in the field of RRIM have been reviewed by different workers [19,20]. A second generation of improved RIM machinary is in operation. It has been claimed that Heneck's parallel-stream-type MP mixing head can satisfactorily handle glass-fiber containing systems [16]. But the real production experience of recently developed techniques, such as mixing and metering, is limited and evaluation of equipment wear has been based on laboratory processing. Much work is still to be carried out before the overall durability of reinforced RIM equipment can be assessed for high-volume automative production.

Reinforced RIM Polyurethanes

A reinforcement is any substance that improves the physical properties when added to the material in reasonable concentrations. The concept of adding inorganic fillers to polyurethanes dates back to the early 1970s. Barytes were used to improve sound absorption properties [21]. Different extenders (which cannot be considered as reinforcement) have been used to cheapen the formulation [22]. Glass has been used to improve some structural properties and also as a straightforward reinforcement for both military [23] and construction [24] use. However, these filled materials were not processed by the RIM technique. Tecnick International was probably the first company that used the RIM process to manufacture toilet seats reinforced with 65% filler and 3% glass fibers [11]. Apart from the processing of pigments, fillers, and low levels of fiber reinforcement, growth of RRIM is rather slow, mainly because of processing difficulties.

As time goes by, new and better ways of processing reinforced RIM polyurethanes have come to light. Equipment has been developed and is being developed for better processing procedures at lower costs.

Various types of reinforcements, such as glass, carbon, aramid fibers, and mica flakes are being considered by the industry as possible considerations for RIM reinforcement, but the majority of published work has concentrated on the use of glass fibers. Presently, two types of glass fibers, hammer-milled and chopped strands, are being used by the RIM industry.

Commercially available hammer-milled fibers exist in a variety of effective aspect ratios. The process of manufacture of these materials produces a distribution of lengths [25], a factor that has to be taken into account in any structure and property investigation. Hammer-milled fiberglass has been used extensively in the automobile industry with significant success, and the work has been reviewed by Ishan [26,27]. Perhaps the only disadvantage of these materials is their high cost.

As compared to the hammer-milled fiberglass, chopped strands are cheap and readily available. Despite this obvious advantage, chopped strands have seen limited application in the polyurethane RIM system because of difficulties in processing.

STRESS–STRAIN PROPERTIES

The most common and widely used method of determining the mechanical properties of solid polymeric materials is to carry out a load-extension measurement using some kind of tensile testing equipment where one end of the test piece is clamped rigidly while the other end is displaced at a constant rate. The actual experimental data derived from these tests are in the form of load extension curves, which are then converted to the desired stress–strain curves. A tensile test provides simultaneously values for Young's modulus, tensile strength, yield strength and strain, ultimate strain, and energy required to break the sample. These values are sensitive to a number of external variables, such as temperature, strain rate, and volume fraction of effective reinforcement. Therefore, for structural applications a design engineer may require information about these variables. Figure 2 shows the stress–strain curves for dry polyurethene samples tested at various temperatures. The behavior is typical of viscoelastic network materials. A given curve is concave to the strain axis until fracture and shows no evidence of yielding. As can be expected, increase in temperature has led to a decrease in stress at a given strain level [28]. Since the range of temperature under study is well below the glass transition temperature ($T_g \approx 130°C$), the polymer molecules are largely immobilized. Therefore, decrease in modulus is not large. Values shown in Table 1 are typical of glassy amorphous polymers. With the increase in temperature there is a decrease in ultimate strength and a corresponding increase in ultimate strain. The effects of moisture on PU (Figure 3 and Table 1) are quite

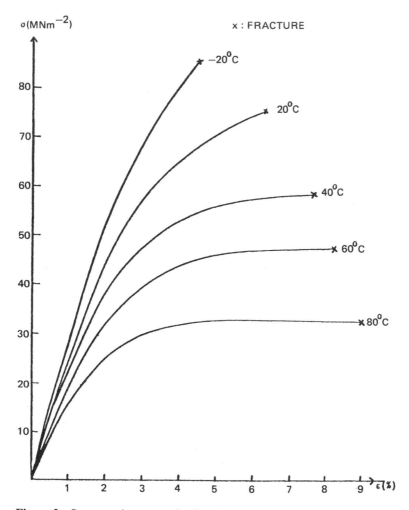

Figure 2 Stress–strain curves of polyurethane at temperature indicated.

small and even lie within experimental error, but largely are in the direction that one would expect. Clemens [29] has shown that Young's modulus of nylon 66 decreases drastically with increase in moisture content. The Young's modulus of nylon 66, in the presence of 9.1% water, drops from 3.03 GN/m^2 to 0.68 GN/m^2. The ineffectiveness of moisture on the PU network as compared to nylon 66 is partly explained by the different amounts of water absorbed, 9.1% for nylon and 2.6% for PU. But it also reflects the advantage of a chemically cross-linked system over a linear thermoplastic which relies entirely on hydrogen bonding. The effect of temperature on dry reinforced PU ($\phi_f = 0.158$) is similar to that of unfilled PU (Figure 4). But the variation in stress–strain properties with temperature, for reinforced PU, are less pronounced compared to the unfilled PU (Table 1). This behavior can be attributed to the contribution of the glass filler, which is relatively less sensitive to temperature.

The stress–strain curve for wet reinforced polyurethane at 20°C (Figure 3 and Table 1) shows a decrease in tensile modulus together with an increase in elongation to break.

Table 1 Tensile Stress–Strain Properties of Unfilled And Reinforced Polyurethane

ϕ_f	Temperature (°C)	Young's Modulus, E (GN/m²)	Ultimate Stress, σ^* (MN/m²)	Ultimate Strain, ϵ_c^* (%)
0	−20	2.73	85.50	4.50
	20	2.34	75.18	6.35
	40	2.17	57.80	7.93
	60	1.89	47.58	8.25
	80	1.70	32.50	9.00
0.045	20	2.75	66.53	3.5
0.088	20	3.21	67.50	3.00
0.126	20	3.70	60.44	2.01
0.158	−20	4.77	55.50	1.2
	20	4.25	53.80	1.31
	40	3.85	50.54	1.50
	60	3.48	46.22	1.57
	80	3.15	42.18	1.70
0 (wet), H₂O = 2.6%	20	2.29	72.11	5.75
0.158 (wet), H₂O = 1.9%	20	3.63	37.00	1.50

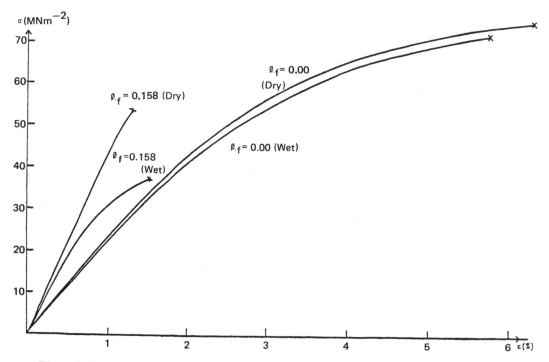

Figure 3 Stress–strain curves of dry and wet samples of unfilled and filled ($\phi_f = 0.158$) polyurethane at 20°C.

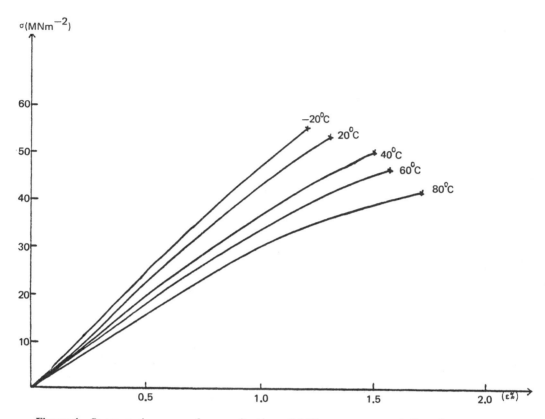

Figure 4 Stress–strain curves of composite ($\phi_f = 0.158$) at temperatures indicated.

The most significant effect of moisture concerns the ultimate tensile strength. The stress at failure of wet reinforced polyurethane almost approaches the stress values of the unfilled polyurethane at that particular strain. The effect of moisture on polyurethane matrix is quite small; therefore, this behavior is due to the dewetting phenomenon. The presence of plasticizers such as moisture in composites results in dewetting at large strain [30]. However, increased filler concentration decreases the value of elongation at which dewetting takes place [31]. Dewetting is a result of creating voids during the stretching of a specimen due to poor interfacial bonding. Filler gives rise to an increase in stiffness over that of the unfilled material, as shown by a steady increase in the tensile modulus. There is a corresponding decrease in the ultimate tensile strength and a decrease in elongation to break (Figure 5 and Table 1).

Tensile Modulus

A number of theoretical equations have been proposed to describe the effect of particulate filler on the tensile modulus [32–35]. The theoretical predictions assume perfect bonding between the matrix and filler. Even if it is not achieved in practice, the equations remain valid because of the mismatch of thermal coefficients of expansion between filler and matrix. Since the matrices usually have a significantly higher thermal coefficient of expansion than the fillers, on cooling down from the fabrication or curing temperature they

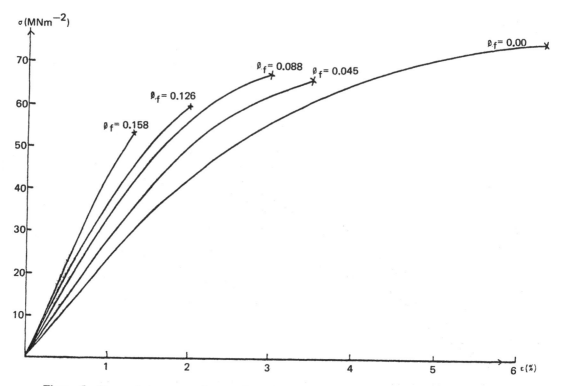

Figure 5 Stress–strain curves of composite and polyurethane at 20°C.

contract around the filler. In addition, the frictional forces between the two phases offer a mechanical restraint preventing motion at the filler–matrix interface during deformation. The equation proposed by Lewis and Nielson [36,37] is generally applied to a rigid filler in a rigid matrix:

$$\frac{E_c}{E_m} = \frac{1 + AB\phi_f}{B\psi\phi_f} \tag{10}$$

where E_c is the composite modulus and E_m is the matrix modulus. The quanity B is defined as

$$B = \frac{(E_f/E_m) - 1}{(E_f/E_m) + A} \tag{11}$$

where A is a constant related to the generalized Einstein coefficient $K\epsilon$ by

$$A = K\epsilon - 1 \tag{12}$$

For randomly oriented rodlike particles the value of $K\epsilon$ depends on the ratio of the length of the particles to their diameter [38]. The factor ψ depends on the maximum packing fraction, ϕ_{max}, of the filler. The necessary boundary conditions are

$$\psi = \frac{1 + \phi_f(1 - \phi_{max})}{\phi_{max}^2} \tag{13}$$

$$\psi\phi_f = 1 - \exp\left(\frac{-\phi_f}{1 - \phi_f/\phi_{max}}\right) \tag{14}$$

The quantity $\psi\phi_f$ can be thought of as a reduced volume fraction. For $\phi_f < 0.1$, it is shown by Figure 6 that there is good agreement between the experimental and theoretical values. However, in the case of higher volume fractions (i.e., $\phi_f > 0.1$), experimental values deviate from theoretically predicted values, but the difference is quite small (ca. 10%). This difference can be attributed to poor dispersion and increased entrapped air. With increased filler concentration, mixing of hammer-milled glass fibers becomes difficult. Poor mixing gives rise to inadequate dispersion and increased void content in the composites. Inadequate dispersion leads to a lower filler–matrix interfacial area, resulting in less interfacial friction and hence a lower tensile modulus. Entrapped air also has a significant deterimental effect on the tensile modulus.

Stress–Strain Analysis

The stress–strain response of short fiber-reinforced composites is a complicated phenomenon and cannot be described by a simple relationship. The complication arises from the possibility of a number of material and geometrical variables which can influence the stress–strain behavior of these materials. Apart from the matrix and filler properties, variables such as fiber length, fiber-length distribution, interface strength, and fiber orientation have a significant influence on the stress–strain properties of short fiber-reinforced composites. Bowyer and Bader [39] have developed a model to describe the stress–strain behavior of such materials. Furthermore, they and other workers [29,40] have applied this model successfully to calculate the interfacial strength and fiber orientation factor of composites from their measured stress–strain curves and fiber length distribution.

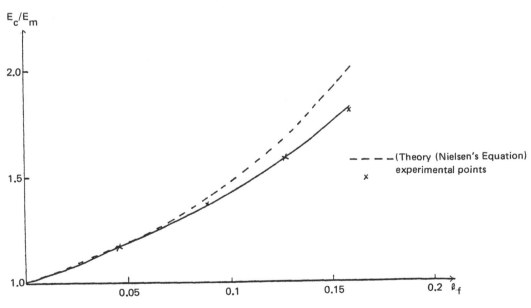

Figure 6 Relative tensile modulus versus ϕ_f for polyurethane composites.

$$\sigma_c = C \left[\sum_{i}^{\bar{l}_i < \bar{l}_\epsilon} \frac{\tau \bar{l}_i \phi_i}{d} + \sum_{j}^{\bar{l}_j \geq \bar{l}_\epsilon} E_f \epsilon_c \phi_f \left(1 - \frac{E_f \epsilon_c d}{4 \bar{l}_j \tau} \right) \right] + E_m \epsilon_c (1 - \phi_f) \tag{15}$$

Equation (15) can be written as

$$\sigma_c = Cx + Cy + z \tag{16}$$

The terms x, y, and z represent the contribution of the subcritical fibers, supercritical fibers, and matrix, respectively, to the composite. $\bar{l}_\epsilon$ is the critical fiber length at a particular strain and may be written as

$$\bar{l}_\epsilon = \frac{E_f \epsilon_c d}{2\tau} \tag{17}$$

Equation (16) may be written as

$$\sigma_c - Z = c(x + y) \tag{18}$$

In Eq. (18) the factor C is assumed to be constant at lower strain values ($\epsilon_c \leq 0.03$). Therefore, a plot of ($\sigma_c - z$) versus ($x + y$) at different values of strain should give a straight line with a gradient c. For a practical system σ_c and z at various strain levels may be obtained from the composite and matrix stress–strain curves. The fiber-length distribution and values of d can be physically determined. The value of E_f can be obtained from the literature. The orientation factor, C, and interfacial shear strength, τ, are, however, generally not known but can be computed from the stress–strain data as follows:

1. A value of 76 GN/m^2 was assumed for modulus E_f [41].

2. The strain values, 0.002, 0.004, 0.006, ... up to fracture strain were selected.

3. The average stresses σ_1, σ_2, σ_3, ... , σ_n, corresponding to these strains are measured from the composite stress–strain data. The matrix stress m was also determined from its stress–strain curve, corresponding to the same composite strain levels. The values of ($\sigma_c - z$) were then evaluated at strain levels mentioned above.

4. A value of τ was assumed and corresponding values of $\bar{l}_\epsilon$ were calculated from Eq. (17).

5. The fiber contribution terms x and y from Eqs. (15) and (16) were calculated using the assumed value of τ, the corresponding value of $\bar{l}_\epsilon$ and fiber-length distribution histogram (Figure 7).

6. The term ($x + y$) was determined at various strain levels.

7. The assumed value of τ was adjusted until the best straight-line plot of ($\sigma_c - z$) versus ($x + y$) was obtained. This value of τ was assumed to be correct, and the orientation factor C was then calculated from the gradient of the straight line.

Although care was taken to follow exactly the same molding procedure for each batch, different values of interfacial shear strength and fiber orientation factor for different volume fractions were expected because of the modification in flow properties with increased filler content. Therefore, values C and τ should be determined independently for all volume fractions tested at room temperature.

It was assumed that variation in temperature does not affect the fiber orientation, and the only adjustable parameter in this case is the interfacial shear strength. The values of τ for the composite tested at temperatures other than room temperature were evaluated

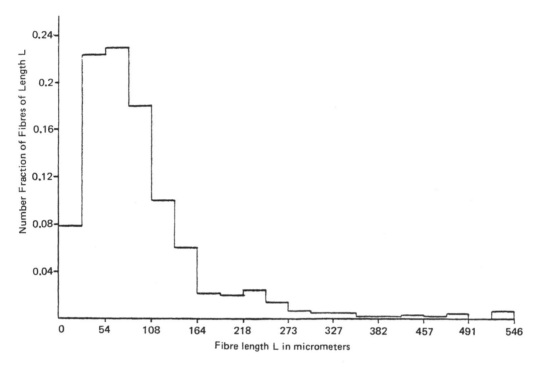

Figure 7 Fiber length distribution.

by adjusting the assumed value of τ until the plot of $(\sigma_c - z)$ versus $(x + y)$ gave the same gradient as obtained for that particular volume fraction at room temperature. This τ value was considered the correct value at this temperature.

From Tables 2 and 3 it is clear that the orientation factor values for all composite volume fractions lie between 0.245 and 0.278. The fiber orientation factor has a value of 1 in the case of fibers aligned parallel to the test direction. The term is reduced to one-third if fibers are randomly dispersed in two dimensions, and if fibers are randomly dispersed in three dimensions, the factor is one-sixth [42]. The values in Table 3 indicate that fibers are neither aligned in any one direction nor do they represent three-dimensional random distribution. However, these values seem closer to the value for two-dimensional random distribution. Small variation in the interfacial shear strength, at constant temperature and

Table 2 τ Values of Composite ($\phi_f = 0.158$) at Various Temperatures

Temperature (°C)	Interfacial Shear Strength, τ (MN/m²)
−20	42
20	35
40	24
60	22
80	18
20 (Wet)	10

Table 3 τ and C Values of Composites at 20°C

Volume Fraction, ϕ_f	Interfacial Shear Strength τ (MN/m^2)	Orientation Factor, C
0.045	31	0.273
0.088	28	0.278
0.126	33	0.245
0.158	35	0.270

moisture content, may arise due to the void content and orientation factor. It increases with the increase in orientation factor and decreases with the increase in void content. Judd and Wright [43] have shown that the interlaminar shear strength of composite material decrease by about 7% for each 1% of voids, up to a total void content of about 4%. Figure 8 shows the effect of temperature on the interfacial shear strength of the composite sample; corresponding values are listed in Table 2. Interfacial shear strength is a temperature-dependent quantity: With the increase in temperature, below the glass transition region, it decreases gradually. The difference in the thermal expansion coefficient of the matrix and the filler results in induced stress in the composite. These stresses develop due to the contraction of the matrix around filler as the composite cools down from the fabrication or curing temperature (the curing temperature of PU and the composites under study is 150°C). When the temperature of the composite is raised again these stresses are relieved, causing a decrease in interfacial strength. Similar trends have been observed in the study of short glass-fiber-reinforced nylon 66 [29,40].

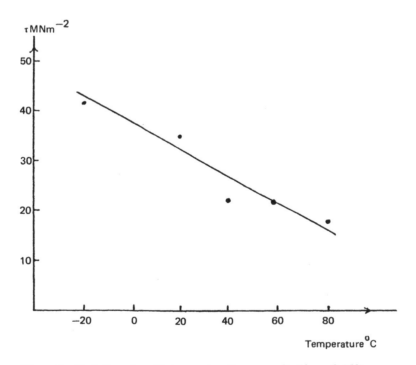

Figure 8 Variation of τ with temperature for composite ($\phi_f = 0.158$).

The effect of moisture on the interfacial shear strength are similar to those of temperature. A drop in interfacial shear strength (from 35 MN/m^2 to 10 MN/m^2) can be attributed to the dewetting effect, in which the adhesion between the filler and matrix phases is weakened because of the presence of moisture. Clemens [29] has reported a decrease in the interfacial shear strength (from 54 MN/m^2 to 16 MN/m^2) of wet short glass-fiber-reinforced nylon 66 at room temperature. It is interesting to note that the magnitude of the effect of moisture on both types of composites is almost the same.

The values of $\bar{l}_e$ critical fiber length at a particular strain level and l_ϵ^*, the critical fiber length at fracture, are listed in Tables 4 to 6. The critical fiber length increases with increased strain. Only at strain level as low as 0.002 are 90% fiber longer than the critical fiber length. At fracture strain, only in the case of $\phi_f = 0.158$, which fails at relatively low strain, are over one-tenth of fibers longer than the critical fiber length.

Table 4 $\bar{l}_\epsilon$ and $\% > \bar{l}_\epsilon$ Values of Composites at 20°C

ϕ_f	0.045		0.088		0.126		0.158	
Composite Strain, ϵ	$\bar{l}_\epsilon \times 10^4$ (m)	$\% > \bar{l}_\epsilon$	$\bar{l}_\epsilon \times 10^4$ (m)	$\% > \bar{l}_\epsilon$	$\bar{l}_\epsilon \times 10^4$ (m)	$\% > \bar{l}_\epsilon$	$\bar{l}_\epsilon \times 10^4$ (m)	$\% > \bar{l}_\epsilon$
0.002	0.29	90.50	0.314	89.11	0.27	92.11	0.26	93.20
0.004	0.58	68.0	0.65	66.0	0.55	69.0	0.52	70.0
0.008	1.18	26.0	1.30	20.0	1.11	27.0	1.04	31.00
0.01	1.47	17.0	1.63	13.0	1.38	15.0	1.30	21.0
ϵ^*	5.14	1.5	4.88	1.70	2.76	4.0	1.71	11.0

Table 5 $\bar{l}_\epsilon^*$ and $\% > \bar{l}_\epsilon$ Values of Reinforced Polyurethane at Various Temperatures ($\phi_f = 0.158$)[a]

					Temperature (°C)				
−20		20		40		60		80	
$\bar{l}_\epsilon \times 10^4$ (m)	$\% > \bar{l}_\epsilon^*$	$\bar{l}_\epsilon \times 10^4$ (m)	$\% > \bar{l}_\epsilon^*$	$\bar{l}_\epsilon \times 10^4$ (m)	$\% > \bar{l}_\epsilon^*$	$\bar{l}_\epsilon \times 10^4$ (m)	$\% > \bar{l}_\epsilon^*$	$\bar{l}_\epsilon \times 10^4$ (m)	$\% > \bar{l}_\epsilon^*$
1.30	20.5	1.71	11.07	3.11	3.20	3.26	3.00	4.30	1.90

[a] $\bar{l}_\epsilon^*$, Critical fiber length at ultimate strain.

Table 6 $\bar{l}_\epsilon^*$ and $\% > \bar{l}_\epsilon$ Values of Both Wet and Dry Composites at 20°C ($\phi_f = 0.158$)

Composite Strain, ϵ	Dry, 20°C, $\tau = 35$ Mn/m^2			Wet, 20°C, $\tau = 10$ MN/m^2		
	0.004	0.008	0.012	0.004	0.008	0.012
$\bar{l}_\epsilon \times 10^4$ (m)	0.52	1.04	1.56	1.82	3.64	5.47
$\% > \bar{l}_\epsilon^*$	70	31.00	18.00	12.0	2.00	0

Table 7   Values of Fracture Energy for
Polyurethane Composites at 20°C

Volume Fraction, ϕ_f	Fracture Energy (MJ/m³)
0.00	4.25
0.045	2.50
0.088	2.19
0.126	1.18
0.158	.734

As test temperature rises, induced stresses in the composite are relieved, resulting in
lower interfacial strength and increased strain at failure, consequently producing an increase
in the critical fiber length and number of subcritical fibers. At 80°C ($\phi_f = 0.158$), less
than 2% fibers are longer than the critical fiber length at the breakpoint. Values of critical
fiber length for dry and wet composite samples are listed in Table 6. At 0.012 strain level
critical fiber length for a dry sample is 1.56×10^{-4} m, and 18.5% fibers are longer than
the critical fiber length, while in the wet composite all fibers are shorter than the critical
fiber length at this strain level. It can be concluded from the discussion above that the
mechanical properties of fiber-reinforced composites may be enhanced by improving either
the interfacial shear strength or by increasing the fiber length, or both. Bowyer and Bader
[39] have suggested that maximum fiber efficiency can be achieved by increasing the fiber
length from l_ϵ to $5l_\epsilon$, a further increase in fiber length being relatively counterproductive.

Processing of longer fibers is difficult and usually results in degradation of fiber length.
Moreover, during handling, fibers may develop some defects. Any defect in fibers affects
the ultimate properties of the composites. The effects vary with the length and diameter
of the fibers. The mean strength of a number of fibers in a bundle decreases with increased
length of the fibers [44]. On the other hand, improvement in interfacial shear strength
reduces the critical fiber length, resulting in enhancement in the mechanical properties at
the expense of overall fiber length, which might have adverse effects.

Toughness

The toughness of a material is the amount of work required to fracture a sample. One
way of determining this property is to measure the area under the stress–strain curves.
The values are listed in Tables 7 and 8; the incorporation of glass fibers in PU results in

Table 8   Values of Fracture Energy for
Wet and Dry Unfilled and Filled ($\phi_f = 0.158$) Polyurethane

Sample Type	Fracture Energy (MJ/m³)
Unfilled dry	4.25
Unfilled wet	3.73
Filled dry	0.734
Filled wet	0.62

decreased ultimate stress and strain, as discussed earlier. Toughness follows the same trend and continues to decrease with increased filler concentration. This decrease is because of the fact that polymers, compared to rigid fillers, are capable of absorbing more energy during fracture. The toughness decreases in the presence of filler, since on a unit volume basis, less polymer is available to absorb energy. The presence of moisture also results in a decrease in the values of toughness, because in this case the decrease in ultimate stress is not matched by an increase in ultimate strain.

DYNAMIC MECHANICAL PROPERTIES

A dynamic test is one in which the polymeric material, unreinforced or reinforced, is subjected to a force that varies sinusoidally with time. The deformation experienced by the material as a result of this force will also be sinusoidal but will lag in phase behind the applied load. The dynamic test yield data concerning the two basic quantities, the dynamic storage modulus and the loss factor or damping factor. The dynamic storage modulus provides a measure of the effective stiffness of the material and is proportional to the peak energy stored and recovered during each cycle of deformation, whereas the loss factor or damping factor is proportional to the ratio of the net energy dissipated per cycle as heat to the peak stored energy. This information is of significant value to the design engineers and also in the study of molecular transitions that occur in the polymeric materials.

The dynamic properties of polymers and composites can be directly related to the use of these materials for controlling vibrations in engineering applications such as machinery, transport, and buildings. When measured over a wide range of temperature, dynamic properties provide a convenient way of studying the different transitions occurring in a polymeric material. Knowledge of these transitions can give insight into structure–property relations and can be related to other important properties, such as impact resistance.

Over a wide range of temperature (or frequency) a polymeric material usually passes through more than one relaxation region and it is often possible to predict the nature and location of the rotational groups by systematically changing the physical or chemical structure of the material [45]. At a fixed frequency the polymeric materials pass through three basic regions of mechanical behavior with increasing temperature: (1) a glass region, (2) a glass–rubber transition region, and (3) a rubbery plateau. At low temperature the amorphous chain conformations are frozen into a rigid network and the polymers are characterized as glassy. In this region they exhibit high modulus ($G' \approx 2$ GN/m^2) and low loss. Some limited movement either in the main chain, or within side groups attached to the chain, is however possible and is responsible for one or more secondary transitions of low magnitude. Occurrence of these secondary transitions can have a significant influence on the impact properties of the glassy polymers [46]. The transition from the glassy to the rubbery region indicates the onset of long-range motions of amorphous polymer chain segments and is characterized by a very large drop in modulus to about $G' = 10^6$ N/m^2 and by a pronounced loss factor peak. The degree of cross-linking can affect the location and magnitude of this transition. In the case of a high degree of cross-linking, the mobility of long-chain segments is restricted and as a result, the transition is shifted to a slightly higher temperature. The inclusion of solid fillers may also have a similar effect on this transition [28,47]. In the rubbery plateau region the polymer behaves like an elastic solid. For non-cross-linked materials the stiffness in this region is due to entanglements, which act as temporary cross-links between the long flexible chains [48]. The modulus of permanently cross-linked rubber is proportional to the degree of cross-linking [49]. The rubbery modulus can be significantly increased by incorporation of filler particles and hard inclusions [36]. This increase is attributed partly to effective cross-linking and partly to modification of the stress distribution in the rubber matrix. To characterize the dynamic properties fully, a number of instruments, capable of being operated over as wide a frequency and/or temperature range as possible, have been developed and used to carry out fundamental studies on different materials. However, for general practical purposes there are two basic types of dynamic tests: (1) the free vibration method (torsion pendulum), in which the test piece is initially deformed, then released, and allowed to oscillate without further input of energy, and (2) the forced vibration method, in which the oscillation of test piece is maintained by external means. This method may be subdivided into those methods operating at resonance (e.g., vibrating reed) and those operating away from the resonance (e.g., Rheovibron).

The Torsion Pendulum

The torsion pendulum used in the study of fiber-reinforced PU composite is simple to use and operates at low frequency (≈ 1 Hz). It is designed on the principles similar to those of Heijboer et al. [50], in which one end of the sample is clamped to a rigid support while the other end is attached to an inertia member that is free to oscillate about the longitudinal axis of the sample. A counterweight is used to remove the tensile load on the sample. The temperature of the test piece is controlled by mounting it in an environmental chamber, and some means of measuring the amplitude of the oscillations without interfering with the damping characteristics of the system is required. Figure 9 illustrates the basic design of the torsion pendulum. The bottom end of a test piece, P, is firmly clamped to a rigid support, and the top end is free to oscillate under the influence of the internal arm, A, which is centrally suspended by a steel torsin wire, w, passing over two pulleys, C, to a counterweight, W. The torsion oscillation of the test piece is initiated by displacing the lever slightly, then rapidly returning it and locking it in its original central position. The decay of sample vibrations is monitored by a capacitative proximeter probe, B. The output from the probe is amplified and monitored on a chart recorder. A thermal chamber, T, is used to determine the dynamic properties at different temperatures. It consists of a fixed rear half and a detachable front half. The front half can easily be dismantled to clamp the test piece. The temperature of the test piece is monitored by means of a thermocouple placed near the bottomend of the sample. Initially, the temperature of the sample is lowered by pouring liquid nitrogen into the thermal chamber, and temperatures down to about $-190°C$ can be attained. The temperature of the thermal chamber is raised by providing controlled heating, and temperatures as high as 200°C can be achieved by means of two built-in heaters.

Theory

When isotropic materials are subjected to shear deformation, their dynamic viscoelastic properties in terms of stress–strain relationship may be expressed in complex form as

$$\sigma^* = G^*\gamma^* \tag{19}$$

where the shear-stress and shear-strain cycles are represented by the real parts of

$$\sigma^* = \sigma_0 \exp[i(\omega t + \delta)] \quad \text{and} \quad \gamma^* = \gamma_0 \exp(i\omega t)$$

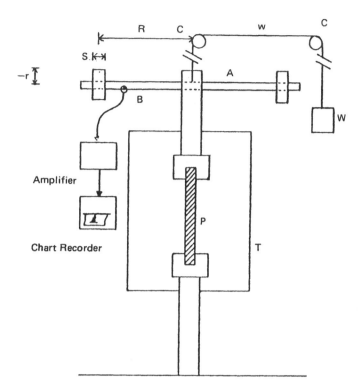

Figure 9 Schematic representation of the torsion pendulum.

respectively. Here σ_0 and γ_0 are the amplitudes of stress and strain cycle, ω is the angular frequency (S^{-1}), t is the time (s) (Figure 10), and $i = (-1)^{1/2}$. The complex shear modulus is thus given by

$$G^* = |G| \exp(i\delta)$$ (20)
$$G^* = |G| (\cos \delta + i \sin \delta)$$

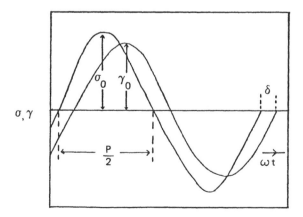

Figure 10 Time dependence of shear stress and shear strain during a low-frequency dynamic test.

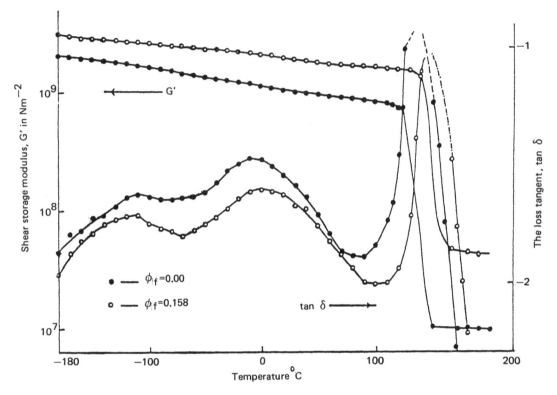

Figure 11 Plots of G' and tan δ versus temperature for dry samples of $\phi_f = 0.0$ and 0.158.

where absolute modulus $|G| = \sigma_0/\gamma_0$. Equation (20) can be resolved as

$$G^* = G' + iG'' \tag{21}$$

where

$$\begin{aligned} G' &= |G| \cos \delta \\ G'' &= |G| \sin \delta \end{aligned} \tag{22}$$

$$\tan \delta = \frac{G''}{G'} \tag{23}$$

G' and G'' are referred to as dynamic shear modulus and shear loss modulus, respectively. The quantity tan δ is termed the shear damping factor.

Figure 11 shows the dynamic behavior of the PU matrix and composite ($\phi_f = 0.158$). PU matrix behaves in a manner that is typical of amorphous polymers. The glassy modulus is ≈ 2.0 GN/m^2 and drops slowly until the vicinity of T_g, when there is a sharp fall in value to 10 MN/m^2 with the temperature.

Curves for tan δ against temperature show three peaks, which according to the prevalent nomenclature can be labeled as α, β, and γ peaks in order of decreasing temperature. It is not possible for highly cross-linked polymers to ascertain the quantitatively exact value of tan δ at glass transition temperatures. In this region tan δ increases too rapidly for accurate measurements, and as damping increases, the number of oscillations decreases to less than 1. Therefore, it was assumed (by extrapolation) that the most prominent α-peak of the PU matrix associated with the glass transition temperature occurs at 133°C. The α-transition in the PU networks originates from the motion of main-chain segments between cross-links. The β-relaxation is a peak occurring at −10 to 0°C. This transition can be attributed to, first, the plasticizing effect of absorbed molecules such as water. Since the test pieces used for dry volume fraction series were throughly dried, the possibility of moisture as a cause for the occurrence of this peak is remote. Furthermore, the addition of water produces only a nominal increase in the magnitude of this peak. The second cause of the transition is the restricted motion of either urethane or phenyl group [51] in the vicinity of allophanate groups. Allophanate groups are formed by the side reaction of free isocyanate groups, particularly at higher extents of reaction after gelation [52,53]. In a study of MDI/LHT240, Cawse [52] has observed such a peak at −6 to −3°C (1 Hz) and attributed it to allophanate cross-links.

The γ-transition is a broad peak occurring at −120 to −100°C and is possibly a result of two mechanisms. The lower end of this peak can be assigned to the methyl side groups [54], while the higher-temperature component can be assigned to the motion of hydrogen-bonded species. McCrum et al. [55] and Saur [51] have reported such peaks in nylon at about −75°C (1 Hz) and −25°C (103 Hz), respectively. Independent studies in PU have revealed the occurrence of such peaks at −88°C (1 Hz) [56], at −63°C (3.6 Hz) [57], at −78°C (1 Hz), and −83 to −75°C (1 Hz) [52]. Since the presence of moisture causes a significant increase in the intensity of this peak, it may be due to some motion of H-bonded urethane groups, such is facilitated when water is present to destroy other intermolecular bonds.

Mechanical damping may increase or decrease with the incorporation of filler [57]. Manson and Sperling [47] suggested that the incorporation of solid elastic filler, at reasonable levels of loading, usually decreases the level of damping. Nielsen [36] proposed a relation accounting for the decrease in damping with the increase in the volume fraction of filler:

$$\tan \delta_c = (1 - \phi_f)\tan \delta_m \qquad (24)$$

It is assumed here that additional damping at the filler–matrix interface is absent and that viscoelastic losses in the filler, as compared with those in the polymer, are negligible, and therefore $\tan \delta_c/\tan \delta_m (1 - \phi_f)$ should equal 1. Figure 12 shows a plot of Eq. (24) and experimentally obtained values of $\tan \delta_c$. All the experimental points lie below the theoretical curve, indicating that the presence of reinforcement has led to a greater reduction in $\tan \delta_c$ than expected by Nielsen's relation. It has been suggested [58] that the increased reduction is due to the formation of a shell of immobilized polymer around each filler particle, restricting any small main-chain motion within this shell. The formation of an immobilized shell around the filler particles is generally attributed to the following factors:

1. Mismatch of the coefficients of thermal expansion, which can cause thermal stresses within the polymer matrix because the matrix tends to squeeze around the filler particles while cooling down from the curing temperature

2. Interaction between the polymer and the filler, which leads to absorption of the polymer on the filler surface, persumably creating some sort of bonding between the filler particles and the polymer matrix

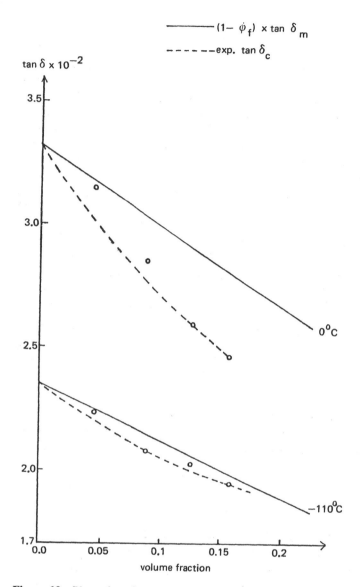

Figure 12 Plots of tan δ versus ϕ_f for dry composites.

Considering the later factor, absorption of the polymer on the filler surface can give rise to a change in glass transition temperature. Adequate reviews are available in literature regarding the shift in glass transition temperature with the incorporation of filler [28,47].

A polymer filler system is basically a heterogeneous system, and the motion of the polymer chain is restrained due to the interaction between the filler, and polymer. The interaction is such that the polymer is absorbed at the surface of the filler particle (formation of hydrogen bonding or primary bonding), and the nature of the absorption is different from that of the original polymer phase. The mechanism is considered to be responsible for reinforcing effects, such as increase in modulus and change in glass transition temperature.

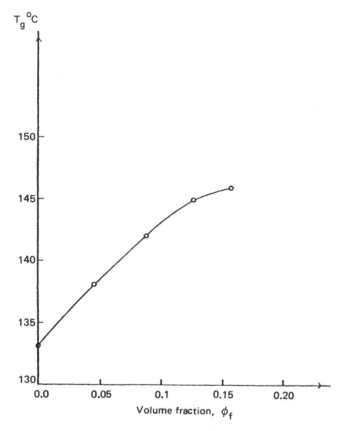

Figure 13 Plots of T_g versus volume fraction.

Figure 13 shows a plot of glass transition temperature against volume fraction. For $\phi_f = 0.158$, an increase of about 13°C over the original matrix was observed, but this increase in the glass transition temperature is not a linear function of volume fraction.

With the increase in temperature, in Figure 11, the values of shear modulus (G') decrease gradually below the glass transition region and then fall sharply when the temperature is raised above the glass transition region. Incorporation of filler has led to an increase in the shear modulus for all temperatures. The increase is less pronounced in the region below the glass transition temperature (around 80%) than in the region above the glass transition temperature, where the relative shear modulus (G_c/G_m) for $\phi_f = 0.158$ has attained a value of about 5.

A plot of relative shear modulus against volume fraction at four different temperatures is shown in Figure 14, which clearly illustrates the trend of increase in the relative shear modulus with temperature increase. The effect of reinforcement on the shear modulus is very similar to that of tensile modulus. Therefore, the same generalized equation that was used for tensile modulus can be applied here for theoretical predictions of relative shear modulus. After changing the symbols to shear notation, the equation becomes

$$\frac{G'_c}{G'_m} = \frac{1 + AB\phi_f}{1 - B\psi\phi_f} \tag{25}$$

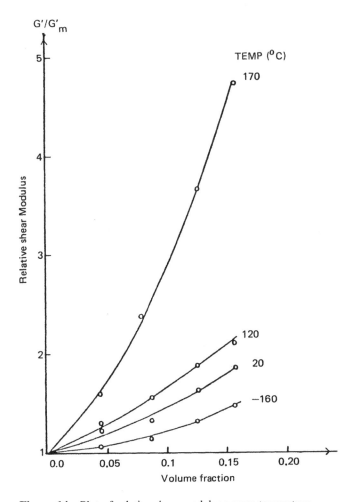

Figure 14 Plot of relative shear modulus versus temperture.

where $A = K - 1$, $B = (G'_f/G'_m - 1)/(G'_f/G'_m + A)$ and $\psi = 1 + \dfrac{(1 - \phi_{max})}{\phi^2_{max}} \phi_f$.

The value of maximum packing fraction ($\phi_{max} = 0.28$) has been determined experimentally [59]. Figure 15 shows a plot of relative shear modulus G'_c/G'_m against temperature, together with the theoretical predictions of Eq. (25). In the glassy region experimental values are in good agreement with the values predicted by Eq. (25), at least for lower volume fractions. In the case of maximum volume fractions ($\phi_f = 0.158$), the theoretical values are higher than those of experimental relative shear modulus values at lower temperatures (-160 to $100°C$), but from 100 to 120°C the agreement between theoretical and experimental values is quite close even for this volume fractions. As the temperature is raised above the glass transition region, the experimental values of the relative shear modulus are much higher than those of theoretical values. Equation (25) does not take into consideration the effects of temperature on the shear modulus of glass fiber G_f and the factor A. The value of G'_c/G'_m is expected to be nearly constant in the region below the glass transition temperature, or at the most may decrease with temperature

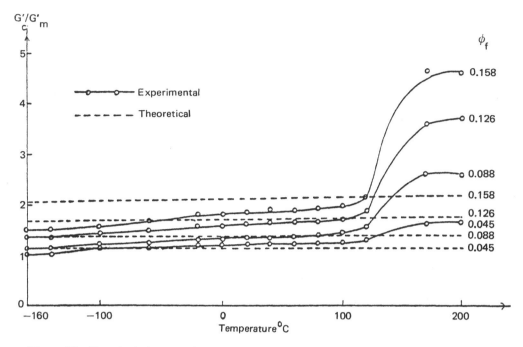

Figure 15 Plot of relative modulus G'_c/G'_m versus temperature.

decrease. But, in practice, the values of G'_c/G'_m often decrease noticeably as the temperature is lowered [28].

Effect of Moisture on Dynamic Behavior

Figure 16 shows plots of tan δ against temperature for unfilled wet and dry matrix. Like the dry sample, the wet samples also show three peaks, the α, β, and γ peaks, in order of decreasing temperature. As mentioned earlier, it was not possible to ascertain the exact quantitative value of tan δ for these polyurethane networks in the glass transition region. Therefore, the effect of moisture on the magnitude of the α-peak (glass transition peak) is not clear. It may be assumed (by extrapolation) that the presence of moisture may tend to increase the intensity of this peak. However, there is a downward shift of about 12° in the temperature at which this peak occurs, which can be attributed to the plasticization phenomenon. The presence of moisture has slightly increased the intensity of the β-peak without lowering the actual temperature of its occurrence. This peak is much broader than the β-peak for dry samples. The possible reason could be that while acting as plasticizer, moisture helps ease the restriction on the mobility of network chains. The effect of moisture on the γ-peak is most pronounced. Moisture also has a significant broadening effect on the γ-peak, but the actual temperature of its occurrence has not changed. As mentioned before, one of the mechanisms responsible for the occurrence of this peak could be the motion of H-bonded urethane groups. The increase in magnitude may be explained by the fact that the presence of water facilitates the motion of H-bonded urethane groups by destroying or weakening the other intermolecular bonds. This fact can be supported by the broadening effect of moisture on this peak. The lower end or high-temperature maximum of this peak

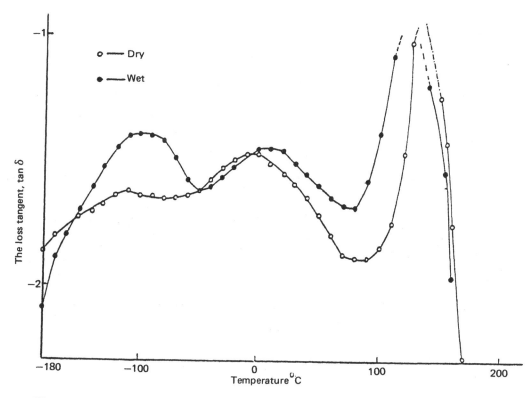

Figure 16   Effect of moisture on tan δ of the unfilled matrix.

is around −80°C, the region where H-bonded urethane groups are believed to be operative in the presence of moisture. Figure 17 shows a plot of tan δ versus temperature for dry and wet polyurethane composites. Apart from lower actual values of tan δ, the effects of moisture on the reinforced polyurethane samples are similar to those observed in unfilled samples. Figure 18 shows a plot of G' against temperature for dry and wet matrix. A similar plot for reinforced dry and wet samples is shown in Figure 19. The effect of moisture on the unfilled and reinforced is almost similar. A little increase in the shear modulus of wet samples over dry samples can be assigned to the contribution of ice, which has a shear modulus much higher than the polymers and so has a reinforcing effect. The shear modulus of both unfilled and reinforced wet samples is less than the dry samples in the rubbery region. Figure 20 shows a plot of relative shear modulus for wet composite together with the theoretically predicted values. The trend is almost similar to that of dry composites.

THERMAL EXPANSION PROPERTIES

The thermal expansion coefficients of polymers are significantly higher than those for metals. In certain applications this does not present a problem and the variations can be accommodated by design and attachment methods. But for those applications where polymers are used to replace metals, or are joined with metals, fabrication would be simplified by reducing the mismatch in the thermal coefficient of expansion. Fillers or

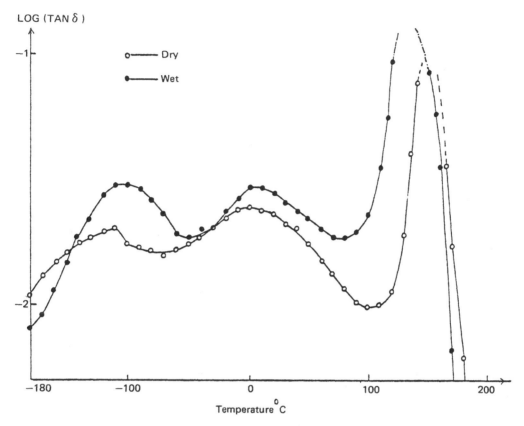

Figure 17 Effect of moisture on tan δ of the reinforced polyurethane $\phi_f = .158$.

reinforcement like glass fibers have much lower thermal coefficients of expansion than polymers; for example, the thermal coefficient of expansion of E-glass is $4.9 \times 10^{-6} \,^{\circ}C^{-1}$ (0 to 100°C), compared with a typical polymer of $10^{-4} \,^{\circ}C^{-1}$ [60]. Therefore, incorporation of such fillers into a polymer can produce a composite system of acceptable thermal expansion behavior as well as achieving enhancement of mechanical properties. In a review of high-modulus RIM systems [61] it has been indicated that the thermal coefficient of expansion of polyurethanes reduces with the incorporation of milled glass fibers. Sufficient dimensional stability can be achieved at 20 to 30% by weight-milled glass fiber loadings regardless of the initial value of the thermal coefficient of expansion or the elastic modulus of base polyurethane system.

Many different methods [62] and variations of methods such as push-rod dilatometers, the twin-telemicroscope method, interferometers, and volumetric methods, have been developed to measure thermal expansion. The choice of method may depend on the material to be measured, the amount of material available, the temperature range of measurement, and the type of information required. The push-rod method [63] for measuring thermal expansion is experimentally simple, reliable, and easy to automate. One of the common sources of error in using this method is the measurement of temperature. All too often the temperature measured is not the temperature of the specimen. This is especially true

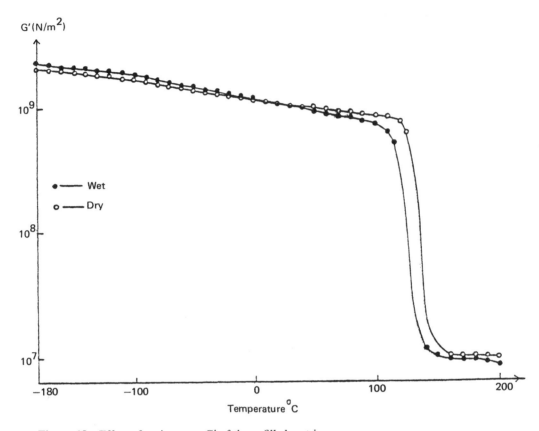

Figure 18   Effect of moisture on G' of the unfilled matrix.

in measurements made while the temperature is changing. This can be minimized by using two thermocouples, attached close to either end of the sample, and by taking care to ensure that their junctions and the specimen are at the same temperature.

Expansion Equipment

Several variations of the push-rod method can be used to measure the thermal expansion of a sample, but the basic principle is essentially the same. The expansion of the sample is transferred out of the heating chamber to an extansometer by means of rods (or tubes) of a stable material. The apparent change in length is calculated from the difference between the extansometer readings at two different temperatures. Details of the expansion equipment are shown in Figures 21 and 22. Equipment is required to follow the change in length of the sample as the temperature is changed and recorded. It is accomplished by fitting a sample holder with a dial gauge to measure the length changes, two thermocouples to measure temperatures, and a heating coil connected to a water circulating device to provide the necessary periodic heat inputs. By using this assembly, the change in length can easily be followed as the temperature is raised. The Circon unit in Figure 21, a thermostatically controlled liquid circulator, pumps the water through the coil surrounding the sample holder and is used to increase the temperature of the sample being

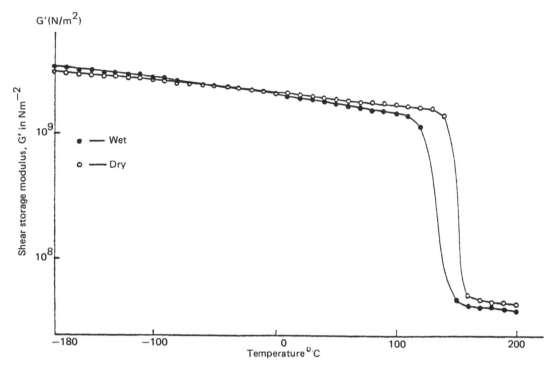

Figure 19 Effect of moisture on G' of reinforced polyurethane ($\phi_f = 0.158$).

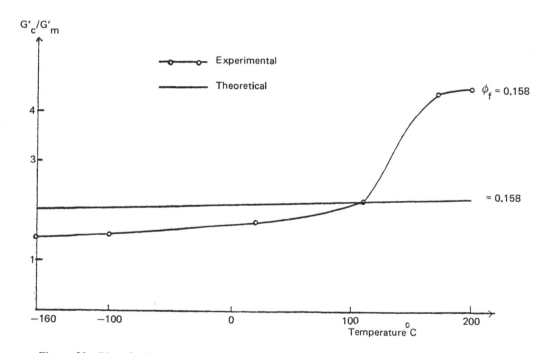

Figure 20 Plot of relative modulus versus temperature for wet composite ($\phi_f = 0.158$).

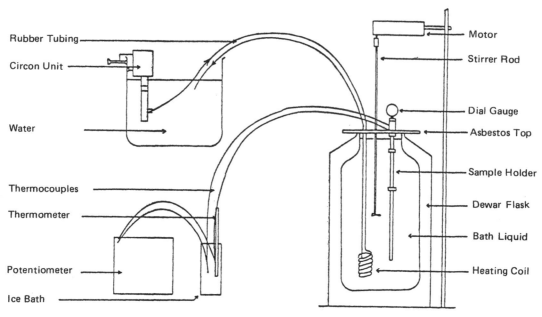

Figure 21 Apparatus for expansion measurements.

tested. After passing through the heating coils, water is reheated in the Circon unit. A digital thermometer, together with two built-in thermocouples, is used to measure the temperature of the bath in which the sample holder is immersed. The thermocouples (T) in Figure 22 are positioned at the opposite ends of the sample (S) to provide an indication of the temperature uniformity of the sample. The sample holding device, with the sample in place, is shown in Figure 21. The sample (s) rests firmly on the base plate (B), which is attached to the stainless steel frame rods (F). The frame rods are fixed to the asbestos or of some suitable material stand (A). A dial gauge (D) supported by an asbestos stand (C) registers the movement of the stainless steel plunger (P). The bottom end of the plunger rests on the top end of the sample. When the device is in use, the section below the bottom of the asbestos top is immersed in a dewar flash containing the bath liquid. Upon heating the apparatus, the frame rods expand so that the base plate moves further from the asbestos top. Since the location of the dial gauge is fixed, this movement causes a decrease in the dial gauge reading. The sample rests on the base plate so that when it is heated, the top of the sample moves toward the asbestos top and thus causes an increase in the dial gauge reading. The expansion of the plunger also causes an increase in the dial gauge reading, since the lower end of the plunger rests on the top of the sample. Therefore, the deflection registered by the dial gauge upon heating the apparatus is the sum of the positive contribution of the sample and the plunger, and the negative contribution of the frame rods.

The plunger and frame rods are cut from the same material and thus should have the same expansion coefficients. The plunger and upper section of the frame rods are exposed to the same environment, and as such, the expansion of the plunger should be identical to that of an equal length of the upper section of the frame rod (W to Y on Figure 22). However, since these expansions register in an opposite manner on the dial gauge the

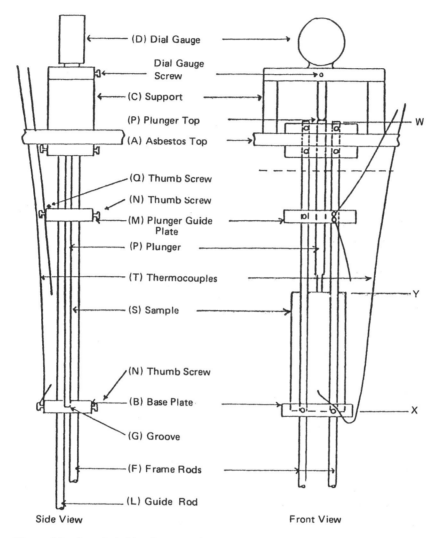

Figure 22 Sample holder for expansion measurements.

effect of one should cancel the other. If this is the case, the dial gauge deflection is simply the difference between the contributions of the polymer sample and that of a section of the frame of the same length as the polymer (Y to X in Figure 22). Therefore, absolute values of the linear coefficient of expansion of the polymer samples can be determined by adding the expansion coefficient of the steel frame to the apparent values of the linear coefficient of expansion of samples. Samples of various lengths and thicknesses can be accommodated by the apparatus. The base plate, normally fixed to the frame rods by screws (N), can be moved to accommodate different lengths of samples by loosening these screws. Two thumb screws (Q) are provided on both the base plate and the plunger guide plate (M) by which separation of the guide rods (L) from the frame rod may be adjusted.

Semiempirical Approach to Predicting Coefficient of Thermal Expansion of Filled Polymer Systems

When thermal stresses arise in solids because of the differences between the expansions of the matrix and filler, the expansion of the composite will be a function of the expansivity and the elastic properties of the individual components of the composite. In general, the thermal behavior of the composite is a complex function of the thermal and elastic properties of the components. A number of equations are available in the literature for calculating the coefficient of thermal expansion of a composite from material constants of the components [64–67]. The various equations often predict quite different values for the coefficient of expansion of a given composite. Some experimental data agree with one equation, while other data agree with a different equation [4]. However, nearly all the equations predict coefficients of expansion which are less than that which would be calculated from the simple "rule of mixtures," because of mechanical restraints of the matrix by the filler particles. In composites where elastic properties of the components are nearly equal, the simple rule of mixtures [68] can, within limits, be applied to predict the thermal coefficients of polymer filler systems. If the filler particles are rodlike in shape and are randomly oriented, the Thomas equation [66], or logarithmic rule of mixtures, can lead to good results. If the constituents of a system occupy volume fraction ϕ_i and have a cubical thermal expansion coefficient β_i, the cubical thermal expansion coefficient β_s of the system is given by

$$\beta_s^c = \sum_i \phi_i \beta_i^c \tag{26}$$

where c is a constant which may vary from -1 to $+1$, depending on the particular system. Thus for a binary system,

$$\beta_s^c = \phi_m \beta_m^c + \phi_f \beta_f^c \tag{27}$$

where ϕ_m and ϕ_f are the volume fraction of the polymer and filler, respectively, and β_m and β_f the cubical thermal expansion coefficient of polymer and filler, respectively. For an isotropic polymer system the cubical expansion coefficient is approximately three times that of the linear coefficient. Hence if α_m and α_f are the linear thermal expansion coefficient of the polymer and filler, respectively, the expression applied to linear coefficient of thermal expansion would be

$$\alpha_s^c = \phi_m \alpha_m^c + \phi_f \alpha_f^c \tag{28}$$

For $c = 0$, Eq. (28) becomes

$$1 = \phi_m + \phi_f \tag{29}$$

For $c = 0$, Eq. (28) becomes

$$\alpha_s = (\phi_m \alpha_m^c + \phi_f \alpha_f^c)^{1/c} \tag{30}$$

Hence Eq. (30) can be used to determine α_s for all values of $c \neq 0$. However, for $c \ll 1$, Eq. (28) assumes a simple form by using Taylor expansion of α_s^c, α_m^c, and α_f^c, around $c = 0$. The Taylor expansion of a function $f(c)$ around $c = 0$ is given by

$$f(c) = f(0) + c\left[\frac{df}{dc}\right]_{c=0} + \frac{c^2}{2!}\left[\frac{d^2f}{dc^2}\right]_{c=0} + \cdots + \frac{1}{n!}\left[\frac{d^nf}{dc^n}\right]^{c=0} \tag{31}$$

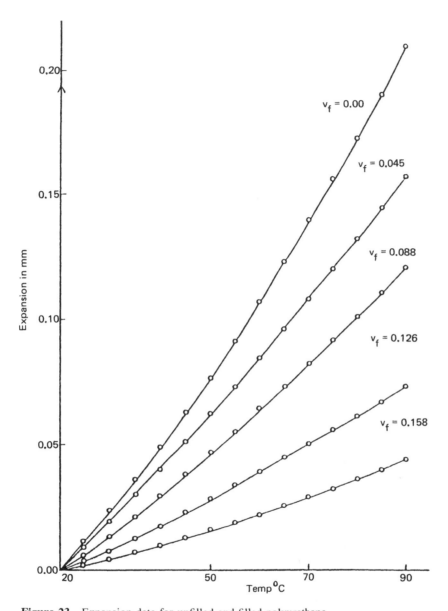

Figure 23 Expansion data for unfilled and filled polyurethane.

If $f(c) = \alpha_s^c$, then

$$\alpha_s^c = 1 + c \ln \alpha_s + \frac{c^2}{2}(\ln \alpha_s)^2 + \cdots \tag{32}$$

where $d\alpha_s^c/dc = \alpha_s^c \ln \alpha_s$.

If the first two terms of the expansion in Eq. (32) and the similar expansions for α_m^c and α_f^c in Eq. (28) and using Eq. (29) give

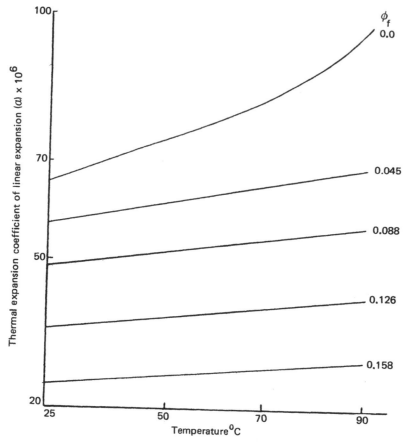

Figure 24 Plot of coefficient of thermal expansion versus temperature.

$$\ln \alpha_s = \phi_m \ln \alpha_m + \phi_f \ln \alpha_f \tag{33}$$

Equation (33) is valid only if $c \ll 1$.

The solution of Eq. (28) for extreme limits (i.e., $c = +1$ and $c = -1$) is for $c = -1$,

$$\frac{1}{\alpha_s} = \frac{\phi_m}{\alpha_m} + \frac{\phi_f}{\alpha_f} + \cdots \tag{34}$$

and for $c = +1$,

$$\alpha_s = \phi_m \alpha_m + \phi_f \alpha_f \tag{35}$$

Expansion data for polyurethane matrix and reinforced polyurethane are shown in Figure 23 and represent a curve rather than a straight line. The corresponding coefficients of thermal expansion for polyurethane matrix and composites are plotted in Figure 24 and listed in Table 9. As the filler concentration increases, the coefficient of linear thermal expansion decreases and tends to become more linear. The filler, E-glass fiber, has a much lower coefficient of thermal expansion than the polyurethane matrix over the range

Table 9 Coefficient of Thermal Expansion (α) of Unfilled and Filled Polyurethane (v_f = 0.00 to 0.158)

Temperature (°C)	Coefficient of thermal expansion, $\alpha \times 10^{6} °C^{-1}$, for volume fraction:				
	0.00	0.045	0.088	0.126	0.158
25	66.21	57.04	48.75	35.72	24.97
30	67.08	58.49	50.00	37.50	25.43
35	68.76	59.70	51.47	38.99	26.16
40	71.26	60.68	52.48	39.69	26.63
45	74.57	61.41	52.92	39.73	27.00
50	77.86	62.06	52.93	39.79	27.32
55	80.24	62.75	53.15	39.87	27.59
60	81.74	63.49	53.58	39.95	27.81
65	82.34	64.29	54.21	40.10	27.99
70	83.07	65.13	54.89	40.34	28.19
75	84.95	66.01	55.47	40.69	28.47
80	87.98	66.91	55.95	41.16	28.83
85	92.16	67.85	56.32	41.80	29.28
90	97.49	68.81	56.59	42.58	29.80

of temperature 20 to 90°C. The coefficient of linear expansion of E-glass is almost constant. Therefore, as the volume fraction of the filler increases, the coefficient of linear thermal expansion becomes less dependent on the temperature. Figures 25 to 27 are plots of the logarithmic rule-of-mixtures equation for different temperatures. For each temperature there is a pair of curves drawn with upper straight portions and the curved lower portions

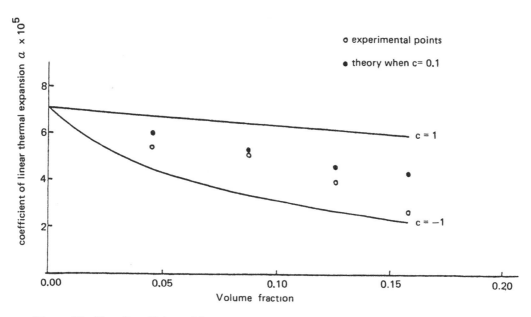

Figure 25 Plot of coefficient of thermal expansion versus volume fraction at 40°C.

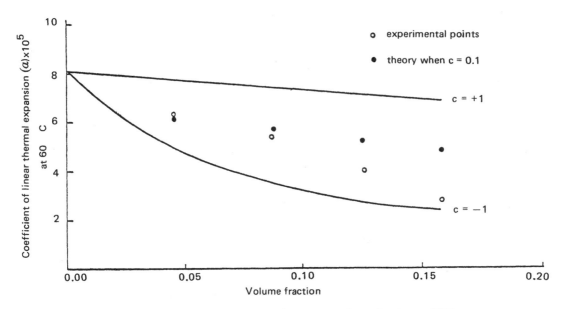

Figure 26 Plot of coefficient of thermal expansion versus volume fraction at 60°C.

representing the theoretical values when the exponent C in Eq. (28) is $+1$ and -1, respectively. The experimental values at all filler volume fraction levels fall within the areas enveloped by their respective theoretical upper and lower limits. Hence the experimental data do follow the theoretical curves. Therefore, a modification of the simple rule of mixtures that would include the volume fraction and thermal expansion of the components

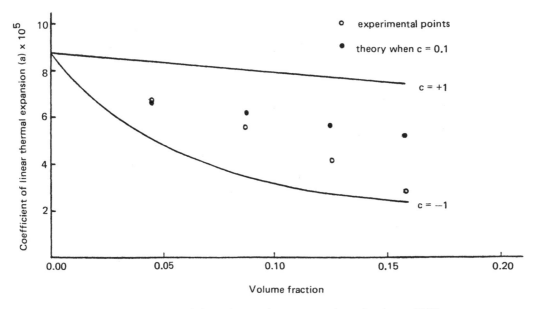

Figure 27 Plot of coefficient of thermal expansion versus volume fraction at 80°C.

of polymer filler system seem plausible as a method for predicting thermal expansion of filled polymeric system.

IMPACT PROPERTIES

Toughness is a quantitative description of how easy or difficult it is to break a specimen. To quantify the parameter and hence compare materials, it is necessary to determine the work done (i.e., the energy required to cause fracture). One way of determining this property is to compute the area under load–extension curves obtained by flexural or tensile tests. However, these values are of limited use if calculated from the area under a conventional low-speed tensile or bending test curve because they apply only to those conditions of tests. In practical situations, design engineers are interested primarily in toughness under conditions of rapid deformations. This has led to the development of various impact tests where the specimens are subjected to rapid blows. These tests are devised in such a manner that the energy required to cause fracture is recorded, rather than force to break. Standard impact tests for plastics can be categorized as follows:

1. *High-speed tensile impact tests* [69]. In these tests one end of the specimen is fixed to a pendulum of the impact testing machine and the other end is clamped to the cross-head. To bring about fracture the moving cross-head is suddenly stopped while the pendulum is still moving.

2. *Pendulum tests* [70,71]. In these tests a blow is delivered on a clamped or supported specimen through the release of the arm or pendulum or hammer and the energy absorbed by the specimen during the fracture is recorded. These tests can be carried out in tension or bending (Charpy or Izod) in a freely supported beam, whereas an Izod test specimen is supported at one end only, cantilever style.

3. *Falling weight tests* [70,72]. These tests are carried out on a rectangular sheet or disk. In these tests the specimen is rested on a support and a striker is dropped from a given height on the specimen at midpoint.

Impact tests provide a ready measure of the energy required to break a specimen. Such information can be quite useful for predicting the behavior of a particular type of material under impact or shock conditions. The main disadvantage of these tests is that they provide information of only one quantity, overall impact energy, which is usually reported in terms of impact strength. Impact strength is defined either in terms of energy per unit width of notch or energy per unit area under the notch required to break the specimen. The measured impact energy contains various energy contributions which cannot be separated. These are due to elastic deformation, viscoelastic process, plastic flow, and deformation of the material at the tip of propagating crack. Nonetheless, this drawback can be overcome by instrumenting the impact machine by attaching a load sensor on the top of the striker. The load time response can then be analyzed with respect to various deformations occurring in the specimen on impact. The overall impact energy or total energy absorbed in the fracture process is made up of two terms, the energy required to initiate a crack and the energy required to propagate that crack. The first term is dependent on random flaws such as scratches and may have a tremendous effect on the toughness and impact strength of the material. These flaws are largely responsible for the scatter often found in impact data on unnotched specimens. To overcome this problem, impact tests are usually carried out on specimens having a central V-shaped notch on the edge opposite the applied force. Notches acts as stress raisers and consequently localize the fracture and limit the spread of the yield zone over a narrow range. Then the total energy required to break the specimen will consist almost entirely of the energy required to propagate the crack, and more reproducible results can be expected. The energy necessary for the unit area crack growth is known as the critical strain energy release rate.

Impact Testing

Conventional Charpy and Izode impact tests, using a limited pendulum apparatus, give only a measure of the total energy involved in the fracture process and give no indication as to the behavior of the material during impact. To obtain more detailed information on the complete impact event, a fully instrumented machine capable of being used for a wide range of polymeric materials and composites is required and is shown in Figure 28. The apparatus is a drop weight device in which the striker may fall under gravity or be propelled at higher velocity by an impulse from a spring-loading system, at velocities in the range 4 to 13 m/s. Information on the complete fracture process occurring during the impact event is gathered in the form of force/time data from a quartz force transducer situated on the striker attached to the cross-head. The output from the transducer is fed through a charge amplifier into a transient recorder, which stores the information over the short duration required for the impact, usually less than 5 ms. This stored force–time information can be displayed continuously as a trace on an oscilloscope, and a permanent record of the impact may be produced by transferring and/or by processing the information in a microcomputer. The computer gives crude graphical representation of the data, and if

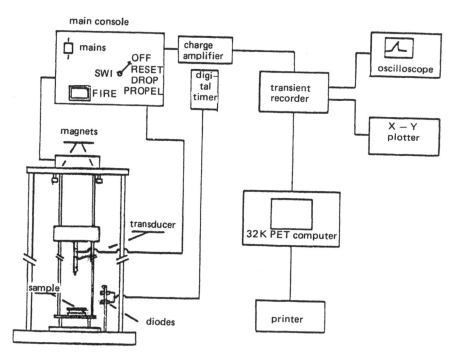

Figure 28 Instrumented impact tester: PET computer and electronics equipment for control of mechanical apparatus and data handling.

properly programmed can compute essential fracture parameters. The cross-head weighs 3.514 kg and is made of welded steel sheet. The movement of the cross-head at the high velocities is facilitated by means of self-lubricating bearings. Before firing, the cross-head is held by the electromagnets and can be fired from the control panel by switching off the power supply to the magnets. The force-time output from the transducer (attached with striker) is fed into the charge amplifier (Kistler 5007). The output from the charge amplifier is fed into a transient recorder (Datalab DL 901), which continuously samples data and stores them in 1024 locations (words), each of 8 bits. The recorder is usually operated in a "pretrigger" mode so that the impact itself acts as a trigger, freezing the memory. Consequently, some preimpact output is recorded, enabling a baseline to be established. The sweep time, over which the 1024 words are sampled, can be varied between 5 ms and 200 s.

The stored trace of force versus time is continuously displaced on an oscilloscope (Telequipment DM64). "Rearming" the transient recorder removes the data from the memory and from the oscilloscope. A permanent output of the impact event may be produced by taking the separate output from the "plot" socket of the transient recorder to an x-y plotter, when the 1024 words are output over a period of 20 s.

To analyze the transferred data, BASIC routines are available that could read words in specified memory locations and carry out the necessary calculations. A flow diagram of the procedure for the analysis and storage of impact data is shown in Figure 29.

Theory of Fracture Mechanics

The fracture energy measured in impact tests depends on the specimen geometry and therefore cannot be regarded as a material property. To overcome this problem, impact strength is often expressed as the specific fracture energy. The specific fracture energy is calculated by dividing the measured fracture energy by the cross-sectional area of the fracture ligament. This method is not entirely satisfactory since the specific fracture energy is strongly dependent on the length of the notch. It decreases with the increase in A/D, the ratio of the notch length to the width of the sample [73,74]. Williams et al. [75–79] have attempted to explain the fracture mechanics approach, which takes into account the notch length and specimen geometry. These workers have used the strain energy release rate G_c, instead of specific fracture energy, as the true parameters that characterizes the fracture toughness of materials.

If a force P is applied to a specimen that contains a crack of length a, and the points of application of force have been displaced by a distance Δ, the extension of the crack will result in a change in the stored elastic energy U of the specimen and an amount of work dw will be done on the system by the external force, and the increase in the stored elastic energy of the system must be the energy G available for the process of fracture. Hence

$$G\,da = d\omega - du \tag{36}$$

Now $d\omega = P\,d\Delta$ for a linear elastic system

$$U = \tfrac{1}{2}P\Delta = \frac{P^2C}{2} \tag{37}$$

where $C = \Delta/P$ is the compliance of the specimen. Substituting Eq. (37) in Eq. (36) yields

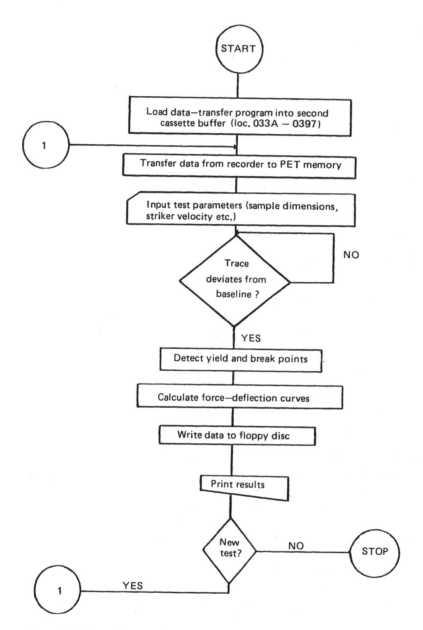

Figure 29 Flow diagram of procedure for the preliminary analysis and storage of impact data.

$$G\,da = P\,d\Delta - \tfrac{1}{2}(P\,d\Delta + \Delta\,dP)$$
$$= \tfrac{1}{2}(P\,d\Delta - \Delta\,dP) \tag{38}$$

Since $C = \dfrac{\Delta}{P}$,

$$\frac{dC}{da} = \frac{P\,d\Delta/da - \Delta\,dP/da}{P^2}$$

whence

$$G = \frac{1}{2}\frac{P^2 dC}{da} \tag{39}$$

According to the linear elastic fracture mechanics (LEFM), the fracture will occur when the value of G is equal to the critical strain energy release rate G_c:

$$G = G_c \tag{40}$$

Combining Eq. (40) and (39), we get

$$G_c = \frac{1}{2}\frac{P^2 dC}{da} \tag{41}$$

and for a specimen of uniform thickness B the critical strain energy release rate G_c is given by

$$G_c = \frac{P^2}{2B}\frac{dC}{da} \tag{42}$$

The critical stress intensity factor K_c is related to G_c as

$$G_c = \frac{K_c^2}{E'} = Y^2 \sigma^2 \frac{a}{E'} \tag{43}$$

where E' is the reduced Young's modulus and is represented by $E/1 - v^2$ for plane strain and Y is a function of a/D.

For three-point bending specimens the stress σ can be represented as

$$\sigma = \frac{3PS}{2BD^2} \tag{44}$$

Substituting Eq. (44) into (43) and equating Eq. (41) with (42),

$$\frac{dC}{da} = \frac{9S^2}{2BD^2 E'}\frac{a}{D}Y^2 \tag{45}$$

or

$$C = \frac{9S^2}{2BD^2 E'}\int Y^2 \frac{a}{D}d\frac{a}{D} + C_o \tag{46}$$

where $C_o = S^3/4E'BD^3 = \Delta/P$ is the specimen compliance at zero crack length. Equation (46) can be written as

$$C = \frac{9S^2}{2BD^2 E'}\left[\int Y^2\left(\frac{a}{D}\right)d\left(\frac{a}{D}\right) + \frac{S}{18D}\right] \tag{47}$$

Now the energy absorbed by the specimen during fracture W^* is expressed as

$$W^* = \frac{1}{2}P\Delta$$

or

$$W^* = \frac{1}{2}P^2 C \tag{48}$$

Substituting the value of C from Eq. (47) into (48) and rearranging yields

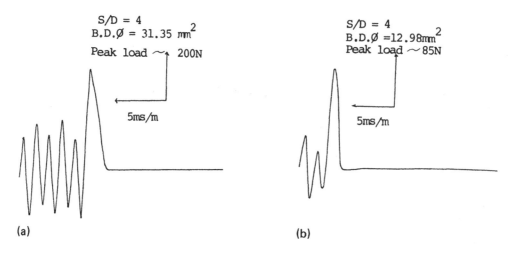

Figure 30 Impact load–time curve for unfilled polyurethane with values of *S/D* and *BD*ϕ as shown.

$$W^* = G_c BD\phi \tag{49}$$

where

$$\phi = \int \frac{Y^2(a/D)\ d(a/D)\ +\ S/18D}{Y^2} \frac{a}{D} \tag{50}$$

B is the specimen thickness and *D* is the specimen width. ϕ is a geometrical correction factor which is determined as a function of *a/D*. A plot of *W** versus *BD*ϕ should give

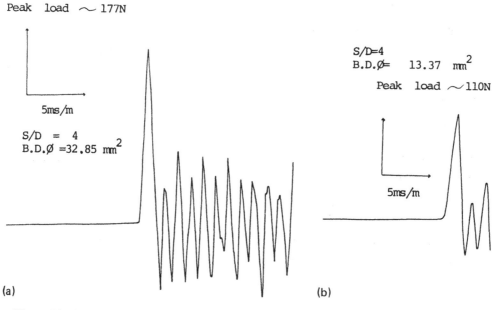

Figure 31 Impact load–time curve for composite of δ$_f$ = 0.045 with values of *S/D* and *BD*ϕ as shown.

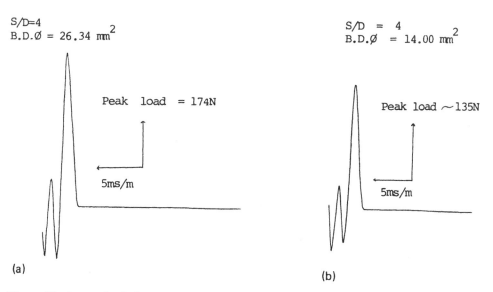

Figure 32 Impact load–time curve for composite of $\delta_f = 0.088$ with values of *S/D* and *BD*ɸ as shown.

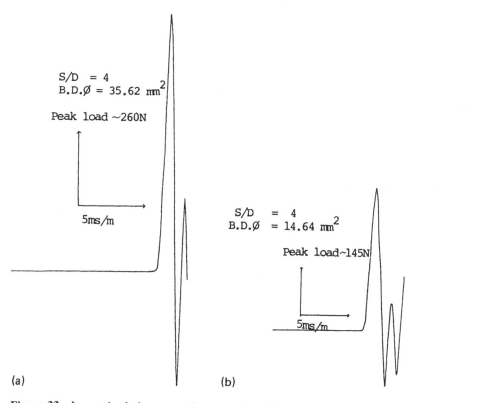

Figure 33 Impact load–time curve for composite of $\delta_f = 0.126$ with values of *S/D* and *BD*ɸ as shown.

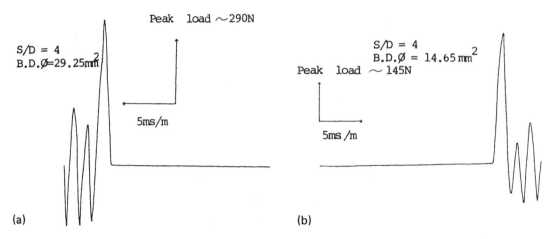

Figure 34 Impact load–time curve for composite of $\delta_f = 0.158$ with values of *S/D* and *BD*ϕ as shown.

a straight line with slope G_c. Here G_c is a constant for any given material and may be used to compare the impact strength of different composite systems. For materials that undergo brittle fracture, it is possible to evaluate the critical strain energy release rate, G_c, as a material parameter which is independent of sample dimension and notch depth.

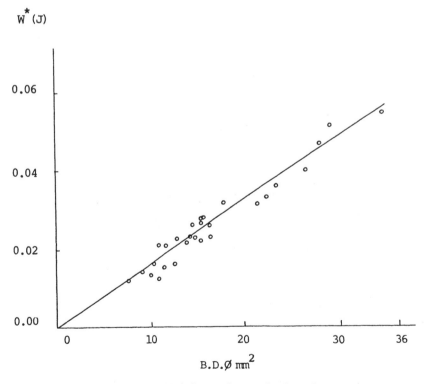

Figure 35 Plot of *W** versus *BD*ϕ for specimens of polyurethane.

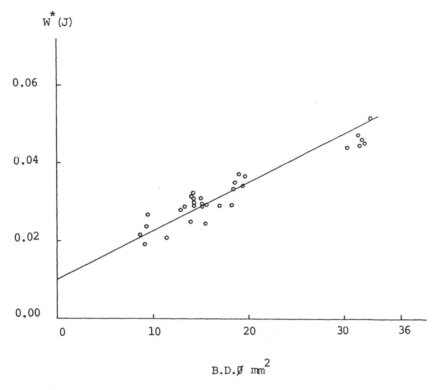

Figure 36 Plot of W^* versus $BD\phi$ for reinforced specimens of $\phi_f = 0.045$.

Critical strain energy release rate, G_c, is evaluated as the slope of a plot of the total fracture energy, W^*, the area under the force–deflection curve, versus the sample geometry factor BD. The parameters B and D are the breadth and depth of the specimen, and ϕ depends on B and D, sample length and notch depth. Thus W^* can be evaluated for a series of samples notched to different depths, giving a series of values of $BD\phi$. Figures 30 to 34 show the representative load–time curves for unfilled and reinforced polyurethane. The time to fracture varies with notch depth: specimens with small notch depth require more time to fracture. Correspondingly, the area under the load–time curve, a measure of the total energy required to fracture a specimen and the sample geometry factor $BD\phi$, also varies with notch depth. Plots of total energy, W^*, against $BD\phi$ for polyurethane matrix and composites are shown in Figures 35 to 39. According to theory, a plot of W^* versus $BD\phi$ should give a straight line passing through the origin. This is the case only for unfilled polyurethane matrix; plots of W^* against $BD\phi$ for composites give a straight line with a positive intercept. This deviation from theory can be attributed to the loss in kinetic energy after specimen is fractured. Every straight line shown in Figures 35 to 39 has a correlation factor better than 0.90. The values of G_c and ratio of composite G_c to matrix G_m are listed in Table 10. Figure 40 shows a plot of G_c versus volume fractions for composites and polyurethane. A plot of ratio of composite G_c to matrix G_m is shown in Figure 41. For the lower volume fraction ($\phi_f = 0.045$) the value of G_c decreases; then there is a gradual increase in the value of G_c and it almost equals the value of G_c for

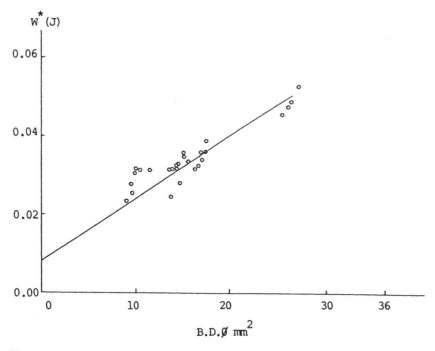

Figure 37 Plot of W^* versus $BD\phi$ for reinforced specimens of $\phi_f = 0.088$.

unfilled polyurethane matrix (for $\phi_f = 0.088$). The increase in G_c continues to increase with further increase in filler concentration.

The impact strength of particulate-filled composites usually decreases with increase in filler concentration of rigid fillers. On the other hand, incorporation of fibers as reinforcing material generally leads to an increase in impact strength, up to a reasonable level of loading, and then must fall again with further increase in concentration of fibers. In the case of composites of discontinuous fibers which have a wide range of fiber lengths, as is the case with the hammer-milled fiber glass used in this investigation, the analysis of impact behavior of composites becomes complicated. The toughening effect of fibers can be attributed to the energy required to fracture the matrix, fiber pullout, resistance to crack propagation, interfacial bond, fiber length, concentration of fibers, and their geometrical organization.

Decrease in the impact strength of polyurethane composites of lower volume fraction ($\phi_f < 0.1$) and subsequent increase in the impact strength of higher-volume-fraction composites suggests that two different types of mechanisms are playing their part in the process of fracture. For lower volume fractions ($\phi_f < 0.1$), fracture takes place primarily by fibers being pulled out of the matrix. In this case, pulling out of fibers is facilitated by the growth of cracks at the fiber ends, particularly fibers that are much shorter than the critical fiber length, and any misaligned fibers are pulled through the matrix before the complete pull out phenomenon is completed. It is also possible that a few fibers themselves may fracture in the process. At higher volume fractions both the cohesive strength of the matrix and the energy required to pull out, contribute to the impact strength of the composite, in accordance with the rule of mixtures. This behavior can be explained by considering the crack propagation phenomenon. At lower volume fractions the number

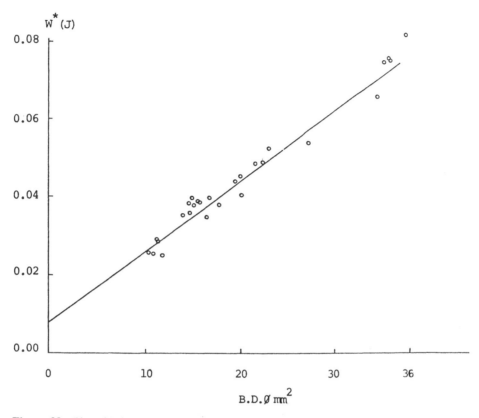

Figure 38 Plot of W^* versus $BD\phi$ for reinforced specimens of $\phi_f = 0.126$.

of fibers in the composite is not sufficient enough to stop, blunt, or deviate the propagation of crack. In other words, the distribution of stresses at the crack tip region are not altered and the crack continues to grow smoothly. As a result, the crack can advance easily through the matrix, leaving fibers bridging the crack. Consequently, the energy required to fracture a specimen is the work done to pull these bridging fibers out of the matrix.

As the volume fraction of fibers increases to a critical level where the number of fibers is sufficient to blunt, deflect, or even stop the propagation of the crack, the distribution of stresses at the crack tip alters, and the crack cannot propagate through the matrix without altering its path. This change of path, however, depends on the particular arrangement of fibers and also on the interfacial bond between the fibers and the matrix. In the case of composites ($\phi_f < 0.1$) the energy required to fracture a specimen will be a combination of energy required to fracture the matrix and the energy required to pull the fibers out of the matrix. Moreover, at higher volume fractions, there is a likelihood of interaction of fibers with each other. When one fiber is surrounded by another, the toughening effect may come from the energy dissipated in frictional sliding of one fiber inside the other during deformation [80]. Phang [40] also observed similar trends while studying the fracture properties of short glass-fiber-reinforced nylon 6,6. Fracture mechanisms can be observed from the fracture surfaces of composites of different volume fractions. The fracture surfaces of lower volume fractions ($\phi_f < 0.1$) will be smooth, indicating a free, easy ad-

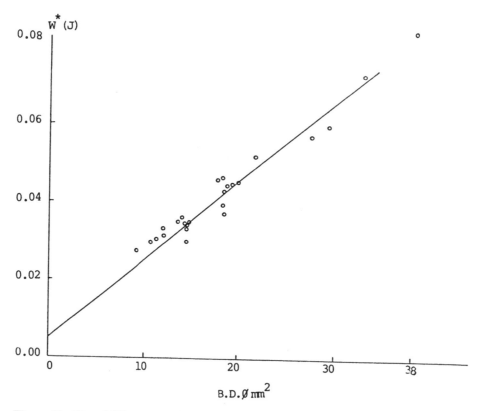

Figure 39 Plot of W^* versus $BD\phi$ for reinforced specimens of $\phi_f = 0.158$.

vance of the crack through the matrix, while the rough, rugged surface of higher volume fractions ($\phi_f > 0.1$) will show the evidence of the contribution of the increased number of fibers.

Other factors that can influence the impact properties of composite materials are the length of fiber, interfacial shear strength, and orientation of fibers with respect to impact loading. For an oriented system, the fracture energy is greater for loading in the direction parallel to the fibers, as compared to the transverse direction, with the highest impact

Table 10 Values of Strain Energy Release Rate, G_c, of Unfilled Polyurethane and Composites

ϕ_f	S/D	G_c(KJ/m^2)	G_c^c/G_c^{m} [a]
0.00	4	1.642	1
0.045	4	1.25	0.761
0.088	4	1.615	0.984
0.126	4	1.823	1.110
0.158	4	2.031	1.237

[a] G_c^c, strain energy release rate of composite; G_c^m, strain energy release rate of matrix.

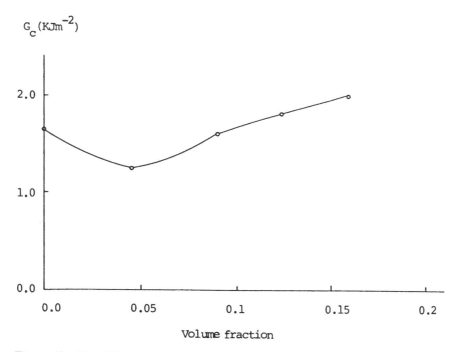

Figure 40 Plot of G_c versus volume fraction for polyurethane composites.

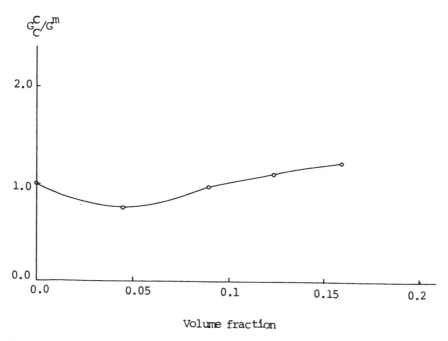

Figure 41 Plot of G_c^c/G_c^m versus volume fraction for polyurethane composites.

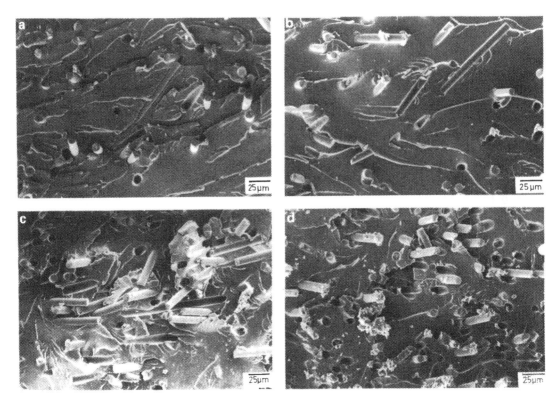

Figure 42 Scanning electron micrographs showing fracture surfaces at the notch tip region of impact specimen of (a) $\phi_f = 0.045$, (b) $\phi_f = 0.088$, (c) $\phi_f = 0.126$, and (d) $\phi_f = 0.158$.

strength being for short fibers of critical length, and for poor, as compared to strong, bonding. For transverse loading, however, a good bonding is preferable.

In our investigations, as has been discussed previously, the orientation factor and interfacial shear strength, at constant temperature, are almost similar for all volume fractions. Therefore, their effect on the impact strength should also be the same for all the different volume fraction composites.

Scanning electron microscopy (SEM) of fractured surfaces of specimens is shown in Figure 42. All micrographs show brittle fracture with no evidence of localized fibrillation. For composites of lower volume fractions, fracture is smooth, while in the case of higher volume fractions (Figure 42), rougher fracture surfaces with a greater deformation of matrix can be observed. None of the micrographs show any evidence of matrix deformation around the fibers or around the voids left by pulling out or removing fibers, as would have been the case if there had been a strong interfacial bond between the fibers and matrix.

REFERENCES

1. *Farbenjabriken Bayer Leverkusen*, 2nd ed., Bayer Plastics, 1959, p. 23.
2. J. H. Saunders, and K. C. Frisch, *Polyurethanes*, Interscience, New York, 1963.

3. P. J. Flory, *Principles of Polymer Chemistry*, Cornell University Press, Ithaca, N.Y., 1953.

4. B. G. Alzner and K. C. Frisch, *Ind, Eng, Chem.*, 51: 715 (1959).

5. F. Mort, *J. Oil Colour Chem. Assoc.*, 45: 95 (1962).

6. R. H. Stengart, "Rigid urethane foam," in *Handbook of Foamed Plastics*, R. J. Bender (ed.), Lake Publishing, Libertyville, Ill., 1965.

7. A. F. Rogers, "Catalysis reactions involved in making flexible urethane foams," *Polymer Conference on Cellular Plastics*, Union Carbide Corporation, Wayne State University, Detroit, Mich., May 1966.

8. K. Kircher and G. Menges, *34th Annual SPE Technical Conference*, Atlantic City, N.J., 1976, p. 481.

9. K. Kircher, S. Schaper, and G. Menges, *34th Annual SPE Technical Conference*, Atlantic City, N. J., 1976, p. 481.

10. J. M. Methven and J. B. Shortall, *Eur. J. Cell. Plast.*, 27 (1978).

11. R. J. Ferrari, in *Reaction Injection Moulding*, W. E. Becker (ed.), Van Nostrand Reinhold, New York, 1979, p. 75.

12. N. P. Suh and S. C. Malguarneva, *Polym. Eng. Sci.*, 2: 17 (1977).

13. U.S. patent 3,924,989, Dec. 1975.

14. H. A. Silverwood, "Performance characteristics of elastomeric bumper fascia," Paper 750007, *SAE Automotive Engineering Congress and Exposition*, Detroit, Mich., Feb. 1975.

15. E. J. Thomson, H. E. Reymore, and A. A. A. Sayigh, *J. Cell. Plast.*, a(1): 35 (1973).

16. *Eur. Plast. News*, Aug. 16, 1988.

17. *Br. Plast. Rubbers*, June 26, 1976.

18. J. M. Buist (ed.), *Developments in Polyurethanes*, Vol. 1, Elsevier Applied Science, Barking, Essex, England, 1978.

19. K. W. Schute, H. Boden, K. Sccl, and C. Weber, *Eur. J. Cell. Plast.*, Apr. 1979, p. 61.

20. J. H. Saunders and K. C. Frisch, *Polyurethane Chemistry and Technology*, Part I, *Chemistry*, Interscience, New York, 1962.

21. USSR patent 459,482 (1972).

22. Ger. offen., 2,351,844 (date unknown).

23. U. S. patent 3,916,060 (1975).

24. Anon., *Mod. Plast. Int.*, June 14, 1974, p. 4.

25. A. B. Ishan, *International Conference on Polymer Processing*, MIT, Cambridge, Mass., Aug. 1977.

26. A. B. Ishan, "Glass fibre reinforced elastomers for automotive applications: a comparison of RIM urethanes and alternative systems," *SAE Automotive Engineering Meeting*, Detroit, Mich., Feb. 1976.

27. A. B. Ishan, *Plast. Eng.*, Feb. 1975.

28. L. E. Nielson, *Mechanical Properties of Polymers and Composites*, Marcel Dekker, New York, 1974.

29. M. L. Clemens, Ph.D. thesis, Department of Polymer Science and Technology, University of Manchester, (UMIST), Manchester, England, 1978.

30. R. Sabia and F. R. Eirich, *J. Polym. Sci.*, 2A: 1909 (1964).

31. L. E. Nielson, *J. Compos. Mater.* 1: 100 (1967).

32. M. Mooney, *J. Colloid Sci.*, 6: 162 (1951).

33. J. E. Ashton, J. C. Halpin, and P. H. Petite, *Primer on Composite Analysis*, Technomic, Lancaster, Pa., 1969.

34. J. C. Halpin and J. L. Kardos, *J, Appl. Phys.*, 43: 2235 (1972).

35. E. Guth, H. Smallwood, *J. Appl. Phys.*, 15: 758 (1944).

36. T. B. Lewis and L. E. Nielsen *J. Appl. Polym. Sci.*, 14: 1449 (1970).

37. L. E. Nielson, *J. Appl. Phys.*, 41: 4626 (1970).

38. J. M. Burgess, *Second Report on Viscosity and Plasticity*, Nordemann, New York, 1938, p. 13.
39. W. H. Bowyer and M. G. Bader, *J. Mater. Sci.*, 7: 1315 (1972).
40. K. W. Phang, Ph.D. thesis, Department of Polymer Science and Technology, University of Manchester (UMIST), Manchester, England, 1981.
41. L. Holliday (ed.), *Composite Materials*, Elsevier, Amsterdam, 1971.
42. H. L. Cox, *Br. J. Appl. Phys.*, 3: 72 (1952).
43. N. C. W. Judd and W. W. Wright, *SAMPLE J.*, Jan./Feb. 1978, pp. 10–14.
44. N. J. Parrot, *Rubber Plast. Age*, 41: 263 (1960).
45. J. Heijboer, *Br. Polym. J.*, 1: 3–14 (1968).
46. J. Heijboer, *J. Polym. Sci.*, C16: 3755 (1968).
47. J. A. Manson and J. H. Sperling, *Polymer Blends and Composites*, Heyden, London, 1976.
48. F. D. Ferry, *Viscoelastic Properties of Polymers*, Wiley, New York, 1961.
49. I. M. Ward, *Mechanical Properties of Solid Polymers*, Wiley-Interscience, London, 1971.
50. J Heijboer, P. Dekking, and A. J. Staverman, in *Proc. 2nd International Congress on Rheology*, V. G. W. Harrison (ed.), Academic Press, New York, 1954, p. 123.
51. J. A. Saur and T. Lim, *J. Macromol. Sci. Phys.*, B13: 419 (1977).
52. J. L. Cawse, Ph.D. thesis, University of Manchester (UMIST), Manchester, England, 1979.
53. T. Yokoyama, *Adv. Urethane Sci. Technol.*, 6: 30 (1978).
54. T. Kajiyama and W. J. Macknight, *Macromolecules*, 2: 254 (1969).
55. N. G. McCrum, B. E. Read, and G. Williams, *Anelastic and Dielectric Effects in Polymer Solids*, Wiley, London, 1967.
56. G. Allen and M. J. Bowden, *Polymer*, 14: 597 (1973).
57. G. L. Ball, *29th Annual Technical Conference SPE*, 1971, p. 71.
58. R. J. Morgan, *J. Mater. Sci.*, 9:1219 (1974).
59. A. Mateen and S. A. Siddiqi, *Mater. Lett.*, 7(3): 110–114 (1988).
60. D. Hull, *An Introduction to Composite Materials*, Cambridge University Press, London, 1981.
61. W. A. Ludwico, *Polym. Plast. Technol. Eng.*, 15(1): 83 (1980).
62. Y. S. Toulloukin, R. K. Kirby, R. E. Taylor, and T. Y. R. Lee (eds.), *Thermophysical Properties of Matter*, Vol. 13, *Thermal Expansion, Nonmetallic Solids*, IFI/Plenum, New York, 1976.
63. G. R. Clausener, Economy considerations for Pushrod—Type Dilatometer, *AIP Conference Proceedings*, No. 3, *Thermal Expansion*, American Institute of Physics, New York, 1972, pp. 51–58.
64. E. H. Kerner, *Proc. Phys. Soc. B.*, 69: 808 (1956).
65. P. S. Turner, *J. Res. Natl. Bur. Stand.*, 37: 239 (1946).
66. J. P. Thomas, *U. S. Dept. Commerce Rep. AD 287826*.
67. R. A. Shpery, *J. Compos. Mater.*, 2: 380 (1968).
68. J. E. Carey, *Thermal Expansion of Filled Epoxy Resins*, Shell Chemical Company, Technical Service Laboratory, Union, N.J. 1956.
69. ASTM D-1822: *Tensile Impact Energy to Break Plastics and Electric Insulating Materials*, American Society for Testing and Materials, Philadelphia.
70. BS-2782: *Materials of Testing Plastics*, Part 3: *Mechanical Properties*, British Standard Institution, 1970.
71. ASTM D-256, *Impact Resistance of Plastics and Electrical Insulating Materials*, American Society for Testing and Materials, Philadelphia.
72. ASTM D-1709, *Test for Impact Resistance of Polyurethane Film by the Free-Falling Dart Technique*, American Society for Testing and Materials, Philadelphia.
73. R. W. Davidge and G. Tappin, *J. Mater. Sci.*, 3: 165 (1968).
74. L. Mascia, *Thermoplastics: Material Engineering*, Elsevier Applied Science, Barking, Essex, England, 1982, p. 255.
75. G. P. Marshall, J. G. Williams, and C. E. Turner, *J. Mater. Sci.*, 8: 949 (1973).

76. E. Plati and J. G. Williams, *Polym. Eng. Sci.*, 15(6): 470 (1975).
77. J. G. Williams, *Polym. Eng. Sci.*, 20: 3 (1977).
78. P. L. Fernando and J. G. Williams, *Polym. Eng. Sci.*, 20: 3 (1980).
79. P. L. Fernando and J. G. Williams, *Polym. Eng.* Sci., 21: 15 (1981).
80. A. Mateen and S. A. Siddiqi, *J. Mater. Sci.* 24: 4516–4524 (1989).
81. R. P. Sheldon, *Composite Polymeric Materials*, Applied Science, Barking, Essex, England, 1982.

6

Steel-Fiber-Reinforced Polymer-Impregnated Concrete

Güngör Gündüz
Kimya Mühendisliği Bölümü
Orta Doğu Teknik Üniversitesi
(Middle East Technical University)
Ankara, Türkiye (Turkey)

DEDICATED TO MY LATE FATHER, MEHMET GÜNDÜZ, THE TEACHER

INTRODUCTION

The great Industrial Revolution introduced two very important products to humankind: machinery and cement. Because of availability, moldability, and price, concrete is the most widely used construction material on the earth. Although cementlike binding agents such as lime, clay, gypsum, and puzzolan were used in the ancient and medieval periods, their mechanical strengths were very poor [1,2]. The compressive strength of these substances would not exceed 5 MPa, which is much lower than the strength of cement paste, which is above 35 MPa.

Hydraulic cement paste and concrete can both be considered as composite materials because of their multiphase nature. The aggregate phase in concrete usually constitutes two-thirds of the weight and gives a type of pseudoductility. The calcium silicates and

273

aluminates in cement give a mixture of amorphous and crystalline material upon hydration. These phases are poorly bounded, primarily by van der Waals forces, so the tensile strength and fracture toughness of cement paste are consequently low. The aggregate–matrix bond is even weaker than the cohesive forces in the paste and is often a critical factor in all cement-based composites. Another weakness is due to the water used in hydrating calcium silicates and aluminates of cement. Some excess water must be used to have workability of cement or concrete. The unreacted water gives rise to porosity in the hardened composite system. Since the aggregate is usually nonporous, all the pores exist in the cement matrix and weaken its strength. In addition, pores make the entrance of deleterious chemicals easy into the matrix and seriously affect the durability of concrete. Therefore, the improvement in mechanical properties and durability of concrete depends mainly on the improvement of (1) the interfacial bond strength between aggregate and cement matrix, and (2) the strength of the cement phase itself.

Although there have been a large variety of methods to improve the strength of concrete, two methods introduced about two decades ago have proved to be very successful in increasing the strength by a few orders of magnitude: (1) polymer impregnation, and (2) steel fiber reinforcement. Both methods have been given serious consideration, and the mechanisms through which they increase the strength of concrete have been well analyzed. These two methods can be used together to further improve the mechanical properties of concrete. Such a composite is then termed *steel-fiber-reinforced polymer-impregnated concrete* (SFRPIC). To understand the mechanisms that lead to very high strength SFRPIC composite, (1) the structure of cement paste with aggregate and/or steel fiber, (2) polymer–pore interaction at different conditions in polymer-impregnated concrete (PIC), (3) improvement of mechanical properties of concrete by steel fiber reinforcement and polymer impregnation, and (4) the effects of parameters such as pore size, types of monomers, polymer loading, impregnation techniques, adhesion between polymer and cement matrix, and interconnection between pores on the interfacial bonds and the mechanical strengths of SFRPIC will be reviewed.

MICROSTRUCTURE OF CEMENT PASTE

The porosity developed in a mature cement paste and the change in porosity during maturing depend strongly on the water/cement (w/c) ratio. The pores existing in a mature paste are of various sizes and shapes. The gel pores (1 to 10 nm in diameter) and capillary or small pores (10 to 100 nm in diameter) are usually caused by the evaporation of unreacted water in hydrolysis. Large pores (>100 nm) are created by entrapped air and occluded voids during casting. These coarse micropores usually form at large w/c ratios, such as 0.5 to 0.7. At low w/c ratios very few coarse micropores form, and the distribution of small pores changes from a maximum at about 50 nm at early ages to about 30 nm at maturity [3]. Pastes prepared by casting and pressing show different structure and composition and thus different strengths [4]. Pressing affects the size of pores and thus also the local gel/crystal ratio, especially in the paste–aggregate interface. Temperature, time, and mode of curing also affect total porosity [5].

Although the toughness of some brittle ceramic materials can be improved by introducing pores, this method does not work in the case of cement paste because of its already existing high porosity. In fact, pores with radii of less than 10 nm have negligible effects on strength [6]. All hydraulic products formed contribute to the strength through the reduction of porosity. Some additives also improve the strength; among these, gypsum is the best known and one of the major additives to clinker. It provides sulfide anion and retards the hydration. It also yields two different hydrates, ettringite and monosulfate; the former causes large increases in solid volume [7].

Calcium hydroxide and ettringite crystals form immediately upon addition of water. Ettringite forms around $C_4A_3\overline{S}$ (calcium aluminum sulfate) particles and changes to long needles during the course of hydration [8–10]. CSH (calcium silicate hydrate) crystals start to form a few hours after hydration and sharply increase after the first day of hydration, causing large decreases in porostiy. C_4AFH (calcium alumina ferrite hydrate) starts to grow at the same time with CSH. The ettringite crystals are larger and thicker than CSH. The framework formed by ettringite crystals begins to disappear 1 day after hydration and is replaced by a similar network of CSH needles. In 1-month-old paste the network of needles is usually identified as CSH needles [13]. They are embedded in hexagonal calcium hydroxide crystals and monosulfate. Some of the calcium hydroxide crystals can grow to large sizes and form panel-shaped crystals.

The dependence of mechanical properties of cement paste on porosity is quite complicated. The dependence can be expressed in the form of some exponential, power, or logarithmic functions [14]. A somewhat linear dependence can be achieved if one considers solid/pore volume ratio rather than pore volume only. The linear dependence holds true both at different w/c ratios and different curing times [15].

The pore shape also have some effect on the strength and fracture energy of cement paste [16,17]. Samples having uniformly shaped and spaced pores have a lower adverse effect on fracture energy [18]. Pores within grains are less detrimental to strength than those at grain boundaries. However, pore shape and composition of CSH may not be independent variables. The chemical composition of hydrated silicates, and hence their density and crystallinity, affect the porosity and thus the strength of cement paste [19]. In general, poorly crystalline and semicrystalline types of CSH give the highest strengths [18].

The role of pores in compressive and tensile (or flexural) strengths may be different. The tensile strength may be controlled by pore and grain size, total porosity, and also by shape of pores. However, in compressive strength pores mostly act to stop crack propagation unless very high stresses are reached [20].

Detailed studies under electron microscope show that crack propagation is a complicated phenomenon, and the models based on fracture mechanics usually oversimplify the problem. Once a crack appeared there may also occur some other branch cracks, but they do not form a continuous network. Some inactive branch cracks are not widened by further enlargement of the main crack. Some discontinuities on the pathway of the main crack may take place [21].

CEMENT–AGGREGATE INTERFACE

The bond between the cement phase and aggregates has an important influence on the behavior of concrete. There are several ways in which a bond may develop between the two phases. Usually, some products may preferentially deposit in the porous zone between the aggregate and cement phase, or epitaxial growth can take place on the surface of the aggregate. The fine cracks on the aggregate may also be filled with the hydrolysis products of cement, and the aggregate may chemically interact with these products.

It is known that the chemical composition of the contact or interfacial zone between cement paste and aggregate surface is different from that of cement paste. The thickness of the interfacial zone is about 30 μm. Since calcium hydroxide and ettringite crystals form immediately after hydrolysis of cement, they adhere on the surface of the aggregate. The ettringite needles are soon enveloped by calcium hydroxide crystals, which run perpendicular to the aggregate surface. This layer is about 2 to 3 μm [12]. Then panel-shaped calcium hydroxide crystals with an average dimension of 10 to 30 μm form in the interfacial zone, while ettringite and CSH needles also form.

The thickness of the interfacial zone tends to be larger for larger aggregates. Its thickness also depends on the shape of the particles. This zone can also be termed the transition zone.

When a carbonate rock or marble is used as aggregate, it interacts with the hydration products of the transition zone [22]. The interaction can be due to chemical reactions or epitaxis growth. A carboaluminate layer forms due to interactions. The improved bond strength is expected to increase the mechanical properties of concrete, but it may not have significant contribution [23]. In fact, an excessive dissolution of carbonate may adversely affect the bond strength [24].

It has been found in a number of research studies that the amount of calcium hydroxide in the interfacial zone decreases with distance. At the surface of aggregate particles, the calcium hydroxide crystals show preferential alignment on the surface of aggregate. This orientation decreases sharply away from the surface and almost lost at the other end of the interfacial zone [25]. The orientation at the interface grows stronger with time [26].

The interfacial zone is weaker than cement paste because of its (1) high porosity, (2) chemical composition, and (3) oriented calcium hydroxide crystals through which a crack easily propagates. Since the interfacial zone is usually the weakest region of concrete, expansive cements modify this zone by producing more ettringite there and a discontinuous calcium hydroxide layer. Hence concrete made using expansive cements usually have higher strengths than those of concrete made using ordinary portland cement [27]. The covering of aggregate particles by certain chemicals before casting can change the composition of the contact layer formed and thus increase the strength [28].

STEEL FIBER REINFORCEMENT

Steel Fiber–Cement Matrix Interface

The composition of the interfacial zone between cement paste and steel fiber looks like that of cement paste and sand aggregate, with a relatively high enrichment of calcium hydroxide in the contact layer. The thickeness of calcium hydroxide changes along the fiber surface with some discontinuities where CSH inclusions are detected [29]. The zone around the steel fiber has less weakness than the bulk of the paste. A Vickers test shows that there is decreased hardness up to 0.75 mm from the steel fiber surface [30]. So high porosity extends to large distances away from the surface of the fiber. The w/c ratio is an essential factor in bond strength. In fact, at a w/c ratio of 0.25, the solids on the fiber are predominantly calcium hydroxide, while at a ratio of 0.35 they are CSH. The increase in the w/c ratio also increases the surface of the fiber covered with solid materials in the first 24 hours of hydration. All calcium hydroxide film is removed from the steel fiber surface at the w/c ratio of 0.25, while less calcium hydroxide is removed from the surface as the ratio increases, and failure takes place through the CSH behind the calcium hydroxide film. After 3 days of hydration the surface coverage of solids on the debonded steel decreases and the bond strengths decrease with the increase in w/c ratio [31].

Bond strength improvement can be achieved by improving both the fiber–matrix adhesion and the cohesive strength of the surrounding region [32]. Some cement additives can improve both adhesion and cohesion forces. However, a very effective way of improving both forces is achieved through polymer impregnation, which gives extensive fiber yielding without slippage at the interface [33].

By the addition of silica fume as low as 5%, the amount of preferentially oriented calcium hydroxide on the contact layer is considerably reduced; thereby denser hydration products are formed, resulting in higher mechanical strength [34]. The type of fiber used also affects the composition and thus the strength of the interfacial zone. Stainless steel, mild steel, copper, and brass show an increasing order of bond strength. it is suggested that the partially acidic character of oxide films on brass an copper causes relatively strong adhesion [35]. The experiments done to measure the bond strength between C_3A (three-calcium aluminate) and iron, copper, and zinc wires show that zinc gives the highest bond strength [36]. When C_3S (three-calcium silicate) is used, iron fiber forms the strongest bond with the cement paste [37]. In pullout tests it was observed that C_3A adhered to the zinc wire, showing high bond strength. Zinc ion probably diffuses easily into C_3A paste.

The compressive strength does not seem to depend on the type of the fiber used, but flexural strength is significantly influenced from the interface bond between different types of fibers and cement paste. However, the increase in improvements is not of much practical value [38].

The temperature of curing also has some effect on the interfacial bond. The low curing temperature results in weaker bonds in the early ages and stronger bonds in the later ages, while the trend seems to be reversed in high-temperature curing [39].

Failure Mechanism

In composite systems the failure of the composite under load depends on the relative stiffness of the matrix and the fiber. In steel-fiber-reinforced concrete, failure takes place at the fiber–matrix interface. The bond strength between the fiber and the matrix is generally determined by bending or by pullout experiments. The load–deflection curve increases almost linearly, passes through a maximum, and gradually decreases. The first-crack strength of the composite is defined as the stress beyond which the curve stops being linear. This point comes before the maximum point, which denotes the ultimate strength. The shape of the load–displacement curve also looks like the load–deflection curve. The maximum point refers to the maximum load carried by the composite.

It was found that the average bond strength at first crack was 3.57 N/mm^2, while the ultimate bond strength was 4.15 N/mm^2 [40]. These close values indicate that composite failure occurs dominantly by fiber pullout from the matrix.

The fracture process can be considered to consist of two stages [40]. The first stage is the progressive debonding of the fibers through the propagation of cracks slowly. Crack propagation stops in the matrix as far as 40 μm away from the fiber. This region is the most porous and the weakest zone of the interface. This may be the reason why most of the surface treatments of fibers do not introduce considerable improvements in strength [41,42]. However, the roughness of the fiber can sometimes necessitate increased debonding load [43]. If the fiber length is greater than a critical value, multiple cracking occurs.

The second stage of fracture corresponds to the complete failure when the interfacial shear stress reaches the ultimate bond strength between the steel fiber and the cement matrix. At this stage the crack propagation is highly unstable, and finally fiber pullout from the surrounding matrix takes place by overcoming all frictional resistances.

Besides the interfacial bond strength, the aspect ratio of the fibers and their concentration are two other important parameters that affect the strength of the composite [44–46]. Some other parameters can also have significant effects on the interfacial bond strength. The pore structure, as well as the shape of hydration products and their spatial distribution, are affected by fibers [47]. The models developed to analyze the fracture of fiber-reinforced concretes have a number of limitations in their applications. The linear fracture mechanics apply to even plain concrete under certain limited conditions. Since the steel fibers give a kind of dissipative behavior to concrete, modeling of its strength and structure is difficult [48]. However, some efforts have been fruitful, and models proposed to clarify the toughening of cement phase by fibers [49–55].

Steel-fiber-reinforced concrete finds a variety of engineering applications [56,57]. For a specified water/cement/sand ratio there exists a certain amount of steel fiber that could be added to the mixture without causing balling [58]. This quantity is around a few percent. It is reported [59] that 4% steel fiber increased the composite strength by negligible amounts, while the flexural strength is increased by about 2.2 times. Steel fiber reinforcement also improves performance at elevated temperatures [60].

POLYMER-IMPREGNATED CONCRETE

Pore-Filling Effect

Impregnation of concrete by polymers makes a very high strength composite material. An incorporation of a small amount of polymer such as 4 to 7%, increases the strength properties three- to six-fold [61,62]. Although concrete has good compressive strength, it has low tensile and flexural strengths and low fracture toughness. The polymer with high tensile strength improves the poor properties of concrete.

There are mainly two important roles of polymer in concrete matrix [63]: (1) to fill the pores, hence to increase the modulus of the cement phase and to improve the strength, and (2) to improve bonding between the cement and aggregate phases [64]. So the extent of pore filling has a predominant effect on the mechanical strength of PIC. Therefore, the monomer to be used must easily penetrate the pores and polymerize there. For this purpose, low-viscosity vinyl monomers such as methyl metacrylate, styrene, acrylonitrile, and their derivatives or mixtures are generally used. To decrease the viscosity of monomers, some solvents can be used in small quantities, but they do not contribute much to the strength. The use of cross-linking agents such as divinyl benzene or trimethylol propane trimetacrylate can contribute to the strength [65].

Polymer impregnation not only improves the mechanical strength but also increases the durability of concrete against severe conditions such as freezing, corrosion, and abrasion [66,67]. As the pores are filled with polymer, the deleterious fluids cannot easily penetrate the matrix, and the durability of the concrete is improved.

Bond Effect

The superior strength of PIC compared to plain concrete is due especially to an increase in the mortar tensile strength. The improved bond strength does not seem to make a contribution to the compressive strength [68]. However, the use of some monomers with reactive groups may increase the bond strength and thus the tensile strength of the composite. It was shown that [69,70] calcium ion makes complexes (i.e., ionomers) with the carboxyl groups of polymers, which are made from acrylates. The monovalent ions, which are rich in pore solutions [71], do not form ionomers [72] and do not make inter- or intrabonds in polymer chains, although interaction with a single carboxyl group is possible. Since ionomers have relatively high thermal stability, C_2S and C_3S constituents of cement have a greater effect on the thermal stability of polymers than do other components [73,74], probably due to larger calcium ion release upon hydration. Increasing the adhesion of aggregate to matrix in PIC can also be achieved by using silane-based molecules, which make complexes with aggregate surface and form covalent bonds with polymers [75].

Methods of Impregnation

To increase monomer loading into the pores, the concrete samples must be well dried in an oven at around 110°C for a few days. The presence of adsorbed water in pores decreases the amount of polymer loading due to hydrophobic effect. The dried samples must then be soaked in monomer for sufficient time, which is usually determined by the thickness of the samples [76]. To get higher amounts of monomer loadings, evacuation can be done prior to soaking or pressure can be applied [64,77]. Monomer loading can also be increased by changing the *w/c* ratio of the concrete before casting [78]. Although the strength of young pastes can be increased substantially, the old pastes cured longer than 6 months possess relatively small pores; hence sufficient impregnation cannot be done, yielding lower increases in strength [79]. The monomer absorbed by the concrete matrix can be polymerized either by thermal catalytic methods using an initiator, or by nuclear radiation, usually gamma rays. Radiation polymerization usually gives better mechanical strength, due to radiation-induced cross-linking. Another advantage of this method is that polymerization can be carried out at room temperature, which limits loss of monomer through vaporization.

In thermal catalytic polymerization 1 to 3% catalyst (or initiator) such as benzoyl peroxide or azobisisobutyronitrile is used. In a thermal catalytic method the polymerization rate is rapid; consequently, the processing time is short. One problem is that a concrete matrix behaves like a molecular sieve, and it may be difficult for large initiator molecules such as benzoyl peroxide (actually dibenzoyl peroxide) to penetrate central regions of concrete [65].

During polymerization some monomer is lost through evaporation from the surface. To minimize these losses, samples can be wrapped by polyethylene sheets and/or immersed in a liquid medium such as water, which does not dissolve the monomer and the polymer.

STEEL-FIBER-REINFORCED POLYMER-IMPREGNATED CONCRETE

Steel fiber reinforcement and polymer impregnation are two different methods used to increase the strength of concrete. When two methods are applied together, they may show an additive effect on the strength properties of concrete [33,80–82]. The properties of this composite were studied and the results are given below.

Preparation of Samples

Specimens used in the experiments were prepared by using a cement/water/aggregate ratio of 1:3:0.6 unless otherwise stated. Curing was done at 95% relative humidity and at 20°C for 28 days. They were then dried at 110°C to dry off all free water. Some of the

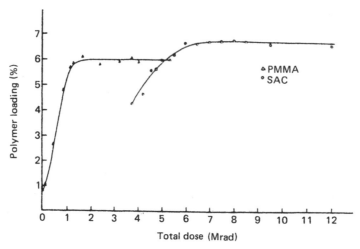

Figure 1 Change of polymer loading with dose. (From Ref. 66.)

dried and cooled specimens were polymerized by ^{60}Co radiation and some by thermal catalytic method using 3% benzoyl peroxide.

Pressure impregnation and evacuation before soaking into monomer were also done. Artificial porosity was created in some of the samples by adding naphthalene while casting. Methyl metacrylate and a styrene–acrylonitrile mixture were used as monomers. The styrene/acrylonitrile ratio was taken as 0.7:0.3, which makes the strongest copolymer [83].

Effect of Pore Filling and Adhesion

The amount of polymer loading can easily be adjusted by using radiation polymerization as seen from Figure 1. The optimum dose needed depends on the type of monomer used. For polymethyl metacrylate (PMMA) it is between 1.3 and 2.0 Mrad. While for styrene–acrylonitrile copolymer (SAC) it is between 7.0 and 8.0 Mrad. For these polymers the maximum polymer loadings are 6.0 and 6.75%, respectively. The relatively high viscosity of methyl metacrylate [65] can cause less penetration into the pores than styrene–acry-

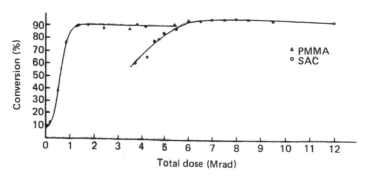

Figure 2 Change of percent conversion with dose. (From Ref. 66.)

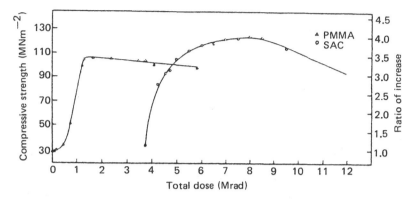

Figure 3 Change of compressive strength of concrete with dose. (From Ref. 66.)

lonitrile. The percent conversion of SAC is larger than that of PMMA, as seen from Figure 2. This may be due to a higher loss of methyl metacrylate into water around the specimens under radiation.

PMMA increases the compressive strength 3.5-fold and SAC about 4-fold at optimum doses, as seen from Figure 3. It is seen that the compressive strength increases sharply with the amount of polymer loaded into the pores. Filling of the pores is the most predominant effect in the improvement of strength. The tensile strengths also show similar changes, as seen in Figure 4. SAC increases the tensile strength 6.1-fold and PMMA increases it 4.1-fold. The ratio of increase in tensile strength is larger than that in compressive strength. In fact, plain concrete has good compressive strength but poor tensile strength. So the higher rate of increase of tensile strength is of real importance.

Irradiation of specimens for longer times results in decreases in strength, as seen from Figures 3 and 4. This may be due to degradation of polymer chains as seen in Figure 5, where PMMA degrades at a faster rate than SAC. The prolonged irradiation of plain concrete samples up to 12 Mrad does not show any change in its compressive strength [66]. So the change of strengths in Figures 3 and 4 can be attributed to the degradation of polymers. Since PMMA degrades faster than SAC does, the rate of decrease in compressive and tensile strengths is also faster in PMMA than in SAC.

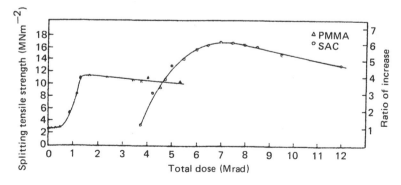

Figure 4 Change of splitting tensile strength of concrete with dose. (From Ref. 66.)

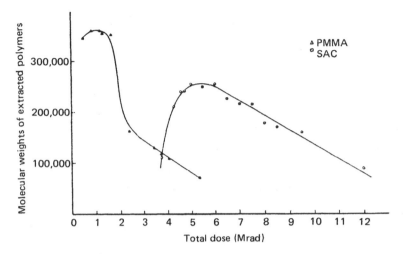

Figure 5 Change of molecular weights of extracted polymers with dose. (From Ref. 66.)

Strengthening of concrete matrix is known to depend highly on pore filling. In fact, from Figures 1, 3, and 4 it is seen that the fast increase of curves in the early period supports this thought. Degradation of polymers at a few Mrad beyond the optimum dose does not cause significant weight losses, but changes molecular weights, a seen in Figure 5. It is known that excessive radiation doses make the polymer brittle, which in turn decreases the adhesion between the polymer and concrete matrix. Consequently, the strength of concrete decreases. So the models developed to express the strength of composites must also have some terms about the effect of adhesion [84].

A close inspection of Figures 3, 4, and 5 shows that maximum strengths were achieved around 7 Mrad, while degradation starts after 5.5 Mrad in SAC-impregnated concrete. So degradation takes place at a faster rate than polymerization after 5.5 Mrad, although polymerization becomes complete around 7 Mrad, as seen from Figures 5 and 2. So in summary one can say that pore filling is the predominant factor in strength, but once the pores are filled the extent of adhesion of polymer to the concrete matrix becomes another important factor [66].

The experimental results discussed so far show that the physical and chemical properties of the monomer has a direct effect on its penetration into pores and also its adhesion to the matrix. Since SAC is superior to PMMA, only SAC will be considered in the forthcoming experimental results.

Effect of Pressure and Vacuum Impregnations on Mechanical Strengths

The pressure impregnation can be carried out under a nitrogen gas atmosphere. Polymer impregnation increases with the applied pressure as shown in Figure 6. The rate of polymer loading decreases significantly after 5 atm and the linear dependence of loading on pressure is lost.

Normally, large pores are filled first with monomer. As pressure increases, small pores also get impregnated. Air molecules entrapped in small pores resist the penetration of

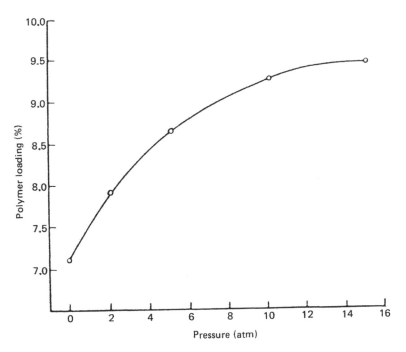

Figure 6   Change of polymer loading with impregnation pressure. (From Ref. 86.)

monomer, decreasing the rate of loading after certain pressure is applied. The tensile strengths of both PIC and SFRPIC increase almost linearly with polymer loading up to 8.5% (i.e., 5 atm pressure), as seen from Figure 7. The dependence of strength on polymer loading becomes less but continues after this point. The dependence of tensile strength on polymer loading after 5 atm implies that the type of pores filled has a significant effect on tensile strength [85]. After 5 atm the small pores, which are difficult to be penetrated by monomer molecules due to viscosity and surface tension restrictions and also possible resistance of air molecules entrapped, are now filled under increased pressure. Filling these small pores seem to have smaller contribution to strength than filling of large pores. To see this effect more clearly, some specimens were evacuated before soaking into monomer. The general trend of curve is shown in Figure 8.

The changes in the compressive and tensile strengths of PIC are shown in Figure 9 and the flexural strength of SFRPIC in Figure 10. The pressure and vacuum impregnations give the same compressive strength. This result verifies the fact that pore filling is the most important factor in compressive strength. The type of pores where polymer goes in is not important in compressive strength. Vacuum impregnation gives lower tensile strength than pressure impregnation for the same amount of polymer loading. In vacuum impregnation the decrease in the resistance of entrapped air enhances capillary action and monomer can easily fill in small pores. The pore size distribution also verifies this fact, as seen from Figure 11. For the same amount of polymer loading, small pores are left unfilled or semifilled in pressure impregnation, while some large pores may not be sufficiently filled in vacuum impregnation. The semifilled large pores act as weak spots in the matrix, resulting in lower tensile strength. In compression tests all pores and the

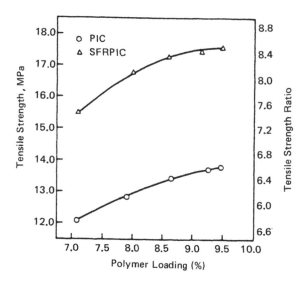

Figure 7 Change of tensile strengths with polymer loading. (From Ref. 85.)

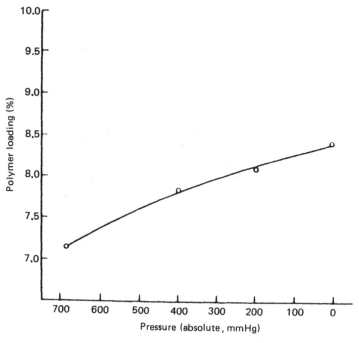

Figure 8 Change of polymer loading with vacuum applied before impregnation. (From Ref. 86.)

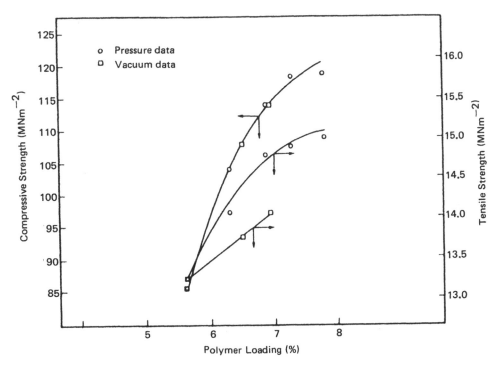

Figure 9 Change of strengths of PIC with polymer loading. (From Ref. 66.)

materials in the matrix are collapsed, while in tensile tests, the materials are removed from each other, increasing the sizes of pores and thus creating cracks in the matrix. Therefore, the semifilled pores help the creation of cracks in tensile strength. The polymer in large pores represents more heterogeneous structure than the polymer in small pores, and the average diameter of polymer in pressure impregnation is larger than the one in vacuum impregnation. So the polymer shows a higher fiber effect in the pressure impregnation case. The polymer in small pores naturally form better adherence to cement paste than the one in large pores. It seems that adherence is not of primary importance. However, for the same type of pores filled with polymer, the adherence helps to improve the strengths further, as discussed earlier for radiation polymerization.

The behavior of flexural strengths of PIC and SFRPIC is similar to the behavior of tensile strength, as shown in Figure 10. Vacuum impregnation yields lower strengths for the same amount of polymer loading. The gap between PIC and SFRPIC curves remains almost constant for both pressure and vacuum impregnation. Polymer generally strengthens the matrix, and steel fibers generally introduce frictional strength. So the increase in the flexural strength caused by polymer does not shadow the contribution of steel fibers very much. This can be seen better in Table 1, which gives the flexural strengths impregnated at atmospheric pressure. The addition of ratios of strengths of SFRC (steel-fiber-reinforced concrete) and PIC with respect to plain concrete is 6.9, which is only 10% higher than that of SFRPIC, which is 6.3.

A close inspection of curves in Figure 10 shows that the effect of impregnation technique on flexural strength is smaller in SFRPIC than in PIC. In SFRPIC, arresting of cracks by

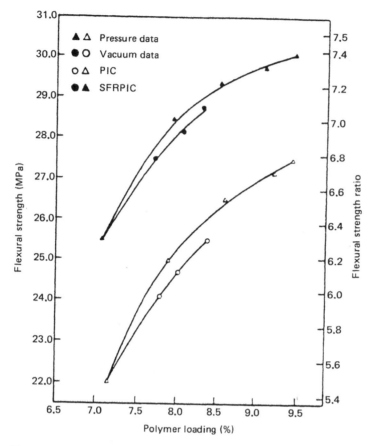

Figure 10 Change of flexural strengths with polymer loading. (From Ref. 86.)

steel fibers can decrease the fiberlike effect of polymer [85]. The strengthening of matrix by polymer has a much larger effect than increased resistive strength by steel fibers, as seen in Table 1. Steel fiber increases the flexural strength 1.48-fold, while polymer increases it 5.42-fold, that is, 3.67 times greater than the former.

The effects of steel fiber reinforcement and polymer impregnation on impact strength are shown in Table 2. The contribution of polymers to impact strength is much higher than that of steel fibers. Matrix strengthening also plays an important role in impact strength. Steel fibers and polymer seem to have an additive effect on impact strength, as seen from the last column. Steel fiber increases the initial cracking load ratio 2-fold, while polymer increases it 7-fold at atmospheric impregnation. In the case of SFRPIC it is increased 9-fold, while failure load is still increased 15-fold, as in the cases of SFRC and PIC. However, the failure load ratio of SFRPIC is increased from 15 to 21. In further polymer loadings, the initial crack load and failure load show similar behavior. These results show that matrix strengthening by polymer and crack arrest by steel fibers both play important roles in the initial crack load and failure load. At 9.5% polymer loading the initial cracking load of SFRPIC increases 12-fold and the failure load increases 28-fold. This shows excellent quality improvement in concrete by steel fiber reinforcement and polymer impregnation.

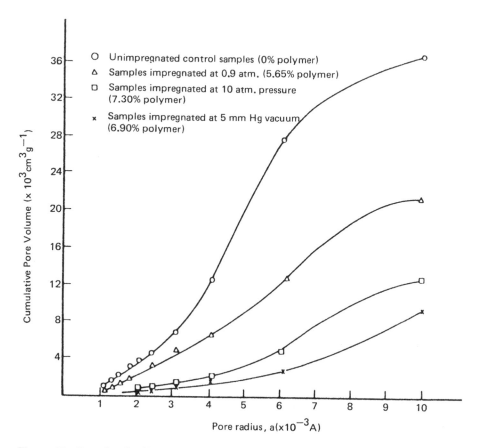

Figure 11 Porosity distribution. (From Ref. 66.)

So far, the strengthening mechanisms of plain concrete prepared according to standards have been taken into consideration. The porosity of the concrete can be changed by changing the *w/c* ratio [78] and adding some volatiles during casting, much as naphthalene, which stays solid in the matrix at room temperature and sublimates at elevated temperature, creating pores behind [87].

Table 1 Flexural Strengths of Various Concretes (Impregnated at Atmospheric Conditions) (From Ref. 86)

Type of Composite Material	Flexural Strength (MPa)	Flexural Strength Ratio
Control	4.07	1
SFRC	5.80	1.48
PIC[a]	22.07	5.42
SFRPIC[b]	25.50	6.30

[a] Polymer loading, 7.16%.

[b] Polymer loading, 7.10%.

Table 2 Changes of Impact Strengths of Various Concretes with Impregnating Pressure (From Ref. 85)

Type of Composite Material	Impregnating Pressure (atm)	Polymer Loading (%)	Number of Blows to Initiate Crack	Number of Blows to Failure	Total Impact Work (N-m)	Toughness (MPa)	Impact Strength Ratio
Control			1	1	4–12	0.20–0.59	1
SFRC			2	3	12–24	0.59–1.44	2–3
PIC	0	7.16	5	6	60–84	2.94–4.12	7–15
	2	7.91	5	6	60–84	2.94–4.12	7–15
	5	8.64	5	7	60–112	2.94–5.50	9–15
	10	9.25	6	7	64–112	4.12–5.50	9–21
	15	9.45	6	7	84–112	4.12–5.50	9–21
SFRPIC	0	7.10	5	7	60–112	2.94–5.50	9–15
	2	8.00	5	7	60–112	2.94–5.50	9–15
	5	8.60	6	7	84–112	4.12–5.50	9–21
	10	9.15	6	8	84–144	4.12–7.06	12–21
	15	9.50	7	8	112–144	5.50–7.06	12–28

The change in compressive and tensile strengths of plain concrete with the *w/c* ratio is given in Figure 12. Both strengths decrease linearly with increased water content, which naturally increases porosity. This, in turn, increases polymer loading, as shown in Figure 13. The change of strength of PIC at different *w/c* values is shown in Figure 14, and the ratio of increase in strength is shown in Figure 15. It is observed that there is a sharp increase in strength in both cases after 60% water content. However, it is seen from Figure 13 that the increase in polymer loading at 65% *w/c* ratio is a linear extension of increase at 60% *w/c* ratio. The sharp increase in strengths at 65% *w/c* therefore cannot be explained only with increased polymer loading. In Figure 16 porosity distributions of both 60 and

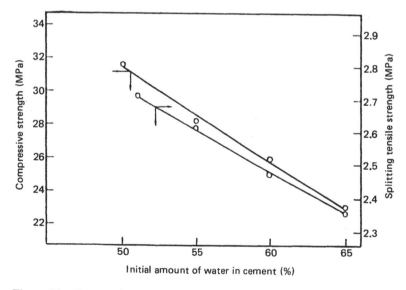

Figure 12 Compressive and tensile strengths of unimpregnated samples. (From Ref. 87.)

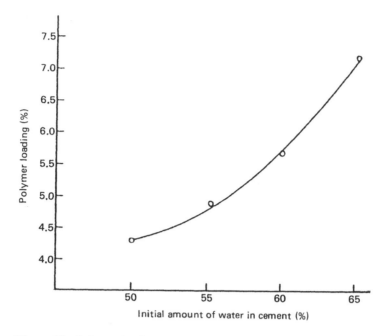

Figure 13 Polymer loading in concrete. (From Ref. 87.)

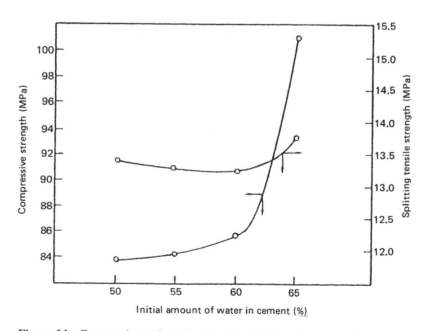

Figure 14 Compressive and tensile strengths of polymer-impregnated samples. (From Ref. 87.)

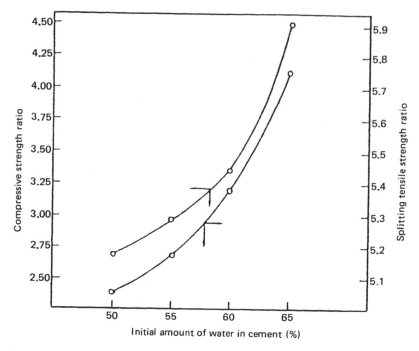

Figure 15 Compressive and tensile strength ratios. (From Ref. 87.)

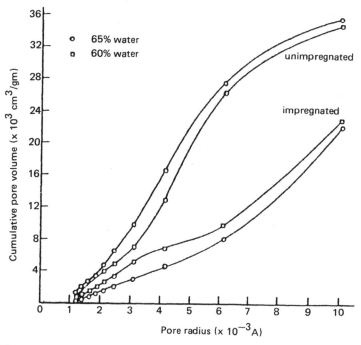

Figure 16 Porosity distribution. (From Ref. 87.)

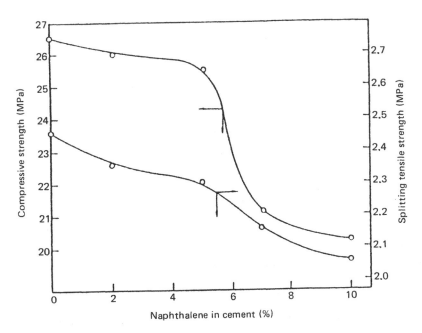

Figure 17 Compressive and tensile strengths of unimpregnated samples. (From Ref. 87.)

65% *w/c* unimpregnated and impregnated samples are shown. The unimpregnated samples with 65% *w/c* naturally has higher amounts of pores, as expected. It is interesting to observe that 65% *w/c* impregnated samples have less number of pores than 60% *w/c* impregnated samples. In 65% *w/c* samples interconnections between pores must have been better achieved, leading to easy access to monomer penetration. This results in better filling of pores with highly interconnected network [87]. In most research done in this field, both pore filling and adhesion have been emphasized as strengthening mechanisms. It is clear that interconnection of pores is another important factor and may be more important than adhesion in the strengthening of concrete by polymer impregnation, as is implied in Figure 14.

The compressive and tensile strengths of unimpregnated and impregnated samples in which additional porosity was created by using naphthalene are shown in Figures 17 and 18, respectively. The strengths of unimpregnated samples decrease with the amount of naphthalene (i.e., amount of pores). There is a sharp decrease after 5% naphthalene, especially in the compressive strength of plain concrete, as seen in Figure 17. The size and structure of pores created by naphathalene above 5% must facilitate the collapse of concrete structure upon compression. However, after 7% naphthalene the decrease in compressive strength again slows down. So the new pores or interconnections created between 5 and 7% naphthalene must play a critical role in the decrease of compressive strength.

In impregnated samples, both strengths, especially the compressive strength, show a maximum around 3% naphthalene. However, the polymer loading increases smoothly, without showing abrupt changes, with the amount of naphthalene [87]. The maximum at 3% naphthalene shows better interconnections in the structure, so that polymer can fill in the pores effectively. This leads to a high-strength impregnated concrete, despite the

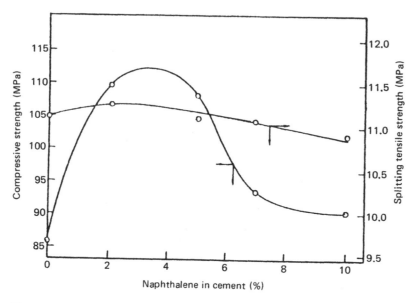

Figure 18 Compressive and tensile strengths of polymer-impregnated samples. (From Ref. 87.)

fact that the strength of plain concrete with 3% naphathalene is lower than that with no naphthalene. The sharp decrease in compressive strength between 5 and 7% naphthalene can be due to the same reasons as for plain concrete.

Interfacial Bond

Interfacial bonding is usually determined by a pullout test. The change of pullout force for plain concrete is shown in Figure 19. The maximum point in each curve indicates the force needed for the steel fiber to break its bonds (i.e., debonding force) from the surrounding matrix. It is a measure of interfacial bond strength [88]. The pullout force (or interfacial bond strength) decreases in general with the increase of porosity (i.e., amount of naphthalene). This can be seen from the lowest curve in Figure 20. As the porosity increases, the amount of lamellar calcium hydroxide crystals increases at the interface. The zone next to the lamellar calcium hydroxide is the most porous zone of the interfacial region. The interfacial bond naturally decreases with the increase in porosity.

The interfacial bond strengths for polymer-impregnated samples are shown in Figure 21. It is seen that except for 8 and 10% naphthalene samples, all others have stronger interfacial bonds than the plain concrete sample without naphthalene, the maximum being at 3% naphthalene. Polymer loadings for 1.5 and 3% naphthalene samples are almost the same [88]. Therefore, 3% naphthalene samples must have a relatively higher number of pores than 1.5% naphthalene samples, even after polymer impregnation. The larger pullout force for 3% naphthalene can be explained in terms of the better interconnections formed between pores, and thus the better polymer network and better adhesion, as mentioned earlier. In high quantities of naphthalene the weakening of cementious phase becomes quite high, so that the polymer cannot compensate for it. However, polymer improves the interfacial bond at all values of naphthalene, as seen in Figures 19 and 20.

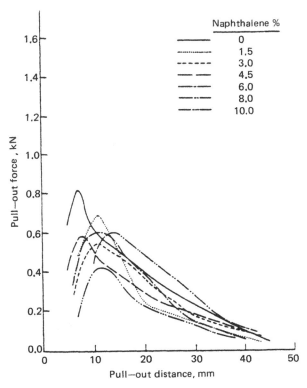

Figure 19 Pullout force in concrete containing no polymer. (From Ref. 88.)

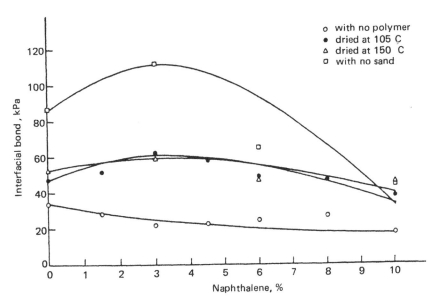

Figure 20 Interfacial bond strength in different concretes. (From Ref. 88.)

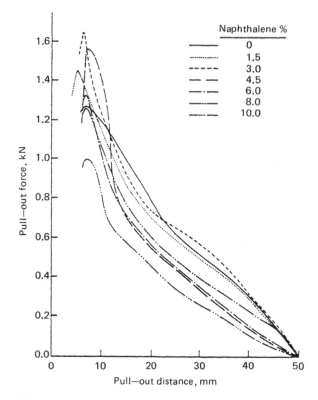

Figure 21 Pullout force in polymer-impregnated concrete. (From Ref. 88.)

To improve adhesion between the polymer and cement phase, the polymer must penetrate very small pores as much as possible. The naphthalene, which might have been trapped by adsorption, may prevent the penetration of monomer. In fact, higher polymer loading was achieved for the samples dried at 150°C than for those dried at 105°C [88]. The change of pullout force for these samples is shown in Figure 22. It is seen that the weakening of cementious matrix due to naphthalene is well compensated by increased polymer loading and thus increased adhesion between the matrix and the polymer. So the interfacial bond strengths tend to get close to each other, as seen from the relatively flat shape of the curve for samples dried at 150°C in Figure 20.

The presence of sand as aggregate and steel fiber in concrete represents a complex system such that the transition zones around aggregates containing lamillar calcium hydroxide and thus the pores formed next to it also effect the transition zone around the steel fiber. To eliminate this effect and see only the interfacial bond between steel fiber and the cementious phase, some specimens were prepared without aggregate (i.e., sand). The change of pullout force in such samples is shown in Figure 23. The interfacial bond is relatively high in such a system, as seen in Figure 20. The matrix of the cementious phase of such a system is quite strong, because there are no aggregates and thus no local weak zones throughout the matrix, except at the steel fiber interface. The weak zone around the fiber must go along the fiber showing a kind of continuity. So the polymer filled in will be highly interconnected in the zone around the fiber. Hence both the strong

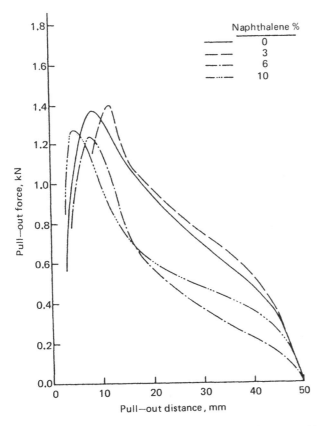

Figure 22 Pullout force in polymer-impregnated concrete (drying temperature 150°C). (From Ref. 88.)

cementious matrix and highly interconnected polymer result in a high interfacial bond in concretes containing no sand aggregate.

At about 10% naphthalene samples the amount of porosity increases to such a high value that the parameters that increase the bond strength make a very low contribution, except the polymer. So the interfacial bond strength becomes similar in all polymer-impregnated cases, as seen in Figure 20.

It was found [89] from electron microscope fractographic examination that the steel fiber pulled out from an ordinary concrete was coated with a thin layer of cementious material. The layer was much thinner in a polymer-impregnated sample. This shows that the adhesive force between the interfacial zone material and the cementious phase is larger than that of between the interfacial zone material and the steel fiber.

The area under each curve in pullout figures gives the pullout work done. Its change is given in Figure 24 for different cases. Using Figures 20 and 24, one can easily calculate that as the amount of naphthalene is changed from 0 to 3%, the interfacial bond strength increases by 30% while the pullout work increases only 7%. So it can be concluded that pullout work is strongly dependent on matrix strength rather than interfacial strength. However, the pullout force and pullout energy are interrelated, so that the curves in Figures 20 and 24 show more-or-less similar behaviors.

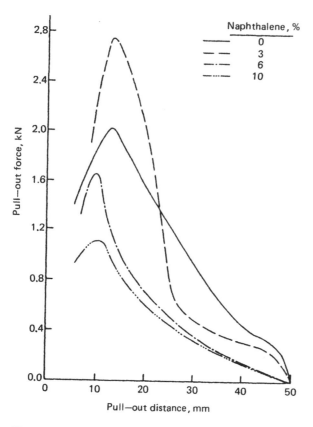

Figure 23 Pullout force in polymer-impregnated concrete containing no sand. (From Ref. 88.)

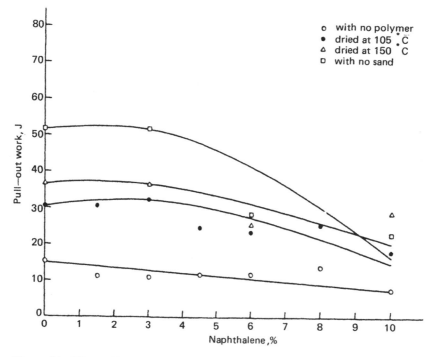

Figure 24 Fiber pullout work in different concretes. (From Ref. 88.)

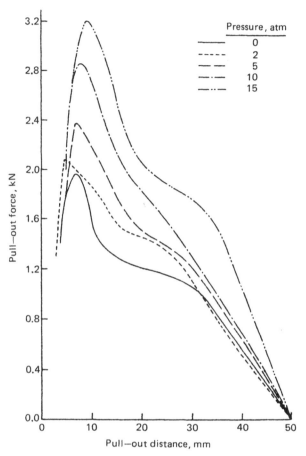

Figure 25 Pullout force in concrete polymer-impregnated under pressure. (From Ref. 88.)

When pressure or vacuum impregnation techniques are used, the pullout force increases with the pressure or vacuum applied, because of higher polymer loadings, shown in Figures 25 and 26. Consequently, the interfacial bond strengths and pullout work also increase, as seen in Figures 27 to 30. The polymer loading at 15 atm is known to be larger than at 5 mmHg vacuum [88]. However, the interfacial bond increases by 70% at 15 atm and 150% at 5 mmHg impregnations, as determined from Figures 27 and 28. Similar behavior is noted in pullout work. At 15 atm it increases by 65%, and at 5 mmHg by 100%, as determined from Figures 29 and 30. The higher percentages of increase of both interfacial bond strength and pullout work in vacuum impregnation is due to the much better adhesion of polymer to the cementious phase. The sharp increase in interfacial bond shows that the size of pores which are evacuated from air molecules below 200 mmHg and then filled by monomer plays an important role in improving the adhesion between the concrete matrix and polymer. However, this is not reflected in mechanical strength. As mentioned earlier for the same amount of polymer loading, vacuum impregnation yields lower tensile and flexural strengths than does pressure impregnation, although compressive strength is not affected by the process technology. So better adhesion does not always mean higher strength. In pressure impregnation coarse polymers are

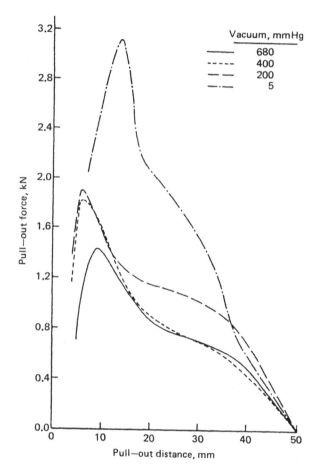

Figure 26 Pullout force in concrete polymer-impregnated under vacuum. (From Ref. 88.)

formed in large pores and in vacuum impregnation fine polymers are formed in small pores. The coarse polymer network in concrete can act as a crack arrestor and can also easily dissipate the energy of applied load, yielding a higher mechanical strength than that of a fine polymer network.

Another observation that can be made from Figures 25 and 26 is that after the sudden drop of the maximum points, there appears a kind of descending plateau region which cannot easily be observed in Figures 21, 22, and 23. In fact, in Figures 25 and 26 the porosity is not increased by naphthalene. So the polymer-loaded fills in already existing pores and the high amounts of polymer start to show a kind of fiber effect. In hybrid-reinforced concretes that contain steel fibers and polymer fibers, load–deflection curves usually have two maxima; the first is a sharp peak showing the debonding force of steel fiber, and the second is a very wide peak showing the increased toughness of the composite, which indicates that after the occurrence of cracking the tensile force is carried by the fibers only, so that a high strength level is also maintained at high strains [90]. Since the amount of polymer in concrete is small and not in the form of regular fibers, a descending plateau is formed in Figures 25 and 26 instead of a new wide peak.

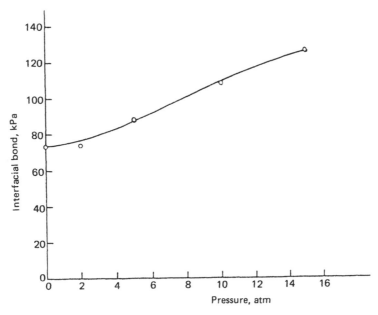

Figure 27   Interfacial bond strength in concrete impregnated under pressure. (From Ref. 88.)

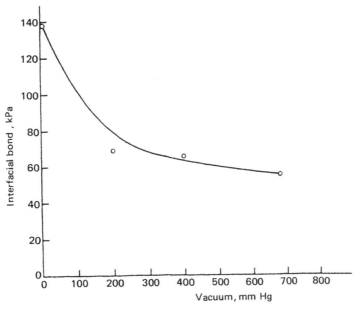

Figure 28   Interfacial bond strength in concrete impregnated under vacuum. (From Ref. 88.)

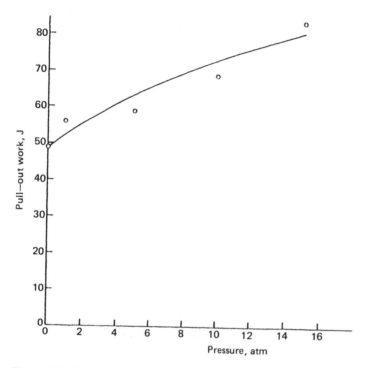

Figure 29 Fiber pullout work in concrete impregnated under pressure. (From Ref. 88.)

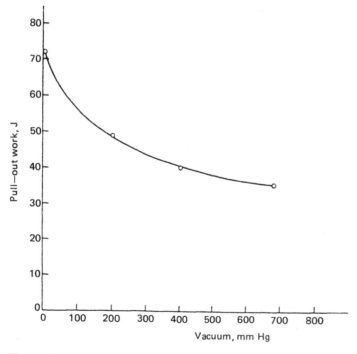

Figure 30 Fiber pullout work in concrete impregnated under vacuum. (From Ref. 88.)

ECONOMICS AND USES

SFRPIC is a new material and it is difficult to estimate its cost until fairly extensive experience with production and service conditions has been established. Monomers are expensive chemicals and make a major contribution to the cost of SFRPIC. Nevertheless, the economic feasibility of emloying PIC as a structural material (e.g., highway structures) has been put forward [91]. If performance and maintenance costs are considered, PIC and SFRPIC can be an alternative in areas where plain concrete may provide an inadequate performance. SFRPIC can find applications in areas where higher strength than that of PIC is aimed. These areas can be tunnel supports, acid- and alkali-resistant floors under heavy load, underground mine support systems, and floors or walls that are subject to impact.

REFERENCES

1. J. Skalny and K. E. Daugherty, *Chemtech*, 2: 38 (1972).
2. A. Lucas and J. R. Harris, *Ancient Egyptian Materials and Industries*, Edward Arnold, London, 1962.
3. H. G. Midgley and J. M. Illston, *Cem. Concr. Res.*, 13: 197 (1983).
4. M. Röbler and I. Odler, *Cem. Concr. Res.*, 15: 320 (1985).
5. A. Bajza and I. Rouseková, *Cem. Concr. Res.*, 13: 747 (1983).
6. I. Odler and M. Röbler, *Cem. Concr. Res.*, 15: 401 (1985).
7. E. S. Jons and B. Osbaeck, *Cem. Concr. Res.*, 12: 167 (1982).
8. K. Ogawa and D. M. Roy, *Cem. Conr. Res.*, 12: 101 (1982).
9. P. K. Mehta, *Cem. Concr. Res.*, 6: 169 (1976).
10. I. Teoreanu and C. Dumitrescu, *Cem. Concr. Res.*, 12: 141 (1982).
11. W. Richartz, *Über die Gefüge- und Festigkeitsentwicklung des Zementsteins, Betontechnische Berichte*, Betonverlag, Düsseldorf, Germany, 1969, p. 67.
12. R. Zimbelmann, *Cem. Concr. Res.*, 15: 801 (1985).
13. S. Chatterji, *Br. Ceram. Trans. J.*, 70: 195 (1971).
14. P. J. Sereda, R. F. Feldman, and V. S. Ramachandran, in *7th International Congress on Chem. Cement*, Vol. I (Principle Reports), Sub-theme VI-1, 1980, p. 3.
15. K. L. Watson, *Cem. Concr. Res.*, 11: 473 (1981).
16. H. F. W. Taylor, *Cem. Concr. Res.*, 7: 465 (1977).
17. G. Verbeck and R. Helmuth, *5th International Symposium on Chem. Cement*, Vol. III, 1968, p. 1.
18. J. J. Beaudoin and R. F. Feldman, *Cem. Concr. Res.*, 15: 105 (1985).
19. V. S. Ramachandran, R. F. Feldman, and J. J. Beaudoin, *Concrete Science*, Heyden, London, 1981, p. 427.
20. R. W. Rice, in *Treatise on Materials Science and Technology*, Vol. 11, *Properties and Microstructure* R. K. MacCrone (ed.), Academic Press, New York, 1977, p. 199.
21. S. Mindess and S. Diamond, *Cem. Concr. Res.*, 10: 509 (1980).
22. P. J. M. Monteiro and P. K. Mehta, *Cem. Concr. Res.*, 16: 127 (1986).
23. M. Saito and M. Kawamura, *Cem. Concr. Res.*, 16: 653 (1986).
24. C. Z. Yuan and W. J. Guo, *Cem. Concr. Res.*, 17: 544 (1987).
25. P. J. M. Monteiro, J. C. Maso, and J. P. Olivier, *Cem. Concr. Res.*, 15: 953 (1985).
26. R. J. Detwiler, P. J. M. Monteiro, H. R. Wenk, and Z. Zhong, *Cem. Concr. Res.*, 18: 823 (1988).
27. P. J. M. Monteiro and P. K. Mehta, *Cem. Concr. Res.*, 16: 111 (1986).
28. R. Zimbelmann, *Cem. Concr. Res.*, 17: 651 (1987).
29. M. N. Al Khalaf and C. L. Page, *Cem. Concr. Res.*, 9: 197 (1979).

30. D. J. Pinchin and D. Tabor, *Cem. Concr. Res.*, 8: 15 (1978).
31. M. Nakayama and J. J. Beaudoin, *Cem. Concr. Res.*, 17: 478 (1987).
32. C. L. Page, *Composites*, 12: 140 (1982).
33. Z. Jamrozy and J. Sliwinski, *Int. J. Cem. Compos.*, 1(3): 117 (1979).
34. P. J. M. Monteiro, O. E. Gjorv, and P. K. Mehta, *Cem. Concr. Res.*, 19: 114 (1989).
35. M. N. Al Khalaf and C. L. Page, *Cem. Concr. Res.*, 9: 197 (1979).
36. C. Tashiro and K. Ueoka, *Cem. Concr. Res.*, 11: 619 (1981).
37. C. Tashiro and S. Tatibana, *Cem. Concr. Res.*, 13: 377 (1983).
38. M. N. Al Khalaf, C. L. Page, and A. G. B. Ritchie, *Cem. Concr. Res.*, 10: 71 (1980).
39. N. Banthia and J. F. Trottier, *Cem. Concr. Res.*, 19: 400 (1989).
40. R. N. Swamy and P. S. Mangat, *Cem. Concr. Res.*, 4: 313 (1974).
41. M. Maage, *Cem. Concr. Res.*, 7: 703 (1977).
42. A. Bentur, S. Diamond, and S. Mindess, *Cem. Concr. Res.*, 15: 331 (1985).
43. D. J. Pinchin and D. Tabor, *Cem. Concr. Res.*, 8: 139 (1978).
44. A. J. Majumdar and V. Laws, *Philos. Trans. R. Soc. London Ser. A*, 310: 191 (1983).
45. R. J. Gray and C. D. Johnson, *Int. J. Cem. Compos. Lightweight Concr.*, 9(1): 43 (1978).
46. P. S. Mangat and K. Gurusamy, *J. Mater. Sci.*, 22: 3103 (1987).
47. R. Sh. Mikhail, M. Abd El Khalik, A. Hassanein, D. Dollimore, and R. Stino, *Cem. Concr. Res.*, 8: 765 (1978).
48. P. Rossi, O. Coussy, C. Boulay, P. Acker, and Y. Malier, *Cem. Concr. Res.*, 16: 303 (1986).
49. D. J. Hannant, D. C. Hughes, and A. Kelly, *Philos. Trans. R. Soc. London Ser. A*, 310: 175 (1983).
50. A. P. Hibbert and D. J. Hannant, *Composites*, 12: 105 (1982).
51. P. Stroven, *Composites*, 12: 129 (1982).
52. V. Laws, *Composites*, 12: 145 (1982).
53. W. Suaris and S. P. Shah, *Composites*, 12: 153 (1982).
54. A. S. Argon, G. W. Hawkins, and H. Y. Kuo, *J. Mater. Sci.*, 14: 1707 (1979).
55. H. Stang and S. P. Shah, *J. Mater. Sci.*, 21: 953 (1986).
56. *Fiber Reinforced Concrete*, Publication SP-44, American Concrete Institute, Detroit, Mich., 1974.
57. C. D. Johnston, *Composites*, 12: 113 (1982).
58. R. Narayanan and A. S. Palanjian, *Concrete*, 17(2): 42 (1983).
59. P. J. Robins and S. A. Austin, *Concrete*, 19(3): 18 (1985).
60. J. A. Purkiss, *Int. J. Cem. Composites Lightweight Concr.*, 6(3): 179 (1984).
61. M. Steinberg, in *Polymers in Concrete*, Publication SP-40, American Concrete Institute, Detroit, Mich., 1973, p. 1.
62. G. W. De Puy and J. T. Dikeou, in *Polymers in Concrete*, Publication SP-40, American Concrete Institute, Detroit, Mich., 1973, p. 33.
63. A. Auskern and W. Horn, in *Polymers in Concrete*, Publication SP-40, American Concrete Institute, Detroit, Mich., 1973, p. 223.
64. E. Tazawa and S. Kobayashi, in *Polymers in Concrete*, Publication SP-40, American Concrete Institute, Detroit, Mich., 1973, p. 57.
65. R. N. Swamy, *J. Mater. Sci.*, 14: 1521 (1979).
66. G. Gündüz, B. Yeter, P. Tuğlu, and I. Ahmed, *J. Mater. Sci.*, 16: 221 (1981).
67. K. R. Kirtania and S. Maiti, *J. Mater. Sci.*, 21: 341 (1986).
68. N. J. Carino, *Cem. Concr. Res.*, 7: 439 (1977).
69. S. Chandra and P. Flodin, *Cem. Concr. Res.*, 17: 875 (1987).
70. J. A. Larbi and J. M. Bijen, *Cem. Concr. Res.*, 20: 139 (1990).
71. K. Andersson, B. Allard, M. Bengtsson, and B. Magnusson, *Cem. Concr. Res.*, 19: 327 (1989).
72. S. Chandra, L. Berntsson, and P. Flodin, *Cem. Concr. Res.*, 11: 125 (1981).
73. T. Sugama and L. E. Kukacka, *Cem. Concr. Res.*, 9: 69 (1979).

74. T. Sugama, L. E. Kukacka, and W. Horn, *Cem. Concr. Res.*, 11: 429 (1981).
75. L. Balewski, M. Wejchan-Judek, and A. Zuk, *Cem. Concr. Res.*, 17: 539 (1987).
76. B. Sopler, A. E. Fiorato, and R. Lenschow, in *Polymers in Concrete*, Publication SP-40, American Concrete Institute, Detroit, Mich., 1973, p. 149.
77. L. E. Kukacka and A. J. Romano, in *Polymers in Concrete*, Publication SP-40, American Concrete Institute, Detroit, Mich., 1973, p. 15.
78. D. G. Manning and B. B. Hope, in *Polymers in Concrete*, Publication SP-40, American Concrete Institute, Detroit, Mich., 1973, p. 191.
79. R. Sh. Mikhail, A. M. Mousa, S. A. Abo-el-Enein, and M. S. Marie, *Cem. Concr. Res.*, 13: 325 (1983).
80. F. Flajsman, D. S. Cahn, and J. C. Phillips, *J. Am. Ceram. Soc.*, 54: 129 (1971).
81. T. Fukuchi, Y. Ohama, H. Hashimoto, and M. Sugiyama, *Proc. Japanese Congress Materials Research*, Nihon University, 1977, p. 163.
82. C. B. Kukreja, S. K. Kaushik, M. B. Kanchi, and P. Jaino, *Indian Concr. J.*, 54: 184 (1980).
83. H. F. Mark, N. G. Gaylord, and M. N. Bikales (eds.), *Encyclopedia of Polymer Science and Technology*, Vol. I, Interscience, New York, 1964, p. 425.
84. A. Muñoz-Escalano and C. Ramos, *Cem. Concr. Res.*, 6: 273 (1976).
85. G. Gündüz and N. Yalçin, *Cem. Concr. Res.*, 16: 793 (1986).
86. G. Gündüz and N. Yalçin, *Compos. Sci. Technol.*, 30: 127 (1987).
87. G. Gündüz and I. Ahmed, *Int. J. Poly. Mater.*, 11: 159 (1986).
88. G. Gündüz, *Compos. Sci. Technol.*, 32: 121 (1988).
89. T. Ertürk and M. Tokyay, *Proc. 4th International Conference on the Mechanical Behaviour of Materials*, Stockholm, Vol. I, Aug. 15–19, 1983, p. 507.
90. K. Kobayashi and R. Cho, *Composites*, 13: 164 (1982).
91. D. J. O'Neil, *Int. J. Polym. Mater.*, 6: 73 (1977).

7

Significance of Defects and Damage in Carbon Fiber/Epoxy Panels

Mustafa Akay
Department of Mechanical and Industrial Engineering
University of Ulster at Jordanstown
Newtownabbey, N. Ireland

INTRODUCTION

Production of continuous fiber-reinforced polymer laminate parts are still heavily manual and stringent; therefore, the integrity of the product relies on the extent of care exercised during manufacturing. Defects are considered to originate from the raw material handling and manufacturing processes, and damage from the postproduction machining, assembly, and service conditions.

Lamination defects include voids and inclusions, kinked fibers, and thermal cracks. Subsequent fabrication can cause handling and machining damage and joint defects. In service, damage can arise from adverse environment (such as excessive loading, heat, moisture, and chemical fluids), careless handling and maintenance, knocks, and scratches. It is popularly accepted that manufacturing defects tend to have a negligible effect on the mechanical performance of carbon fiber/epoxy resin composites, whereas service-induced damage (particularly due to impact) can cause significant reductions.

Deficiencies in the appearance and performance of composites can also arise from the susceptibility of the fiber to structural flaws and the resin to aging. Furthermore, developments in resin formulations may improve certain properties at the expense of others.

In this chapter the emphasis is on the consequences of prepreg aging, variation of production conditions, hot/wet environmental conditioning, low-energy impact, and pin bearing for various laminate systems and sandwich panels. These aspects are expanded on below.

A realistic estimation of the useful life of carbon fiber/epoxy prepregs is essential for reasons of fabrication, economics, environment, and most important, product integrity. Prepregs consist of a layer of fiber arrangement impregnated with a partially reacted mixture of monomers either in solvent or as a solvent-free medium. The mixture will continue to react and lose its solvent prior to product fabrication, to a degree determined by the storage, handling, and environmental conditions, mainly temperature. Accordingly, the raw material suppliers recommend that the material be transported and stored refrigerated (mostly at $-18°C$) and used up within the specified shelf life. Out-of-refrigerator shelf life (referred to as "out-time") is of particular importance in its influence on tack, and some adverse influence on the design capabilities must also occur, even if slowly. A prepreg is tacky and flexible when handled above its glass transition temperature (T_g). Loss of tack can be tolerated in the making of flat laminate but is especially a problem in the manufacture of complex shapes involving drape and where tack facilitates placement. Therefore, definition of useful life or overage for a prepreg depends on the enduse of a specific resin–fiber system. For the type of prepreg investigated here (i.e., unidirectional carbon fiber/epoxy of 37 wt % resin content), about 500 h is recommended as the handling life. The practice of advanced composites users is even more stringent, particularly on work contracted out: for example, beyond 400 h out-time at 25°C or 1000 after bagging, the material may be stipulated as unacceptable. Sanjana [1] has studied the aging behavior of a Hercules AS/3501–6 unidirectional carbon fiber/epoxy system with resin contents of 42 and 35 wt % under various heat and humidity conditions and ambient. The work has identified the prepreg tack as the critical property and has aimed to correlate loss of tack with aging using dynamic mechanical analysis (DMA), dielectric analysis, and TTW (a time–temperature integrating indicator). All these techniques have been shown successfully to monitor the age of the prepreg.

Performance of a fiber–polymer composite depends inherently on structural features, but then notably on the production history and service environment. In epoxy resin systems the processing parameters—temperature, pressure, and time—influence the extent of curing (conversion of reactive resin to solid thermoset) and intra- and inter-prepreg ply consolidation, hence of void content and uniformity of fiber packing. The process cycle is also determined with due regard to economics, but the integrity of the formed component must not be compromised. Accordingly, the technology of molding must become more of a science and less of an art. Recent work by Purslow and Childs [2] has examined an autoclave cure cycle in terms of the resin properties, viscosity, and gel time, for a carbon fiber (CF)/epoxy and emphasized the importance of parameter interactions. A computer simulation of cure for a glass fiber/epoxy has been developed by Morrison and Bader [3]. This work using a resin flow model [4] describes rheological behavior and examines the influence of the permeability of the bleeder cloth and the prepreg fabric.

Relevant environmental considerations have been partially investigated: the effects of chemicals [5] fluids, fuel, radiation [6], and in particular, moisture and heat. Absorbed moisture can act as a plasticizer, causing a reduction in glass transition temperature (T_g) [7–10] and in matrix-controlled mechanical properties [11]. Furthermore, the moisture ingress causes dimensional instability, although a relaxation of frozen-in thermal tensile stress is also associated with this phenomenon [12–14]. Carbon fiber is not itself affected by water [15], but there may be an affect on resin fiber interface. Moisture-induced deterioration is aggravated by temperature, and under thermal-spiking conditions (when the temperature can reach the T_g value—already depressed by moisture) it assumes critical proportions [11,16,17].

Composites can be vulnerable following low-energy impacts, and in application areas such as aerospace structures, the damage induced can be critical. The residual strength–deformation behavior of impact-damaged composites has been reported by several authors [18–25]. The reduction in the compressive strength of graphite–epoxy systems following low-velocity impact is outlined in Refs. 21 and 22. The impact damage zone has been shown to be V-shaped: no visual damage at the point of impact but severe damage, in the form of cracks and delamination, at the back surface. A comparison of the CF/epoxy and CF/PEEK systems is given in Refs. 20 and 23. Several authors [18,19] have shown that at least some aspects of damage can be predicted by appropriate theoretical analysis.

Sandwich structure with advantageous properties can be prepared using a low-density cellular core with shear rigidity and solid skins with significant in-plane compressive and tensile load-bearing properties. Such a structure combines a very high level of bending stiffness with low weight, a combination demanded by transport applications—from advanced aerospace to automotive. Honeycomb is widely employed as a core material in the aerospace industry. Rigid polymeric foams can be a feasible alternative. Foam does not provide a planar connection between the skin panels and some structural efficiency is lost; the strength and stiffness-to-weight coefficients are reduced, particularly with traditional foams of PVC and PU. Some new generation closed-cell rigid foams, such as Rohacell (based on polymethacrylimide), do, however, have superior mechanical and thermal properties, and the isotropic nature of the foam can provide wider design options. Furthermore, Rohacell lends itself to a single autoclave operation for prepreg lamination and sandwich bonding. This is not always successful with honeycomb cores because at the levels of pressure required for satisfactory laminate consolidation, cell-wall buckling and laminate surface sagging into the cells can occur. An ongoing concern for end users, such as the aerospace industry, is the low-energy impact sensitivity of composite structures. Accordingly, considerable work has been undertaken on prepreg laminates [18–25]. To date, most of the work on sandwich structures has been on metallic and nonmetallic honeycomb cores with unidirectional carbon fiber/epoxy skins [26–28]. Recent papers [29–30] have focused on the low-velocity impact damage of sandwich panels consisting of various CF/epoxy resin skins with cores of Nomex (aromatic polyamide paper) honeycomb and Rohacell foam.

Bolted joints can be a feasible alternative for the assembly of laminate-type structures based on composite materials. They result in higher stress concentration than would be the case with bonded joints but enable easy inspection, facilitate assembly, and provide reliability at an economic cost. Current applications include aerospace structures and the filament-wound rocket booster case of the *Challenger* space shuttle [31]. Successful adoption of bolted joints in composites depends on a clear understanding of pin bearing strength. Limited information is available in review articles [32–34] and specific research papers dealing with unidirectional CF/epoxy laminates. The effect of ply layup has been well covered [35–39]. The influence of geometric features [i.e., specimen width (w) and thickness (t), end distance (e), and hole diameter (d)] has been covered extensively [33,36,40,41]. There is a general consensus that for a given layup, minimum e/d and w/d ratios must be provided; otherwise, the potential bearing strength will not be realized. Work [35,36,39] on the effects of lateral constraint has shown that bearing strength increases gradually and reaches a plateau value with increasing clamping pressure via bolt torque. Various failure modes have been identified [32,34,36,42]: end-section bearing, net-section tension, edge-section shear, end-section cleavage, and tension–cleavage combination. Some mixture of these modes may, of course, occur [40]. Failure can be defined as the maximum load sustained [36] or as a deformation of the hole. The definition of bearing strength in relation to fixed diametric strain as in ASTM D953–87 is open to question. Clearly, strength should be defined in relation to permanent damage within the joint, and the deformation to create damage is a function of both the tolerances of the hole and the stress–deformation behavior of a particular material system. To date, the volume of work has not adequately covered woven carbon fiber composite systems, and scarcely any work has been published on pin bearing strength under fatigue loading [43].

EXPERIMENTAL PROCEDURES

Materials and Production

Flat laminate panels were prepared from the following prepregs:

1.   8-Harness (8-HS) satin fabric of 37% by weight resin content (wrc): carbon fiber (CF)/epoxy (Cycom 985 resin and T300 3k fiber)
2.   8-HS satin fabric of 37% by wrc: CF/epoxy (Fiberite 934 resin and T300 3k fiber)
3.   5-HS satin fabric of 40% by wrc: CF/epoxy (Fiberite 984 resin and T300 3k fiber)
4.   5-HS satin fabric of 40% by wrc: CF/toughened epoxy (Fiberite 974 resin and T300 3k fiber)
5.   Plain-weave (PW) fabric of 40% by wrc: CF/epoxy (cycom 985 resin and T300 3k fiber)
6.   PW fabric of 40% by wrc: CF/epoxy (Fiberite 934 resin and T300 3k fiber)
7.   PW fabric of 40% by wrc: CF/toughened epoxy (Cycom 1010 resin and T300 3k fiber)
8.   Unidirectional (UD) tape of 37% by wrc: CF/epoxy (Cycom 985 resin and T300 3k fiber)
9.   UD tape of 37% by wrc: CF/epoxy (Fiberite 934 resin and T300 3k fiber)
10.   UD tape of 37% by wrc: CF/toughened epoxy (Cycom 1010 resin and T300 3k fiber)

The prepreg stacks were debulked, bagged, and then autoclave-cured, normally at 179°C and 600 kPa, for 2 h. The heating rate was maintained at 2°C/min, with cooling at 3°C/min.

To examine the influence of curing conditions on laminate behavior, prepregs in material 1 were cured under alternative conditions as follows: temperatures, pressures, and times of (i) 179°C, 300 kPa, 2 h; (ii) 179°C, 600 kPa, 2 h; (iii) 179°C, 600 kPa, 4 h; and (iv) 189°C, 600 kPa, 2 hr. Eight-ply stacks of prepregs (0° layup) produced panels of approximately 3.15 mm thickness. Conditions (ii) are the standard for load-bearing aircraft components.

A roll of prepreg in material 8 was removed from cold storage and taken out of its seal upon reaching room temperature. The material was then kept on a shelf under ambient conditions and used periodically to prepare laminated panels representing a range of out-times from 100 to 930 h. Each panel consisted of 14 plies in 0° layup and produced 2.2 mm thickness.

Test pieces were cut parallel to the warp direction in fabric laminates and longitudinal (0°) and transverse (90°) in UD laminates, using a high-speed, diamond-tipped rotary cutter. A set of test pieces were hot/wet conditioned by full immersion in water at 70°C for 21 days and resulted in approximately 1.3 and 1.5 wt % moisture uptake in fabric and UD laminates, respectively.

Cellular core sandwich panels were prepared from the following materials:

Panel A:	Skins:	prepregs 1 and 5
	Core:	Rohacell 51 WF foam of 51 kg/m³ density
Panel B:	Skins:	prepregs 2 and 6
	Core:	Nomex honeycomb of 48 kg/m³ density and 3.1 mm cell opening
Panel C, D, and E:	Skins:	prepregs 2 and 6
	Core:	Rohacell 71 WF (71 kg/m³) for panels C and E and Rohacell 110 WF (110 kg/m³) for panel D

Panels A, B, C, and D had a preproduction core thickness of 12.5 mm, and panel E, a core thickness of 25 mm. The layup sequence was (45°/90°/90°/90°/45°) for the five-ply 8-HS face skin and (90°/45°/90°) for the three-ply PW back skin, giving 1.97 and 0.63 mm thickness, respectively.

The foam was precleaned with an oil-free air jet to remove dust and debris, and the honeycomb was degreased in a trichloroethylene vapor bath. The foam was annealed as recommended by the manufacturers to improve the thermal stability. Debulked and bagged stacks of prepregs and core, interdispersed with grade 10, FM300 Cyanamid adhesive film (epoxy-based), were cured and bonded in an autoclave at 179°C for 2 h, under pressures of 300 kPa for panels A and B, and 600 kPa for panels C, D, and E. All the panels were inspected by C-scan in a water immersion tank. Those with a void level in the skins of 2% and above were discarded. Samples for low-energy impact, compression, and impact-puncture tests were cut 76 mm wide and 127 mm long with a high-speed diamond-tipped cutter.

Laminates, cured under conditions (ii) with layup details presented in Table 1, were cut into coupons of approximately 194 mm length and 30 mm width for pin bearing tests. Centrally located holes of 4.84 mm diameter and 35 mm end distance were machined, and the specimen end to be gripped was bonded with aluminum alloy tabs.

Test Procedures

Flexure and short-beam shear (interlaminar shear) tests were conducted in accordance with ASTM D790M and ASTM 2344, where 10-mm-wide test pieces were loaded in three-point bending on a support span of 5× specimen thickness for interlaminar shear strength (ILSS) and on a 80-mm span for flexural tests. A cross-head displacement rate of 1 mm/min was used to load the specimens. The central loading pin and outer support rods were 6 mm in diameter.

Table 1 Details of Pin-Bearing Specimens

Specimen[a]	Prepreg	Ply Nos.	Layup	% 0° Ply
α	5	14	$(0°,90°)_{14}$	50
β	5	14	$[(\pm45°)/(0°,90°)]_7$	25
γ	1	8	$[(0°,90°)/(\pm45°)/(0°,90°)/(\pm45°)]_s$	25
δ	1	8	$(\pm45°)_2/(0°,90°)_4/(\pm45°)_2$	25
ε	6	12	$[(0°,90°)/(\pm45°)_2/(0°.90°)/(\pm45°)_2]_s$	16.7
1	8	18	$[0°/45°/-45°/0°/45°/-45°/0°/45°/-45°]_s$	33.3
2	10	18	$[0°/45°/-45°/0°/45°/-45°/0°/45°/-45°]_s$	33.3
3	8	12	$[0°/45°/-45°/0°/45°/-45°]_s$	33.3
4	10	12	$[0°/45°/-45°/0°/45°/-45°]_s$	33.3
5	9	18	$[0°/45°/-45°/0°/45°/-45°/0°/45°/-45°]_s$	33.3
6	9	12	$[0°/45°/-45°/0°/45°/-45°]_s$	33.3

Source: Adapted from Ref. 43.

[a] Specimen ε was 35 mm wide, and the remainder were 30 mm.

Compression tests were conducted at 1 mm/min loading speed, employing a Celanese fixture (see ASTM D3410 for details), on 10-mm-wide and at least 100-mm-long test pieces with an unsupported surface of 10 mm × 10 mm. Dynamic properties were determined at 1 Hz and a 4°C/min rate of heating.

Fracture toughness tests were carried out in a three-point loading mode on a 40-mm span and using an instrumented falling weight machine at a 1-m/s impact speed. Bars of 80 mm length and 10 mm depth were milled with a 45° × 0.25 mm cutter in a range of notch depths/bar depth ratios ranging from 0.03 to 0.46 and impacted on edge. The fracture properties G_c and K_c were determined from triangular force–deflection traces using the analysis based on energy calibration factor (ϕ) and stress-intensity calibration factor (Y) [44,45], so that

$$G_c = U(BD\phi)^{-1} \quad \text{and} \quad K_c^2 = y^2\sigma^2 a$$

where U is the energy absorbed to failure and σ is the failure stress. B, D, and a are the breadth, depth, and notch depth, respectively, of the specimen. ϕ and Y are functions of a/D.

A falling-weight impactor with a hemispherical striker of 12.7 mm diameter was used for the low-energy tests. The samples were either clamped (using 2.5 N · m torque) between a pair of 15-mm-thick rings of 50 mm internal and 90 mm external diameter, or freely supported (unclamped) on one of the rings. Incident impact energies in the range 1 to 16 J were generated by changing the mass of the drop weight and keeping the drop height constant at 0.7 m. Laminates consisting of prepregs 3, 4, 5, and 7, representing ordinary and toughened epoxy systems, cut as 100-mm-wide and 550-mm-long strips, were clamped and impacted at 50-mm spacings. Test pieces for the subsequent compression and flexure tests (50 × 100 mm) were cut from these damaged strips of laminates. The impact energy was limited so that visible cracks stopped at least 5 mm short of the test-piece edge. Compression specimens were supported along their edges by antibuckling guides and loaded at 1 mm/min. Tests were conducted on undamaged specimens, using both the antibuckling device and a Celanese rig.

In sandwich panels, the face skin (8-HS) was impacted and the extent of damage in

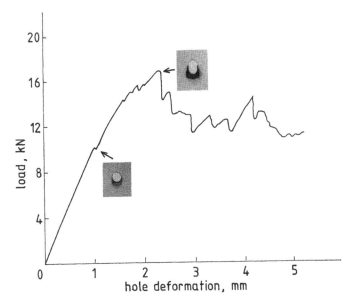

Figure 1 Typical bearing load–displacement curve (showing damage at initial failure and at maximum load). (From Ref. 73.)

the face and backskins was measured by C-scan. Through-thickness skin damage was examined in greater detail by sectioning, mounting, and polishing a number of samples. Edge-on compression tests were conducted as described in Ref. 30.

Static pin-bearing tests were conducted with reference to ASTM D953–87 and the CRAG [46] test methods. All of this work was conducted at room temperature using an Instron floor model machine in tensile mode. Hole displacement was measured using an extensometer. Load was applied at 1 mm/min via 4.76-mm-diameter pins. Stress values, load/dt, were determined corresponding to 4% hole displacement (as per ASTM D953–87), initial discontinuity in a load–displacement trace (Figure 1, and maximum load. All the specimens were prepared with fingertight lateral constraint, and specimens 5 and 6 were further examined over a range of constraints, corresponding to pin torque values of 0, 2, 4, 6, and 8 N · m. The upper limit of the applied torque was dictated by pin thread stripping. The constraint pressure (σ_{press}) has been identified elsewhere [36,43] and based on the pin torque (T) and the constrained area employed, σ_{press} (MPa) = 6.17T. Fatigue (sinusoidal load-amplitude) tests on an Instron servohydraulic machine were conducted at 3 Hz using a base load of 0.2 kN and a range of preset maximum loads. The tests were continued to a pin displacement of 2 mm (40% hole deformation); this value corresponds to the deformation experienced under maximum static bearing load. Specimens that were highly prone to tensile failure had fractured within this limit.

RESULTS AND DISCUSSION

Prepreg Aging

The mechanical behavior of material 8 was investigated at out-times well beyond its assumed mechanical life (750 h). Figures 2 to 7 (where the data represent an average of

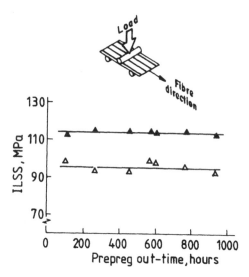

Figure 2 ILSS versus prepreg out-times. ▲, As-cured; △, H/W conditioned test pieces. (Adapted from Ref. 68.)

five tests) show that there is no deterioration in the properties of the as-cured and hot/ wet (H/W) conditioned specimens over a wide range of prepreg out-times, although the expected aging is evident from the dynamic mechanical analysis. Figure 8 shows a reduction in tan δ–peak height and an increase in tan δ–peak temperature (popularly accepted as T_g); this behavior is consistent with increased curing and/or cross-linking [47–49]. The increment of change in T_g seems to be too small ($\simeq 5°C$) to affect the mechanical behavior.

Figure 7 shows that with fibers oriented normal to the line of impact, the notched specimens on the whole produced greater impact strength (defined as energy to crack initiation/cross-sectional area of specimen at notch). This is attributed to the failure mode of the specimens, the notched ones frequently exhibiting an interface failure within the tensile zone (see Figure 9a). It would therefore appear that a notch in unidirectional-fiber

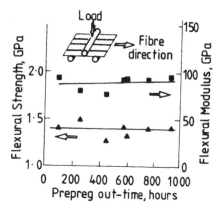

Figure 3 Flexural strength (▲) and flexural modulus (■) versus prepreg out-time for as-cured test pieces. (Adapted from Ref. 68.)

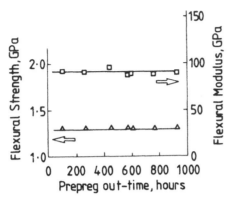

Figure 4 Flexural strength ($\triangle$) and flexural modulus ($\square$) versus prepreg out-time for H/W conditioned test pieces. (Adapted from Ref. 68.)

laminates acts not only as a stress raiser but does in fact advantageously influence the crack failure path, resulting in higher total failure loads.

In light of the experienced mechanical integrity tolerance to prepreg aging, it is worth indicating that the loss of tack and flexibility can be counteracted partially by heating to no more than 60°C, to facilitate fabrication and to avoid wastage.

Curing Conditions

The dynamic mechanical properties (Figures 10 to 13) showed that compared to the standard setting (ii), the more severe curing conditions (iii) and (iv) generated increases of approximately 10°C in T_g. The static flexural modulus also improved by 10%. However, autoclave conditions (iii) and (iv) produced significant variations in strength properties.

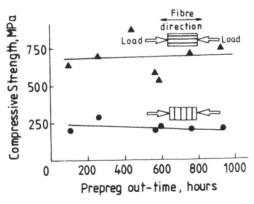

Figure 5 Compressive (Celanese) strength versus prepreg out-time for longitudinal ($\blacktriangle$) and transverse ($\bullet$) test pieces. (Adapted from Ref. 68.)

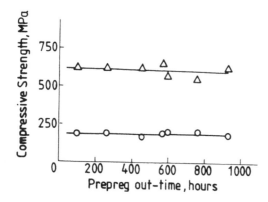

Figure 6 Compressive (Celanese) strength versus prepreg out-time for H/W conditioned longitudinal (△) and transverse (○) test pieces. (Adapted from Ref. 68.)

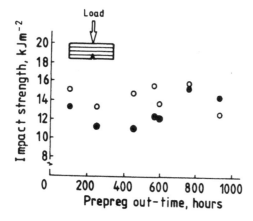

Figure 7 Impact strength at failure initiation versus prepreg out-time for notched (○) and unnotched (●) longitudinal as-cured test pieces. (Adapted from Ref. 68.)

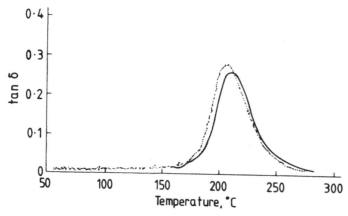

Figure 8 Dynamic mechanical data for longitudinal test pieces. . . . , 100-h, and ——, 800-h prepreg out-times. (Adapted from Ref. 68.)

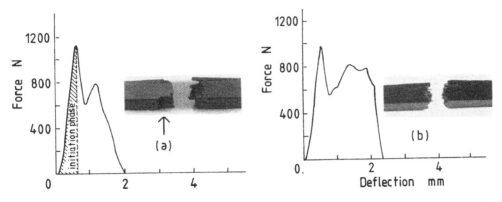

Figure 9 Force–deflection traces and fracture modes for edge-on impacted longitudinal (a) notched and (b) unnotched test pieces. (The arrow locates interface failure at the tip of the notch.) (Adapted from Ref. 68.)

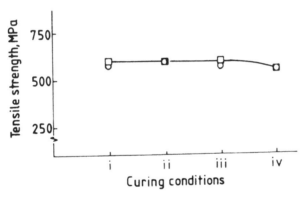

Figure 10 Tensile strength versus curing conditions. □, As-cured; ○, H/W conditioned. (Adapted from Ref. 50.)

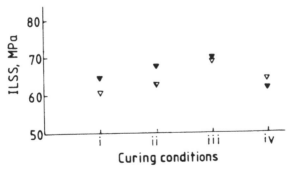

Figure 11 ILSS versus curing conditions. ▼, As-cured; ▽, H/W conditioned. (Adapted from Ref. 50.)

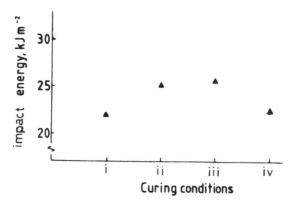

Figure 12 Impact strength at crack initiation versus curing conditions for as-cured test pieces. (Adapted from Ref. 50.)

The higher cure-temperature route (iv) caused reductions in tensile strength (8%), inter-laminar-shear strength (9%) and fracture energy (10%), and a slight improvement in compressive strength (2%). The reasons are perhaps a combination of structural variations, such as cross-link density distributions and the extent of frozen-in thermal stresses as the curing regime alters. The origins and nature of thermal stresses/strains are covered, among others, in Refs. 12 to 14 and 50.

The C-scans (Figure 14) and optical micrographs (Figure 15) show that setting (iii) produced the best and that of (i) the worst laminate, the longer hold-time enabling maximum consolidation with no evidence of voids. The inference is that there exists sufficient interply movement, well beyond the gel point, to drive off final traces of the void. This would be at odds to popular assertion that the resin is immobile beyond the gel point [2].

ILSS shows a maximum (Figure 11) corresponding to the setting (iii); compared to the standard setting (ii), the improvement in ILSS is 4%. Ultrasonic scans show that laminates produced at setting (ii) contained some voids, but there was no evidence of voids when using setting (iii). The interrelationship between void contents and ILSS has

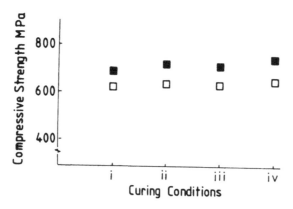

Figure 13 Compressive strength versus curing conditions. ■, As-cured; □, H/W conditioned. (Adapted from Ref. 50.)

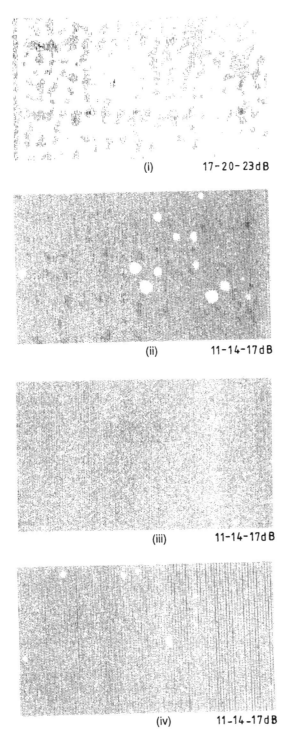

(i)   17-20-23dB

(ii)   11-14-17dB

(iii)   11-14-17dB

(iv)   11-14-17dB

Figure 14 C-scans of laminates (i), (ii), (iii), and (iv). (Adapted from Ref. 50.)

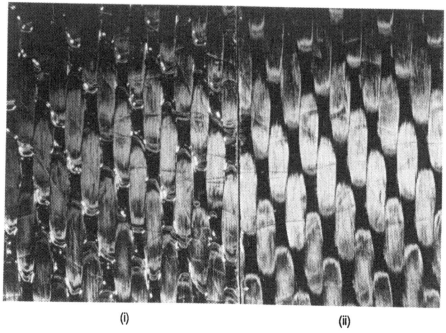

Figure 15 Optical micrographs of laminates (i) and (ii). (Adapted from Ref. 50.)

been identified elsewhere. Judd and Wright [51] indicate a reduction of 7%, corresponding to a void content 1%.

The quality of the resin–fiber interface influences dynamic mechanical properties: the damping behavior, as measured by E'', seems to be sensitive to changes in the quality of the interface. There is an increase in E'' peak value (Figure 16) when laminates (i) and (iii) are compared. This trend is attributed to the improved interface and therefore improved transfer of stress between resin and fiber, so that the resin is not shielded from the applied load and the damping is not obscured. This inference is based on the fact that, normally, an increase in tan δ–peak and E''–peak temperatures, as induced by increased curing/cross-linking, is accompanied by a reduction in the magnitude of the peak values [47].

Hygrothermal Behavior

Figures 2 to 6 and 10 to 13 show the influence of hot/wet conditioning on mechanical properties. Approximately 1.5 wt % moisture uptake had no significant affect on the tensile strength and the flexural modulus (fiber-dominated properties), but there was a reduction in flexural strength. More significant deterioration was observed in resin-dominated properties of compressive strength and ILSS. This arises mainly from the moisture plasticization, whereby the reduced friction within the matrix, particularly between the matrix and the fibers, limits effective stress transfer and hence results in poor stress distributions over the test piece. The mositure-weakened matrix and interface give rise to either delamination or to a more localized form of deformation and failure, depending on the loading mode. In the compression mode, moisture penetration results in a split-type failure (Figure 17). In flexure the presence of mositure inhibits delamination (Figure 18b).

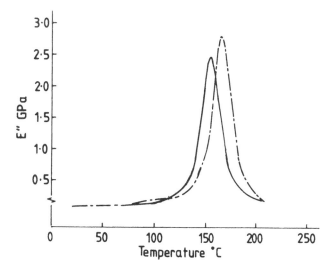

Figure 16 Dynamic loss modulus (E'') versus temperature. ——, Laminate (i); —·—, laminate (iii). (Adapted from Ref. 50.)

One of the most dramatic effects of moisture ingress is the extent of reduction in T_g. Figure 19 shows the drop in T_g and in tan δ–peak height with moisture plasticization. The magnitude of moisture-induced T_g depression is, however, a rather disputed subject, concerning mainly the prediction models: their origins and the values chosen for the parameters that are included in these predictions [9,10,52–54].

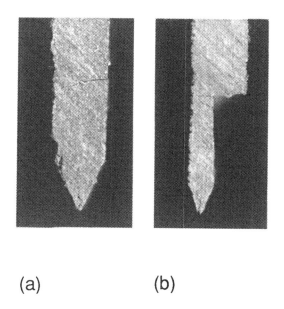

(a)   (b)

Figure 17 Failure modes in compression for transverse UD test pieces (side views); typical for (a) as-cured and (b) H/W conditioned. (Adapted from Ref. 68.)

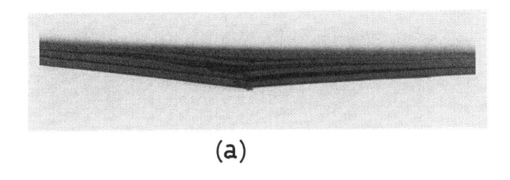

(a)

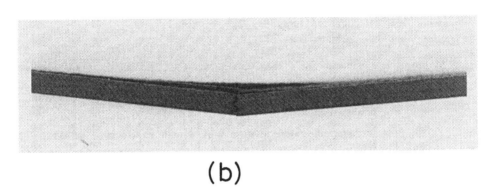

(b)

Figure 18 Failure modes in flexure for longitudinal UD test pieces (side views): typical for (a) as-cured and (b) H/W conditioned. (Adapted from Ref. 68.)

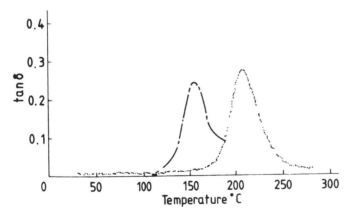

Figure 19 Dynamic mechanical data for longitudinal UD test pieces of 250-h prepreg out-time. . . . , As-cured; —·—, H/W conditioned. (Adapted from Ref. 68.)

Table 2 Predicted Values of Moisture-Induced T_g Depression (ΔT_g) Using Different Models and Data

Model Type	Parameter Values (source references)				Predicted ΔT_g (°C)
	α_2 (°C^{-1})	$\Delta\alpha$ (°C^{-1})	ΔC_{p2} (J g^{-1} °C^{-1})	ΔC_{p1} (J g^{-1} °C^{-1})	
Eq. (1)	2.1×10^{-4} at 23°C (Ref. 54)	4.8×10^{-4} (WLF constant)			8
	8.4×10^{-4} at 100°C (Ref. 54)	3.78×10^{-4} (Ref. 10)			35
Eq. (2)			1.94 (Refs. 63, 65)	0.31[a]	70
			1.94 (Refs. 63, 65)	0.16[a]	114

Source: Adapted from Ref. 68.
[a] The values were obtained from the empirical relationship [69] of $(T_g + 273) \times \Delta C_p = 78.1$ to 146.2.

The Kelley–Bueche prediction [58] for the compositional dependence of T_g based on free volume theory [59] states that

$$T_g = \frac{V_1 \Delta\alpha T_{g1} + V_2\alpha_2 T_{g2}}{V_1 \Delta\alpha + V_2\alpha_2} \qquad (1)$$

where $V_{1,2}$ is the volume fraction of polymer and diluent, respectively, α_2 is the coefficient of cubic expansion of the diluent in a nonglassy phase, and $\Delta\alpha$ is the change in the coefficient of cubic expansion of the polymer of T_g. Equation (1) has been applied to water-conditioned epoxy systems with varying degrees of success [9,10,57].

A thermodynamic theory [60,61] based on entropy considerations has been modified for systems where the component T_g values are well apart, as in epoxy–water, and gives the T_g of the system as [62,63].

$$T_g = \frac{w_1 \Delta C_{p1} T_{g1} + W_2 \Delta C_{p2} T_{g2}}{w_1 \Delta C_{p1} + w_2 \Delta C_{p2}} \qquad (2)$$

where subscripts 1 and 2 refer to polymer and diluent, respectively, w is the weight fraction, and ΔC_p is the incremental change in specific heat at glass transition.

Table 2 shows the application of these predictions to the experimental data of $T_{g1} = 207$°C and $\Delta T_g \simeq 50$°C (extracted from Figure 19) at a moisture uptake of 4.1% (assuming the composite moisture weight gain of 1.5% to be totally absorbed by the matrix). The T_g of water was taken to be -138°C [64,65] and the specific gravity of epoxy as 1.27. Equation (1) underestimates the T_g depression, and Eq. (2) considerably overestimates it. Although the prediction due to Eq. (2) will improve if the moisture retained within the resin–fiber interface and within interlaminar defects can be isolated and accounted for.

One of the justifiable criticisms leveled at the use of these models is the choice of the values for the material parameters [52]. For example, the value of ΔC_p for the polymer ought to be measured, although this in itself is fraught with difficulties. The technique commonly employed for such a measurement is DSC; which is a rather indefinite and

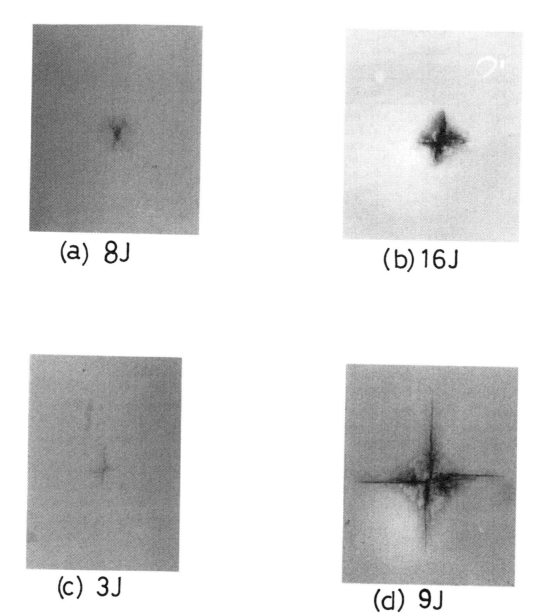

Figure 20 X-radiographs of damaged test pieces of 5-HS satin weave Fiberite materials: (a, b) CF/toughened-epoxy; (c, d) CF/epoxy. (Adapted from Ref. 25.)

subjective method, at best providing data with large scatter, particularly with highly cross-linked networks [52,53,66,67]. For convenience it will be worth seeking a parameter of the same origin as the heat capacity but easier to measure. The data obtained from DMA are very definitive and reproducible, and the dynamic mechanical properties have the same sensitivity to molecular features as the heat capacity (e.g., tan δ–peak height and C_p [66] both decrease as cross-link density increases). Accordingly, DMA data may be incorporated

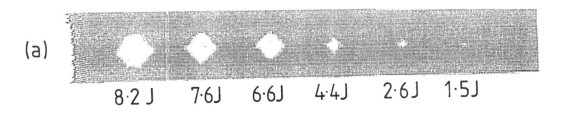

(a)

8·2 J   7·6J   6·6J   4·4J   2·6J   1·5J

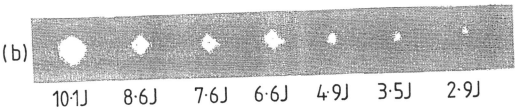

(b)

10·1J   8·6J   7·6J   6·6J   4·9J   3·5J   2·9J

Figure 21   C-scans of damaged test pieces of PW Cycom materials: (a) CF/epoxy; (b) CF/toughened epoxy. (Adapted from Ref. 25.)

into a modified Couchmann–Karasz relationship such that K in the following equation becomes a function of tan δ–peak height [68].

$$T_g = \frac{w_2 T_{g2} + K w_1 T_{g1}}{w_2 + K w_1}$$

It will further be noted that the magnitude of the reduction in tan δ-peak height, due to moisture ingress, corresponds to the decrease in matrix-dominated mechanical properties: ILSS and compressive strength.

Postimpact Residual Strength in Laminates

Figures 20 and 21 show examples of impact damage, in the form of delaminations and resin–fiber cracking. The radiographs in Figure 20 show that the toughened system arrests and limits the crack propagation. The crack is confined to the delaminated area in the case of a toughened system but is ahead of the delaminated region in an untoughened system. Figure 21 shows that for the same impact energy the damage area is greater for the untoughened material.

Residual compressive strength changes rapidly at first and then more gradually as the incident impact energy increases; see Figures 22 and 23. Undamaged test pieces invariably failed at the clamps; consequently, additional tests were conducted using the Celanese method. Failures were then mainly within the gauge length and gave strength values consistently higher (approximately 1.5-fold; see Figure 22). This improved performance is explained by avoiding clamp-induced premature failure and by the different mode of failure due to the change in specimen dimensions. Figure 24a shows, for a test using the Celanese rig, a generation of multiple delaminations which diverts and helps arrest the propagation of shear failure.

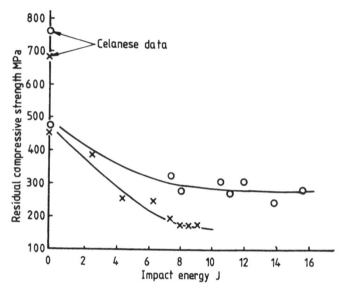

Figure 22 Residual compressive strength versus incident impact energy for 5-HS satin weave Fiberite. ○, CF/tougened epoxy; X, CF/epoxy (10-ply laminates of 3.4 mm thickness). (Adapted from Ref. 25.)

Residual flexural strength values for the woven system are presented in Figures 25 and 26 and show a leveling off in deterioration as the impact energy increases. In the toughened systems, the deterioration is less and, moreover, becomes significant only beyond some threshold of impact energy. This is particularly evident in Fiberite systems, but the comparison in this case is perhaps unfair, since Fiberite 984 is claimed to have high environmental resistance rather than toughness. Flexural failure occurred by through-

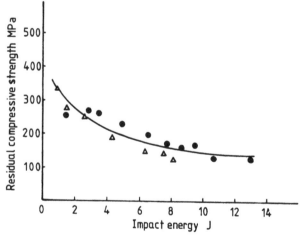

Figure 23 Residual compressive strength versus incident impact energy for PW Cycom. ● CF/toughened epoxy; △, CF/epoxy (10-ply laminates of 2.2 mm thickness). (Adapted from Ref. 25.)

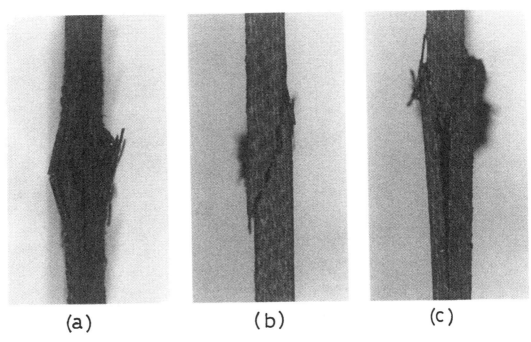

Figure 24 Failure modes in compression for the 5-HS woven systems. (Adapted from Ref. 25.)

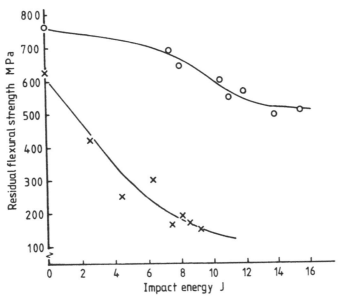

Figure 25 Residual flexural strength versus incident impact energy for 5-HS satin weave Fiberite. ○, CF/toughened epoxy; X, CF/epoxy (10-ply). (Adapted from Ref. 25.)

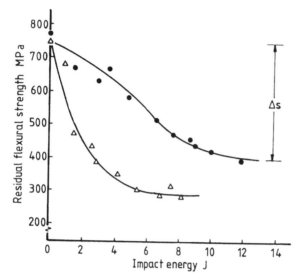

Figure 26 Residual flexural strength versus incident impact energy for PW Cycom. ●, CF/ toughened epoxy; △, CF/epoxy (10-ply). (Adapted from Ref. 25.)

thickness cracks running across the specimen width and causing a final split, but in toughened systems delamination (Figure 27) was pronounced (i.e., greater strain energy input is required for failure in the controlled rather than sudden failure mode).

There is evidence of a relation between the total drop in residual flexural strength and the fracture toughness, K_c. The fracture data are presented in Figures 28 and 29. The material with higher K_c is more damage tolerant and accordingly, retains a higher percentage

Figure 27 Side view of failed flexural test piece. (Adapted from Ref. 25.)

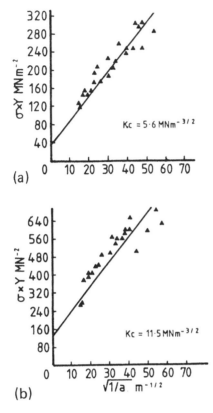

Figure 28 Fracture data for 5-HS satin weave Fiberite: (a) CF/epoxy; (b) CF/toughened epoxy (10-ply). (Adapted from Ref. 25.)

of its initial strength. In fact, for corresponding toughened and untoughened systems, K_c (untoughened) $\cdot K_c^{-1}$ *(toughened)* $\simeq \Delta s$ *(toughened)*. Δs^{-1} (untoughened) where Δs is the total drop in flexural strength as defined in Figure 26.

Repaired specimens which contained significant impact damage (incident impact energy $\geq$ 6 J), employing either autoclave or heating blanket repair methods, were examined under compression. Considerable strength was regained, repaired specimens exhibiting approximately 80% of the undamaged strength. Figure 30 illustrates some of the repairing steps. Detailed information on the techniques of repairing and evaluation of composite repairs is presented in Ref. 70.

Postimpact Residual Strength in Sandwich Panels

Impact-Induced Damage

The nature of the low-energy impact-induced damage was dictated mainly by the type of material, panel thickness, and the level of incident energy. The damage was confined to the face skin and the core; there was no detectable damage in the back skin. This concurs with previous work [29], where the panels had also been supported peripherally, and differs from the work [71] where the whole panel had been placed on a rigid surface.

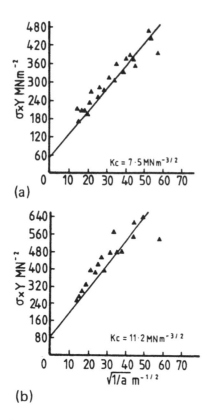

Figure 29 Fracture data for PW Cycom: (a) CF/epoxy; (b) CF/Toughened epoxy (10-ply). (Adapted from Ref. 25.)

Figure 31 shows the area of damage on the face skin as a function of incident impact energy, and Figure 32 shows the associated C-scans. Damage was detected by C-scan at a 3-J incident impact level but was only visible beyond 5 J. Through-thickness micrographs (Figure 33) of the damaged skin and part of the honeycomb core show the progression of delaminations from the inner plies.

Figure 31 shows that the damaged area increases in a sigmoidal pattern with increasing energy and indicates impact levels for incipient and maximum damage. The maximum damage area for 12.5-mm-thick panels is either comparable with, or less than, the projected area of the striker (123 mm²). This is at variance with the behavior of sandwich structures of unidirectional carbon fiber/epoxy skins [26,29]. The damage area was greater ($\simeq$ 140 mm²) in the thicker panel. The thicker panels being stiffer, less of the impact energy is expended in total panel deformation and rather more in localized deformation and damage. The differences in the bending stiffness is illustrated by instrumented impact photography [72] (see Figure 34).

A comparison of the 12.5-mm-thick panels with respect to the core material suggests that the Rohacell foam is more effective in containing the skin damage. Obviously, such a comparison is misleading, since in the case of panel A the skins consisted of a different prepreg system, Cycom 985, which gives a tougher laminate, as will be shown subsequently. The skins of panels B and C were from the same prepreg, Fiberite 934, but panel

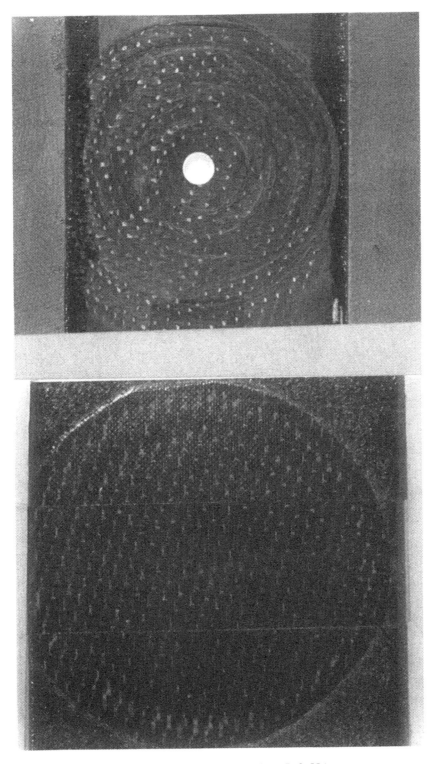

Figure 30 Stepped and repaired area. (Adapted from Ref. 55.)

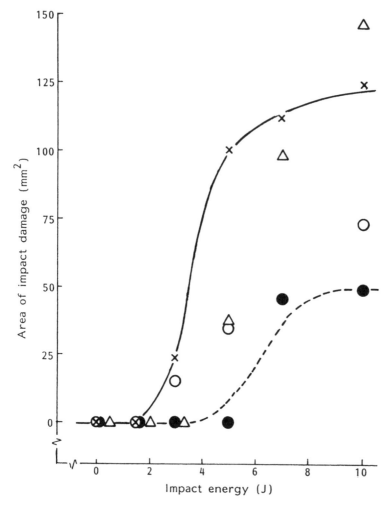

Figure 31   Face-skin damage area (by C-scan) versus incident impact energy for panels A (○), B (X), C (●), and E (△). Adapted from Ref. 30.)

C was produced with a greater autoclave pressure (twice as great), leading to improved laminate consolidation.

Production-Induced Defect

It will be appreciated that honeycomb panels of enhanced performance could be fabricated by employing a two-step operation: the prepregs would be laminated at appropriate pressures and then bonded to the honeycomb at a much lower pressure, to avoid cell-wall buckling and localized laminate sag. The low autoclave pressure adopted in the present single-step operation resulted in an inferior laminate surface and, possibly, poor consolidation of prepreg plies [50].

The manufacturers of Rohacell claim that single-step fabrication can be successful. In the present work problems were encountered in the production of Rohacell 51 panels at

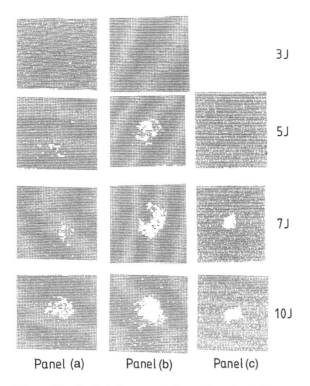

3 J

5 J

7 J

10 J

Panel (a)   Panel (b)   Panel (c)

Figure 32 Typical C-scans. (Adapted from Ref. 30.)

the standard autoclave pressure of 600 kPa; the panels were distorted and were of non-uniform thickness, due to core collapse. Uniform thickness panels were produced only by reducing the pressure to 300 kPa. Rohacell 51 has a density comparable to that of the 48-kg/m^3 Nomex honeycomb and hence provides the same advantages in terms of lightness, but the suppliers do not recommend it for aircraft applications. The higher grades of Rohacell foam did provide uniform panels at an autoclave pressure of 600 kPa, but some localized cell-wall collapse and cell coalescence occurred. Figure 35 shows that cell coalescence results in a layer of large voids immediately beneath the skin–core adhesive interface. The magnitude of the voids was reduced in the higher-density samples. Foam collapse was also evident by the reduced thickness of Rohacell panels relative to the honeycomb panels: by between 0.5 and 2 mm, depending on the production pressure and the foam density.

Postimpact Residual Compressive Behavior

Figures 36 and 37 show that, as would be expected, compressive load to failure for undamaged Rohacell panels increases with foam density and/or core thickness. The load to failure reduces asymptotically as the incident impact energy level increases. It is also clear that Rohacell 51 panels had limited residual capacity. Rohacell 71 and Nomex honeycomb panels exhibited identical capacity, and no further improvement was observed with Rohacell 110. An increase in core thickness provided no significant improvement, since as outlined previously, increased thickness led to more severe skin damage. A

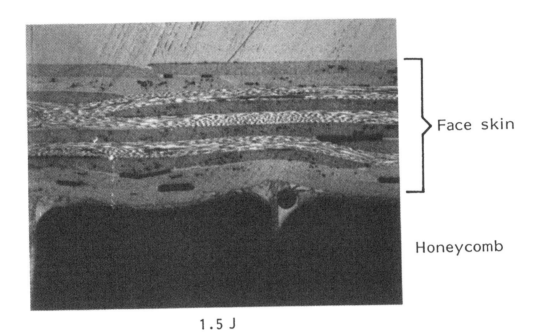

1.5 J

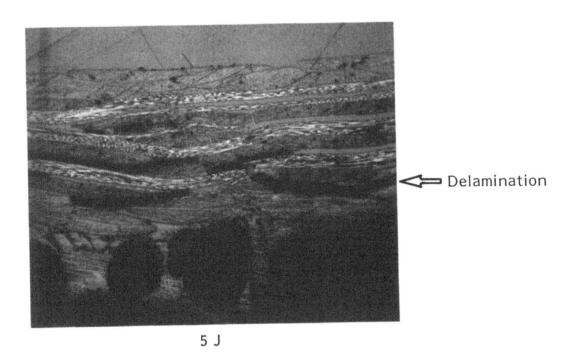

5 J

Figure 33 Micrographs of impact-damaged sections for a honeycomb core sample. (Adapted from Ref. 30.)

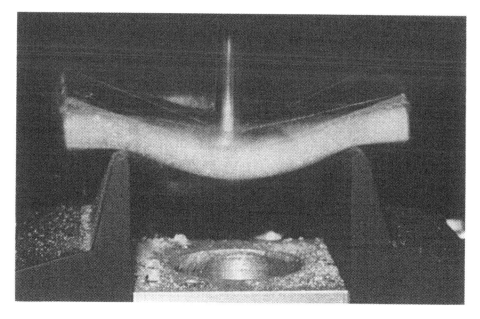

(a)

(b)

Figure 34 Impact bending of (a) panel C and (b) panel E.

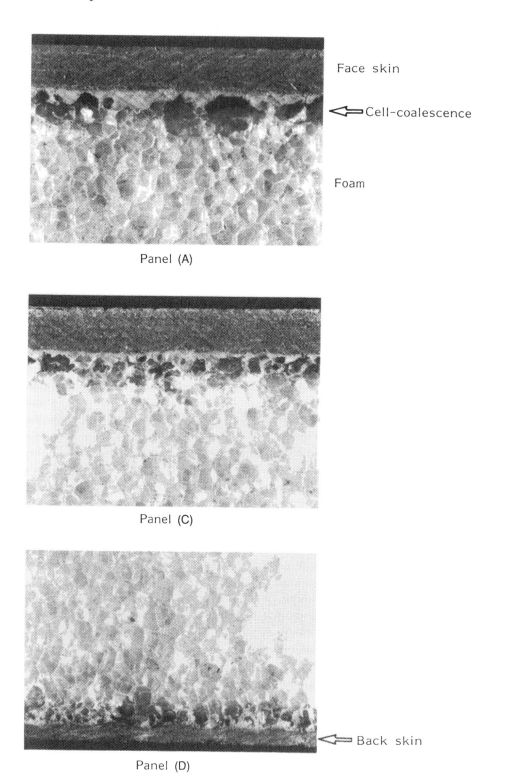

Face skin

⇐ Cell-coalescence

Foam

Panel (A)

Panel (C)

⇐ Back skin

Panel (D)

Figure 35 Samples with Rohacell foam core, showing cell-coalescence. (Adapted from Ref. 30.)

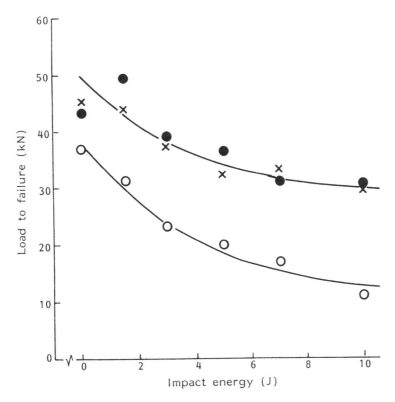

Figure 36 Compressive load to failure versus incident impact energy for unclamped samples. ○, Panel A; X, panel B; ●, panel C. (Adapted from Ref. 30.)

comparison of Figures 36 and 37 shows that the manner of supporting panels (i.e., clamped or unclamped), does not influence subsequent compressive behavior.

Under compression, the failure developed from the damaged skin and moved across the core to the unbroken skin. There were, however, distinct features of deformation and/ or failure, depending on the core type. In the honeycomb test pieces, the second skin was either visibly intact (Figure 38a) or showed a crack at a 45° shear plane relative to the crack in the first skin (Figure 38b). There was no visible skin–core separation, and the type of localized honeycomb tear shown in Figure 38c was an exception.

Rohacell-core panels exhibited one of two forms of failure: (1) a crack along the skin and (2) crushing across the panel cross section, to an extent determined by the density of the foam. Rohacell 51 showed very pronounced tearing (Figure 39a) with occasional, complete skin–core separation, but less so at the higher densities. The crack in the opposite skin usually occurred directly opposite that in the damaged skin and via a through-thickness crack in the foam (Figure 39c). In some test-pieces the second skin showed no failure, and these corresponded to limited through-thickness cracks in the most dense foam (Figure 39d) and with severe along-the-skin core tear in the lower-density foams (Figures 39a and b). The higher strength of the Rohacell 110 foam and the premature splitting of the Rohacell 55 foam helped avoid skin failure, albeit in the latter case the integrity of the panel was destroyed by the magnitude of core failure.

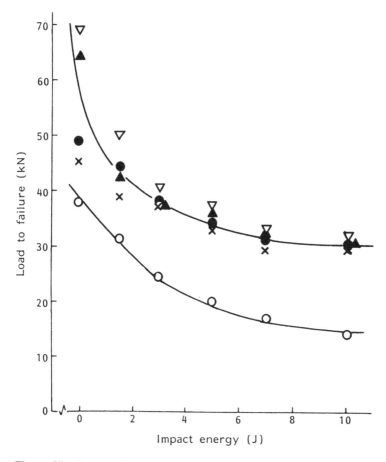

Figure 37 Compressive load to failure versus incident impact energy for clamped samples. ○, Panel A; X, panel B; ●, panel C; ▲, panel D; ∇, panel E. (Adapted from Ref. 30.)

The ease of along-the-skin foam splitting, especially in low-density foams, was a reflection of the magnitude of the production-induced voiding. This clash between the desirable properties of panel lightness and foam integrity could be resolved by using an appropriate foam formulation. The aim should be to achieve a multitier cell-size distribution with smaller cells near the surfaces (i.e., a variant of integral-skin foam).

Impact Puncture Behavior

Figure 40 shows the load–deflection traces of through-thickness puncture of 8-HS fabric laminates and the sandwich panels of 12.5-mm core thickness. It can be seen that the Cycom prepreg system has superior impact properties. A comparison of the load–deflection traces for laminates and their counterparts as panel skin show that the latter absorbs greater energy to fracture. This is attributable to the support provided by the core, and particularly obvious is the improved resistance to crack propagation; that is, the load drops immediately after crack initiation in the case of single laminates, whereas it continues to increase into the propagation phase in the failure of panel skins.

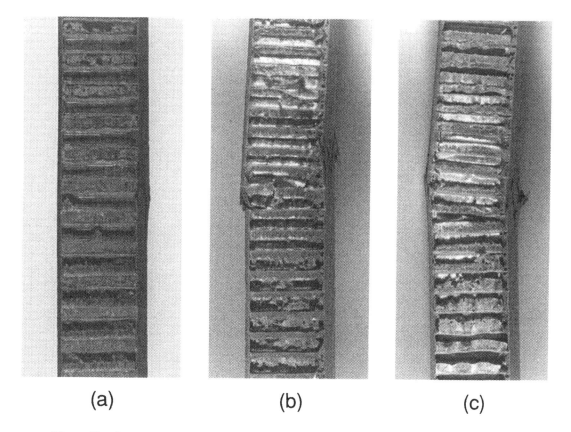

Figure 38 Compression-failed honeycomb core samples. (Adapted from Ref. 30.)

The load–deflection traces (Figure 40b and c) clearly show the resistance offered by the cores. The foams provide greater resistance than the honeycomb, to a degree determined by the foam density. Impact loads sustained by the various cores are extracted from the load-deflection traces and presented in Figure 41.

Sectioning test pieces through the punctured holes (examples shown in Figure 42) enabled an assessment of skin–core separation. Honeycomb specimens showed no visible separation, and the Rohacell ones exhibited varying degrees of separation, caused by cracking of the foam, depending on the density—complete separation with the lowest and no visible separation with the highest densities.

Static Pin-Bearing Behavior

Resistance to bearing will be described as the maximum stress (σ_m) sustained by the specimen, the stress at initial failure (σ_i), and the stress at which the bearing hole is deformed 4% of its diameter.

Influence of Prepreg Type and Layup

Figure 43 shows the maximum bearing stress for the laminate systems, constrained only by finger tightening. In general, woven laminates supported higher stresses than UD

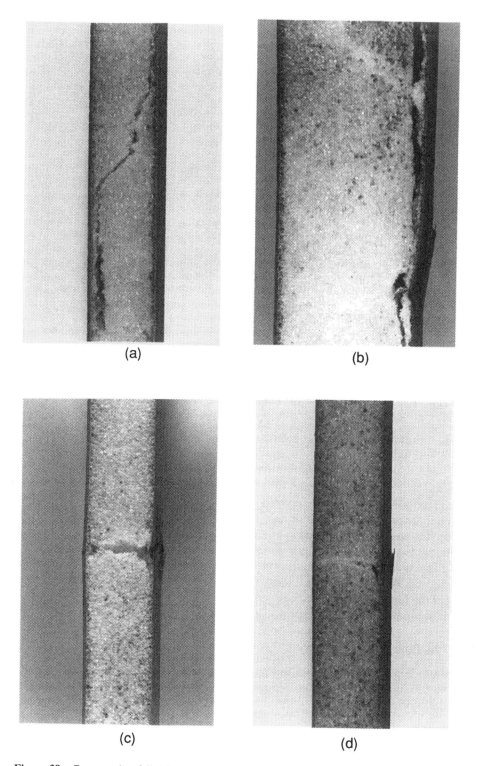

Figure 39 Compression-failed foam core samples: (a) panel A; (b) panel E; (c, d) panel D. (Adapted from Ref. 30.)

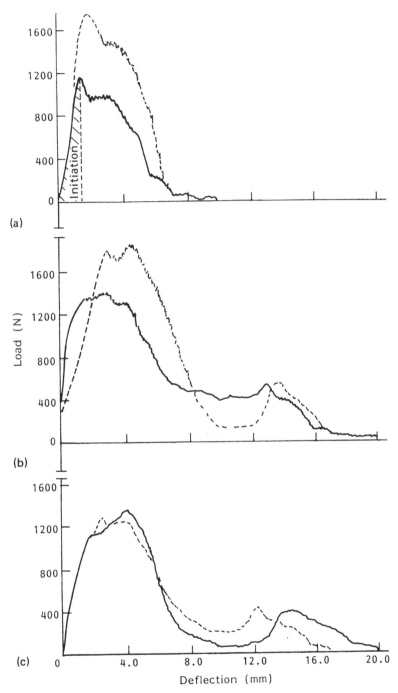

Figure 40 Instrumented impact load–deflection traces for (a) ——, Fiberite; ---, Cycom laminates; (b) ——, panel D; ---, panel A; (c) ——, panel B; ---, panel C. (Adapted from Ref. 30.)

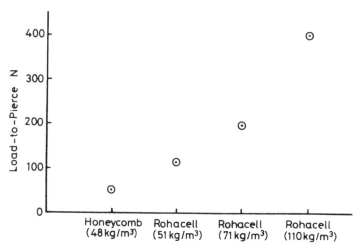

Figure 41 Impact load sustained by various core types.

laminates, depending on the layup and geometric factors. The adoption of all 0°/90° fiber orientation (as in specimen α) generates high stress concentrations and causes premature failure in tension.

The resin systems designated as Cycom 985 and Fiberite 934 are considered to be similar, but Cycom 1010 is a toughened epoxy and increases the ductility of the composite and hence reduces the likelihood of high stress concentrations. Accordingly, specimen 2 produced the highest value for σ_m.

The failure modes are indicated on Figure 43. Most specimens underwent a significant degree of bearing failure, except the (0°, 90°) PW specimen, which failed in tension following limited bearing failure, and the 8-HS specimen, which exhibited a mixture of bearing, cleavage, and shear-type failures.

The woven laminates, particularly PW types, and as would be expected, the UD laminates with the higher number of plies showed greater resistance to the initiation of permanent damage (Figure 44). The low σ_i values experienced with the toughened epoxy system probably result from the greater ductility of the matrix. Toughened epoxies are excellent in containing damage, but they can suffer initial damage more readily under certain loading conditions.

In general, the UD specimens exhibited slightly higher stress levels than the woven specimens at 4% strain. This is as would be expected, since UD laminates are inherently stiffer, suffer no weaving-induced fiber distortion, and in this case, there was significant 0°-ply content (Table 1). At this low level of strain, the specimen suffered no visible damage or any damage that could be identified by x radiography. It follows that in regard to material integrity, the significance of the ASTM recommendation referred to earlier is open to question.

Influence of Lateral Constraint

Laminates 5 and 6, which were investigated over a range of lateral constraint (Figures 45 and 46), show that the bearing stress gradually increased with increasing constraint

Panel (B)

Panel (D)

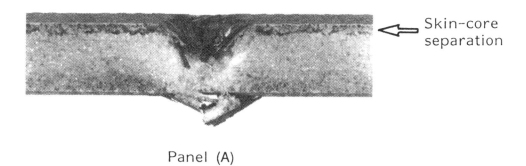

Skin-core
separation

Panel (A)

Figure 42 Sample sections showing impact puncture. (Adapted from Ref. 30.)

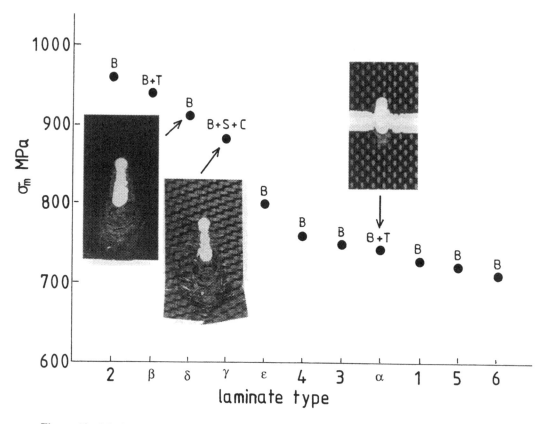

Figure 43 Maximum bearing stress (σ_m) for various laminate types. (Types of failure: B, bearing; T, tension; C, cleavage; S, shear.) (Adapted from Ref. 43.)

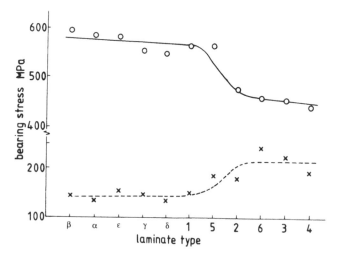

Figure 44 Bearing stress for various laminate types. ○, σ_i; X, stress at 4% pinhole deformation. (Adapted from Ref. 43.)

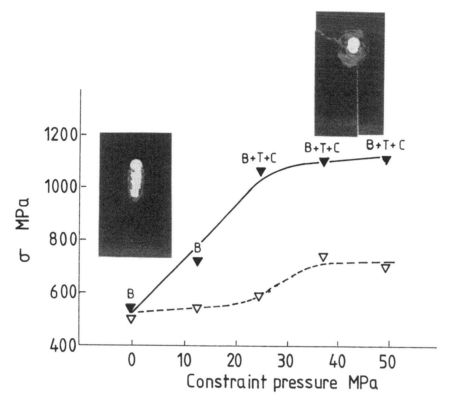

Figure 45 Influence of constraint pressure on (▼) σ_m and (▽) σ_i for specimen 5. (Adapted from Ref. 43.)

pressure and leveled off above 25 MPa (specimen 5). The associated increases for specimens 5 and 6, respectively, were approximately 118% in σ_m and 36% in σ_i for specimen 5 and 275% and 250% for specimen 6. These magnitudes are consistent with previously published work [35,39]. Thinner laminates benefit to a greater degree from the introduction of constraint pressures, since they are more prone to local buckling.

A transition in the failure mode was noted and coincided with the change in bearing stress pattern (Figure 45). The thicker laminate, specimen 5, exhibited a bearing-type failure only up to a 25-MPa constraint; beyond this, limited initial bearing failure was followed by a final failure in the combined tension–cleavage mode. It would seem that effective inhibition of laminate "brooming" requires a certain minimum constraint level. The transition in the failure mode occurred at 46 MPa for the thinner laminate (specimen 6), and the increase in bearing stress showed no leveling off up to this point. The failure mode changed from 100% bearing failure to limited bearing–tension failure along 45° fiber alignment.

An assessment was made of the correlation between static strengths, as represented by σ_m and σ_i, and fatigue behavior. Figure 47 shows comparisons between specimens of

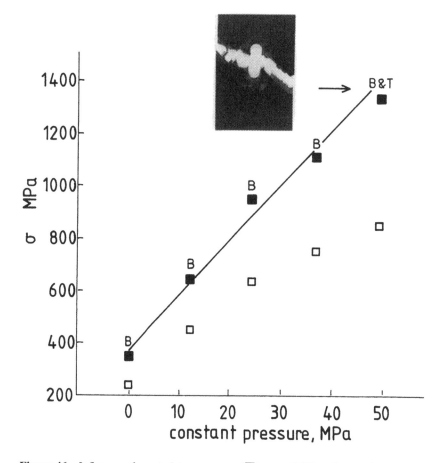

Figure 46 Influence of constraint pressure on (■) σ_m and (□) σ_i for specimen 6. (From Ref. 56.)

types 5 and 6 with similar σ_m but different σ_i values. Figure 48 shows comparisons between specimens α and β with different σ_m but similar σ_i. It can be seen that for $N \geq 10^5$, σ_i controls the fatigue performance; the fatigue strength approaches the value σ_i. Further work would be needed to confirm the extrapolation for $N > 10^5$ in Figure 47.

ACKNOWLEDGMENTS

The materials were kindly donated by Fiberite Europe, Cyanamid Fothergill UK, and Short Bros. plc, Belfast. The help of Short Bros., particularly in the production of laminates, is greatly appreciated. Special gratitude is due to Mrs. Dorothy Savage and Miss June Johnston, who transformed all sorts of manuscripts into very high quality typescript.

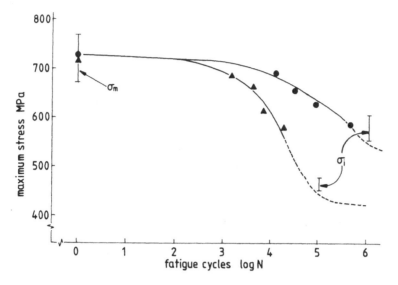

Figure 47 *S–N* curves for specimen 5 (●) and specimen 6 (▲). (Adapted from Ref. 43.)

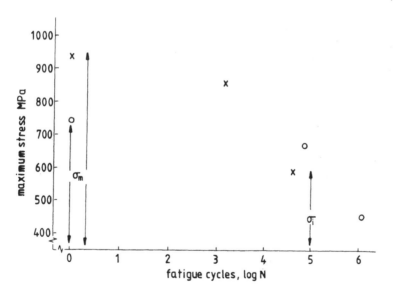

Figure 48 *S–N* curves for specimen α (○) and specimen β (X). (Adapted from Ref. 43.)

REFERENCES

1. S. N. Sanjana, *SAMPE J.* 5 (1980).
2. D. Purslow and R. Childs, *Composites*, 17: 127 (1986).
3. C. E. Morrison and M. G. Bader, *Composites*, 20: 9 (1989).
4. G. S. Springer, *J. Compos. Mater.*, 16: 400 (1982).
5. F. R. Jones, J. W. Rock, and J. E. Bailey, *J. Mater. Sci.*, 18: 1059 (1983).
6. R. F. Dickson, C. J. Jones, T. Adam, H. Reiter, and B. Harris, *Proc. First International*

6. R. F. Dickson, C. J. Jones, T. Adam, H. Reiter, and B. Harris, *Proc. First International Conference on Acoustic Emission from Reinforced Composites*, Paper A-4, SPI, New York, 1983.

7. R. Delasi and J. B. Whiteside, *Advanced Composite Materials: Environmental Effects*, ASTM STP-658, ASTM, Philadelphia, 1978, p. 2.

8. P. Peyser and W. D. Bascom, *J. Mater. Sci.*, 16:75 (1981).

9. E. L. McKague, J. D. Reynolds, and J. E. Halkias, *J. Appl. Polym. Sci.*, 22: 1643 (1978).

10. C. E. Browning, G. E. Husman, and J. M. Whitney, *Composite Materials: Testing and Design*, ASTM STP-617, ASTM, Philadelphia, 1977, p. 481.

11. T. A. Collings and D. E. W. Stone, *Compos. Struct.*, 3: 341 (1985).

12. T. A. Collings and D. E. W. Stone, *Composites*, 16: 307 (1985).

13. W. W. Right, *Composites*, July 1981, p. 201.

14. H. T. Hahn, *J. Compos. Mater.*, 10: 266 (1976).

15. J. Aveston, A. Kelly, and J. M. Sillwood, Proc. ICCM3 (Paris), *Advances in Composite Materials*, Vol. 1, Pergamon, Oxford, 1980, p. 556.

16. E. L. McKague, J. E. Halkias, and J. D. Reynolds, *J. Compos. Mater.*, 9: 2 (1975).

17. C. E. Browning, *SAMPE 22nd National Symposium*, Dan Diego, Calif. 1977, p. 365.

18. R. Jones, W. Broughton, R. F. Mousley, and R. T. Potter, *Compos. Struct.*, 3: 167 (1985).

19. K. M. Lal, *J. Rein. Plast. Compos.*, 2: 226 (1983).

20. S. M. Bishop, *Compos. Struct.* 3: 295 (1985).

21. M. D. Rhodes, J. G. Williams, and J. H. Storness, *34th Annual Technical Conference, Reinforced Plastics/Composites Institute*, 1979.

22. V. S. Avva, *ASTM STP-808*, p. 140 (1982).

23. D. R. Carlisle and D. C. Leach, *15th National SAMPE Technical Conference*, 1983.

24. G. Dorey, AGARD Conference Proc. *163*, Paper 8, 1975.

25. M. Akay, *Compos. Sci. Technol.*, 33:1 (1988).

26. D. W. Oplinger and J. M. Slepetz, *Foreign Object Impact Damage to Composites*, ASTM STP-568, ASTM, Philadelphia, 1975, p. 30.

27. M. D. Rhodes, *NASA TM-78719* (Oct. 1978).

28. A. V. Sharma, *Test Methods and Design Allowables for Fibrous Composites*, ASTM STP-734, ASTM, Philadelphia, 1981, p. 54.

29. M. L. Bernard and P. A. Lagace, *Proc. American Society for Composites, 2nd Technical Conference*, University of Delaware, Technomic, Lancaster, Pa., 1987, p. 167.

30. M. Akay, *Composites*, 21: 325 (1990).

31. S. R. Swanson and J. S. Burns, *31st International SAMPE Symposium*, 1986, p. 1070.

32. T. H. Tsiang, *Compos. Tech. Rev.*, 6: 74 (1984).

33. E. W. Godwin and F. L. Matthews, *Composites*, July 1980, p. 155.

34. J. R. Vinson, *Polym. Eng. Sci.*, 29: 1333 (1989).

35. T. A. Collings, *RAE Tech. Rep. 75127*, 1975.

36. T. A. Collings, *Composites*, Jan. 1977, p. 43.

37. T. A. Collings, *Composites*, July 1982, p. 241.

38. T. A. Collings and M. J. Beauchamp, *Composites*, 15:33 (1984).

39. P. A. Smith and K. J. Pascoe, *Compos. Struct.* 6: 1 (1986).

40. P. A. Smith, K. J. Pascoe, C. Polak, and D. O. Stroud, *Compos. Struc.*, 6: 41 (1986).

41. T. A. Collings, *Joining Fibre-Reinforced Plastics*, F. L. Matthews (ed.), Elsevier Applied Science, Amsterdam, 1987.

42. R. L. Ramkumar, *ASTM STP-734*, p. 376 (1981).

43. M. Akay, *ECCM-4*, Stuttgart, Germany, Sept. 1990.

44. W. F. Brown and J. E. Srawley, *ASTM STP-410* (1966).

45. E. Plati and J. G. Williams, *Polym. Eng. Sci.*, 15: 470 (1975).

46. P. T. Curtis (ed.), *CRAG Test Methods*, RAE TR 85099, RAE, Farnborough, England, Nov. 1985.
47. L. E. Nielsen, *Mechanical Properties of Polymers and Composites*, Vol. 1, Marcel Dekker, New York, 1974, Chap. 4.
48. M. Akay, S. J. Bryan, and E. F. T. White, *J. Oil Colour Chem. Assoc.*, 56: 86 (1973).
49. T. Murayama, *Dynamic Mechanical Analysis of Polymeric Material*, Elsevier, Amsterdam, 1978, Chap. 3.
50. M. Akay, in *Interfacial Phenomena in Composite Materials*, F. R. Jones (ed.), Butterworth, London 1989, p. 155.
51. N. C. W. Judd and W. W. Wright, *SAMPE J.*, 14: 10 (1978).
52. P. Peyser and W. D. Bascom, *J. Mater. Sci.*, 16: 75 (1981).
53. H. G. Carter and K. G. Kibler, *J. Compos. Mater.*, 11: 265 (1977).
54. L. S. A. Smith and V. Schmitz, *Polymer*, 29:1871 (1988).
55. G. Wright, M. Sc. dissertation, University of Ulster at Jordanstown, Ireland, 1987.
56. M. Akay, work to be published.
57. R. J. Morgan and J. E. O'Neal, *Poly. Plast. Technol. Eng.*, 10: 49 (1978).
58. F. N. Kelley and F. J. Bueche, *J. Polym. Sci.*, 50: 549 (1961).
59. M. L. Williams, R. F. Landel, and J. D. Ferry, *J. Am. Chem. Soc.*, 77: 3701 (1955).
60. P. R. Couchman and F. E. Karasz, *Macromolecules*, 11: 117 (1978).
61. P. R. Couchman, *Polym. Eng. Sci.*, 24: 135 (1984).
62. G. ten Brinke, F. E. Karasz, and T. S. Ellis, *Macromolecules*, 16: 244 (1983).
63. T. S. Ellis and F. E. Karasz, *Polymer*, 25: 664 (1984).
64. C. A. Angell, J. M. Sare, and E. J. Sare, *J. Phys. Chem.*, 82: 2622 (1978).
65. M. Sugisaki, H. Suga, and S. Seki, *Bull. Chem. Soc. Jpn.*, 41: 2591 (1968).
66. P. Moy and F. E. Karasz, in *Water in Polymers*, ACS Symposium Series 127, P. S. Rowland (ed.), American Chemical Society, Washington, D.C., 1980, paper 30.
67. R. P. Kambour, C. L. Gruner, and E. E. Romagosa, *J. Polym. Sci. Polym. Phys. Ed.*, 11: 1879 (1973).
68. M. Akay, *Compos. Sci. Technol.*, 38: 359 (1990).
69. R. Simha and R. F. Boyer, *J. Chem. Phys.*, 37: 1003 (1962).
70. H. Brown (ed.), *Composite Reparis*, SAMPE Monograph 1, Covina, Calif., 1985.
71. W. G. J. 't Hart, National Aerospace Laboratory, Amsterdam, Dec. 1981.
72. D. Barkley, M. Akay, work to be published.

8

Delamination Onset and Growth in Composite Laminates

Lin Ye
Department of Engineering Mechanics
Xian Jiaotong University
Xian, Shaanxi, China

INTRODUCTION

The use of graphite–epoxy composites in the design of secondary structures of aircraft (such as flight control surfaces) has become well established. These components, developed to replace aluminum structures on existing transport aircraft, have provided a weight reduction of up to 25%. These materials are also used in military aircraft for primary structures such as wings and tails, and weight saving has reached an average of 34% [1]. The application of composite materials to primary structures of commercial aircraft is still under development. However, using first-generation epoxies, the damage tolerance requirements result in nonoptimum structures which do not utilize the maximum weight-saving potential of composite materials [2,3].

349

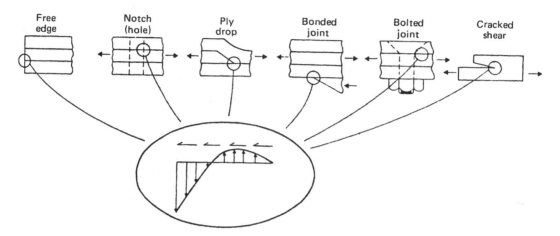

Figure 1   Sources of out-of-plane loads from load path discontinuities. (From Ref. 7.)

When a laminated composite structure is subjected to a static or cyclic load, damage initiation and propagation occur in a series of events, whereby individual modes develop and interact with each other [4]. Due to the anisotropic and heterogeneous nature of the composite materials, the possibility of damage modes is large [5,6]. However, in a composite laminate, there are three principal failure modes—intraply cracking, interlaminar delamination, and fiber failure—which play a role in affecting mechanical properties of the composite laminate. Among the three principal damage modes, delamination has been proven to be the most predominant and life-limiting failure mechanism [7–17].

Delamination may develop during manufacturing due to nonoptimum curing or the introduction of foreign bodies. It may also result from impact damage or from the three-dimensional interlaminar stresses at stress-free edges, ply drop-offs, bonded joints, and other discontinuities (Figure 1). Even in the absence of such discontinuities, delamination can result from applied compressive loading, which may cause local or global buckling but also produces internal transverse tensile and shear stresses that can cause delamination.

Since delamination may significantly influence the structural properties and performance, its behavior needs to be identified thoroughly and the techniques for its delay or prevention need to be developed. Some approaches for prevention or delay of delamination being considered are improvement in the current resins by toughening them or developing new tougher resins [18–22], adhesive interleaving [23], braiding [24], three-dimensional weaving [25], through the thickness stitching [26], and novel design to reduce interlaminar stresses [27]. The damage tolerance design methods for composite structures depend on full knowledge of damage initiation and growth, involving inspection and measurement of damage accumulation, modeling of damage growth, detailed stress analysis, development of failure criteria, and so on.

The purpose of this section is to review the state of art in delamination, highlighting reasons for its onset and growth, methodologies used to describe the onset and growth, and characterizations of delamination resistance. The major topics covered here are (1) interlaminar stresses, (2) delamination onset and initiation, (3) delamination growth, and (4) characterization of delamination resistance.

INTERLAMINAR STRESSES

Many analytical efforts to date have attempted to devise a rational methodology which generally describes the onset and growth of delamination. Most of such work has been to determine the interlaminar stresses in the boundary layer region near straight or curved free edges, where delamination occurs. The occurrence of interlaminar stresses is considered to result from the presence and interactions between geometric discontinuities of the composite laminates and the mismatch in engineering properties through the laminate thickness. The investigation by Herakovich [28] indicated that good correlations exist between interlaminar stresses and the mismatch in engineering properties [i.e., mismatch in Poisson's ratio (ν_{xy}) and coefficient of mutual influence ($\eta_{xy,x}$) between plies].

The first approximate solution of interlaminar stresses was proposed by Puppo and Evensen [29], based on a laminate model with interlaminar normal stress being neglected in the laminate. The approach, which consisted of three interlaminar stress components, was first made by Pipes and Pagano [30]. A great variety of methods have been proposed to calculate interlaminar stress. The representative methods are the finite difference method by Pipes and Pagano [30], the boundary layer theory by Tang and Levy [31], the finite element method by Wang and Crossman [32], the Raleigh–Ritz method by Pagano [33], the hybrid finite element method by Spilker and Chou [34], the complex variable stress potentials by Wang and Choi [35], and the global–local model by Pagano and Soni [36]. However, because of the complexities, most of the studies were restricted to the composite laminates with straight edges. There is no efficient solution available for the prediction of interlaminar stresses in composite laminates with arbitrary curved edges. The method that can be used is that of three-dimensional finite elements [37], which require large amounts of computer storage and is very time consuming.

In this chapter we describe a methodology [38] that can be used to deal with interlaminar stresses in composite laminates with arbitrary curved edges. From the following model, it is also easy to understand why interlaminar stress occurs near free edges of composite laminates.

Formulation

Consider a symmetric composite laminate with arbitrary curved edges subjected to in-plane loading (Figure 2). The laminate is composed of n plies and the thickness of each ply is t_i. The thickness of the laminate is $2h$. For convenience, the origin of Cartesian axes is chosen as an arbitrary point in the midplane of the curved edges of the laminate. The x-axis is tangential to the intersection of the midplane and the edge, the y-axis is taken as the direction normal to the edge, and the z-axis is perpendicular to the midplane.

The following assumptions were used: (1) each lamina is described by a macroscopically homogeneous orthotropic material; (2) the material is linear elastic under a generalized Hooke's law and the displacement is small; (3) an initially stress-free field with negligible body force is assumed; and (4) the laminate thickness $2h$ is small compared to an in-plane characteristic length (such as the width of laminate or the diameter of a hole).

The dimensionless variable $\eta = z/h$ is introduced. The basic governing equations for kth ply are as follows [31,39]:

The equilibrium equations are

$$\sigma_{5,3} + h(\sigma_{1,1} + \sigma_{6,2}) = 0$$
$$\sigma_{4,3} + h(\sigma_{2,2} + \sigma_{6,1}) = 0 \qquad (1)$$
$$\sigma_{3,3} + h(\sigma_{5,1} + \sigma_{4,2}) = 0$$

Hereafter, the contracted notation is used, and a comma before a subscript denotes partial differentiation with respect to the space coordinate associated with the subscript.

The generalized Hooke's law is

$$\boldsymbol{\epsilon} = \mathbf{S}\boldsymbol{\sigma} \qquad (2)$$

where

$$\mathbf{S} = \begin{bmatrix} S_{11} & S_{12} & S_{13} & 0 & 0 & S_{16} \\ S_{12} & S_{22} & S_{23} & 0 & 0 & S_{26} \\ S_{13} & S_{23} & S_{33} & 0 & 0 & S_{36} \\ 0 & 0 & 0 & S_{44} & S_{45} & 0 \\ 0 & 0 & 0 & S_{45} & S_{55} & 0 \\ S_{16} & S_{26} & S_{36} & 0 & 0 & S_{66} \end{bmatrix} \qquad (3)$$

The compatibility equations are

$$\begin{aligned}
\epsilon_{1,22} + \epsilon_{2,11} - \epsilon_{6,12} &= 0 \\
\epsilon_{1,33} &= h\epsilon_{5,13} - h^2\epsilon_{3,11} \\
\epsilon_{2,33} &= h\epsilon_{4,23} - h^2\epsilon_{3,22} \\
\epsilon_{6,33} &= h(\epsilon_{4,13} + \epsilon_{6,23}) - 2h^2\epsilon_{3,12} \\
2\epsilon_{1,23} - \epsilon_{6,13} &= h(\epsilon_{5,21} - \epsilon_{4,11}) \\
2\epsilon_{2,13} - \epsilon_{6,23} &= h(\epsilon_{4,12} - \epsilon_{5,22})
\end{aligned} \qquad (4)$$

The boundary conditions are

$$^k\sigma_i = \begin{cases} 0 & k = 1, n \text{ (the top and bottom surfaces)} \quad i = 3, 4, 5 \\ {}^{k+1}\sigma_i & k = 1, 2, \ldots, n - 1 \text{ (interface)} \end{cases} \qquad (5)$$

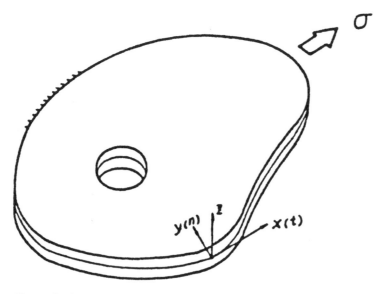

Figure 2 Symmetric composite laminate with curved edges under in-plane loads.

Along the free edge

$$\sigma_i = 0 \qquad i = 2, 4, 6 \tag{6}$$

Interior Problem

By means of asymptotic expensions, the stress vector is expended in a power series of h [31]:

$$\boldsymbol{\sigma} = \sum_{n=0}^{N} \mathbf{g}^n h^n \tag{7}$$

Substituting Eq. (7) into Eqs. (1) to (4) gives the following zeroth-order solutions:

$$\sigma_i^\circ = 0 \qquad i = 3, 4, 5 \tag{8}$$

$$\epsilon_i^\circ = \epsilon_i^\circ(x,y) \qquad i = 1, 2, 6 \tag{9}$$

$$\sigma_i^\circ = \mathbf{C}^{-1} \epsilon_i^\circ \qquad i = 1, 2, 6 \tag{10}$$

The corresponding equilibrium equations can be expressed as

$$N_{xx,x}^\circ + N_{xy,y}^\circ = 0 \tag{11}$$

$$N_{xy,x}^\circ + N_{yy,y}^\circ = 0 \tag{12}$$

where

$$\mathbf{N}^\circ = \int \boldsymbol{\sigma}^\circ \, d\eta \tag{13}$$

Compared with the basic equations of the plane problem in elasticity, it is noted that for the zeroth-order solution, governing equations (8)–(13) correspond to the classical plane stress problem. Hence the zeroth-order interior problem coincides with classical lamination theory (CLT). For some typical shapes of orthotropic or anisotropic plates, such as a rectangular plate or an infinite plate with an elliptical hole subjected to uniform in-plane far-field stress, the plane stress solutions are available. For the laminate with arbitrary curved edges, the plane stress solutions can be determined with numerical method based on the CLT. The strains in the laminate can be obtained from the resultant strain relations. Then the stresses in each ply may be calculated from the ply stress–strain law,

$$\epsilon_i = \frac{C_{ij} N_j^\circ}{2h} \qquad i, j = 1, 2, 6 \tag{14}$$

$$^k\overline{\sigma}_i = {}^k Q_{ij} {}^k \epsilon_j = {}^k Q_{ij} \epsilon_j^\circ \qquad i, j = 1, 2, 6 \tag{15}$$

where C_{ij} is the laminate compliance tensor evaluated from the CLT, and $^k Q_{ij}$ is the reduced stiffness tensor for each ply. Hence for the zeroth-order solution, the stresses in the composite laminate are determined by the CLT. However, this solution usually cannot satisfy the stress boundary conditions in each ply along the free edge; it can only satisfy the stress boundary conditions in an integrated manner.

$$N_i^\circ = \int \overline{\sigma}_i^\circ \, d\eta = 0 \qquad i = 2, 6 \tag{16}$$

To satisfy the stress boundary conditions exactly, an additional stress field is superposed, that is, the boundary layer stresses.

Boundary Layer Problem

The additional stress field defined in boundary layer region must satisfy the following stress boundary conditions along the edge:

$$\sigma_i = -\bar{\sigma}_i^{\circ} \qquad i = 2, 6$$
$$\sigma_4 = 0 \tag{17}$$

An equation similar to Eq. 16 can be obtained:

$$\int \sigma_i \, d\eta = 0 \qquad i = 2, 6 \tag{18}$$

From Saint-Venant's principle, the additional stress σ is an effect in a local region frequently referred to as the boundary layer region, and it can also be considered to be induced due to mismatch of stress field determined by the CLT at the free edge of the composite laminate. Far from the edge,

$$\lim_{y \to \infty} \sigma = 0 \qquad i = 1, 2, \ldots, 6 \tag{19}$$

Now introduce the dimensionless variable $\xi = y/h$. By asymptotic expansions, the stress vector is expended in a power series of h:

$$\sigma = \sum_{n=0}^{N} \mathbf{f}^n h^n \tag{20}$$

Substituting Eq. (20) into Eqs. (1) to (4) yields the following main governing equations of zeroth-order solutions:

Equilibrium equations:

$$\sigma_{6,2}^{\circ} + \sigma_{5,3}^{\circ} = 0$$
$$\sigma_{2,2}^{\circ} + \sigma_{4,3}^{\circ} = 0 \tag{21}$$
$$\sigma_{3,3}^{\circ} + \sigma_{4,2}^{\circ} = 0$$

Compatibility equations:

$$\epsilon_{2,33}^{\circ} + \epsilon_{3,22}^{\circ} = \epsilon_{4,23}^{\circ}$$
$$\epsilon_{6,23}^{\circ} - \epsilon_{5,22}^{\circ} = 0$$
$$\epsilon_{6,33}^{\circ} - \epsilon_{5,23}^{\circ} = 0 \tag{22}$$
$$\epsilon_{1,33}^{\circ} = \epsilon_{1,22}^{\circ} = \epsilon_{1,23}^{\circ} = 0$$

Compared with the governing equations of three-dimensional elasticity, it is very important to notice that no quantities in the equations above are functions of the x-direction. This means that the zeroth-order boundary layer theory corresponds to a fictitious straight-edge problem: If the laminate is of straight, free edges and is subjected to in-plane uniform extension, the equations obtained here are identical to those derived by Pipes and Pagano in 1970 [30], and in this case these equations are exact ones. Hence for the composite laminate with arbitrary curved edges, the determination of approximate three-dimensional stresses in the boundary layer region can be separated into a series of problems on the dispersed locations (origin of the coordinate) along the edge. Each of them consists of a fictitious straight-edge laminate whose principal directions coincide with the normal and tangential directions (Figure 3). When these locations are sufficient in number, the continuous distributions of the three-dimensional stresses (including interlaminar stresses) in the boundary layer region can be obtained.

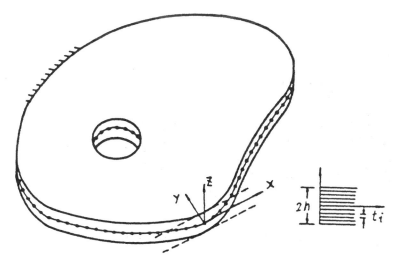

Figure 3 Zeroth-order solution of stresses in boundary layer region (it consists of a series straight-edge problems).

Hence, from the methodology described herein, the determination of stress field in the composite laminate subjected to in-plane loading can be divided into two parts. One is the in-plane stress domain field predicted by the CLT; another is the boundary layer stress field in the boundary layer region. Interlaminar stresses only exist in the latter, and they can be considered as the direct products due to the matching of the in-plane stresses evaluated by the CLT. From the superposition of the in-plane stresses and the boundary layer stresses, the stress field in the composite laminate can be obtained.

Some Basic Characteristics of Interlaminar Stresses

Consider a symmetric composite laminate with straight edges subjected to in-plane uniform loading. The geometry and coordinate are shown in Figure 4. No quantities depend on x-direction. A structural element is taken out as in Figure 4. The lower surface of the element coincides with the bottom surface of the laminate, and the upper surface is an arbitrary plane parallel to the x-y plane. The transverse dimension b of the element is sufficiently large so that Eq. (19) can be satisfied. Together with the boundary equations (17), the equilibrium equations of the element can be expressed as

$$\Sigma F_x = 0: \quad \int_{z=c} \tau_{xz} \, dy + \int_{y=0} \bar{\tau}_{xy} \, dz = 0 \tag{23a}$$

$$\Sigma F_y = 0: \quad \int_{z=c} \tau_{yz} \, dy + \int_{y=0} \bar{\sigma}_{yy} \, dz = 0 \tag{23b}$$

$$\Sigma F_z = 0: \quad \int_{z=c} \sigma_{zz} \, dy \qquad = 0 \tag{23c}$$

The three equations above define some important characteristics of interlaminar stresses.

Interlaminar Normal Stress, σ_{zz}

Equation (23c) indicates that for any z-location through the laminate thickness, σ_{zz} as a function of y must cross the y-axis at least once for its integral to be zero. Far from the

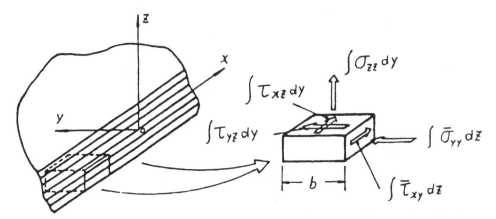

Figure 4 Integral equilibrium of a structural element on the edge of a laminate.

edge, σ_{zz} approaches zero, and the stress field defined by the CLT is obtained. If σ_{zz} does not have these characteristics, the solution cannot be reasonable.

Interlaminar Shear Stress, τ_{yz}

For any balanced angle-ply $(\theta/-\theta)_{ns}$ laminate, due to this typical construction, the transverse stress $\bar{\sigma}_{yy}$ in each ply evaluated by the CLT is everywhere equal to zero:

$$\bar{\sigma}^i_{yy} = 0 \tag{24}$$

Equation (23b) reduces to

$$\int \tau_{yz}\, dy = 0 \tag{25}$$

This equation implies that τ_{yz} is either identical zero or crosses the y-axis at least once in each ply of the angle-ply laminate. It should be noted that this behavior is not limited to angle-ply laminate but is valid for any laminate for which Eq. (24) is true.

Interlaminar Shear Stress, τ_{xz}

For any balanced cross-ply $(0/90)_{ns}$ or $(90/0)_{ns}$ laminates, the in-plane shear stress $\bar{\tau}_{xy}$ in each ply evaluated by CLT is everywhere equal to zero,

$$\bar{\tau}^i_{xy} = 0 \tag{26}$$

Equation (23a) reduces to

$$\int \tau_{xz}\, dy = 0 \tag{27}$$

This equation implies that τ_{xz} is either identically zero or crosses the y-axis at least once in each ply of the cross-ply laminate. It is also valid for any laminate for which Eq. (26) is true. Equations (23) to (27) can be used as a critical standard to examine whether a solution for interlaminar stresses is reasonable.

Approximate Approaches

For thick realistic structural laminates, exact evaluation of interlaminar stresses tends to be expensive. Hence several approximate solutions have been developed [40–42,38]. The approximate solution by Pipes and Pagano [40] for the normal component of interlaminar stress state, σ_{zz}, may give an idea of free-edge delamination prone and resistant stacking sequences. This ability of differentiating the stacking sequences offers promise that the approximation when employed with a suitable failure criterion can accurately predict edge delamination. Accordingly, Lagace [41] and Ye [38] proposed two approximate solutions which have the ability to perform the analysis of thick laminates (with 100 plies or more). Another useful approximate method has been proposed by Whitney [42] to estimate the free-edge interlaminar stresses. This method gives the interlaminar stresses, which are close to the exact estimates.

It should be noted that although approximate solutions are available, the usual method used in the research work is the finite element method [32,43–51]. One reason is probably due to the fact that the approximate solutions lack enough examples to support their accuracy. Additionally, using these numerical techniques, it has been relatively easy to incorporate the effect of moisture and temperature on the interlaminar stresses [52] as well as the effect of material nonlinearity [53,54].

FREE-EDGE DELAMINATION ONSET

Regardless of the nature of loading, free-edge delamination is believed to be driven by the interlaminar stresses that exist near the laminate edges. Basically, there are two approaches to the prediction of delamination onset, one based on fracture mechanics and another on mechanics of material strength.

Fracture Mechanics Approach

The fracture mechanics method used the assumption of delamination as an edge crack at the interface and thus determines either stress intensity factor, K, or the strain energy release rate, G. Due to the complexity in evaluation of the "free-edge interlaminar stress intensity factor" [35], the most popular method is the strain energy release rate approach.

In 1977, Rybicki et al. [55] proposed that edge delamination can be modeled as an interfacial crack near the free edge and analyzed by the finite element method. They determined that the strain energy release rate remained nearly constant during propagation, which indicates that the critical strain energy release rate, G_c, may be a material property independent of geometry and is hence a candidate parameter for predicting delamination growth. Additionally, they suggested an approach to predict the delamination onset or initiation by introducing a characteristic length of the flaw equivalent to one-ply thickness. The application to a $(\pm 30/\pm 30/90/\overline{90})_s$ showed the success of the methodology in predicting onset of delamination.

The work by O'Brien and his colleagues [56,57] has been central in developing and extending this methodology. O'Brien [58] has given a simple method based on the laminate theory to predict the strain energy release rate. For the case of a laminated coupon under uniaxial tensile load, a simple equation to measure the strain energy release rate was derived as

$$G = \frac{\epsilon^2 t}{2}(E_{\text{lam}} - E^*) \tag{28}$$

where t is laminate thickness and ϵ is the strain level; E_{lam} is the extensional stiffness of undelaminated laminate and E^* is the stiffness of the laminate completely delaminated along one or more interfaces given by

$$E^* = \frac{\sum_{i=1}^{m} E_i t_i}{t} \tag{29}$$

with E_i being the stiffness of the sublaminate of thickness t_i. The critical strain energy release rate, G_c, can be determined by measuring the strain level at which delamination onset occurs. Accordingly, the onset of delamination in a laminate occurs at the critical strain, ϵ_c, given by

$$\epsilon_c = \sqrt{\frac{2G_c}{t(E_{lam} - E^*)}} \tag{30}$$

O'Brien has used Eq. (30) to predict the onset of delamination successfully in various laminates with brittle and toughened-matrix resins. The predicted behavior is seen to agree well with the test data. A few of their results are presented in Figure 5, where the critical strain for delamination onset at the 0/90 interface for $(+45_n/-45_n/0_n/90_n)_s$ for $n = 1$, 2, 3, is shown.

Whitney and Browning [59] modified Eq. (30) to account for any discrepancies between theoretically and experimentally determined stiffness of undamaged laminate, that is,

$$\epsilon_c = \sqrt{\frac{2G_c}{t(1 - E^*/E_{\text{lam}})\bar{E}_{\text{lam}}}} \tag{31}$$

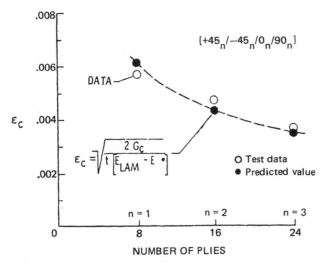

Figure 5 Edge delamination onset prediction in 0/90 interface of T300/5208 laminates. (From Ref. 56.)

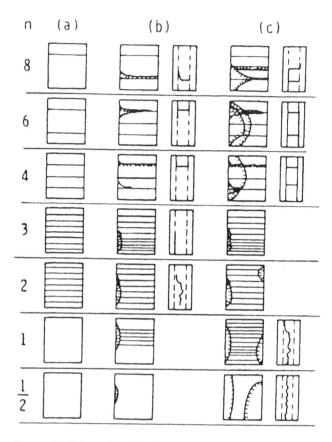

Figure 6 Schematic of the fracture sequence in the $(\pm 25/90_n)_s$ laminates: (a) just prior to edge delamination; (b) subsequent to edge delamination; (c) just prior to final failure. (From Ref. 60.)

where $\overline{E}_{\mathrm{lam}}$ is experimentally determined laminate stiffness. Additionally, E^* should be determined by the stiffness of the sublaminates, which are usually unsymmetric rather than symmetric sublaminates.

One important observation is that in some cases there are significant interactions between delamination and matrix ply cracks. As shown schematically in Figure 6, $(\pm 25/90_n)_s$ laminates may be segregated into three categories based on their fracture behavior: thin ($n = \frac{1}{2}$ and 1), thick ($n = 2$ and 3), and very thick ($n = 4$, 6, and 8) $90°$ layers. The initial fracture of the thin $90°$ layer laminates is a free-edge delamination along the midplane of the laminate. The thick $90°$ layer laminates showed transverse ply cracking as their first failure mode, followed by free-edge delamination along the 25/90 interface. Finally, the very thick $90°$ layer laminates exhibited transverse ply cracking as their first failure mode, followed by a combination of free-edge delamination and transverse-crack-tip delamination. These fracture modes are illustrated in Figure 7. Based on these modes, Wang and his colleagues [61–64] have developed a methodology to describe the onset and growth of delamination. First, the energy release rate as a function of crack length is represented by a set of shape functions [61] which were obtained by the finite-element crack-closure procedure. Then according to different fracture modes, the fracture criteria

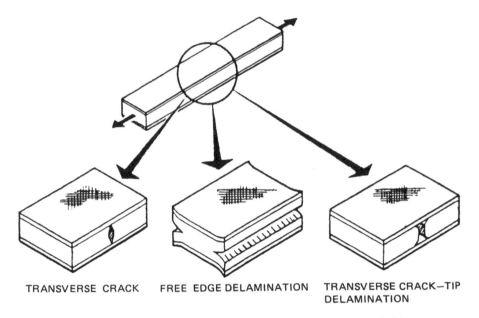

TRANSVERSE CRACK   FREE EDGE DELAMINATION   TRANSVERSE CRACK–TIP
DELAMINATION

Figure 7 Schematic of fracture types in $(\pm 25/90_n)_s$ laminates. (From Ref. 66.)

are used to define the onset of delamination. For example, the analysis of combined mode I–mode II crack growth, such as the 25/90 interface and transverse-crack-tip delamination problems, is based on application of a mixed-mode fracture criterion advanced by Wu [65] of the form

$$\sqrt{\frac{G_I}{G_{Ic}}} + \frac{G_{II}}{G_{IIc}} = 1 \tag{32}$$

where G_{Ic} and G_{IIc} are the critical strain energy release rates for mode I and mode II crack growth, respectively. Finally, the analysis of a combined transverse-crack-tip delamination (TCTD) and free-edge delamination (FED), as exhibited by the very thick 90° layer laminates, requires a three-dimensional solution. A method was introduced by Law [66] to superimpose two-dimensional solutions. This superposition method is illustrated by Figure 8. Since these two forms of crack growth are orthogonal (the transverse-crack-tip delamination grows in the 0° or loading direction and the free-edge delamination grows in the 90° direction transverse to the loading), it was proposed that the energy release rates be treated as vector quantities such that

$$G_{tot} = \sqrt{G_{TCTD}^2 + G_{FED}^2} \tag{33}$$

where G_{TCTD} and G_{FED} refer to the energy release rate for the transverse-crack-tip and free-edge delamination problems, respectively. This "rule" is then applied to each fracture mode individually and substituted into the mixed-mode fracture criterion given in Eq. (33). However, this superposition is not based on rigorous analysis and is thus a replacement for the three-dimentional analysis required to describe this complex problem.

Another important concept proposed by Wang [64] was the "effective" material flaw distribution (Figure 9), which enables definition of the initial crack length a as a random

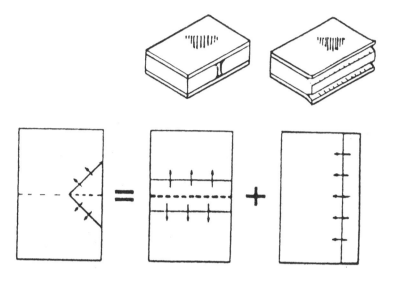

--- **TRANSVERSE CRACK**   **DELAMINATED AREA**

Figure 8 Superposition of transverse-crack-tip and free-edge delamination. (From Ref. 66.)

variable. This concept theorizes that a characteristic distribution of "effective" flaws can be determined for a given representative volume of the basic ply material; each of the effective flaws has a definite size as well as location. Although hypothetical in nature, the effective flaws are characterized so as to represent the aggregate effect of local macroflaws and imperfections at the ply level. In this context, each effective flaw is further assumed to act like a small crack. Then each of the effective flaws can individually become

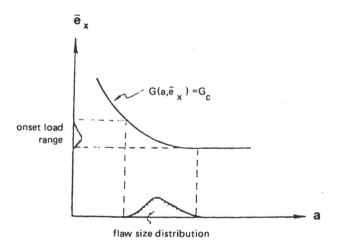

Figure 9 Rationale for predicting onset load of delamination according to fracture mechanics criterion. (From Ref. 64.)

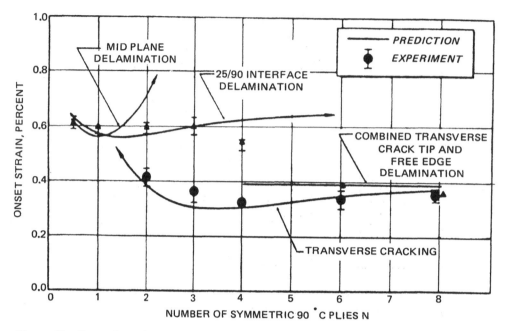

Figure 10 Comparison of predicted onset strains and experiments. (From Ref. 66.)

a sublaminate crack at some critical stress level, and they form collectively multiple sublaminate cracks in the course of loading.

The results of applying the mode I and mixed mode I–mode II fracture models by Law [66] to the ($\pm 25/90_n$)$_s$ laminates are displayed in Figure 10. Good agreement is demonstrated between the theory and experiments. Mode I delamination at the free edge is the initial predicted mode in the laminate when $n = \frac{1}{2}$ and 1, while transverse cracking in the 90° layers is the initial predicted fracture mode for $n = 2$. Free-edge delamination for the laminates of $n = 2$ and 3 is predicted to occur as a mixed mode I–mode II fracture at the 25/90 interface subsequent to transverse cracking. Finally, for $n = 4$, 6, and 8, a combined transverse-crack-tip and free-edge delamination along the 25/90 interface is the predicted fracture mode. Each of these predicted fracture modes is as observed in the experiments.

O'Brien [67] has proposed a analytical model to determine delamination onset strains for these locally combined transverse-crack-tip and free-edge delaminations. In his study the composite laminate is an elastic body containing a matrix ply crack through the thickness of n 90° plies, with delaminations forming at the matrix crack tip and growing in the ply interfaces (Figure 11). For simplicity, the strain-energy release rate associated with the growth of delamination from a single matrix crack is considered. The strain energy release rate associated with the growth of delamination from a matrix ply crack can be expressed as [67]

$$
\begin{aligned}
G &= \frac{P^2}{2mw^2}\left(\frac{1}{t_{LD}E_{LD}} - \frac{1}{tE_{lam}}\right) \\
&= \frac{\epsilon^2 E_{lam}^2 t^2}{2m}\left(\frac{1}{t_{LD}E_{LD}} - \frac{1}{tE_{lam}}\right)
\end{aligned}
\tag{34}
$$

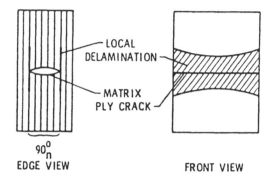

Figure 11 Schematic of local delamination growing from matrix cracks. (From Ref. 67.)

where E_{LD} is the modulus of the locally delaminated region, t_{LD} the thickness of the locally delaminated region, and m the number of delaminations growing from the matrix ply crack. For the case illustrated in Figure 11, $m = 2$. The critical strain ϵ_c for delamination onset predicted by using Eqs. (30) and (34) for different sources of delaminations is shown in Figure 12. Good correlation between the analytical predictions and experimental results is observed.

Material-Strength Mechanics Approach

The material-strength mechanics approach involves the detailed stress analysis near the free edge used in conjunction with a failure criterion similar to Nuismer and Whitney's [68] point and average stress failure or Wu's [69] strength criteria for composites.

Kim and Soni [70] used the point stress and average stress criteria to predict delamination onset in composite laminates such as $(\pm 30_n/90_n)_s$, $(0_n/\pm 45_n/90_n)_s$, and $(0_n/90_n/\pm 45_n)_s$

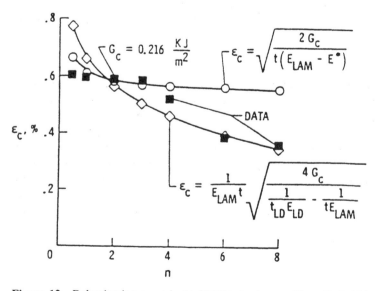

Figure 12 Delamination onset in $(\pm 25/90_n)_s$ laminates. (From Ref. 67.)

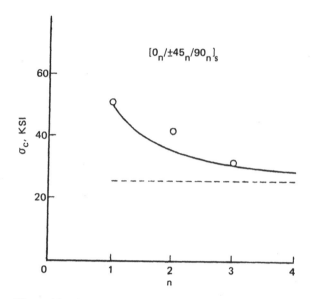

Figure 13 Comparison of predictions and experiments for onset of delamination. ———, average stress criterion; — — — —, maximum point stress criterion; ○, experiments. (From Ref. 70.)

for several values of n. Their studies have shown the point stress criterion to be conservative, but the average stress failure criterion predicts the behavior, which agrees well with the test data. For the average stress criterion, they have used a characteristic distance, a^*, as equivalent to a ply thickness over which the stress is averaged. Figure 13 shows a few of their results.

However, the point or average stress approaches by Kim and Soni [70] are suitable only when delamination onset is governed by σ_{zz}. If the delamination occurs under the influence of shear and normal stress, the approach can be modified using a criterion similar to that of Wu [69], defined as follows: "the onset of delamination will occur when at a characteristic distance a^* from the edge, the stress state satisfies the following failure criterion":

$$F_i \sigma_i + F_{ij} \sigma_i \sigma_j = 1 \tag{35}$$

where F_i and F_{ij} are strength tensors whose components can be expressed in terms of the material principal strengths, and σ_i and σ_j are the stress components at the interface at a point a^* away from the edge. Herakovich et al. [71] used this approach for the prediction of delamination onset in angle-ply laminated composites. The tensor polynomial [Eq. (35)] was evaluated near the free edge for each finite element to determine the location of initial failure in various laminate configurations. In addition, the individual terms of the polynomial were examined to identify modes of failure. It has been shown by these studies that for angle-ply $(\pm \theta)_s$ laminates $\theta = 15°$ is the most critical angle-ply laminate, with the shear stress τ_{xz} dominating failure. It is interesting to note from these studies that as the fiber angle is increased above 15°, the failure mode of $(\pm \theta)_s$ laminates shifts to mixed shear (τ_{xz} and τ_{yz}), mixed shear and normal (τ_{xz}, τ_{yz}, and σ_z), and finally, to transverse tension (σ_y).

Another modified approach to include all three interlaminar stresses was proposed by Brewer and Lagace [72]. Each stress is normalized by its respective ultimate, and it is

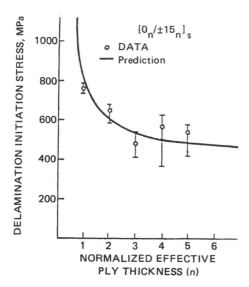

Figure 14 Predicted and actual delamination initiation stresses for $(O_n/\pm 15_n)_s$ laminates. (From Ref. 72.)

postulated that failure occurs when the sum of the squares of these normalized components equals 1:

$$\left(\frac{\overline{\sigma}_{zz}}{Z^t}\right)^2 + \left(\frac{\overline{\sigma}_{1z}}{Z^{s1}}\right)^2 + \left(\frac{\overline{\sigma}_{2z}}{Z^{s2}}\right)^2 = 1 \tag{36}$$

where the overbar denotes the average stress components defined by

$$\overline{\sigma}_{ij} = \frac{1}{x_{\text{avg}}} \int_0^{x_{\text{avg}}} \sigma_{ij} \, dx \tag{37}$$

where x_{avg} is the critical length over which the values are averaged. The normal tensile ultimate is denoted by Z^t, while the interlaminar shear strengths are represented by Z^{s1} and Z^{s2} for the σ_{1z} and σ_{2z} stresses, respectively. Shown in Figure 14 are the experimental delamination onset stresses and theoretical correlations for $(0_n/\pm 15_n)_s$ laminates.

It has been noted [21] that the average stress approach like that in Eq. (37) is necessary since the strength properties of composite materials (such as Z^t and Z^{si}) are valid only over a finite material volume. This viewpoint is similar to the concept of "effective modulus."

At present it is difficult to make the comparison between two methodologies—fracture mechanics or strength-based—for the prediction of delamination onset. However, some researchers [73,21] believed that the reasonable approach is strength-based.

Delamination Onset near Curved Edge

O'Brien has extended his energy-release model for delamination onset [58] to the composite laminates with an open hole. A simple technique [74] was proposed for estimating the interlaminar stresses at the boundary of the hole, which involved modeling several discrete locations around the hole as straight free edges with the ply orientations rotated by an

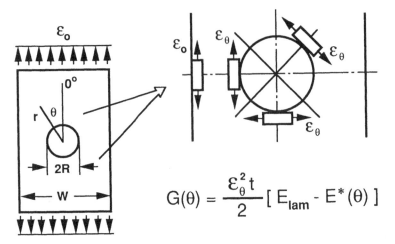

$$G(\theta) = \frac{\varepsilon_\theta^2 t}{2} [E_{lam} - E^*(\theta)]$$

Figure 15 Estimation of G distribution around the hole boundary. (From Ref. 74.)

appropriate angle (Figure 15). The circumferential strain, calculated from an elasticity solution, was substitued into Eq. (28) to generate the G distribution around the hole boundary. For the quasi-isotropic laminates

$$G(\theta) = \frac{\epsilon_\theta^2 t}{2} [E_{lam} - E^*(\theta)] \tag{38}$$

$$= (1 - 2 \cos 2\theta)^2 \epsilon_0^2 \frac{t}{2} [E_{lam} - E^*(\theta)]$$

Hence, for a given laminate of thickness t with a remote tensile strain ϵ_0, G will vary with θ due to the variation of circumferential strain amplitudes as well as the variation in laminate modulus after delamination. Equation (38) was then used to identify regions around the hole boundary where G values were high and hence where delamination was likely to occur. The predicted and measured delamination remote strains are listed in Table 1. It can be seen that there is a good correlation between the predictions and experimental observations.

Table 1 Comparison of Predicted and Measured Delamination Onset Remote Strains

	Layup[a]	Interface	θ (deg)	$(\epsilon_0)_c$
Predicted	A	90/45	± 60	0.00582
	B	45/90	-80	0.00424
	B	90/-45	-80	0.00572
Measured	A	90/45	± 60	0.00420
	B	45/90 or 90/-45	-80	0.00300
	B	45/90 and 90/-45	-80	0.00570

Source: Ref. 74.

[a] A, $(0/90/\pm45)_s$; B, $(45/90/-45/0)_s$.

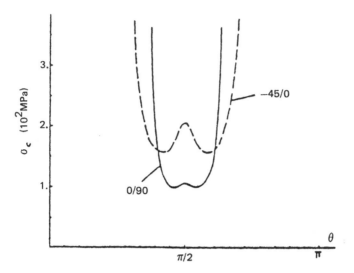

Figure 16 Critical load levels for delamination to initiate at each location along the hole edge, at 0/90 and $-45/90$ interface. (From Ref. 75.)

Based on the methodology discussed earlier for estimation of interlaminar stress near curved edges, a strength criterion for delamination onset along the curved edge was proposed by Ye [21,75]: When $\sigma_{zz} > 0$,

$$\left(\frac{\bar{\sigma}_{zz}}{Y}\right)^2 + \left(\frac{\bar{\tau}_{Lz}}{S}\right)^2 + \left(\frac{\bar{\tau}_{Tz}}{S_T}\right)^2 = 1 \tag{39}$$

and when $\sigma_{zz} \leq 0$,

$$\left(\frac{\bar{\tau}_{Lz}}{S}\right)^2 + \left(\frac{\bar{\tau}_{Tz}}{S_T}\right)^2 = 1 \tag{40}$$

where L denotes the fiber direction; Y and S are the tensile strength and shear strength of a lamina, respectively; S_T is the shear strength in the plane perpendicular to fibers. The stress components in Eqs. (39) and (40) are defined as the average stresses over a distance from the edge. Figure 16 shows the remote critical stress levels predicted by the model for delamination onset at each location along the hole edge at 0/90 and $-45/90$ interfaces in a graphite–epoxy ($\pm 45_2/0_2/90_2$)$_s$ laminate. The possible location of delamination onset at the 0/90 interface is in the range 70° to 110°, and at the $-45/90$ interface, in the range 60° to 120°. Comparison with the delamination states recorded in x-ray radiographs has shown a good correlation between predictions and experimental observations [75].

DELAMINATION GROWTH

Delamination growth depends on the stress state of the crack tip, which is governed by the mixed-mode stress intensity factors K_I, K_{II}, K_{III}, or the strain energy release rates G_I, G_{II}, and G_{III}. Basically, delamination growth behavior can be modeled in two ways, according to the feature of loading.

Table 2 Stress Intensity Factors K_i for Edge Delamination in $(\pm\theta)_s$ Angle-Ply Graphite–Epoxy Laminate under Uniform Strain, ϵ_x^a

$\pm\theta$ (deg)	K_I	K_{II}	K_{III}	K_{III}/K_I
15	0.095	0.01334	−5.009	−52.73
30	0.256	0.0366	−4.022	15.71
45	0.148	0.0152	−1.425	9.63
60	0.2225	0.0149	−0.1951	0.88
75	0.0069	−0.000324	0.0899	13.02

Source: Ref. 76.

a K_i (MPa$\sqrt{m}$) are scaled by $10^3\epsilon_x$.

Delamination Growth under Static Loading

It has been found that delamination growth may be dominated by only one or two modes. For example, the analysis by Wang [76] for edge delaminations at the interface or angle-ply $(\pm\theta)_s$ laminates shows that K_{III} is considerably higher than K_I and K_{II} (Table 2), and edge delamination is dominated by K_{III} (i.e. the tearing mode). On the other hand, O'Brien's [5] analysis for the edge crack at the $-30/90$ interface of $(\pm30/\pm30/90/\overline{90})_s$ laminate shows that the mode III contribution is negligible.

The growth of edge delamination is a stable fracture process in laminates subjected to tensile loading [55,56], (i.e., the applied load has to be increased to force the delamination to grow). Such a growth has been characterized by O'Brien [56] using the concept of crack growth resistance curves (R-curve) whereby G_R versus delamination growth curves were evaluated. The initial value of G_R represents the critical G_c of the delamination onset. O'Brien considered that G_c is a material property and is independent of ply orientation. This is contrary to the observations of Johannesson and Blikstad [77] for delamination of angle-ply laminates. Their studies have shown that the G_c is strongly dependent on the ratio K_{III}/K_I for angle-ply laminate and obtained a relationship in the form

$$G_c \propto \sqrt{1 + \left(\frac{K_{III}}{K_I}\right)^2} \tag{41}$$

The mixed-mode fracture analysis of off-axis unidirectional graphite–epoxy composites by Wang et al. [78] also shows that the mixed-mode material fracture toughness measured in terms of the critical energy release rate $G_{(I,II)c}$ is highly dependent on the G_{II}/G_I ratio.

The mixed-mode delamination growth is not observed to follow a single propagation law. Various laws have been used by different investigators to correlate their test data. The simplest mixed-mode delamination propagation law is defined as follows: "The delamination growth occurs when the total strain energy release rate G_T ($= G_I + G_{II} + G_{III}$) reaches a critical value G_c"; that is,

$$G_T = G_I + G_{II} + G_{III} = G_c \tag{42}$$

where G_I, G_{II}, and G_{III} are the mode I, II, and III strain energy release rate, respectively. However, Eq. (42) may describe mixed delamination growth for the materials only when $G_{Ic} = G_{IIc} = G_{IIIc}$. Rybicki et al. [55] and O'Brien [56] have used Eq. (42) to describe their mixed-mode delamination growth.

For most graphite–epoxy composites, $G_{Ic} \ll G_{IIc}$, and thus Eq. (42) may not adequately describe the delamination growth. A more appropriate relation to describe the delamination growth is considered as

$$\left[\frac{G_I}{G_{Ic}}\right]^m + \left[\frac{G_{II}}{G_{IIc}}\right]^n = 1 \tag{43}$$

where the exponents m and n have been found to have different values for different cases.

Johnson and Mangalgiri [79] and Jurf and Pipes [80] show that $m = n = 1$ agrees with their test data for mixed-mode delamination growth. It can be seen from Figure 17 that a straight-line relation does a good job of fitting the test data for different material systems [79]. On the other hand, Law [66] has used Eq. (43) with $m = \frac{1}{2}$ and $n = 1$ to fit the test data for free-edge delamination behavior in $(\pm 25/90)_s$ laminates. Whitcomb [14,81,82] considered Eq. (43) with $m = n = 2$ and Eq. (41) as $G_I = G_{Ic}$ to describe the delamination growth of a laminate under uniaxial compression. Johnson and Mangalgiri [79] pointed out that Eq. (42) is a reasonable static failure criterion for many tough resin systems, while Eq. [43] may be appropriate for brittle materials.

Hahn [83] has proposed another fixed-mode crack propagation law as

$$(1 - g)\left[\frac{G_I}{G_{Ic}}\right]^{1/2} + g\left[\frac{G_I}{G_{Ic}}\right] + \left[\frac{G_{II}}{G_{IIc}}\right] = 1 \tag{44}$$

where $g = G_{Ic}/G_{IIc}$

For composites with epoxy matrices, $g \approx 0.1$ to 0.2. For negligible g, Eq. (44) is similar to Eq. (43) with $m = \frac{1}{2}$, $n = 1$, which was used by Law [66].

On the basis of morphology of fracture surfaces, a geometrical fracture criterion was suggested by Hahn and Johannesson [84] for mixed-mode fracture as

$$G_c = \alpha_1 + \alpha_2 \sqrt{1 + \left(\frac{K_{II}}{K_I}\right)^2} \tag{45}$$

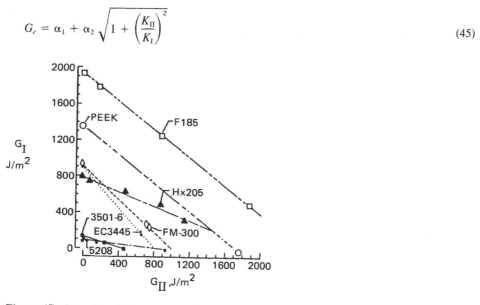

Figure 17 Mixed-mode fracture toughness. (From Ref. 79.)

where α_1 may be considered to be the critical strain energy release rate due to fiber–matrix debonding, and α_2 is related to the surface energy of the resin, γ_m, by

$$\alpha_2 = 2\gamma_m(1 - V_f) \tag{46}$$

where V_f is the fiber volume fraction. The constants α_1 and α_2 were determined to be 61 and 3.1 J/m for delamination of angle-ply laminates of T300/934 graphite–epoxy by Johannesson and Blikstad [77].

Donaldson [85,86] compared various crack propagation laws for graphite–epoxy and graphite–PEEK composites. He has shown that mixed-mode delamination of graphite–epoxy can be described by

$$G_c = - e^{-(C_1M + C_2)} + C_3 \tag{47}$$

where

$$M = \sqrt{1 + \left(\frac{G_{II}}{G_I}\right)\frac{\sqrt{E_L}}{E_T}} \tag{48}$$

and $C_1 = 0.25$, $C_2 = -6.31$, and $C_3 = 503.3$. E_L and E_T are the longitudinal and transverse moduli for the laminate, respectively. However, he suggested that the delamination growth for graphite–PEEK was described by the relation

$$\frac{G_I}{G_{Ic}} + \left[\frac{G_{II}}{G_{IIc}}\right]^{3/2} = 1 \tag{49}$$

Russell and Street [87] also studied various delamination propagation laws and demonstrated that mixed-mode delamination growth may be described by Wu's [69] tensor polynomial criterion. They modified the original Wu's criterion, which was from the plane stress condition to plane strain conditions, and was more appropriate for delamination. The plane strain model was found to fit well the experimental data for mixed-mode delamination.

Delamination Growth under Cyclic Loading

The growth of cracks in metals under cyclic loading can be described by a function of the stress intensity factor range ΔK [88],

$$\frac{da}{dN} = \mathrm{f}(\Delta K) \tag{50}$$

with the most widely used form being the empirical power law function,

$$\frac{da}{dN} = C(\Delta K)^n \tag{51}$$

An analogous power law equation in terms of the strain energy release rate was proposed by O'Brien [56] for one-dimensional delamination propagation. Considering the behavior at constant load ratio R,

$$\frac{da}{dN} = C(G_{max})n \tag{52}$$

where G_{max} is the maximum total strain energy release rate corresponding to the cyclic loading. Figure 18 shows da/dN as a function of G_{max} for free-edge delamination growth in $(\pm 30/\pm 30/90/\overline{90})_s$ graphite–epoxy laminate [56]. It can be seen that an excellent

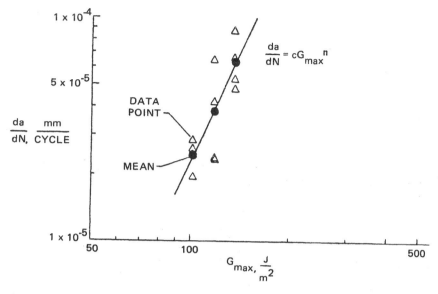

Figure 18 Power law curve fit for *da/aN* as a function of G_{max}. (From Ref. 56.)

correlation was achieved for the data. Some other researchers have also used Eq. (52) to describe delamination growth in graphite–epoxy composites [89] as well as in composite adherends [90,91].

However, Eq. (52) considers only the driving force for delamination growth, and for simplicity, assumes a constant resistance. Although this is quite valid for most materials, it should be modified to represent the changing delamination resistance in laminates. Poursartip [92] proposed that Eq. (52) should become, in terms of the maximum strain energy release rate and constant load ratio R,

$$\frac{da}{dN} = f(G_{max}, G_R) \tag{53}$$

and the equivalent power law relation to Eq. (52) would then be

$$\frac{da}{dN} = C\left(\frac{G_{max}}{G_R}\right)^n \tag{54}$$

Mixed-mode delamination growth is potentially influenced by all the strain energy release rate components (G_I, G_{II}, and G_{III}). In the delamination growth rate expressions in Eqs. (52) and (54), individual mode contributions to *da/dN* are not addressed explicitly. To separate these contributions, Ramkumar and Whitcomb [93] proposed that the contribution of mode I alone to delamination growth rate was described by

$$\frac{da}{dN} = C_1\left(\frac{G_{Imax}}{G_{Ic}}\right)^{n_1} \tag{55}$$

and the contribution of mode II to delamination growth rate by

$$\frac{da}{dN} = C_2\left(\frac{G_{IImax}}{G_{IIc}}\right)^{n_2} \tag{56}$$

and then the mixed-mode delamination growth rate, assuming G_{III} effects to be negligible, was assumed:

$$\frac{da}{dN} = C_1 \left(\frac{G_{Imax}}{G_{Ic}}\right)^{n_1} + C_2 \left(\frac{G_{IImax}}{G_{IIc}}\right)^{n_2}$$ (57)

where

$$G_{Imax} + G_{IImax} = G_{max}$$ (58)

and C_1, C_2, n_1, and n_2 are constants.

Wang et al. [64] have noted that the quantity G in Eqs. (52) to (58) already includes the laminate stacking sequence, lamina thickness, laminate shape features, the individual flaw geometry, the nature of the applied load, and so on. Furthermore, the quantity G/G_c represents the crack driving force relative to the material's resistance. Hence the fatigue constants C and n are expected to be material dependent only.

Two-Dimensional Delamination Growth Problems

In the preceding sections, the energy release rate method has been applied to simulate a number of delamination growth problems. However, delamination was confined to grow in a one-dimensional, self-similar manner. Under this assumption, the kinematics of the delamination propagation is much simplified, with the delamination crack front being represented by a point known as the crack tip, and the crack size having the magnitude a. This simplification has made it possible to calculate energy release rate function G by a number of techniques.

Generally speaking, one-dimensional delamination growth is a very special case and happens only under ideally controlled conditions. Most problems encountered in practice are localized events usually caused by local defects. The growth of localized delamination is almost always multidirectional in nature.

For example, consider a laminate having a small through-hole. Upon loading, localized delamination may be induced near the curvilinear edge of the hole. In this case the crack front will be some line contour, and the associated growth will be two-dimensional. The instantaneous crack size a is now the delaminated area, which will have both a magnitude and a definite shape contour.

In a separate effort, Wang et al. [64] developed a three-dimensional finite element routine in which Irwin's crack-closure concept was incorporated to calculate the crack front energy release rate along a prescribed delamination contour. For the simulation of crack growth in a finite element analysis using a stiffness derivative or virtual crack extension, Parks [94] and Hellen [95] have shown that the crack extension affects only a few elements near the crack tip. Making use of this observation, Mahishi and Adams [96,97] developed a finite element technique for estimation of the local elastic strain energy release rate in the presence of plasticity. Their results demonstrated that the method is very effective in predicting stable delamination growth.

EVALUATION OF DELAMINATION RESISTANCES

The success of any analytical prediction of delamination growth depends on the use of the right critical interlaminar fracture energies, which must be determined experimentally using proper test methods. In addition, a correct assessment of interlaminar fracture toughness will aid in material development, screening, selection, and design. During the

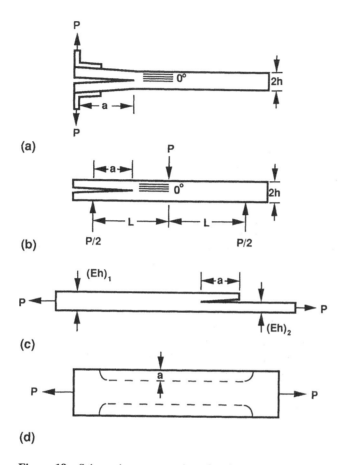

Figure 19   Schematic representation of various test specimens for IFT of composite laminates: (a) double cantilever beam (DCB) specimen; (b) end notched flexure (ENF) specimen; (c) cracked lap shear (CLS) specimen; (d) free edge delamination (FED) specimen.

past decade, a great deal of research work has been conducted on the development of test methods for evaluation of interlaminar fracture toughness (IFT). The most common tests used for IFT characterization in different modes are illustrated schematically in Figure 19.

A delamination can grow under pure mode I (opening or peel), through various combinations of mode I and II (sliding or shear) or mode III (tearing) to pure mode II or mode III. Hence IFT characterization is needed for mode I, mode II, mode III, and mixed mode.

The most commonly used test methods for IFT characterization are based on the global energy balance of fracture mechanics. The strain energy release rate in the test specimens is calibrated by means of a relation between G and compliance C, which is frequently referred to as the *compliance method*:

$$G = \frac{P^2}{2B} \frac{\partial C}{\partial a} \tag{59}$$

where a is the crack length, B the width of the specimen, P the applied load, and C the compliance, defined by the slope of the load (P) versus displacement (δ) curve for the specimen, that is,

$$C = \frac{\delta}{P} \tag{60}$$

Mode I Testing

The most commonly used test for IFT characterization in mode I is the DCB test. The DCB test was applied previously for testing adhesives. Extensive investigation has been performed by Devitt et al. [98] and others [89,93,99–101] to establish the DCB as a standard test for the measurement of composite material resistance to interlaminar crack growth. There are two basic configurations for the DCB specimens, the constant width [99] and the width-tapered DCB (WTDCB) [102]. The width of the tapered specimen is designed so that a/B is constant. The strain energy release rate derived from the analysis of a WTDCB specimen is independent of crack length, and the crack grows under constant load. Therefore, monitoring of the crack length during the test is eliminated [103–106]. However, the fabrication of the tapered composite specimen is much more complicated than the straight-sided one.

A DCB (Figure 19a) is tested under displacement-controlled conditions and P versus δ curves are obtained for various crack lengths (a) which are then used to obtain C and $\partial C/\partial a$. The value of G_{Ic} may then be determined using the critical load P_c for the expansion of delamination and the values of $\partial C/\partial a$ in Eq. (59).

The DCB specimen may be considered as two sufficiently long cantilevers, jointed at one end (i.e., the crack tip). The compliance derived from load–displacement relationships using simple beam analysis in plane stress is

$$C = \frac{2a^3}{3EI} \tag{61}$$

where E is the elastic modulus and I is the moment of inertia of the cross section.

However, the exact description of the compliance is slightly different from the simple beam theory, due to the finite length of the beam and some rotation components. These may be taken into account by replacing the exponent 3 in Eq. (61) by n and writing [107]

$$C = \frac{a^n}{k} \tag{62}$$

where k is a material-geometrical constant. Both n and k must be determined empirically.

The interlaminar fracture toughness in mode I, G_{Ic}, may thus be evaluated as

$$G_{Ic} = \frac{P_c^2}{2B} \frac{nC}{a} = \frac{P_c n \delta}{2Ba} \tag{63}$$

Fracture toughness may also be derived by the area method [99]. This method is based on delamination of the area, ΔA, enclosed within the loading and unloading load–displacement curves in a DCB test, and the increment Δa of a new crack length, and calculating G_{Ic} by

$$G_{Ic} = \frac{1}{B} \frac{\Delta A}{\Delta a} = \frac{1}{2B\Delta a} (P_1 \delta_2 - P_2 \delta_1) \tag{64}$$

where P_1 and δ_1 are the load and displacement for crack length a, and P_2 and δ_2 are the load and displacement for crack length $a + \Delta a$. Note that the right-hand side of Eq. (64) is for linear material only. An average value of G_{Ic} may be determined by measuring P_1, P_2, δ_1, and δ_2 for a series of N crack extensions of length Δa. Thus

$$G_{Ic} = \frac{1}{2BN\,\Delta a} \sum_{i=1}^{N} (P_{1i}\delta_{2i} - P_{2i}\delta_{1i}) \tag{65}$$

The beam relation [Eq. (61)] can be used as long as the shear deformations can be ignored and the deflections are small; otherwise, the corrections for these should be applied [99,100]. However, Williams [108, 109] has pointed out that an end correction should be made to consider the slope and deflection at the cantilever root, which results in an increment in the crack length. Hence Eq. (61) is replaced by

$$C = \frac{2(a + \Delta)^3}{3EI} \tag{66}$$

where Δ can be obtained by the analytical method [108] or by fitting the test data [109]. Whitcomb [110] has considered a nonlinear DCB using beam theory, the crack being corrected to compensate for the nonlinear behavior of the specimen. Mall and Johnson [111] tested several adhesive systems using the DCB specimen; linear and geometric nonlinear finite element analyses were performed but no significant difference was found between them. The maximum difference in the computed compliances from linear and nonlinear analyses was 5% for the maximum debond length investigated (200 mm) at its critical load. The experimental values of compliance were within ±5% of those given by a geometric linear finite element analysis.

Mode II Testing

Russell and Street [112] introduced the end-notched flexural (ENF) fracture specimen for pure mode II testing. The test specimen is essentially a three-point flexure specimen with an embedded through-delamination placed at the laminate midplane, where interlaminar shear stresses are greatest (see Figure 19b).

In a beam loaded in flexure as in the ENF test, a state of almost pure shear prevails at the edge of the artificial midplane delamination. The relationship between the compliance C and the delamination length a is more complicated than for the DCB test, but may be found by conjugate beam analysis or by the deflection of cantilever beams [113]. The expression for this compliance is

$$C = C_0\left(1 + \frac{3a^3}{2L^3}\right) + C_s \tag{67}$$

where $C_0 = L^3/6EI$ is the compliance of the reference specimen with no delamination ($a = 0$), L is the half-span, and c_s is the contribution of the shear stresses to overall compliance.

Neglecting C_s, substituting of Eq. (67) into Eq. (59) yields the expression for G_{IIc}:

$$G_{IIc} = \frac{9a^2 P_c^2 C_0}{2B(2L^3 + 3a^2)} \tag{68}$$

Gillespie et al. have investigated the validity of the ENF test for pure mode II testing [114]. Linear elastic, two-dimensional, finite element stress analysis of the ENF specimen was performed using a four-node plane stress element procedure. The strain energy release rate was evaluated using the crack closure and compliance techniques. Results from both techniques demonstrate that the ENF fracture specimen is a pure mode II test.

The major difficulty in designing a pure mode II ENF specimen is in preventing any crack opening without introducing excessive friction between the crack faces. Mode II crack propagation causes sliding of the crack surfaces; friction between the crack surfaces opposes the sliding and is consequently an energy-absorbing mechanism in addition to the energy dissipation in the creation of new crack surfaces. However, it has been noted [112–114] that the error in the value of G_{IIc} using the ENF test due to friction is small. Recently, Gillespie et al. [115] refined the analysis of ENF specimen using the finite element method, including the effects of local shear deformation around the crack tip (not accounted for by beam theory). This resulted in higher G_{II}. This study has revealed that inclusion of shear deformation in the derivation of G_{II} by beam theory improves the results. Yet the discrepancy between finite element and beam theory solutions could be on the order of 20 to 40% for typical graphite composites. Therefore, they have suggested an improved data reduction scheme which retains the simplicity of beam theory and accuracy of the finite element method [115].

In general, IFT in mode II (G_{IIc}) may be derived from mixed-mode tests; in fact, this was the only way to evaluate G_{IIc} until a pure mode II test was developed. The crack lap shear (CLS) test was used mainly for mode II mixed-mode tests [89,93,111,112,114]. This test develops a mixed mode, with mode II dominating. Wilkins [89] was the first to use the CLS specimen successfully to measure mixed-mode IFT in composites.

There are some other mixed-mode tests from which G_{IIc} may be determined: the end-notched cantilever beam (ENCB) [116], the cantilever beam enclosed notch (CBEN) [116], the end-loaded split laminate (ELS) [117], a modified Arcan test (a single edge notch composite specimen tested in an Arcan fixture) [118], and uneven loading of a double cantilever beam type of specimen [119].

Mixed-Mode Testing

Using the CLS specimen shown in Figure 19c, the P–δ curves may be obtained for various crack lengths and $\partial C/\partial a$ determined. The substitution of P_c and $\partial C/\partial a$ in Eq. (59) gives the value of G_c. G_c may also be obtained using [79]

$$G_c = \frac{P_c^2}{2B^2}\left[\frac{1}{(Eh)_2} - \frac{1}{(Eh)_1}\right] \tag{69}$$

where the subscripts 1 and 2 refer to the sections indicated in Figure 19c.

Design of the specimen geometry results in a variety of mixed-mode conditions. Johnson [120] reports that $0.6 > G_I/G_{II} > 0.2$, based on finite element analysis of different geometries.

The edge delamination tension (EDT) test proposed by O'Brien [56,57] is a simple tension test for measuring the mixed-mode IFT of composites. Laminates are loaded in tension to develop high interlaminar tensile and shear stresses at their free edges, causing delamination. The layup of the specimen is designed to yield the lowest delamination onset strain to measure a given G_c. The two layup configurations commonly used for EDT are the 11-ply ($\pm 30/\pm 30/90/\overline{90}$)$_s$ and the 8-ply ($\pm 30/0/90$)$_s$ laminates (other layups were also investigated [58,121]). For these laminates, a noticeable change in the slope of the load versus deflection curve occurs at the onset of edge delamination with the strain denoted as ϵ_c. The strain at delamination onset is substituted into a closed-form equation

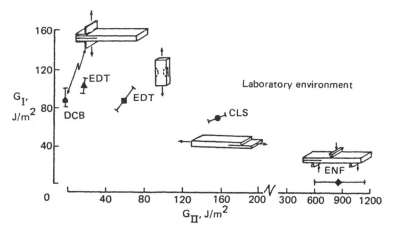

Figure 20 Interlaminar fracture toughness of T300/5208 composites. The data spread is the maximum and minimum values. (From Ref. 120.)

[Eq. (28)] for the strain energy release rate, G, to obtain the critical value G_c for edge delamination.

Since the EDT specimen is a mixed-mode test, mode separation analysis is required. Finite element analysis (and virtual crack extension techniques) [116] of the two EDT layups indicate that the $(\pm30/\pm30/90/\overline{90})_s$ layup consists of 57% G_I due to interlaminar tension, whereas the $(\pm30/0/90)_s$ layup consists of nearly 90% G_I. In both cases the remainder of G is due to G_{II}, resulting from interlaminar shear. The total G, represented by $G_I + G_{II}$, reaches the value predicted by Eq. (28) when the delamination has grown a very small distance from the edge. In addition, the existing analysis has been modified [122] to account for the influence of residual thermal and moisture stresses on strain energy release rate for edge delamination. The EDT test is one of the standard NASA tests [123] used to measure the IFT of toughened resin composites.

There are some questions regarding the validity of the EDT test. It would seem that the relationship for G_c in Eq. (28) would depend on the size and shape of delamination as well as the position along the delamination.

Johnson and Mangalgiri [120] have compared the G_{Ic} and G_{IIc} values obtained using various test specimens for T300/5208 composite material as depicted in Figure 20. From this figure, DCB and ENF appear to be most suitable for characterizing mode I and II delamination behavior, respectively. The figure shows G_{IIc} to be about 8 to 10 times G_{Ic}. On the other hand, EDT and CLS can only give the total G_c for fracture.

Mode III Testing

There is really no standard test method available for tearing mode fracture toughness (i.e., C_{IIIc} in delamination). Ripling et al. [124] have used a tapered double cantiliver beam specimen with a scarf joint for the study of modes I and III. By increasing the scarf angle they could achieve a higher percentage of tearing. This appears to be a visible specimen for studying mode III behavior. Donaldson [125,126] has successfully determined G_{IIIc} by the use of a DCB specimen loaded in the tearing mode for AS4/3502 graphite–epoxy laminates. He has designated the test specimen as a split cantiliver beam (SCB) test.

REFERENCES

1. A. M. James and E. Williams, *Lockheed Horizon*, pp. 31–43 (1986).
2. H. L. Bohon, in *Tough Composite Materials*, NASA CP-2334 pp. 421–445 (1984).
3. J. P. Sandifer, *NASA CR-166091* (1983).
4. K. L. Reifsnider, E. G. Henneke, W. W. Stinchcomb, and J. L. Duke, in *Mechanics of Composite Materials*, *Recent Advances*, Z. Hashin and C. T. Herakovich (eds.), Pergamon Press, Elmsford, N.Y., 1983, pp. 339–390.
5. R. D. Jamison, K. Schulte, K. L. Reifsnider, and W. W. Stinchcomb, *ASTM STP-836*, pp. 21–55 (1984).
6. C. K. H. Dharan, *J. Eng. Mater. Technol.*, 100: 233–247 (1978).
7. G. Dorey, *AGARD LS-124, pp. 6.1–6.11 (1982).*
8. N. J. Pagano and R. B. Pipes, *Int. J. Mech. Sci.*, 15: 679 (1973).
9. J. G. Bjeletich, F. W. Crossman, and W. J. Warren, in *Failure Modes in Composites*, Vol. VI, ASME, Washington, D.C., 1979, p. 118.
10. C. E. Harris and D. H. Morris, *ASTM STP-864*, p. 153 (1985).
11. D. O. Stalnaker and W. W. Stinchcomb, *ASTM STP-674*, pp. 620–641 (1979).
12. K. L. Reifsnider, E. G. Henneke, and W. W. Stinchcomb, *AFML TR-76-81, Part IV (1979).*
13. A. L. Highsmith, W. W. Stinchcomb, and K. L. Reifsnider, *ASTM STP-836*, pp. 194–216 (1984).
14. J. D. Whitcomb, *J. Compos. Mater.*, 15: 403–426 (1981).
15. S. N. Chatterjee, R. B. Pipes, and R. A. Blake, Jr., *ASTM STP-836*, pp. 161–174 (1984).
16. J. M. Hopper, E. Demuts, and G. Milizianto, *AIAA/ASME/ASCE/AHS 25th Structures, Structural Dynamics and Materials Conference*, Palm Springs, Calif., May 1984, pp. 15–20.
17. C. S. Frame and G. Jackson, *AGARD CP-355*, pp. 21.1–21.7 (1986).
18. C. Y. Kam and J. V. Walker, *ASTM STP-937*, pp. 9–22 (1987).
19. D. L. Hunston, R. J. Moulton, N. J. Johnston, and W. D. Bascom, *ASTM STP-937*, pp. 74–97 (1987).
20. S. M. Bishop, *Compos. Struct.*, 3: 295–318 (1985).
21. L. Ye, *Compos. Sci. Technol.*, 33: 257–277 (1989).
22. W. D. Bascom, R. Y. Ting, R. J. Moulton, C. K. Riew, and A. R. Siebert, *J. Mater. Sci.*, 16: 2657–2664 (1981).
23. R. E. Evans, J. E. Masters, and J. L. Courter, *Advanced Composites*, Conference Proc., American Society for Metals, Dearborn, Mich., Dec. 1985, pp. 249–257.
24. S. S. Yau, T. W. Chou, and F. K. Ko, *Composites*, 17: 227–232 (1986).
25. F. K. Ko and C. M. Pastare, *ASTM STP-864*, pp. 428–439 (1985).
26. L. A. Mignery, T. M. Tan, and C. T. Sun, *ASTM STP-876*, pp. 371–385 (1985).
27. L. M. Lackman and N. J. Pagano, *AIAA Paper 74-355 (1974).*
28. C. T. Herakovich, *J. Compos. Mater.*, 15: 336–348 (1981).
29. A. H. Puppo and H. A. Evensen, *J. Compos. Mater.*, 4: 204–220 (1970).
30. R. B. Pipes and N. J. Pagano, *J. Compos. Mater.* 4: 538–548 (1970).
31. S. Tang and A. Levy *J. Compos. Mater.*, 9: 42–52 (1975).
32. A. S. D. Wang and F. W. Crossman, *J. Compos. Mater.*, 11:92–106 (1977).
33. N. J. Pagano, *Int. J. Solids Struct.*, 14: 385–400 (1978).
34. R. L. Spilker and S. C. Chou, *J. Compos. Mater.*, 14: 2–20 (1980).
35. S. S. Wang and I. Choi, *ASME J. Appl. Mech.*, 49: 549–560 (1982).
36. N. J. Pagano and S. R. Soni *Int. J. Solids Struct.*, 19: 207–228 (1983).
37. E. F. Rybicki and D. V. Schmueser, *J. Compos. Mater.*, 12: 300 (1978).
38. L. Ye and B. X. Yang, *J. Reinf. Plast. Compos.*, 7: 179–198 (1988).
39. P. Bar-Yoseph and T. H. H. Pian, *J. Compos. Mater.*, 15: 225–239 (1981).
40. R. B. Pipes and N. J. Pagano, *ASME J. Appl. Mech.*, 41: 668–672 (1974).

41. P. A. Lagace and C. Kassapoglou, *ASME J. Appl. Mech.*, 53: 744 (1986).
42. J. M. Whitney, *ASTM STP-521*, pp. 167–180 (1973).
43. J. D. Whitcomb and I. S. Raju, *J. Compos. Mater.*, 17: 492–507 (1983).
44. J. D. Whitcomb, I. S. Raju, and J. G. Goree, *Comput. Struct.*, 15: 23–37 (1982).
45. I. S. Raju and J. H. Crews, Jr., *Comput. Struct.*, 14: 21–28 (1981).
46. J. D. Whitcomb and I. S. Raju, *ASTM STP-876*, pp. 69–94 (1985).
47. L. Ye, *Int. J. Solids Struct.*, 26: 331–351 (1990).
48. J. R. Yeh and I. G. Tadjbakhsh, *J. Compos. Mater.*, 20: 347 (1986).
49. E. F. Rybicki and A. T. Hooper, *AFML TR-73-100* (1973).
50. W. M. Lucking, S. V. Hoa, and T. S. Sankar, *J. Comos. Mater.*, 17: 188–198 (1984).
51. K. Ericson, M. Presson, L. Carlsson, and A. Gustavsson, *J. Compos. Mater.*, 18: 495–506 (1984).
52. A. S. D. Wang and F. W. Crossman, *J. Compos. Mater.*, 11: 300–312 (1977).
53. C. T. Sun and J. K. Chen, *J. Compos. Mater.*, 21: 969–985 (1987).
54. L. Ye and B. X. Yang, *Acta Mater. Compos. Sin.*, 4(2): 44–51 (1987).
55. E. F. Rybicki, D. W. Schmueser, and J. Fox, *J. Compos. Mater.*, 11: 470–487 (1977).
56. T. K. O'Brien, *ASTM STP-775*, pp. 140–167 (1982).
57. T. K. O'Brien, N. J. Johnston, D. H. Morris, and R. A. Simonds, *SAMPE J.*, 18(4): 8–15 (1982).
58. T. K. O'Brien, *ASTM STP-836*, pp. 125–142 (1984).
59. J. M. Whitney and C. E. Browning, *ASTM STP-836*, pp. 104–124 (1984).
60. F. W. Crossman and A. S. D. Wang, *ASTM STP-775*, pp. 118 (1982).
61. A. S. D. Wang, *Proc. International* Conference on Composite Materials, Paris, Vol. 1, 1980, p. 170.
62. A. S. D. Wang and F. W. Crossman, *J. Compos. Mater.*, Suppl. Vol., p. 71 (1980).
63. F. W. Crossman, J. Warren, A. S. D. Wang and G. E. Law, *J. Compos. Mater.*, Suppl. Vol., p. 88 (1980).
64. A. S. D. Wang, M. Slomiana, and R. B. Bucinell, *ASTM STP-876*, pp. 135–167 (1985).
65. E. M. Wu, *Univ. Ill. T&AM Rep. 275* (1973).
66. G. E. Law, *ASTM STP-836*, pp. 143–160 (1984).
67. T. K. O'Brien, *ASTM STP-876*; pp. 282–297 (1985).
68. R. J. Nuismer and J. M. Whitney, *ASTM STP-593*, pp. 117–142 (1975).
69. E. M. Wu, *Composite Materials*, Vol. V, L. J. Broutman (ed.), Academic Press, New York, 1974, pp. 191–247.
70. R. Y. Kim and S. R. Soni, *J. Compos. Mater.*, 18: 70–80 (1984).
71. C. T. Herakovich, A. Nagarkar and D. W. O'Brien, in *Modern Developments in Composite Materials and Structures*, J. R. Vinson (ed.), ASME, New York, 1981, pp. 53–66.
72. J. C. Brewer and P. A. Lagace, *J. Compos. Mater.*, 22: 1141 (1988).
73. P. A. Lagace, *Material Sciences for the Future*, Society for the Advancement of Material and Process Engineering, 1986, pp. 738–749.
74. T. K. O'Brien and I. S. Raju, *AIAA Paper 84-0961* (1984).
75. L. Ye, *J. Reinf. Plast. Compos.*, 8: 79–91 (1989).
76. S. S. Wang, *J. Compos. Mater.*, 17: 210–223 (1983).
77. T. Johannesson and M. Blikstad, *Report of Linkoping Institute of Technology*, Linkoping, Sweden, 1984.
78. A. S. D. Wang, *Advances in Composite Materials*, *Proc. International Conference on Composite Materials*, Vol. 3, Paris, 1980, p. 170.
79. W. S. Johnson and P. D. Mangalgiri, *ASTM STP-937*, pp. 295–315 (1987).
80. R. A. Jurf and R. B. Pipes, *J. Compos. Mater*, 16: 386–394 (1982).
81. J. D. Whitcomb, *ASTM STP-836*, pp. 175–193 (1984).
82. J. D. Whitcomb, *Compos. Sci. Technol.*, 25: 19–48 (1986).
83. H. T. Hahn, *Compos. Technol. Rev.*, 5: 26–29 (1983).

84. H. T. Hahn and T. Johannesson, *Mechanical Behavior of Materials VI*, ICM4, Stockholm, Vol. 1, 1983, pp. 431–488.
85. S. L. Donaldson, *Composites*, 16: 103–112 (1985).
86. S. L. Donaldson, *Comos. Sci. Technol.*, 28: 33–44 (1987).
87. A. J. Russell and K. N. Street, *ASTM STP-876*, pp. 349–370 (1985).
88. D. Broek, *Elementary Engineering Fracture Mechanics*, Noordhoff, Leyden, The Netherlands, 1978.
89. D. J. Wilkins, J. R. Eisenmann, R. A. Camin, W. S. Margolis, and R. A. Benson, *ASTM STP-775*, pp. 168–183 (1982).
90. S. Mall and W. S. Johnson, *ASTM STP-893*, pp. 322–334 (1986).
91. S. Mall, W. S. Johnson, and R. A. Everett, Jr., *NASA TM-84577* (1985).
92. A. Poursartip, *ASTM STP-937*, pp. 222–241 (1987).
93. R. L. Ramkumar and J. D. Whitcomb, *ASTM STP-876*, pp. 315–335 (1985).
94. D. M. Parks, *Int. J. Fract.*, 10: 487–502 (1974).
95. T. K. Hellen, *Int. J. Numer. Methods Eng.*, 9: 187–207 (1975).
96. J. M. Mahishi and D. F. Adams, *ASTM STP-876*, pp. 95–111 (1985).
97. J. M. Mahishi, *Eng Fract. Mech.*, 25: 197–228 (1986).
98. D. F. Devitt, R. A. Schapery, and W. L. Bradley, *J. Compos. Mater.*, 14: 270–284 (1980).
99. J. M. Whitney, C. E. Browning, and W. Hoogesteden, *J. Reinf. Plast. Compos.*, 1: 297–313 (1982).
100. P. E. Keary and L. B. Ilcewicz, *J. Compos. Mater.*, 19: 154–177 (1985).
101. J. M. Whitney, *Compos. Sci. Technol.*, 23: 201–219 (1985).
102. J. M. Scott and D. C. Phillips, *J. Mater. Sci.*, 10: 551–562 (1975).
103. S. M. Lee, *J. Compos. Mater.*, 20: 185–196 (1986).
104. W. D. Bascom, J. L. Bitner, R. J. Moulton, and A. R. Siebert, *Composites*, 11: 9–18 (1980).
105. I. M. Daniel, I. Shareef, and A. A. Aliyi, *ASTM STP-937*, (1987).
106. M. L. C. E. Verbrugenn, *UTH M-513 (1984)*.
107. J. P. Berry, *J. Appl. Phys.*, 34: 63 (1963).
108. J. G. Williams, *Compos. Sci. Technol.*, 35: 367–376 (1989).
109. S. Hashemi, A. J. Kinloch, and J. G. Williams, *J. Mater. Sci. Lett.*, 8: 125–129 (1989).
110. J. D. Whitcomb, *J. Compos. Technol. Res.*, 7: 65–67 (1985).
111. S. Mall and W. S. Johnson, *NASA TM-86355 (1985)*.
112. A. J. Russell and K. N. Street, *ASTM STP-876*, pp. 349–370 (1985).
113. L. A. Carlsson, J. W. Gillespie, and R. B. Pipes, *J. Compos. Mater.*, 20: 594–604 (1986).
114. J. W. Gillespie, L. A. Carlsson, R. B. Pipes, R. Rothschilds, B. Trethewey, and A. Smiley, *NASA CR-176416 (1985)*.
115. J. W. Gillespie, L. A. Carlsson, and R. B. Pipes, *Compos. Sci. Technol.*, 27: 177–197 (1986).
116. T. K. O'Brien, in *Tough Composite Material, Recent Developments*, NASA Langley Research Center, Hampton, Va., Noyes Publications, Park Ridge, N.J., 1985, pp. 14–27.
117. W. M. Jordan and W. L. Bradley, *ASTM STP-937*, pp. 95–114 (1987).
118. R. A. Jurf and R. B. Pipes, *J. Compos. Mater.*, 16: 386–394 (1982).
119. W. L. Bradley and R. N. Cohen, *ASTM STP-876*, pp. 389–410 (1985).
120. W. S. Johnson and P. D. Mangalgiri, *NASA TM-87571 (1985)*.
121. L. Ye, *Composites*, 20: 275–281 (1989).
122. T. K. O'Brien, I. S. Raju, and D. P. Garber, *NASA TM-86437* (1985).
123. *Standard Tests for Toughened Resin Composites*, rev. ed., NASA RP-1092 (1983).
124. E. T. Ripling, J. S. Santner, and R. B. Crosley, *J. Mater. Sci.*, 18: 2274–2282 (1983).
125. S. L. Donaldson, in Proc. *6th International Conference on Composite Materials*, July 1987, pp. 3.233–3.242.
126. S. L. Donaldson, *Compos. Sci. Technol.*, 32: 225–249 (1988).

Phenolic Fiber-Reinforced Thermoplastics

Eiichi Jinen
Mechanical and Systems Engineering
Kyoto Institute of Technology
Kyoto, Japan

INTRODUCTION

Thermoplastics have begun to be considered more often as possible materials for use in the field of industrial plastics, and many attempts to find a new processing method for production are under way. The use of thermoplastics as alternatives for other materials seems to be growing in industry. According to statistics, both the production and consumption of thermoplastics are gradually on the increase. The reason for this tendency seems to be an increasing demand for lightweight structures or parts, especially for material used in automobiles and aircraft. Another reason for increased production is the fact that process control and the control of production stages are easier than in the case of thermosetting plastics. In polymeric composites, one of the aims of reinforcing the mechanical properties of the resin matrix by mixing it with some other material is to produce a brittle matrix polymer, an improvement in toughness, and for a ductile matrix polymer, an increase in stiffness. In either case, from a macroscopic viewpoint, the purpose is to incorporate reinforcing materials while maintaining good adhesive conditions at the interface between the matrix and the fiber when in service. A nonflammable phenolic fiber with the trade name Kynol [1] is commonly used to reinforce materials. The reason for preferring this fiber is its low density compared to that of other reinforcing materials, as well as other characteristics, such as its performance at high temperatures. So an improvement in the physical properties of material reinforced by this fiber can be expected with respect to both the thermal properties of the resin matrix and a favorable specific modulus at high temperatures. In this section we discuss the reinforcement effects of phenolic fibers with regard to the heat distortion and temperature dependence of both the dynamic modulus and the loss modulus of the thermoplastics polystyrene, polypropylene, and nylon 66.

PHENOLIC FIBER

The phenolic resin fiber called Kynol [1] was invented by Dr. Economy in 1968 and developed by Carborundum Co. Ltd., which is famous for the production of ceramic and abrasive materials. The original aim of this work was an attempt at fiberfication of phenolic precuser fiber resin to a fibrous material for use as the precursor of a carbon fiber that could be employed in the aerospace construction, and for the formation of fibers that were flameproof and could withstand very high temperatures. Before the invention of the new material, Economy had established the possibility of getting an unresolved fiber from the fiberfication of phenolic resin by means of a bridge formation with formaldehyde. In the early stages he designed the material for use as a precusor (a raw fiber) of a carbon fiber. However, the varied characteristics of the material and the possibility of the industrial production of a flameproof fiber were recognized, and a developmental project for the production of the new fiber was begun. Of the two types of phenolic resins generally used as raw materials, resoles (thermosetting resins) and novolacs (thermoplastic resins) [2], Economy was interested in the latter because it is soluble in organic solvents and can be melt-spinned into a textile fiber. The principal barriers to the invention of this new fiber was the curing method, using heat with moisture present to yield formaldehyde and ammonia, which acts as the catalyst, as well as how to form the bridge and provide a

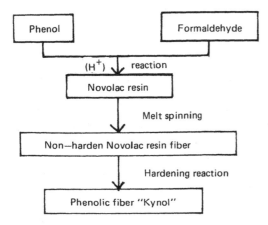

Figure 1 Flow sheet for production of the phenolic fiber Kynol.

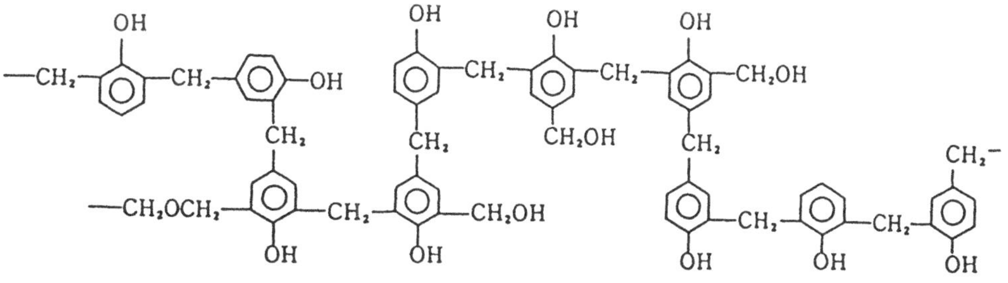

Figure 2 Chemical structure of Kynol.

three-dimensional molecular structure. The problem was solved by finding a skillful method that achieved a critical reaction and balancing by restriction with an acid- and heat-controlling system [3]. However, development of the production technology was not completed, and was transferred to Japan. In 1972, study of the technical improvements in production of the raw fiber and basic study of the phenolic resin raw material were begun using a small-scale pilot plant. GUN-EI Chemical Industry Co. Ltd. is now the sole worldwide supplier of Novoloid and produces many types of raw fibers.

Structure and Properties of Phenolic Fiber

Figures 1 and 2 show a flow sheet for the production of Kynol [4] and its chemical structure [5], respectively. Figure 3 also indicates the functional groups that have been found in small amounts in the fiber molecule by means of instrumental analysis. The characteristics and functional abilities of the phenolic fiber (Table 1) have been induced from those chemical structures, that is, from the existence of several functional groups in the molecule. These groups consist of carbon, hydrogen, and oxygen elements, and the structure is cross-linked, three-dimensional, and has low crystallinity. The phenolic fiber's molecular characteristics give it the following special characteristics [6]: nonmeltability and non-shrinkability at high temperatures, low gas toxicity when burned (Table 2 and Figure 4), high heat and chemical resistance (Figure 5 and Table 3), high heat insulation (Table 4) and nonbrittlenes at low temperatures (Table 5), minimal smoke generation when burned, light in weight, and with good chemical adhesion to other matrix resins. When this fiber is used as a reinforcing material for resin composite systems, the flammability of the composite in the burning state will be restricted [7–10] there will be a decrease in mechanical

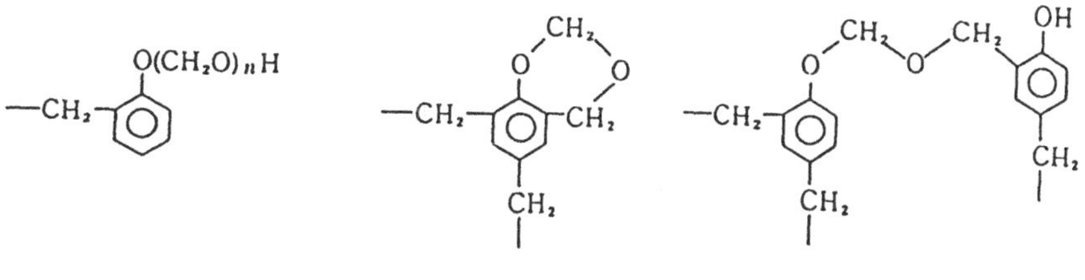

Figure 3 Other functional groups in the Kynol molecule. 1, Hemiacetal; 2, linkaged acetal compound; 3, acetal compound.

Table 1 Characteristics and Mechanical
Properties of Kynol Fiber

Color	Gold
Diameter	14–33 (2–10 denier[a])
Fiber length	0.2 mm endless
Specific gravity	1.27
Tensile strength	1.3–1.8 g/d (15–20 kgf/mm^2)
Elongation	10–50
Modulus	300–400 kgf/mm^2
Loop strength	2.2–3.1 g/d
Knot strength	1.1–1.5 g/d
Moisture regain	6% (at 20°C, 65% RH)

[a] A count number for estimated fiber linear density; numbers
are weight in grams per 9000 m.

Table 2 Flammability
of Organic Fiber

Fiber	LOI[a]
Kynol	30–34
Aramid	28–31
Polychlal	28
Modacryl	26
Wool	24
Polyester	22
Acryl	20
Cotton	19

[a] Limited oxgen index: an
index for estimating degree
to which a fiber is flameproof or flame-retardant.

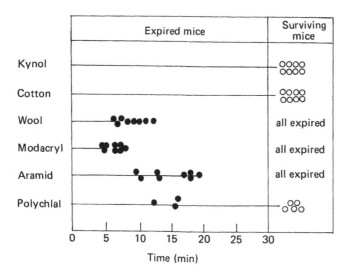

Figure 4 Mouse test for toxicity of combustion gases.

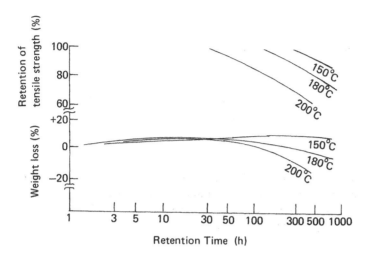

Figure 5 Characteristics of heat resistance in air.

Table 3 Tensile Strength Loss[a] of Organic Fibers After Bleaching Test (100 h at 25°C)

| Organic Fiber | Bleaching Agent (Concentration) | | | | |
	$NaClO_4$ (50)	Na_2SO_3 (10)	H_2O_2 (35)	Cl_2 (0.3)[b]	NaClO (12)[b]
Kynol	A	A	B	A	A
Aramid (*para* type)	B	A	A	D	D
Aramid (*meta* type)	B	A	A	D	B
Polyimideamide	C	A	C	D	D
Polybenzimidazole	A	A	A	D	D
Polynosic	A	A	A	D	D

[a] A, Less than 5% strength loss; B, 6–15% strength loss; C, 16–30% strength loss; D, 70–100% strength loss.
[b] Available chlorine content.

Table 4 Thermal Conductivity[a]

Material	Specific Gravity (g/cm³)	Temp. (°C)	Thermal Conductivity (kcal/m·h·°C)
Kynol felt	0.01	−40	0.024
	0.01	20	0.035
	0.01	40	0.039
Glass fibers	0.04	20	0.035
Glass wool	0.06	20	0.036
Rock wool	0.10	20	0.032

[a] ASTM C-177 method.

Table 5 Retention of Properties After Exposure at Very Low Temperature

Treating Conditions	Tenacity (g/d)	Elongation (%)
None	1.34	34
−44°C, 1000 h	1.39	47
−139°C, 6 h	1.30	39
−196°C, 6 h	1.30	40

toughness in a low-temperature atmosphere. These characteristics will promote a demand for the use of such systems in the interiors of transportation vehicles such as automobiles and airplanes, because the most important considerations regarding materials used therein are the fact that they are flameproof and lightweight.

SPECIMENS

Test specimens are molded to a ASTM A-type specimen in which each resin pellet (about 4 mm) contains a 40 wt % phenolic fiber (12 to 15 $\mu\phi$) or such a fiber mixed with resin pellets to gain from each fiber the content of the material produced by an injection machine under normal conditions. The resin matrices utilized are the popular ones: polystyrene, polypropylene, and nylon 66. The fiber content of these specimens in weight percent is for polystyrene, 0, 2, 4, 5, 11, 21, and 22%; for polypropylene, 0, 1, 2, 3.5, 4, 5, 18, and 33%; and for nylon, 0, 2, 5, 13, 14, 23, and 39%. Moreover, to find or compare the effects of reinforcement by short fibers and fiber powders for the two former types of material, a specimen with a 20 wt % content is also prepared. Molded specimens are tested after conditioning for 6 months in a standard air-conditioned room, so that their moisture content seems to be close to equilibrium. The ASTM A-type specimen is available for use in the measurement of heat distortion, because the standard rules require that the specimen be 126 mm (5 in.) in length, 13 mm (½ in.) in depth, and any width beginning at 3 mm (⅛ in.) The dimension tolerance should be on the order of ±0.13 mm (0.005

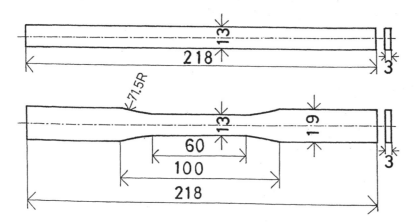

Figure 6 Configurations of the material and the specimen (dimensions in mm).

in.) of the length of the specimen (Figure 6). Specimens are obtained by cutting the molded specimens with a milling machine to the accuracy noted above.

HEAT DISTORTION

Equipment

The heat distortion temperature method is used to find the yield point when a specimen has been deflected by 0.25 mm (0.010 in.) [11]. The equipment used for this purpose is constructed essentially as shown in Figure 7 and consists of a number of parts. The specimen is held by metal supports positioned 100 mm (4 in.) apart, allowing a load to be applied on top of the specimen midway between the supports. The contact edges of the supports and the piece that applies the load are rounded to a radius of 3 mm. The support has provisions for inserting a thermometer to a point within 3 mm of the specimen. A bar of rectangular cross section is tested as a simple beam while a load is applied at its center to give maximum fiber stress (maximum bending stress), which is given by the following simple formula:

$$P = \frac{2Sb^2}{3L} \quad \text{or} \quad P' = \frac{P}{9.80665}$$

where P is the load in newtons (or lbf); S the maximum fiber stress in the specimen: 1820 kPa (264 psi), 910 kPa (132 psi), or 455 kPa (66 psi); b the width of the specimen in meters (or in.); d the depth of the specimen in meters (or in.), L the width of span between the supports, 0.1 m (4 in.); and P' the load in kgf when P in newtons has been measured.

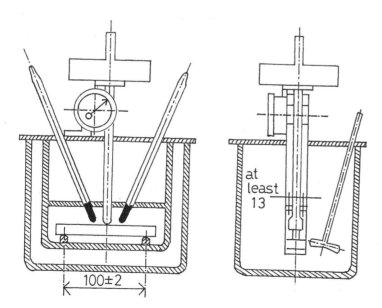

Figure 7 Apparatus for deflection temperature test.

Test Procedure

When using a multirack bath (such as Custom Scientific Instruments, Inc.'s Model CS-107 with an insulated stainless steel bath and either an automatic solid-state programmed temperature control for a 2°C/min temperature rise or a manual temperature control), all specimen supports are adjusted with an Invar calibration bar ($\frac{1}{2}$ in.) across the anvils. All supports should be checked as follows. Place a calibration bar across the anvils, then lower the weight rack with the pressure foot on the calibration bar. The pressure foot must be at right angles to the length of the bar, with a deflection arm adjustment screw located in the dial indicator. Regulate the screw adjustment so that it deflects the dial indicator to its midpoint of travel. This permits plus or minus deflection. Zero the indicator by rotating the face. Then install the specimen across the anvils with the 1.27-cm ($\frac{1}{2}$-in.) dimension vertical. Lower the weight rack and regulate the screw adjustment for a zero reading on the indicator. Place the weight carefully for the desired fiber stress on the weight rack and insert a holder into the temperature bath. The initial temperature of the bath is generally 20 to 23°C, but if it can be proved that a higher initial temperature will not affect the test results, the higher temperature may be used provided that it is not less than 30°C below the expected temperature of deflection under load. Five minutes after the load has been applied, the deflection measuring device, the dial needle, is set to zero and the temperature of the bath is raised as described in the preceding section and watched visually until a 0.25-mm (0.010-in.) deflection occurs. The heat distortion temperature of each specimen is averaged from the results of four to six specimens.

Polystyrene

Dependence of Heat Distortion Temperature on Fiber Content

The results for maximum fiber stresses of 455 kPa ($\times$), 910 kPa ($\bigcirc$), and 1820 kPa ($\triangle$) are shown in Figure 8. The dashed symbols in this figure indicate the results for the specimen, which is reinforced by a 20 wt % fiber powder content. These results will be used to compare the reinforcing effects of fiber or fiber powder and/or to consider the dependence on fiber length of the reinforcement used in the matrix. In the first case improvement in the heat distortion temperature for the range 3 to 5°C is performed by reinforcement with only 2% of the fiber, which yields a temperature gain of 14.4 to 8.4°C by mixing 22% of the fiber under the given maximum fiber stress conditions. This improvement corresponds to the range 18.3 to 11.5% for the matrix results. However, the results for reinforcing with a 20 wt % fiber powder content, do not show the improvement that the results using fiber show. In this case, the reinforcing effect corresponds to that of mixing with 4 to 5% fiber weight content only for the given maximum fiber stress conditions; that is, the reinforcing effect of the powder is one-fourth that of the fiber. The second point is that it is obvious that the heat distortion temperature and fiber content show a clear linear relationship, except for the matrix specimen. The meaning of the gradient of these lines implies the importance of the temperature rise with increased unit fiber content. Also, the magnitude of the gradient in these lines shows a maximum fiber stress dependence on the specimen in three places with regard to bending conditions. That is, the gradient decreases with increased maximum fiber stress. So, to find the stress dependence of the three lines, the correspondences of each value are plotted on the graph using a normal scale. This result is shown in Figure 9. It is clear that stress dependence on the increased temperature per unit fiber content seems to indicate a linear relationship.

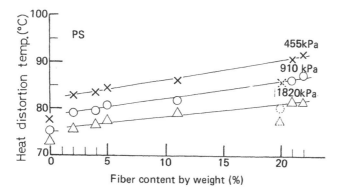

Figure 8 Fiber content dependence of heat distortion temperature for PS. The dashed symbols indicate the result of a specimen reinforced with 20% fiber powder.

These characteristic features of the reinforcing effect of a phenolic fiber may be due to the thermal properties of the matrix and the interaction between the fiber and the matrix.

Dependence of Heat Distortion Temperature on Maximum Fiber Stress

Figure 10a shows the relationship on the normal scale for each level of fiber content. This relation is obviously nonlinear, and its benefit is that changes in the location of the curved line and in its gradient can be determined. The former change is due to an increase in the reinforcing effect with increased fiber content and is self-evident. The latter is a change in the mechanical properties of the material under the test conditions and seems to be due to structural changes in the material due to the increased fiber content. Moreover, the reinforcing effect of fiber seems superior to that of fiber powder. This result is easy

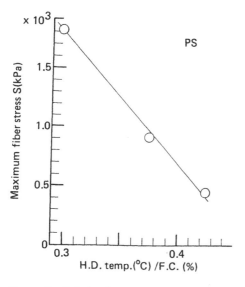

Figure 9 Relation between maximum fiber stress *S* (kPa) and increased temperature values upon increasing 1% fiber content (the gradient of the line in Figure 8) for PS.

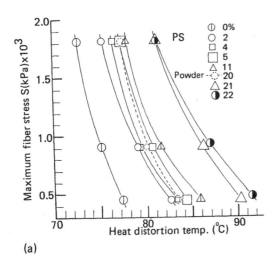

(a)

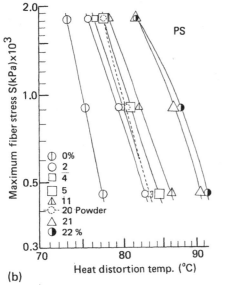

(b)

Figure 10 Heat distortion temperature of maximum fiber stress S (kPa) for PS: (a) plotting on normal scale; (b) log-log scale. Dashed line and symbols indicate the results of the specimen reinforced with 20% fiber powder.

to understand from the location of the curve; that is, the curved line exists in the area between the 4 and 5% lines.

To clarify these relations, data for the maximum fiber stress dependence of the heat distortion temperature are plotted on a log-log sheet. Figure 10b shows the results of these relations. It is obvious that they are shown by a linear line except for the results for higher fiber contents. The gradient of the linear lines implies a stress dependence on heat distortion. It is noticeable that the magnitude of the gradient decreases with increased fiber content. Also, in the case of reinforcement by fiber powder, the gradient of the line is larger than

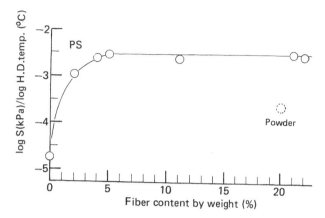

Figure 11 Relation between log S (kPa)/log heat distortion temperature (°C) and fiber content for PS.

those for low fiber content, despite the fact that the heat distortion temperatures of these specimens show nearly the same value. This result indicates that the difference in fiber stress dependence and its properties seems to depend on fiber length. However, the linearity of the relations disappears at higher levels of fiber content, and the gradient of the curve becomes more inclined. The dependence of the gradient of these lines or the incline of the curve of the fiber content are plotted on the graph in Figure 11 using a normal scale. The figure shows that dependence of the gradient on fiber content appears within a 5% range. No change can be found in the outer range, and thus the relation indicates a saturation phenomenon. On the other hand, in the case of reinforcement by a 20 wt % fiber powder, the gradient corresponds to the result for about a 1% fiber content.

Polypropylene

Dependence of Heat Distortion Temperature on Fiber Content

The results plotted on a normal scale for maximum fiber stresses of 455 kPa ($\times$), 910 kPa ($\bigcirc$), and 1820 kPa ($\triangle$) are shown in Figure 12a. The dashed line indicates the results of reinforcement by a 20 wt % fiber powder. In this case, a tendency that differs from the polystyrene results can be observed. The heat distortion temperature was below the value of the matrix under a low maximum fiber stress. Moreover, under these fiber stress conditions, the relation is obviously nonlinear, and a stronger tendency can be recognized under low fiber stress conditions. For the former investigation, to recover the temperature reduction from the value of the matrix, about 10% and 5% more fiber must be used for the maximum fiber stress levels 455 kPa and 910 kPa, respectively.

In injection molding of a crystalline polymer, when processing for solidification of a given form, the temperature changes from above the melting point of a crystalline to below it, through cooling of the die. In general, the volume change in the crystalline polymeric material via a phase change is larger than for a noncrystalline material. Polypropylene is a crystalline polymer and its die shrinkage is 1 to 2.5%. Under room-temperature conditions the specimen will contain some nonequilibrium portion due to residual stress and/or some unevenness in the structure due to rapid cooling from the melting state to the solid state. However, with a raise in temperature in the process for

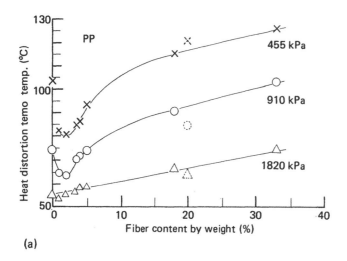

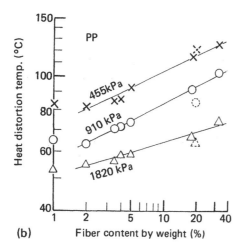

Figure 12 Fiber content dependence of heat distortion temperature for PP: (a) plotting on a normal scale; (b) log-log scale. Dashed symbols indicate the results of the specimen reinforced with 20% fiber powder.

measuring heat distortion, a reverse phenomenon, a volume change, will be observed; and if the temperature is high enough, a change in the morphological structure, such as recrystallization, may occur. Such a change will take place more readily at higher temperatures than at lower ones. The unevenness in local deformation areas behaves like those in which local stress concentration has been experienced. As a whole, deformation of the specimen will be expected to have a larger value than in the case of the matrix specimen. A reduction in fiber content dependence may be explained by the foregoing interpretation. In fact, reduction in the phenomenon of distortion temperature depends more closely on temperature than on maximum fiber stress. Another reason for such an interpretation of the phenomenon is that the degree of adhesion or problem in the interface between the polypropylene and phenolic fiber should be noted. Figure 12a seems to indicate a mutual interaction between the volume change of the matrix and the adhesive conditions

[12]. In the case of higher maximum fiber stress levels, tough a down traces can be found in the range of a few percent a reduction of temperature disappear and the relation clearly seems to be linear. The results for reinforcement by fiber powder cannot be expected to match those for reinforcement by fiber except in the 455 kPa case. This tendency is similar to the results for polystyrene; however, at low maximum fiber stress levels, it seems that reinforcement by fiber powder is effective.

To clarify the nonlinearity of the relationship between the heat distortion temperature and the fiber content, all data except for the value of the matrix are plotted on a log-log scale. The result is shown in Figure 12b. The plots that indicate the correspondence of the relation under three different maximum fiber stress conditions seem to form on a straight line except for that of 1% fiber content. That is, for a fiber content over 2%, the relation can be expressed as an exponential function, and the gradient of the line indicates some dependence on the maximum fiber stress. From the former experimental results, this fact implies that the reinforcing effect in polypropylene is nonlinear and seems to be due to the thermal properties of the matrix. With regards to the latter, for a maximum fiber stress of 455, 910, and 1820 kPa, the magnitude of the gradient for each line is 0.48, 0.50, and 0.3, respectively.

Dependence of Heat Distortion Temperature on Maximum Fiber Stress

As shown in Figure 12a, dependence on the fiber content indicates a reduction from the value of the matrix in the range of low fiber content. This tendency also depends on the maximum fiber stress. For the maximum fiber stress the value is 1820 kPa, but in the lower temperature range, this phenomenon disappears completely. So it is assumed that the maximum fiber stress dependence on heat distortion temperature also shows a complicated relation with regard to the fiber content by an expression such as that shown in Figure 10a using a normal scale. In fact, in the region of low fiber content, the relation can be indicated by curved lines; however, these curves do not form in order of fiber content directly across from each other. Then these complicated relations are plotted on a log-log scale such as that shown in Figure 10b. The results are shown in Figure 13. In

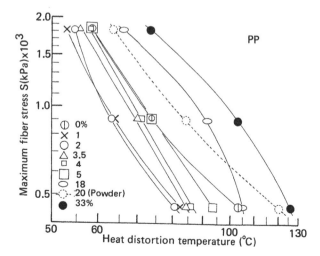

Figure 13 Heat distortion temperature dependence of maximum fiber stress S (kPa) by log-log scale for PP. Dashed line indicates the results of the specimen reinforced with 20% fiber powder.

this figure, the relation cannot be expressed by a straight line for all the fiber contents; however, it should be noted that the nonlinearity of these curves can be changed by changing the fiber content. That is, the curve for the matrix specimen is concave on the right side. However, the degree of this tendency decreases with a mixture of 1 to 2% fiber. For the results of a 4 or 5% specimen, the relation is nearly linear. With a higher fiber content, the relation reverses and shows a convex curve on the right side. In other words, when maximum fiber stress dependence on the heat distortion temperature is described on a log-log scale, the differential coefficient of the second order of these curves changes with the fiber content. This tendency seems to be caused by the change in mechanical properties of the material due to the structural change caused by raising the temperature.

A second interesting result is as follows: As shown by the dashed line in Figure 13, the specimen reinforced with fiber powder shows an inverse tendency compared with the results for the normal specimen; the shape of the curve is similar to those of the matrix and the 1 to 2% specimens. It is obvious from these results that the mechanical properties

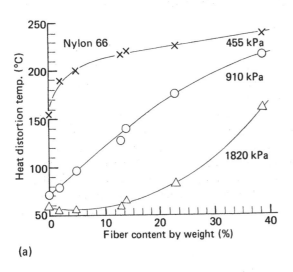

(a)

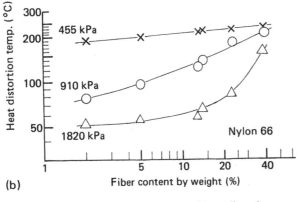

(b)

Figure 14 Fiber content dependence of heat distortion temperature for nylon 66: (a) plotting on a normal scale; (b) log-log scale.

of the specimen relate to fiber length, and the geometry of the curve indicates the character of the structural factor.

Nylon 66

Dependence of Heat Distortion Temperature on Fiber Content

The relationship plotted on a normal scale and described by a parameter is the maximum fiber stress, shown in Figure 14a. In the case of nylon 66, it is noteworthy that the dependence relates more closely to maximum fiber stress than do the results for the two thermoplastics discussed earlier. The character of the relationship changes drastically with an increase in maximum fiber stress. That is, for a low fiber stress level, a rise in temperature can be seen at a low fiber content, which increases moderately with increased fiber content. A saturation phenomenon is indicated by this tendency. At the middle fiber stress level, the heat distortion temperature change shows over a very wide temperature range, and the gradient of the curve is moderate. However, the character of the relationship is the same as in the results for the low fiber stress level in which the curve is convex at the top. But a change in this tendency can be recognized at a higher fiber stress level. That is, the character of the curve changes in form from convex to concave, so it seems that the relation becomes linear at some fiber stress level between the two.

To confirm the character of these curves, the correlative relationships of the data are plotted on a log-log scale. The result is shown in Figure 14b. From this figure it is clear that at the low fiber stress level, the relation can be described by a linear line and the magnitude of the gradient is very small. So at this level the relationship can be expressed exponentially. However, the linearity of this relationship decreases with increased fiber stress level, and the gradient of the curve increases with increased fiber stress level.

Dependence of Heat Distortion Temperature on Maximum Fiber Stress

The relationship described on a log-log scale by a variable parameter has the fiber content shown in Figure 15. A characteristic and interesting tendency of the relationship can be

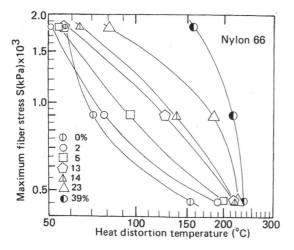

Figure 15 Heat distortion temperature dependence of maximum fiber stress *S* (kPa) by log-log scale for nylon 66.

recognized from this figure. On the whole, the relationships are nonlinear, especially in the matrix specimen. This tendency shows a considerable variation in degree. When reinforced with a fiber, however, the tendency decreases with increased fiber content and a nearly linear relation can be found in some fiber contents between 5 and 13%. The reason for mentioning this may be recognized in the changing shape of the curve. That is, a characteristic change can be found by comparing the results for a low fiber content with those for a higher content: the shape of the curve changes from concave to convex. Moreover, the results for the higher-content (39%) heat distortion temperature show only a weak dependence.

TEMPERATURE DEPENDENCE OF THE DYNAMIC PROPERTIES

Equipment

A Brüel & Kjaer instrument is employed for determination of the resonance frequency and measurement of the internal damping for a rectangular specimen, as shown in Figure 6. To determine the temperature dependence of the dynamic modulus and the loss factor, the test jig, which is a complex modulus apparatus, should be used in an environmental chamber to control the temperature during measurement.

The temperature dependence of the dynamic properties for the phenolic fiber used as the reinforcing material was measured by a Rheovibron DDV-II-EA of Toyo Baldwin Co. Ltd. The measurement principle of the machine is not reasonance but forced vibration, using a longitudinal sine wave with constant frequency. Therefore, to define the transmission path as a fiber length, weak tension must be applied to a multifilament specimen. The phase difference between the forced signal for exciting the specimen and the small deformation is used to determine the loss factor of the specimen.

Test Procedure

The expression of a complex modulus of elasticity or dynamic modulus is given by

$$E^* = E'(1 + jd)$$

where d, the loss factor, $= \tan \delta$, where δ is the phase shift caused by friction. The amplitude of vibration is plotted as a function of frequency. From such a curve, at a resonance peak, the loss factor is calculated as

$$d = \frac{\Delta f_n}{f_n}$$

where Δf_n is the bandwidth at the half-power points (3-dB points) and f_n is the resonance frequency. The index n is the order of resonance, or mode number.

The dynamic modulus E' can be found from the resonance frequency and the mechanical dimensions of the bar [13]:

$$E' = 48\pi^2\rho \left(\frac{l^2}{h} \times \frac{f_n}{K_n^2} \right) \qquad \text{Pa (N/m}^2\text{)}$$

where l, h, and ρ are free length (m), thickness (m), and material density (kg/m^3), respectively. K is a coefficient dependent on the resonant mode number and damping method, with both ends free or clamped; $K_1 = 4.73$, $K_2 = 7.853$, $K_3 = 7.855$. In this experiment the latter mode is employed in the vibration state, using a free length of 16 to 18 cm. Measurements have been made in the temperature range 25 to 200°C, with the atmospheric temperature kept within ±0.5°C for the low-temperature range and ±1°C for the higher-temperature range during the heating process.

Regarding the dynamic tensile properties of the phenolic fiber, the complex modulus E^* can be determined as the ratio of the maximum excited stress ($\dot{\sigma}_{max}$) to that of dynamic strain ($\dot{\epsilon}_{max}$), that is,

$$|E^*| = \sqrt{E'^2 + E''^2} = \frac{\dot{\sigma}_{max}}{\dot{\epsilon}_{max}}$$

$$E' = |E^*| \cos \delta, \qquad E'' = |E^*| \sin \delta, \qquad \frac{E''}{E'} = \tan \delta$$

where E' and E'' are the dynamic modulus and loss modulus of the fiber, respectively. Measurement of the temperature dependence on the dynamic and loss moduli for a multifilament (995 denier) phenolic fiber has been made under the following conditions: frequency of forced excitation, 110 Hz; amplitude of deformation, 32 μm; additional tension for a given static strain in the specimen, 105 grf; specimen length, 40 mm; and density, 1.29×10^3 kg/m³ (measured by a density gradient tube after conditioning). This value is also used to calculate the cross-sectional area of the multifilament, because it can be determined as the product of the denier value multiplied by the density.

Phenolic Fiber

Measurement plots of the temperature dependence on dynamic modulus E', loss modulus E'', and tan δ are shown in Figure 16 using a normal scale [14]. From the results of E',

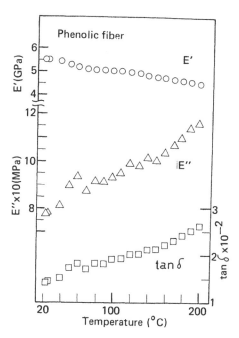

Figure 16 Temperature dependence on dynamic modulus E' (○) or loss modulus E'' (△) and tan δ (□) of phenolic fiber by a longitudinal wave of 100 Hz.

it is easily seen that the fiber is a suitable material for reinforcing thermoplastic resin. That is, the change in E' is only about 20%, which corresponds to the range of temperature change from 23°C to 200°C and shows a moderate decreasing tendency. However, if the plot is checked again carefully, a slight dispersion can be found in the temperature region around 60°C. The trace of this weak dispersion can easily be found by comparing it with the plot for E''. The data follow almost a linear relationship on a normal scale except for this small area of weak dispersion. In the case of temperature dependence on E'', it is easy to find a dispersion that is of fair size in the same temperature region as the one for E'. Moreover, a slight dispersion can also be found in the temperature region around 125°C.

Polystyrene

Temperature Dependence of E' and E''

The results of plotting on a normal scale for fiber weight contents of 0, 11, and 21% are shown in Figure 17. For the dependence of E' on the resin matrix, the property of the E' value shows a rapidly falling tendency. However, such a tendency is improved by mixing the resin with 11% of the fiber. At this fiber content, the reinforcing effect of the fiber with regard to the dependence of E' is superior in the high-temperature region. That is, over 90°C, E' values fall rapidly with about a 20°C temperature difference in the plot of the matrix and the 11% specimens, which can be found by comparing them at the same E'' value. As shown in the preceding section, this phenomenon seems to be due to the good reinforcing properties of the fiber in the high-temperature region. To improve the properties in the lower-temperature region, an increase in fiber content up to 21% was found necessary.

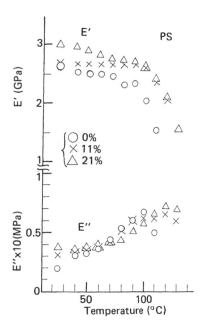

Figure 17 Temperature dependence on E' or E'' of phenolic-fiber-reinforced PS. Fiber content by weight: 0% ($\bigcirc$), 11% ($\times$), and 21% ($\triangle$).

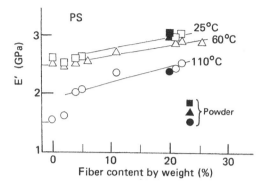

Figure 18 Fiber content dependence on E' of phenolic fiber reinforced PS for 25°C (□), 60°C (△), 110°C (○).

The change in dependence on E'' due to the change in fiber content can be seen in the regions close to room temperature and above 90°C. A large change is especially evident in the latter case. For the result of the matrix, the peak of dispersion is found at about 100°C; however, the peak of the reinforced specimen extends to a broad shape and the peak temperature of the dispersion shifts to the higher-temperature side with increased fiber content.

Dependence of E' *on Fiber Content*

The results of plotting on a normal scale for the 25, 60, and 100°C data are shown in Figure 18. It is clear from the figure that fiber content dependence shows a nearly linear relation in the low-temperature region. Moreover, the location of these plots shifts to a lower value on the E' scale when the temperature is raised from 25°C to 60°C, although the magnitude of the gradient for the two lines seems to be almost identical. In the high-temperature region, however, this linear relationship deviates at two points for the matrix specimen and the specimen with the lower fiber content. Except for these two points in the low fiber content, the relationship is also nearly linear, but the magnitude of the gradient is twice that of the magnitude for the two lines in the lower-temperature region.

Dependence of tan δ on Fiber Content

Variations of tan δ with fiber content at 25, 40, 80, 100, and 120°C are shown in Figure 19. In this figure a rise in temperature can be seen in two regions. In both the temperature region over 100°C and the neighboring one at rom temperature, the fiber content dependence is almost imperceptible on the whole, because the gradient of the plots cannot be recognized or only a slight negative slope is present in the higher-temperature region. By contrast, in the plots for 80°C and 100°C, the relation is nonlinear, with an especially strong tendency toward nonlinearity in the region of lower fiber content.

Effect of Filler Morphology on Dynamic Properties

Although the fiber contents of the specimen used in comparing these properties does not exactly accord with each other, the arraying plots for E' or E'' and tan δ of the 20% fiber powder and 21% fiber are shown in Figure 20a and b, respectively. For the results of E'

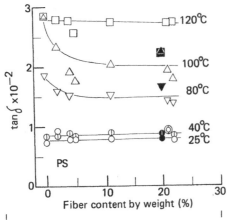

Figure 19 Fiber content dependence on tan δ of phenolic-fiber-reinforced PS for 25°C (○), 40°C (◑), 80°C (▽), 100°C (△), and 120°C (□).

in the temperature region below 110°C, the points for the specimen with powder occupy a higher level than those for the specimen with fiber. However, at 120°C, the E' value for the reinforced specimen is obviously determined. So the results seem to show that the reinforcing effect of the fiber acts as a more effective resistance to deformation of the material in the higher-temperature region than in the lower one. However, the difference in reinforcing effect due to fiber size can easily be seen in the temperature dependence on E''. That is, the temperatures that appear at the peak of plots for the powder and fiber specimens are at about 100°C and 120°C, respectively. The location of the peaks obviously

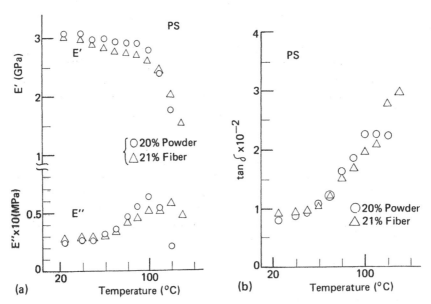

Figure 20 (a) Temperature dependence of E' and E'' for phenolic-fiber-reinforced (△) and fiber-powder-reinforced (○) PS; (b) temperature dependence of tan δ.

differs, and in the latter case shifts to the higher-temperature side, but the curve of the plot is broader than in the former case.

The relation for tan δ also shows such characteristic differences, as does the temperature dependence of E'', due to different fiber sizes. In the temperature region below 70°C, the values of the two specimens are almost in accord. However, in the temperature region from 70 to 110°C, the plot for the powder specimen occupies a higher location than does the plot for the fiber specimen. These results for effective reinforcement by fiber powder seem to be related to the difference in the conditions of dispersion and the adhesive state of the powder and/or those for the fiber in the matrix resin.

Polypropylene

Temperature Dependence of E' *and* E"

The results of plotting on a normal scale for fiber weight contents of 0, 5, 18, and 33% are shown in Figure 21. For the matrix specimen, the temperature dependence of E' shows a decreasing tendency with a rise in temperature until about 70°C. In regions above 70°C, the E' value indicates a nearly constant or slightly increasing tendency. By contrast, for the 5% specimen the points are usually located lower at each given temperature than those for the matrix specimen. For the temperature region below 110°C, this tendency disappears at over 18% of the fiber; however, a reinforcing effect cannot be expected at temperatures above 100°C.

The change in temperature dependence of E'' due to the change in fiber content is small, and these values are plotted in a limited, narrow range. Also, the relation for tan δ is almost linear and the value increases with an increase in temperature. The magnitude

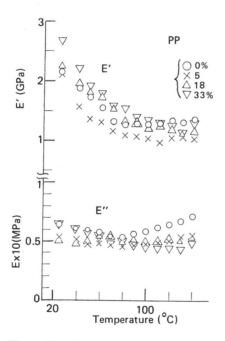

Figure 21 Temperature dependence of E' and E'' for phenolic-fiber-reinforced PP. Fiber content by weight: ○, 0%; ×, 5%; △, 18%; ▽, 33%.

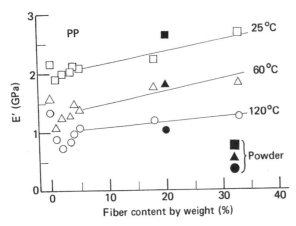

Figure 22 Fiber content dependence of E' for phenolic-fiber-reinforced PP at 25°C (□), 60°C (△) and 120°C (○).

of the gradient for these plots decreases with a gradual increase in the fiber content of the specimen.

Dependence of E' *on Fiber Content*

The results plotted on a normal scale for the temperatures 25, 60, and 120°C are shown in Figure 22. A particular phenomenon can be observed in the low-fiber-content specimen: despite the fiber mixture, the value of E' in the results at 25°C shows a slightly lower value for the matrix specimen. However, the phenomenon of a lower E' value is accompanied by a large change in its size. The amount of this decrease increases with increased temperature and the minimum appears to be at about 2% fiber. This decrease recovers with a fiber mixture of over 5%. For a specimen of over 5%, the E' value seems to increase with increased fiber content and the relation is nearly linear. The magnitude of

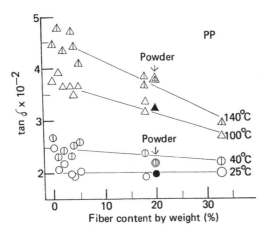

Figure 23 Fiber content dependence of tan δ for phenolic-fiber-reinforced PP at 25°C (○), 40°C (◑), 100°C (△), and 140°C (△).

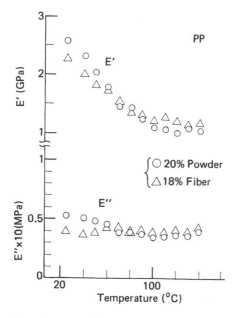

Figure 24 Temperature dependence of E' and E'' of phenolic-fiber-reinforced ($\triangle$) and fiber-powder-filled ($\bigcirc$) PP.

the slope of the plot for a specimen at 25°C and one at 60°C is larger than for one at 120°C.

Dependence of tan δ on Fiber Content

The data for 25, 40, 100, and 140°C are shown in Figure 23. It is obvious that the order of the plots changes with the magnitude of the temperature, which is the principal parameter of the relation. The fiber content dependence seems to be shown as a linear relation, the magnitude of the slope increasing with a rise in temperature, but at room temperature almost no dependence is shown. The phenomenon of the decrease in value of tan δ with a fiber content of 2 to 3% shows a similar tendency in the relation for the dependence of E'.

Effect of Filler Morphology on E' and E''

The temperature dependence of E' and E'' is shown in Figure 24. For the dependence of E', the location of the plot for the two specimens changes positions according to the boundary temperature. That is, the E' value for the powder specimen is superior to that of the fiber in the lower-temperature region, and vice versa at the boundary point. Such a boundary temperature seems to exist in the region from 70 to 80°C. The same relation for the dependence of E'' and the turning-point temperature region can also be found, and this result is similar to the one for the polystyrene resin.

Nylon 66

Temperature Dependence of E' and E''

The results plotted on a normal scale for fiber contents of 0, 5, 13, 23, and 39 wt % are shown in Figure 25. The largest magnitude of the negative slope in the array of points

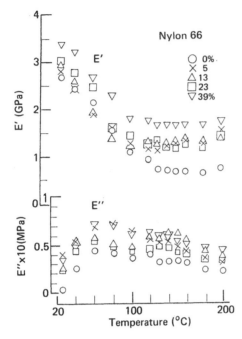

Figure 25 Temperature dependence of E' and E'' for phenolic-fiber-reinforced nylon 66. Fiber content by weight: ○, 0%; ×, 5%; △, 13%; □, 23%; ▽, 39%.

seems to occur in the temperature region from 60 to 70°C. The temperature relates to the second glass transition temperature of the resin. For the reinforced specimen, the plots usually occupy a higher position than those of the matrix in all temperature regions. So in this case a better reinforcing effect can be expected than in the case of polypropylene. The improvement in the utility of the tensile properties by using fiber with the resin can be recognized especially in the high-temperature region. However, this extends to the lower-temperature region with increased fiber content.

For the relation of E'', the character of the dependence for the reinforced specimen is basically similar to that for the matrix specimen. The peak of the array of points for the matrix specimen seems to be in the region from 60 to 80°C. This result agrees with the one for the dependence of E'. For changes in the E'' value, the gradient of the array of points shows a rapid increase in the lower-temperature regions, which indicates a peak in dependence. By contrast, a moderate decreasing tendency can be seen in the higher-temperature region. Such general dependence can also be recognized in the reinforced material.

Dependence of E' *on Fiber Content*

The results plotted on a normal scale for the temperatures 25, 60, and 120°C are shown in Figure 26. The relationship is a fairly linear one in the room-temperature region, and the results for the other two temperatures indicate the same tendency. However, in the higher-temperature region, a difference or some changes from the results for 25°C can be seen in a band for a 0 to 2% specimen. That is, a lowering or decreasing and an increasing of the E' value can be recognized at 60°C. In this temperature region, a nearly

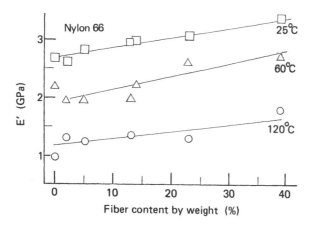

Figure 26 Fiber content dependence of E' for phenolic-fiber-reinforced nylon 66 at 25°C ($\square$), 60°C ($\triangle$), and 120°C ($\bigcirc$).

linear relationship is evident in the higher-fiber-content specimen than in the lower one. Also, the largest magnitude in the slope of plots can be seen at 60°C; the plot for the other two temperatures shows almost the same slope magnitude.

REFERENCES

1. Carborundum Co., BPI 256,924 (Dec. 15, 1971); U.S. patent 3,650,102 (Mar. 21, 1972); J. Economy et al, *Appl. Polym. Symp.*, 21: 81 (1973).
2. F. W. Billmeyer, Jr., *Textbook of Polymer Science*, Wiley, New York, 1962, p. 465.
3. R. Fujii, *Sen-i Gakkaishi*, 33: P-387 (1974).
4. T. Iizuka, *Kobunshi Kako* (*Polym Process.*), 38: 193 (1989).
5. M. Koyama, *Proc. 25th Symposium on Thermosetting Resins*, 1975, p. 32.
6. T. Iizuka and Y. Arita, *Netsukasoseijushi* (*Thermal Setting Resin*), 9: 38 (1988).
7. K. A. Scott, *The Flammability of Plastics*: *Method of Testing*, Technical Review 27, Rubber and Plastic Research Association of Great Britain, Shawbury, England, 1965; also *RAPRA Bull.*, 168 (1965).
8. L. B. Allen and L. N. Chellis, Chap. 11 in *Flammability Tests*, Vol. 2, J. V. Schmity (ed.), Interscience, New York, 1966.
9. L. A. Ashton, Trans., *J. Plast. Inst. London*, 32(98) (1964).
10. *Ignition Properties of Plastics*, ASTM D-1929–68, or BS 476, Part 4 (1970); *Method of Testing Plastics*, BS-2782, Method 508A (1970).
11. *ASTM D-648–72* (1972).
12. E. Jinen, *J. Appl. Polym. Sci.*, 36: 773 (1988).
13. Brüel & Kjaer, *Instructions and Applications*, No. 3939 (1968).
14. E. Jinen, *Compos. Sci. Technol.*, 33: 19 (1988).

10

Surface Treatment of Wood Fibers with Silane Coupling Agents and Their Effect on Properties of the Composites

R. G. Raj and B. V. Kokta
Center for Research in Pulp and Paper
University of Quebec at Three Rivers
Quebec, Canada

INTRODUCTION

The use of coupling agents (silanes and titanates) is a well-known method to improve the mechanical properties of short-fiber-filled thermoplastics [1,2]. Although silane coupling agents are generally used in glass fibers, they are also found to be effective on particulate minerals in filled polymers [3,4]. The surface modification of inorganic fillers with coupling agents greatly improves the adhesion between the matrix and filler. In addition, the surface-coated fillers showed improved melt viscosity, dispersion, and flow during molding of the compounds [5]. It is interesting to note that the same coupling agent can be used in different matrix phases such as thermosets, thermoplastics, and elastomers. Han et al. [6] reported an increase in mechanical properties of a polypropylene–calcium carbonate system using a titanium coupling agent. Pretreatment of a glass fiber with suitable silane coupling agents produced a significant increase in mechanical properties of polypropylene filled with glass fiber [7].

A number of theories have been advanced on the function of coupling agents. It is generally accepted that the organofunctional group of the silane coupling agent forms a covalent bond with the matrix resin, while hydroxyl groups on silicon form oxane bonds with the filler surface. The performance of the silane coupling agent depends on the choice of the organofunctional group [8], the nature of the filler/matrix [9,10], and the method of application [11]. Most of the work above has focused on silane coupling agent reactivity with an inorganic filler. The use of various renewable organic fillers, such as wood fiber as a filler/reinforcing agent in thermoplastic polymers, has received considerable attention in recent years. Surface treatment using coupling and wetting agents is the key factor in the use of these fillers in thermoplastic matrices [12,13]. In the present work, silane coupling agents were used to precoat the wood fibers to improve the adhesion between the fiber and polyethylene matrix. Different silanes with vinyl (silane A-172), methacryloxy (silane A-174), and amino functional (silane A-1100) groups were used. The influence of silane treatment of fiber on mechanical properties of the composites is discussed.

EXPERIMENTAL PROCEDURE

Materials

The polymer used was a high-density polyethylene (HDPE 2907) supplied by Dupont Canada Inc. Melt index: 5.0 dg/min; density: 0.960 g/cm^3. Wood flour (aspen) and chemithermomechanical pulp (CTMP) of aspen were used as fillers. The physical properties of fillers were specified in an earlier publication [14]. The following silane coupling agents (Union Carbide), with different organo functional groups, were used to precoat the wood fibers:

1. Vinyltri(2-methoxyethoxy)silane (silane A-172)
2. γ-Methacryloxypropyltrimethoxysilane (silane A-174)
3. γ-Aminopropyltriethoxysilane (silane A-1100)

Precoated Fiber

The wood fibers were precoated with silane coupling agents before compounding with the polymer. The following procedure was used for silane A-172 and silane A-174 treatment: 25 g of fiber was placed in a flask to which 150 mL of carbon tetrachloride was added, followed by the addition of 1% of dicumyl peroxide and 2% of silane A-172 or silane A-174. The mixture was refluxed for 3 h with continuous stirring. After cooling, the carbon tetrachloride was evaporated and the fibers were dried at 60°C for 24 h. In the case of silane A-1100, a two-stage mixing procedure was used. In the first stage, a mixture of wood fiber, silane A-1100 (2%), and dicumyl peroxide (1%) was refluxed for 3 h in carbon tetrachloride. In the second stage, HDPE (5%) was dissolved in p-xylene to which maleic anhydride (2%) and benzoyl peroxide (0.05%) were added and then refluxed for 3 h. After that the contents of the first stage were added and the entire mixture was again refluxed for 2 h. At the end, the fibers were cooled to room temperature, washed with distilled water, and dried in an oven at 105°C for 12 h.

Fabrication of Composites

The compounding of precoated wood fiber and polymer was done in a C. W. Brabender roll mill. The roller temperature was kept at 160°C during mixing. The polymer–fiber mixture was added gradually to the heated rolls and then mixed thoroughly for 5 to 7 min to obtain uniform mixing of the fiber in the polymer. The resulting mixture was allowed to cool to room temperature and then ground to mesh 20 for compression molding. The concentration of wood fiber in the samples varied from 0 to 30 wt % of fiber.

Compression molding was done in a Carver laboratory press (temperature 165°C, pressure 3.2 MPa). Dog-bone-shaped tensile specimens (ASTM D-638, type V) were obtained after the samples were slowly cooled to room temperature with the pressure maintained during the process.

Mechanical Tests

Tensile properties of the composites were studied in an Instron Model 4201. Before testing, the samples were conditioned at 23°C and 50% relative humidity for 24 h. A minimum of six samples were tested in each series. Tensile strength and elongation were measured at peak load. Modulus was calculated at 0.1% elongation. The tensile results were obtained with the help of a HP86B computing system using the Instron 2412005 general tensile test program. The Izod impact strength (unnotched) of the composites, a minimum of four specimens in each series, was determined in a TMI 43–01 impact tester. The coefficients of variation of the reported properties were less than 6.4%.

RESULTS AND DISCUSSION

Effect of Silane Concentration

Wood flour (aspen) precoated with silane A-172 and silane A-174 coupling agents produced a significant increase in tensile strength compared to untreated fiber (no coupling agent) composites. Relative tensile strength as a function of silane concentration, at different filler loading, is shown in Figure 1. In HDPE filled with silane-precoated fiber, the tensile strength increased steadily at all concentrations of fiber compared to untreated fiber composites. The variation in silane A-172 and silane A-174 concentration from 2 to 6% produced only a slight gain in tensile strength at 20 and 30% filler loadings. The improvement in tensile strength appears to be greater, at 20 and 30% filler loading, in silane A-174-precoated wood flour composites. The results suggest that the degree of adhesion at the fiber–matrix interface is higher in the composites containing silane A-174-precoated fibers than silane A-172-precoated fibers. The effect of the organofunctional group on the silane coupling agent is dependent on its ability to react with the polymer matrix. The bonding of the organofunctional group of the silane with the polymer matrix can take place in several forms. It can form a copolymer, an interpenetrating polymer network, or diffuse into the polymer matrix and cross-link at the fabrication temperature [1]. The performance of the composite is affected strongly by the resin morphology at the interface. It is possible that the presence of vinyl and carbonyl groups in silane A-174 can lead to effective bonding with the polyethylene.

Figure 2 shows the effect of silane concentration on relative tensile modulus, at different fiber loading, of the composites. It can be seen that tensile modulus increased linearly with filler concentration and that the variation in silane concentration had only a marginal effect on modulus. The slight improvement in modulus at 20 and 30% filler loadings in silane A-172- and silane A-174-precoated fiber composites can be attributed to reduced fiber agglomeration and an improved fiber dispersion in the polymer matrix. The degree of adhesion between the fiber and matrix affects the stiffness of the composite less than the tensile strength. It is understandable, therefore, that tensile modulus was not much affected by fiber treatment. The modulus in a short-fiber-filled composite is more sensitive to stiffness of the fiber, concentration of the fiber, fiber aspect ratio (L/D), and orientation of the fiber [15,16].

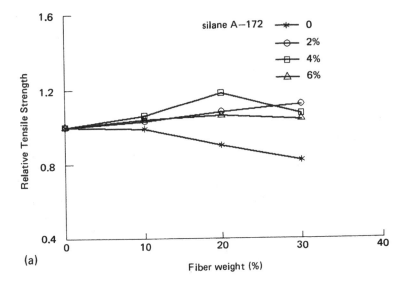

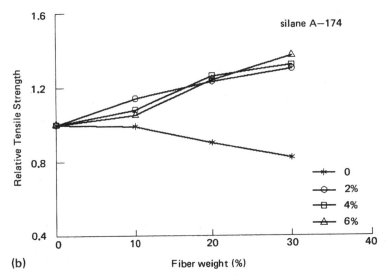

Figure 1 Relative tensile strength as a function of filler concentration for (a) HDPE–silane A-172- and (b) silane A-174-coated wood flour composites.

The effect of silane concentration on elongation of the composites, at different fiber loading, is shown in Figure 3. The addition of fiber had a negative effect on elongation of the composites. The results show that the elongation decreased drastically with the increase in fiber loading and was not greatly affected by the variation in silane concentration. One possible explanation is that the viscoelastic behavior of the matrix is greatly affected by an increase in the stiffness of the system. This means that with the addition of fiber greater stiffening of the matrix can be achieved, but not without a considerable loss in

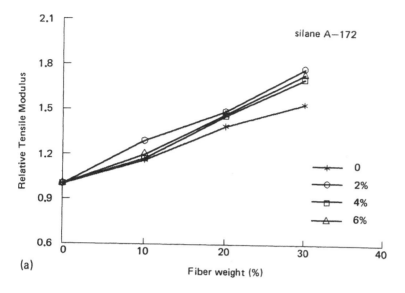

(a)

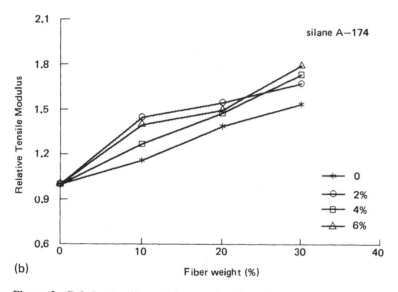

(b)

Figure 2 Relative tensile modulus as a function of filler concentration for (a) HDPE–silane A-172- and (b) silane A-174-coated wood flour composites.

matrix elongation. Figure 4 shows the effect of silane concentration on fracture energy (area under the stress–strain curve) of the composites at various filler loadings. The fracture energy decreased steadily with the increase in fiber loading in the composites. It can be seen from the results that HDPE filled with silane A-172-precoated fibers, at 20 and 30% fiber loadings, are less effective than silane A-174-precoated fiber composites in absorbing the energy during fracture. As mentioned earlier, this may be due to a higher degree of

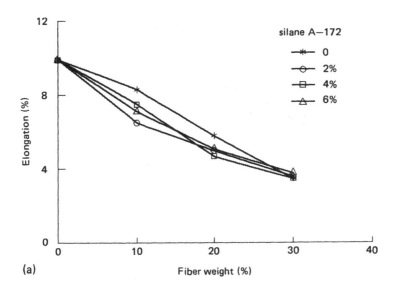

(a)

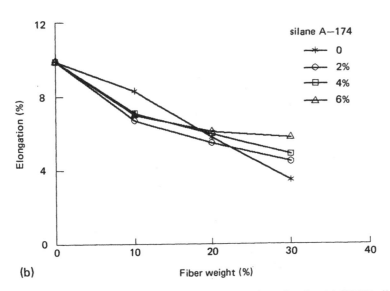

(b)

Figure 3 Elongation as a function of filler concentration for (a) HDPE–silane A-172- and (b) silane A-174-coated wood flour composites.

adhesion in silane A-174-precoated fiber–HDPE composites, which increases the toughness of the system as measured by the fracture energy.

Comparison of Silane Treatments

Silanes with different organofunctional groups (vinyl, methacryloxy, and amine) were used to precoat the wood fiber in order to examine the influence of silane treatment on

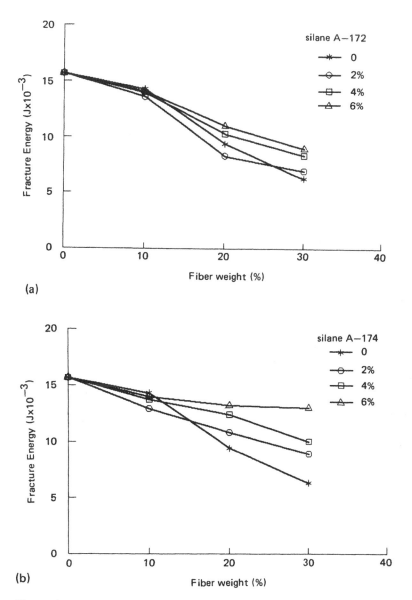

Figure 4 Fracture energy as a function of filler concentration for (a) HDPE–silane A-172- and (b) silane A-174-coated wood flour composites.

mechanical properties of the composites. The percent change in tensile strength of HDPE filled with different silane-treated fibers (2% silane) is given in Figure 5. It is clear that silane treatment produced a significant increase in tensile strength as a result of improved adhesion between the fiber and matrix. The increase in tensile strength is higher in the samples containing silane A-174-precoated fibers, as the fiber loading increased from 10 to 30% in the composites. Of the three silanes used, silane A-174, which carries a

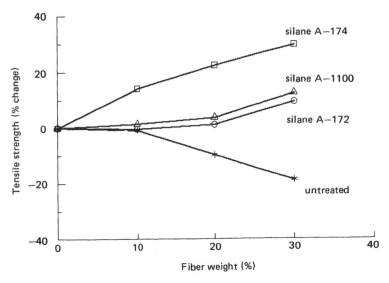

Figure 5 Effect of different silane treatments on tensile strength of HDPE–wood flour composites.

methacryloxy functional group, seems to have a higher degree of reactivity of the silanol groups with the fiber hydroxyls than that of vinyl (silane A-172) and amino (silane A-1100) functional groups, although one would expect improved adhesion with amino functional silane because of its self-catalytic nature for bonding with the filler surface. However, this was not observed in the present case. It is possible that the silanols, formed by the hydrolysis of the alkoxy groups of the silanes, can self-condense to form a thick layer of oligomeric silanol deposition in th fiber–matrix interface which can greatly reduce the efficiency of bonding.

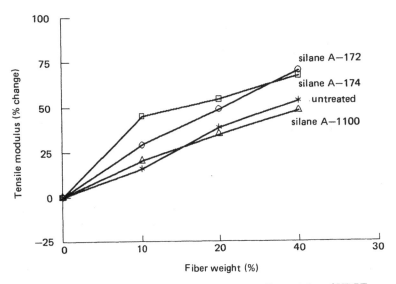

Figure 6 Effect of different silane treatments on tensile modulus of HDPE–wood flour composites.

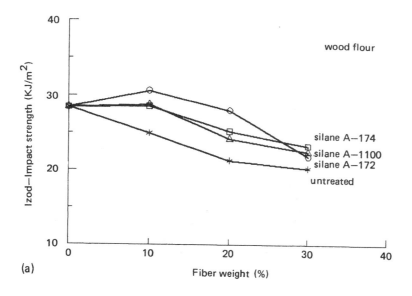

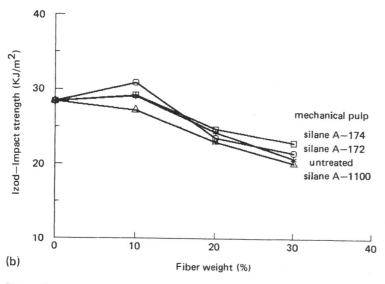

Figure 7 Effect of different silane treatments on Izod impact strength (unnotched) of (a) HDPE–wood flour and (b) mechanical pulp composites.

The effect of silane treatment on the tensile modulus of the composites (at different fiber loading) is presented in Figure 6. It can be seen that HDPE filled with silane A-172- and silane A-174-coated fibers (2%) produced higher tensile modulus than silane A-1100-treated and untreated fibers at the same fiber loading in the composites. This may be attributed to improved wetting of the fibers as a result of better dispersion of the silane coupling agent over the fiber surface. A more uniform coverage of the filler surface is an important factor for the effective transfer of stress at the fiber–matrix interface. Figure 7 shows the Izod impact strength as a function of filler concentration for HDPE composites

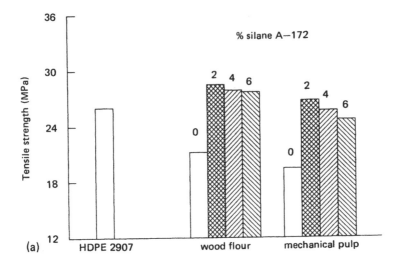

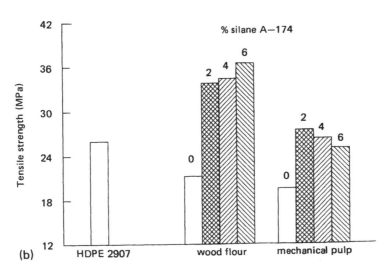

Figure 8 Comparison of tensile strength of (a) HDPE–silane A-172- and (b) silane A-174-coated wood flour and mechanical pulp composites.

containing 2% silane A-172-, A-174-, and A-1100-coated wood flour/mechanical pulp. It can be seen that an increase in fiber loading tends to lower the impact strength of the composites. HDPE filled with silane-coated wood flour composites produced a slightly higher impact strength than that of untreated wood flour composites. Silane coupling agent with a vinyl functional group (silane A-172) slightly increased the impact strength of the composites (at 10 and 20% fiber loading). However, in samples containing silane-coated mechanical pulp fibers, no such trend was observed. The impact strength remained relatively unaffected regardless of fiber treatment, particularly at higher filler concentrations. It was observed by Woodhams et al. [17] that the ductile deformation of the matrix is affected by the increase in the concentration of wood fiber and reduces the ability to absorb the energy during fracture.

Comparison of Mechanical Pulp and Wood Flour

The type of wood fiber used as a filler may affect the properties of the thermoplastic composites. Wood pulps are produced predominantly by chemical or mechanical pulping processes. The yield and cost of the wood pulp varies accordingly. In the present study, hardwood (aspen) pulp fibers produced using a chemithermomechanical process (CTMP) were used. The conditions of preparation were described in an earlier publication [16]. The CTMP process produces high-yield, low-cost pulp. The wood pulp fibers obtained using the foregoing process are better defiberated than wood flour obtained by grinding wood chips in a laboratory grinder (C. W. Brabender). However, the CTMP fiber has less lignin, which may aid fiber dispersion, particularly at higher concentrations of filler. Composites of HDPE containing different concentrations (0 to 6%) of silane A-172- and silane A-174-coated CTMP aspen/wood flour (aspen) fibers were prepared (30% filler loading). The results presented in Figure 8 show that both the silane A-172- and silane A-174-coated wood flour composites produced higher tensile strength than that of mechanical pulp (CTMP) composites. The best improvement in tensile strength was observed at 2% silane concentration. The results were somewhat surprising since one would expect the defibrated fibers obtained by CTMP process to perform better than wood flour as a reinforcement. One possible explanation is the decrease in bonding ability of CTMP fibers due to the loss of hemicellulose and lignin during the pulping process. The increase in tensile strength in wood flour composites is attributed to a higher degree of interaction with the coupling agent that increases the bonding at the fiber–matrix interface. However, further studies are necessary to determine the degree of interaction of the silane coupling agent with different wood fibers. Tensile modulus of silane A-172- and silane A-174- coated mechanical pulp composites is found to be similar to that of wood flour composites (Figure 9). The data also show that the increase in silane concentration has little effect on tensile modulus of the composites.

Fracture Analysis

The fractured surface of the specimens (tensile test) were shown in Figure 10. A comparison of SEM micrographs shows that silane A-174-coated wood flour is well bonded to the HDPE matrix (Figure 10b), but it is clearly not the case in untreated fiber composites (Figure 10a), where there is poor bonding between the fiber and polymer matrix. In the latter case, the failure of the material is caused primarily by debonding between the fiber and matrix, and the appearance of fiber pullout is clear on the surface. Figure 11 shows the fractured surface (impact test) of untreated and silane A-174-treated wood flour composites, respectively. They are characterized by extensive fiber pullout and fiber–matrix separation. In both cases the samples had poor impact strength and the matrix–fiber interface is not effective in stopping crack propagation through the matrix.

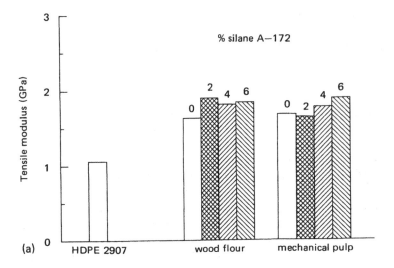

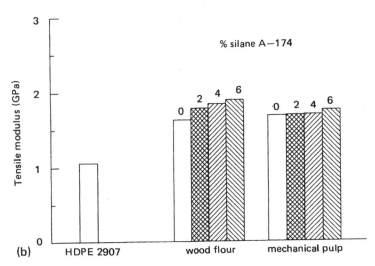

Figure 9 Comparison of tensile modulus of (a) HDPE-silane A-172- and (b) silane A-174-coated wood flour and mechanical pulp composites.

CONCLUSIONS

Wood fibers coated with silane coupling agents significantly improved the tensile strength of wood-fiber-filled HDPE composites compared to untreated fiber composites. Silane with methacryloxy functional group seems to have higher reactivity with fiber/polymer, as seen from higher tensile strength, than silanes with vinyl or amino functional groups. The silane treatment had little influence on modulus of the composites. Regardless of silane treatment, a decrease in elongation and fracture energy was observed with the increase in fiber loading. HDPE filled with wood flour produced higher tensile strength than mechanical pulp fibers at the same filler loading in the composites.

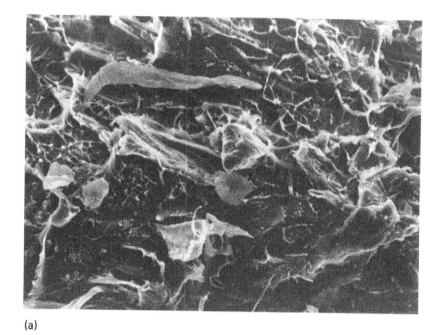

(a)

(b)

Figure 10   Fractured surface (tensile test) of (a) HDPE-untreated and (b) 2% silane A-174-treated wood flour composite (30% fiber weight).

(a)

(b)

Figure 11 Fractured surface (impact test) of (a) HDPE-untreated and (b) 2% silane A-174-treated wood flour composite (30% fiber weight). Magnification 25 ×.

REFERENCES

1. E. P. Plueddemann, *Silane Coupling Agents*, Plenum Press, New York, 1982.
2. A. T. DiBenedetto and A. D. Wambach, *Int. J. Poly. Mater.*, 1: 159 (1972).
3. M. S. Boaira and C. E. Chaffy, *Polym. Eng. Sci.*, 17(10): 715 (1977).
4. P. Vondracek, M. Capka, and M. Schatz, *J. Appl. Polym. Sci.*, 24: 1619 (1979).
5. S. J. Monte and G. Sugarman, *Plast. Compound.*, 1: 56 (1978).
6. C. D. Han, C. Sanford, and H. J. Yoo, *Polym. Sci. Eng.*, 18: 11 (1978).
7. J. G. Marsden, *J. Appl. Polym. Sci.*, 107 (1970).
8. K. Dawes and R. J. Rowely, *Plast. Rubber Mater. Appl.*, 3(1): 23 (1978).
9. T. Nakatsuka, H. Kawaski, K. Itadami, and S. Yamashita, *J. Appl. Polym. Sci.*, 24: 1985 (1979).
10. M. W. Ranney, S. Berger, and J. G. Marsden, in *Interfaces in Polymer Matrix Composites*, P. Pleuddemann (ed.), Academic Press, New York, 1974.
11. H. Ishida and J. L. Koenig, *J. Colloid Interface Sci.*, 64(3): 565 (1978).
12. L. A. Goettler, U.S. patent 4,376,144 (1983).
13. B. V. Kokta, R. Chen, C. Daneault, and J. L. Valade, *Polym. Compos.*, 4: 229 (1983).
14. R. G. Raj, B. V. Kokta, G. Grouleau, and C. Daneault, *Polym. Plast. Technol. Eng.*, 29(4): 339 (1990).
15. C. Klason, J. Kubat, and H.-E. Strömvall, *Int. J. Polym. Mater.*, 10: 159 (1984).
16. A. D. Beshay, B. V. Kokta, and C. Daneault, *Polym. Compos.* 6(4): 261 (1985).
17. R. T. Woodhams, G. Thomas, and D. K. Rodgers, *Polym. Eng. Sci.,* 24(15): 1166–1171 (1984).

11

Mössbauer Spectroscopic Characterization of the Mixed Oxides Containing Iron Ions

Svetozar Musić
Laboratory for the Synthesis of New Materials
Rudjer Bošković Institute
Zagreb, Yugoslavia

INTRODUCTION

In the two last decades scientists and engineers have paid considerable attention to chemistry and physics of the mixed oxides. These materials have already found application in a great variety of technical fields. Mixed oxides are characterized by specific electric, magnetic, piezoelectric, or other physical properties. Their specific physical behavior in relation to other materials is a consequence of their specific chemistry and structures. Mixed oxides can be prepared using different experimental methods, such as coprecipitation, sol-gel preparations, solid-state preparations, ceramic sintering, film growth on different substrates, and so on.

In a narrow sense the term *mixed oxides* describes metal oxides which contain more than one metal cation in the structure. They can exist as solid solutions or with known sites for each metal cation in the crystal structure. In a broad sense *mixed oxides* also describes mixtures of metal oxides: for instance, the corrosion products of iron (mixtures of different oxyhydroxides, oxides, and ferrites), inorganic oxide catalysts, and so on.

In many cases the chemical and physical characterization of mixed oxides is a very difficult task, due to their complexity. Scientists and engineers use very different instrumental techniques to characterize the mixed oxides: for instance, x-ray diffraction, neutron diffraction, Mössbauer spectroscopy, infrared (IR), Fourier transform IR, and Raman spectrocopy. Mössbauer spectroscopy is a very useful technique for the characterization of mixed oxides containing iron or other Mössbauer active nuclides (e.g., Sn, Sb, Eu, Dy, etc.) and in some cases, when several techniques are used, Mössbauer spectroscopy can serve as a "judge" for the conclusion.

In this chapter the fundamentals of the Mössbauer effect and the hyperfine interactions are described. The application of Mössbauer spectroscopy is illustrated on selected examples of mixed oxides containing iron ions. Attention is paid to the mixed oxides, which have application in electrical engineering and electronics. The possibilities of Mössbauer spectroscopy in the research of high-T_c superconductors are also presented. The literature references are primarily from the last decade.

MÖSSBAUER SPECTROSCOPY

Discovery of the Mössbauer Effect

The phenomenon of the emission of a gamma ray from an atomic nucleus and its resonance absorption by an identical nucleus without any loss of energy, due to the recoil, is known as the *Mössbauer effect*. In fact, the recoil energy produced as a result of gamma-ray emission is taken up by the entire crystal lattice. In 1957, Rudolf Mössbauer discovered the recoil-free absorption of the 129-keV gamma ray by ^{191}Ir in iridium metal [1,2]. The gamma-ray source was ^{191}Os. The transmission of the gamma ray decreased unexpectedly as the temperature was lowered from 370 K to 90 K. For this discovery, Rudolf Mössbauer received the Nobel Prize for Physics in 1961. After the discovery of the Mössbauer effect, many other nuclides, ^{57}Fe among them, were found to show the resonant absorption of the gamma ray. The Mössbauer effect was the basis for the development of a new spectroscopic technique called *Mössbauer spectroscopy*. In the literature, the alternative name *nuclear gamma resonance* (NGR) *spectroscopy* can also be found. Mössbauer spectroscopy has already made important contributions to research in physics, chemistry, biochemistry, metallurgy, and mineralogy. The theory and practice of Mössbauer spectroscopy have been treated extensively in several books [3–8]. Therefore, only the basic facts about Mössbauer spectroscopy of importance for the reader of this chapter are discussed.

Principles of the Mössbauer Effect

Let us consider the energy processes involved in gamma-ray nuclear resonance. Assume a free nucleus with excited and ground nuclear energy states and their energy difference, E_0. When the nucleus goes from the excited state to the ground state, a gamma ray of energy E_γ is emitted, and the nucleus receives the recoil energy E_R. From the law of conservation of energy, the transition energy E_0 must be equal to the sum

$$E_0 = E_\gamma + E_R \tag{1}$$

Assuming that the nucleus has mass M and is originally at rest, it will gain velocity v_R by the emission of the gamma ray. The recoil energy E_R is given by the equation

$$E_R = \frac{M}{2} v_R^2 = \frac{(Mv_R)^2}{2M} = \frac{p_n^2}{2M} \tag{2}$$

where p_n is the momentum of the nucleus. The momentum of the gamma ray is given by the equation

$$p_\gamma = \frac{E_\gamma}{c} = \hbar k \tag{3}$$

where c denotes the velocity of light, k the magnitude of the wave vector of the gamma ray, and $\hbar$ is Planck's constant divided by 2π. It follows from the momentum principle that the momentum p_γ of the gamma ray and the momentum of the nucleus undergoing recoil must be equal in magnitude and opposite in sign. Since the recoil considered here is rather small compared with the rest of energy Mc^2 of the emitting system, the problem can be treated in a nonrelativistic manner. From the conservation of the momentum follows

$$E_R = \frac{p_\gamma^2}{2M} = \frac{E_\gamma^2}{2Mc^2} \approx \frac{E_0^2}{2Mc^2} \tag{4}$$

The replacement of E_γ with E_0 in Eq. (4) will not change the value E_R significantly. For $E_0 \approx E_\gamma \approx 10^4$ eV and $M = 100$, a typical value of $E_R \approx 5 \times 10^{-4}$ eV is obtained. The emission of the gamma ray from a nucleus moving with a velocity v along the gamma-ray propagation pathway will shift the energy of the gamma ray in accord with a first-order linear Doppler effect, and this is given by

$$E_D = \frac{v}{c} E_\gamma \tag{5}$$

If the nucleus moves in different directions in relation to velocity v, during gamma-ray emission, the resulting linear Doppler effect is

$$E_D = \frac{p_n p_\gamma}{M} = \frac{v}{c} E_\gamma \cos \alpha \tag{6}$$

where α is the angle between the momentum vectors of the moving nucleus p_n and the photon p_γ. Accordingly, in the real system the energy consideration during gamma-ray emission must also include the energy of the Doppler effect, owing to the thermal motion of nucleus. In the present case the conservation of energy is described by

$$E_0 = E_\gamma + E_R - E_D \tag{7}$$

In the process of gamma-ray absorption, the signs of E_R and E_D in Eq. (7) are reversed.

Considering the energy of gamma rays (photons) emitted by excited atoms, it was found that they do not have exactly the same energy. This is a consequence of the Heisenberg uncertainty principle, according to which the product of the uncertainty of energy (ΔE)

and uncertainty of time (Δt) is on the order of magnitude of $\hbar$, where $\hbar$ is Planck's constant h divided with 2π:

$$\Delta E \cdot \Delta t \geq \hbar \qquad (8)$$

In other words, the Heisenberg principle says that no measurement or observation of the two conjugate variables can be made with higher accuracy than that stated by relation (8). Applying this principle to the resonance theory, the following equation can be written:

$$\Delta E = \Gamma = \frac{\hbar}{\tau} = \frac{0.69\hbar}{t_{1/2}} \qquad (9)$$

where Γ is the natural line width of the source emission or absorption line, τ the mean lifetime of the excited state, and $t_{1/2}$ its half-lifetime ($t_{1/2} = \tau \ln 2$). The line width is defined as the full width at half maximum of the peak. The first excited state of ^{57}Fe (14.4 keV) has a value of $t_{1/2} \approx 10^{-7}$ s and $\Gamma \approx 5 \times 10^{-9}$ eV. If the excited state of nucleus has a very small mean lifetime, τ, the energy of this state is very uncertain, i.e., it cannot be measured exactly. If the ground state has a long mean lifetime, τ, its energy level is well defined. The characteristic energy distribution for the excited state is shown schematically in Figure 1. When the nuclei are embedded in real lattices, the effect of recoil and Doppler broadening, due to the thermal oscillation, must be taken into consideration. Figure 2 shows schematically the energy distribution of the emission line and the corresponding absorption line. We note that the energy required for absorption is higher than E_0 by an amount E_R, because the momentum and energy must be considered in the absorption process. Part of the energy absorbed by the nucleus goes for its recoil. For gamma rays generated by nuclear decay, the energy $E_D < E_R$ at standard temperature, and consequently the overlap of the emission and absorption spectra, which represents the possibility of the resonance absorption, is negligible. In the case of optical transitions, where the transition energies are of the order of 1 eV (the energy $E_D \sim E_R$), there is appreciable overlap of emission and absorption spectra.

From previous discussion and the classical analogy it is obvious that it is possible to observe appreciable nuclear resonance if we avoid the effects of recoil and Doppler broadening. The probability of such recoil-free emission of gamma rays and the corresponding absorption is known as the recoil-free fraction, f, which is given by

$$f = \exp\left(-k^2 <\mathbf{x}^2>_T\right) \qquad (10)$$

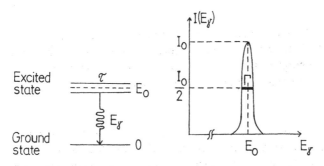

Figure 1 Natural line width. The energy of the excited level with a mean lifetime τ has a width $\Gamma = \hbar/\tau$.

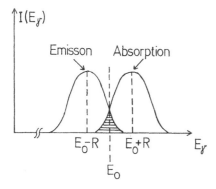

Figure 2 Emission and absorption lines are Doppler broadened and a small overlap (shadowed) exists. The Doppler-broadened lines are much less peaked than the natural ones.

where k is the wave vector of the gamma ray, $<x^2>_T$ the mean square displacement, and $<\cdot>_T$ denotes the thermal average. If the Debye model is used to describe the solid, the recoil-free fraction is

$$f = \exp\left\{-\frac{3E_R}{2k_B\Theta_D}\left[1 + 4\left(\frac{T}{\Theta_D}\right)^2 \int_0^{\Theta_D/T} \frac{1}{e^x - 1}\,dx\right]\right\} \tag{11}$$

where E_R is the free nucleus recoil energy, k_B is Boltzmann's constant, Θ_D is the Debye temperature of the lattice, and x is the position of the emitting nucleus. Recoil-free fraction is usually called the Debye–Waller factor.

The probability of observing the effect increases with decreasing temperature, and the larger effects are observed in rigidly bound solids (i.e., in those that possess a high Debye temperature). Recoil-free emission and absorption have an energy distribution characterized by natural line width Γ and are centered at E_0, thus providing a strong resonance effect. Typically, Γ/E_0 has a value of about 10^{-11} to 10^{-16}, and a very small change in the energy level of the absorbing nucleus will prevent resonant absorption from taking place. This is the key to the importance of Mössbauer spectroscopy, because such small energy differences can be caused by chemical and physical changes in the absorber environment. It is immediately seen that Mössbauer spectroscopy can be used as an extremely sensitive tool for examining such changes in materials.

Mössbauer effect was observed for many elements. However, only a limited number of elements is suitable for routine Mössbauer spectroscopy. The most used were the nuclides of iron, tin, antimony, iodine, europium, and dysprosium. It is relatively easy to work with these nuclides. They show strong resonant fractions, have line widths suitable for hyperfine studies, and represent different regions of the periodic system of elements.

Measurement of the Mössbauer Effect

So far, the principles of the Mössbauer effect have been discussed and now we deal with the question of its experimental measurement. In principle, to arrange a Mössbauer experiment the following systems are needed: a source of gamma rays and the corresponding absorber in the solid state, a driving mechanism to produce a slight change in the gamma-ray energy, and a detector with a multichannel analyzer.

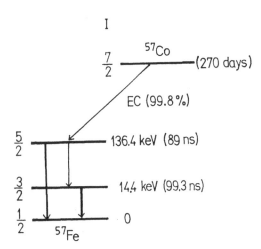

Figure 3 Decay scheme of ^{57}Co.

A gamma-ray transition suitable for the study of Mössbauer effect must satisfy certain conditions. To a great extent, the decay scheme and nuclear parameters determine the usefulness of the radionuclide as a Mössbauer source. The lifetime of the excited state must not be very short, because in accordance with the uncertainty principle, its energy spread is great. The nuclear transition must be to the ground state of the Mössbauer nucleus or it will not be possible to carry out reabsorption of the gamma ray. The upper limit of the gamma-ray energy is near 200 keV, to avoid the excitation of the lattice vibrations. The excited state is usually formed by α- or β-decay, but it can also be formed by other nuclear processes. The parent must have a half-life of at least a few hours to allow measurements to be made. The most commonly used Mössbauer nuclide is ^{57}Fe. The source is ^{57}Co, which decays by K-capture into the 136.4-keV level of ^{57}Fe, and after 89 ns this level emits a gamma ray of 122 keV. The 14.4-keV level of ^{57}Fe decays after 99.3 ns to the ground state, and this transition is generally used to observe the Mössbauer effect in ^{57}Fe. The Mössbauer effect has also been obtained using the 136.4-keV transition. The decay scheme of ^{57}Co is shown in Figure 3.

In most cases the ^{57}Co source is embedded in the metal matrix (Pt, Pd, Rh, etc.) and the absorber being studied is sampled. The absorber thickness plays a significant role in quantitative analysis by Mössbauer spectroscopy. The effective absorber thickness t_A for a single-line absorber is given by the relation

$$t_A = \sigma_0 f_A n_a d_A \alpha_a \tag{12}$$

where σ_0 is the maximum absorption cross section, f_A the recoil-free fraction in the absorber, n_a the number of atoms per cubic centimeter of the particular element, d_A the physical thickness of the absorber in centimeters, and α_a the nuclidic abundance of the resonance nuclide. In the emission Mössbauer spectroscopy, the chemical and physical environment of the source is changing while the absorber properties remain constant.

An essential part of the Mössbauer spectrometer is a driving unit capable of producing slight energy changes of the gamma ray generated in a recoil-free emission. The slight energy change in the gamma ray is produced by imparting a velocity v to the source or absorber, and due to the Doppler effect the corrected energy E is given by

$$E = E_\gamma \left(1 + \frac{v}{c} \right) \tag{13}$$

where c denotes the velocity of light. In this way the extra velocity will either reduce or increase the gamma-ray energy just enough to match the energy of the level exactly and allow resonant absorption to take place. For practical reasons the function of velocity is better to use than the function of energy. A generally accepted method of measurement of the Mössbauer effect is to use a multichannel analyzer whose memory stores are syncronized with the relative velocity normally imparted to the source. The source moves cyclically, so that the velocity range (or equivalent of energy) required is covered several times per second. To ensure a linear energy scale, the velocity of the source is adjusted so that it is linear with respect to time and that the analyzer spends equal time in each channel. The transmission geometry, with absorption producing a relative drop in counting rate, is used primarily in Mössbauer spectroscopy. For some experiments, particularly those in surface science, the use of scattering geometry is favored. In the scattering experiment the absorption corresponds to a relative increase in counting rate and the Mössbauer spectrum is simply inverted. Measurements at low temperatures will reduce the proportion of events that excite vibrations in the crystal lattice.

Three types of detectors are generally used in measurement of the Mössbauer effect: scintillation detectors, proportional counters, and lithium-drifted Ge and Si detectors. The gamma-ray detector selected depends on the type of Mössbauer experiment. Efficiency, resolution, and other properties of detectors must be considered separately for different Mössbauer sources.

Hyperfine Interactions

Earlier it was shown that Mössbauer spectroscopy is characterized by very high energy resolution. This is the basis for the measurement of very small energy shifts and splittings of nuclear levels caused by hyperfine interactions, i.e., interactions of the atomic nucleus with electric and magnetic fields. The three hyperfine interactions usually measured are the isomer shift, electric quadrupole interaction, and magnetic dipole interaction. On the basis of the parameters of hyperfine interactions, it is possible to study lattice, electronic, and magnetic properties of the matter in solid state. These interactions will be illustrated using the ^{57}Fe nucleus, which is naturally present in the nuclidic composition of iron.

Isomer Shift

The electrostatic energy shift of a nuclear level as a consequence of the finite size of nucleus and the penetration of electronic charge into the nuclear volume can be calculated by taking the difference between the electrostatic interaction of a hypothetical point nucleus and one of actual radius R. If we assume that the nuclear charge distribution is uniform within a sphere with given radius and zero outside this sphere, and that the electronic charge density is uniform over the nuclear volume, the energy change, δE, due to the penetration of electrons into the nuclear volume is given by

$$\delta E = \frac{2\pi}{5} Ze^2 R^2 |\psi(0)|^2 \tag{14}$$

where Ze is the nuclear charge and $|\psi(0)|^2$ the finite density at the nuclear site. Generally, the nuclear radius is different for each energy level, and for the actual case of ^{57}Fe in

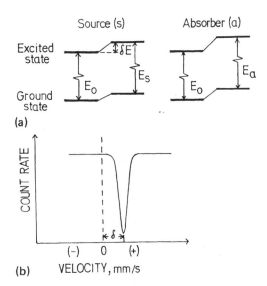

Figure 4 Origin of isomer shift: (a) electric monopole interaction shifts the nuclear energy levels without lifting the degeneracy; (b) resultant Mössbauer spectrum (schematic).

which the Mössbauer transition takes place between the nuclear ground state and the first excited state the energy change δE is given by

$$\delta E = \frac{2\pi}{5} Ze^2 |\psi(0)|^2 \, (R_e^2 - R_g^2) \tag{15}$$

where R_e and R_g represent the nuclear excited and ground state mean charge radii. Since $(R_e^2 - R_g^2)$ is a constant for a given Mössbauer transition, an energy shift can be observed if the source and absorber have a different value for $|\psi(0)|^2$. This energy shift, usually called the isomer shift δ, is described by

$$\delta = E_a - E_s = \frac{2\pi}{5} Ze^2(R_e^2 - R_g^2) \, [|\psi_a(0)|^2 - |\psi_s(0)|^2] \tag{16}$$

where the subscripts a and s refer to "absorber" and "source." In ^{57}Fe, for the 14.4-keV level the R_e^2 is greater than R_g^2. δ is normally measured in velocity units which are proportional to the energy change. The value of δ is positive when the s-electron density in the absorber is smaller than in the source, and the center of the measured spectrum is then shifted toward a positive velocity. $|\psi(0)|^2$ increases with the ionicity of an iron ion, owing to the shielding effect of the $3d$ electron at iron. Mössbauer spectroscopy can serve for the measurement of the ionicity or covalency of the chemical bond. The chemistry of Mössbauer active element will change the isomer shift. The origin of isomer shift is shown schematically in Figure 4.

Electric Quadrupole Interaction

In our discussion of the electric monopole interaction and isomer shift, it has been assumed that the nuclear charge distribution is uniform and spherically symmetrical. However, the nuclear charge distribution can deviate more or less from spherical symmetry and deviation may be different in each state of excitation. This produces an electric field gradient (EFG)

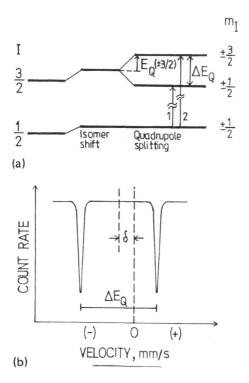

Figure 5 Quadrupole splitting for ^{57}Fe with spin $I = \frac{3}{2}$ in the excited state: (a) $I = \frac{3}{2}$ level splits into two sublevels by electric quadrupole interaction; (b) resultant Mössbauer spectrum (schematic).

at the nucleus site. The quadrupole splitting arises from the interaction between the nuclear quadrupole moment eQ and the principal component of the diagonalized electric field gradient tenzor $V_{zz} = \partial^2V/\partial z^2$ at the nucleus site. The nuclear state splits into sublevels with the eigenvalues

$$E_q = \frac{eQV_{zz}}{4I(2I - 1)} \left[3m_I^2 - I(I + 1) \right] \left(1 + \frac{\eta^2}{3} \right)^{1/2} \qquad (17)$$

The asymmetry parameter η is given by

$$\eta = \frac{V_{xx} - V_{yy}}{V_{zz}} \qquad \text{with} \quad |V_{zz}| \geq |V_{yy}| \geq V_{xx}$$

$V_{zz} + V_{yy} + V_{xx} = 0$ and thus $0 \leq \eta \leq 1$. m_I is the nuclear magnetic spin quantum number, with values $m_I = I, I - 1, \ldots, -I$. I is the nuclear spin quantum number.

The electric quadrupole interaction splits the first nuclear excited state of ^{57}Fe ($I = \frac{3}{2}$) into two sublevels with the eigenvalues

$$E_q = \pm \frac{1}{4} eQV_{zz} \left(1 + \frac{1}{3}\eta^2 \right)^{1/2} \qquad (18)$$

This situation is shown schematically in Figure 5. The resultant Mössbauer spectrum is known as a *quadrupole doublet*. Generally, the contributions to the electric field gradient at ^{57}Fe arise from an asymmetric distribution of the surrounding atoms or ligands. The nonspherical distribution of valence electron over $3d$ orbitals of Fe atom will contribute to the electric field gradient. High-spin Fe^{3+} and low-spin Fe^{2+} compounds have small, temperature-independent quadrupole splittings in almost all cases. Compounds with high-spin Fe^{2+} have much larger quadrupole splittings, which are markedly dependent on temperature. High values of the quadrupole splittings for Fe^{2+} ions can be used for identification of many Fe(II) compounds.

Magnetic Dipole Interaction

The third important contribution to hyperfine interactions is the nuclear Zeeman effect. The interaction of the nuclear magnetic dipole moment μ with a magnetic field H at the site of the nucleus will split the nuclear level with spin $I(I > 0)$ into $(2I + 1)$ sublevels with the eigenvalues

$$E_m = - \frac{\mu H m_I}{I} = - g_N \beta_N H m_I \tag{19}$$

where m_I is the nuclear magnetic quantum number with the values $m_I = I, I - 1, \ldots , -I$, β_N the nuclear Bohr magneton, and g_N the nuclear Landé splitting factor. In the case of ^{57}Fe, the first excited state has a spin $I = \frac{3}{2}$ and the ground state has a spin $I = \frac{1}{2}$. The result of the magnetic splitting are four sublevels of the excited state and two sublevels of the ground state. The allowed transitions in ^{57}Fe are characterized by the selection rule $\Delta m = 0, \pm 1$, and six possible transitions will produce Mössbauer spectrum with six absorption lines. The magnetic splitting of nuclear levels (nuclear Zeeman effect) at ^{57}Fe nucleus is shown in Figure 6. In this illustration the electric quadrupole splitting is zero. For a thin absorber in which the magnetic moments are randomized, for example, in a powder sample, the six-line spectrum exhibits relative peak intensities of approximately 3:2:1:1:2:3. The source of the magnetic field may be either ''internal,'' as in the case of ferromagnetic and antiferromagnetic materials, or it may be an externally applied magnetic field. From Eq. (19) it is seen that the energy difference between sublevels is directly proportional to the magnitude of the magnetic field. By taking the ratio of the splitting between the outermost peaks (peaks 1 and 6 in Figure 6) of unknown material and the standard compound, the internal magnetic field of an unknown compound can be calculated. α-Iron with a known internal field of 330 kOe is commonly used as a standard. The magnetic dipole interaction (nuclear Zeeman effect) measured by Mössbauer spectroscopy has contributed significantly to the fundamental and applied studies of iron oxides and other iron compounds. The Mössbauer parameters of internal magnetic field (or hyperfine magnetic field, HMF) are used for the identification of magnetically ordered phases in materials of complex composition.

Effect of Particle Size

In Mössbauer spectroscopy, particularly in the study of hyperfine interactions of iron oxides, it is very important to know the influence of fine particle size on the magnetic hyperfine spectrum. Generally, fine particles (20 to 200 Å) of iron oxides do not show magnetic ordering below the Néel temperature of the bulk phase. In this case the magnetic splitting spectrum can be obtained by decreasing the temperature of measurement. For example, fine particles of α-Fe_2O_3 (ca. 100 Å) exhibit a central doublet in the Mössbauer spectrum at room temperature instead of the six-line spectrum characteristic of bulk α-Fe_2O_3 [9]. This effect, known as *superparamagnetism*, is due to the process of collective reorientation of the magnetic moment direction in fine particles. If fine particles are cooled

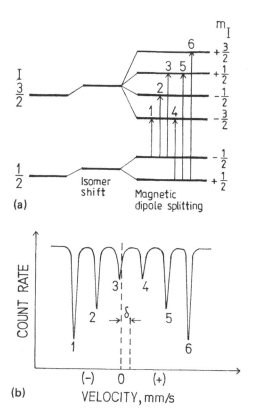

Figure 6 Magnetic splitting for ^{57}Fe with spin $I = \frac{3}{2}$ in the excited state and $I = \frac{1}{2}$ in the ground state without electric quadrupole splitting: (a) $I = \frac{3}{2}$ level splits into four sublevels and $I = \frac{1}{2}$ into two sublevels; (b) resultant Mössbauer spectrum (schematic).

below the magnetic ordering temperature, the electron spins will tend to go from one easy direction to another in a single domain. Above the temperature of magnetic ordering the magnetic splitting collapses because of the fast spin relaxation, and the resultant paramagnetic Mössbauer spectrum consists of one or two lines. Near the temperature of magnetic ordering, intermediate relaxation times produce spectra with broad lines. In a large crystallite, magnetic ordering temperature will not change, except under the influence of external magnetic field. The superparamagnetic relaxation time τ is a function of temperature, and most scientists assume exponential temperature dependence of the relaxation time as originally suggested by Néel [10]:

$$\tau = \tau_0 \exp\left(\frac{KV}{kT}\right) \tag{20}$$

where τ_0 is on the order of 10^{-10} s, KV is the anisotropy energy barrier for a particle of volume V, k is Boltzmann's constant, and T is the temperature. The observation of relaxation spectra below the magnetic ordering temperature may be evidence for the presence of fine particles. The temperature at which the fine particle converts to bulk behavior will depend on the size of this particle. This phenomenon has been used for average particle-size measurement on the basis of superparamagnetic spectra.

BACKGROUND ON FERRITES

Ferrites is the general name for a group of mixed oxides containing iron ions, which are characterized by specific electric and magnetic properties. They have an appreciably higher (by six orders of magnitude) electric resistance and a lower saturation magnetization than those of silicon steels and permalloys. This reduces eddy-current losses and makes the ferrites suitable for making cores and coil chokes in electronic and other electrical devices. Advanced ferrites found very important applications as magnetic recording materials [11,12].

Ferrites can generally be considered as the reaction products between Fe_2O_3 and other metal oxides. Their formula, in the general case [13], is $(Me_2^{k+}O_k^{2-})_{m/2}(Fe_2^{3+}O_3^{2-})_n$, where Me is a metal cation with valency k, and m and n are full numbers. In addition to ferrites containing oxygen anions, there are ferrites in which oxygen anions are substituted with anions of fluorine, chlorine, sulfur, selenium, or tellurium.

On the basis of their structural properties, the ferrites can be divided in four subgroups [13]:

1. Spinel ferrites with the structure of natural mineral spinel, $MgAl_2O_4$

2. Hexagonal-type ferrites with the structure of mineral magnetoplumbite, $PbFe_{7.5}Mn_{3.5}Al_{0.5}Ti_{0.5}O_{19}$

3. Garnet-type ferrites with the structure of mineral garnet, $Ca_3Al_2(SiO_4)_3$

4. Orthoferrites with the structure of mineral perovskite, $CaTiO_3$

The chemical formula of spinel ferrites can be written in the form $(Me_2^{k+}O_k^{2-})_{m/2}$ $(Fe_2^{3+}O_3^{2-})_n$ ($k = 2$, $m = 1$, $n = 1$), or $MeFe_2O_4$. Me^{2+} represents cations such as Fe^{2+}, Co^{2+}, Ni^{2+}, Mn^{2+}, Zn^{2+}, Cu^{2+}, Cd^{2+}, or Mg^{2+}. In this group of ferrites there are also $Li_{0.5}^+Fe_{0.5}^{3+}Fe_2^{3+}O_4$ and γ-Fe_2O_3. For practical applications, ferrites with spinel structure, which contain more than one divalent cation, are more important than simple ferrites, $MeFe_2O_4$.

The main structural feature of the oxide spinels is a close-packed cubic oxygen sublattice in which cations are placed in tetrahedral and octahedral interstices. Cations occupy one-eighth of tetrahedral sites and one-half of octahedral sites, corresponding to the $8a$ and $16d$ positions for space group $Fd3m$. In the normal spinels all divalent cations occupy tetrahedral sites, while iron cations are in octahedral sites. On the other hand, in inverse spinels divalent cations are in octahedral sites, while iron cations are equally distributed between tetrahedral and octahedral sites. In the intermediate case the spinel formula can be written as follows [14]:

$$(Me_{1-\delta}Fe_\delta)\,[Me_\delta Fe_{2-\delta}]O_4 \tag{21}$$

where δ is the degree of inverse character of the spinel, which depends on temperature:

$$\frac{(1-\delta)(2-\delta)}{\delta^2} = \exp\left(-\frac{\Delta p}{kT}\right) \tag{22}$$

The change in lattice energy Δp is a consequence of divalent and trivalent cation exchange between octahedral and tetrahedral sites. This means that using different thermal treatment of spinel ferrites, it is possible to change the cation distributions in spinel structures.

Hexagonal ferrites comprise a number of structural types. Barium and strontium ferrites, which are of very high coercive force and have magnetoplumbite structure (M-type), are

Table 1 Structural Properties of Some Hexagonal Ferrites

Ferrite	Type	Composition of Elementary Cell	Number of Layers in Hexagonal Elementary Cell	c^a (nm)	Theoretical Density (g/cm³)	Molecular Mass[b]
$BaFe_{12}O_{19}$	M	RSR*S*	10	2.32	5.28	1112
$Ba_2Me_2Fe_{12}O_{22}$	Y	$(TS)_3$	18	3 · 1.45	5.39	1408
$BaMe_2Fe_{16}O_{27}$	W(MS)	RS_2R*S_2	14	3.28	5.31	1575
$Me_2Fe_4O_8$	S	S			5.24	232
$Ba_2Me_2Fe_{28}O_{46}$	M₂S	$(RSR*S_2^*)_3$	36	3 · 2.8	5.29	2636
$Ba_3Me_2Fe_{24}O_{41}$	Z(MY)	RSTSR*TS*T*	22	5.23	5.33	2520
$Ba_4Me_2Fe_{36}O_{60}$	M₂Y	RSR*S*T*S*	16	3.81	5.31	3622

Source: Ref. 13.

[a] Assuming that c is not dependent on the type of Me^{2+} cation.

[b] Molecular mass calculated for the case $Me^{2+} = Fe^{2+}$.

produced industrially in large amounts and further processed to give permanent magnets for different applications. The most important hexagonal ferrites and their properties are shown in Table 1.

The magnetoplumbite structure (M-type) contains 10 close-packed layers on traversing the unit cell along the c-axis. Four oxygen layers alternate with an ordered BaO_3 layer. The iron ions are distributed in oxygen octahedral, tetrahedral, and trigonal–bipyramidal sites in such a way as to give alternating spinel (S) and hexagonal (R) blocks along the c-axis, as shown in the review by Gleitzer and Goodenough [15]. The BaO_3 layers contain trigonal–bipyramidal iron, so that the stacking sequence RSR*S gives five distinguishable iron sites (the asterisk designates inversion of the structural block for 180°).

The structure of $Ba_2Fe_{14}O_{22}$ (Y-type) contains a double BO_3 layer, which forms a T block, interleaved with the same S blocks in the stacking sequence TSTSTS. In the T block the iron ions occupy tetrahedral and octahedral positions. The iron ions are in tetrahedral positions and six iron ions are in octahedral positions within the same T block.

By using different combinations of the blocks S, R, T, as well as the blocks S*, R*, T*, it is possible to describe all other structures of the hexagonal ferrites.

The general formula of garnet-type ferrites can be written in the form $(Me_2^{k+}O_k^{2+})_{m/2}$ $(Fe_2^{3+}O_3^{2-})_n$ ($m = 6, n = 5, k = 3$), or in the form $Me_3Fe_5O_{12}$. Me^{3+} represents cations such as Y^{3+}, Sm^{3+}, Gd^{3+}, Tb^{3+}, Dy^{3+}, Ho^{3+}, Er^{3+}, Tm^{3+}, Yb^{3+}, and Lu^{3+}. La^{3+}, Ce^{3+}, Pr^{3+}, and Nd^{3+} have no tendency to form simple garnet-type ferrites. However, La^{3+}, Ce^{3+}, Pr^{3+}, and Nd^{3+} in a mixture with other rare earth cations form solid solutions with garnet structures. In garnet-type ferrites the rare earth cations and Fe^{3+} can be substituted by other cations, such as two- and five-valent cations at molar ratio 2:1, as shown for compound $\{Y_{3-2x}^{3+}Ca_{2x}^{2+}\}[Fe_2^{3+}]$ $(Fe_{3-x}^{3+}V_x^{5+})O_{12}$. Garnet-type ferrites contain mainly oxygen anions. However, the oxygen anions can be partially or completely substituted with fluorine; for instance, in the compounds $Me_{3-x}^{3+}Ca_xFe_5O_{12-x}F_x$ ($Me^{3+} = Y^{3+}$, Sm^{3+}, Gd^{3+}, Dy^{3+}, Er^{3+}, etc.), $Me_3Me_{5-x}^{3+}Me_x^{2+}O_{12-x}F_x$ ($Me = Y$, Gd; $Me^{3+} = Fe^{3+}$, Ga^{3+}, Al^{3+}; $Me^{2+} = Fe^{2+}$, Ni^{2+}, Co^{2+}), $Y_{3-3x}Ca_{3x}Fe_{5-x}Me_x^{5+}O_{12-x}F_x$ ($Me^{5+} = Sb^{5+}$, V^{5+}, Nb^{5+}), and $Na_3Li_3Me_2^{3+}F_{12}$ ($Me^{3+} = Al^{3+}$, V^{3+}, Cr^{3+}, Fe^{3+}, Sc^{3+} etc.).

Table 2 Crystallographic Parameters (nm) for Orthoferrites of Some Rare Earth Cations and Yttrium

	Orthorhombic Cell			Monoclinic Pseudocell		
Orthoferrite	a	b	c	$a = c$	b	$\beta,°$
$GdFeO_3$	0.5346	0.5616	0.7668	0.3877	0.3834	92.8
$EuFeO_3$	0.5371	0.5611	0.7686	0.3884	0.3843	92.5
$SmFeO_3$	0.5349	0.5592	0.7711	0.3885	0.3856	92.0
$NdFeO_3$	0.5441	0.5573	0.7753	0.3895	0.3877	91.4
$PrFeO_3$	0.5495	0.5578	0.7810	0.3912	0.3905	90.8
$LaFeO_3$	0.5556	0.5565	0.7862	0.3932	0.3931	90.2
$YFeO_3$	0.5280	0.5592	0.7602	0.3845	0.3801	93.3

Source: Ref. 13.

Orthoferrites with the perovskite structure $CaTiO_3$ can be presented with general formula $(Me_2^{k+}O_k^{2-})_{m/2}(Fe_2^{3+}O_3^{2-})_n$ ($n = 1$, $m = 2$, $k = 3$), or $MeFeO_3$, where $Me = Y$, La, Ce, Pr, Nd, Sm, Eu, Gd, Tb, Er, and so on. Table 2 shows crystallographic parameters (nm) for orthoferrites of some rare earth cations and yttrium.

SPINEL OXIDES

Lithium Ferrites

The alkali ferrites can be prepared by electrochemical methods or by solid-state reactions between Fe_2O_3 and alkali carbonate at high temperatures. El-Shobaky and Ibrahim [16] studied solid-state reactions between α-Fe_2O_3 and Li_2CO_3. The experimental results showed that the mixture of α-Fe_2O_3 and Li_2CO_3 in a molar ratio 1:2 yielded α-$LiFeO_2$ at temperatures $\geq 500°C$. The unreacted portion of Li_2CO_3 decomposed at 735°C, forming Li_2O, which dissolved into the α-$LiFeO_2$ structure. The dissolved Li_2O increased the thermal stability of α-$LiFeO_2$ and prevented its phase transformation into any other type of lithium ferrite, with further increase in temperature. On the other hand, Fe_2O_3 and Li_2CO_3 in an equimolar ratio yielded α-$LiFeO_2$ at $\geq 500°C$, which transformed in part into β-$LiFeO_2$ and β-$LiFe_5O_8$ at 700 to 1000°C. At 1100°C, β-$LiFe_5O_8$ and an important portion of $LiFeO_2$ transformed to well-crystallized α-$LiFe_5O_8$.

The mechanical treatment of materials by grinding is a usual technique for increasing the reactivity of solids. Fernandez-Rodriguez et al. [17] investigated the effects of grinding on a stoichiometric mixture $LiOH \cdot H_2O$ and α-$FeOOH$. At the beginning, a phase structurally related to disordered α-$LiFeO_2$ was formed. On the basis of the asymmetrically broadened diffraction lines at $2\theta \approx 45°$, it was also concluded about the presence of the tetragonal metastable phase, β-$LiFeO_2$, which further transformed to α-$LiFeO_2$. After 21 h of grinding, the disordered cubic structure, α-$LiFeO_2$, was dominant. Small amounts of unreacted particles of γ-$FeOOH$, as well as α-$LiFe_5O_8$, were also found.

Thackeray et al. [18] investigated the structural properties of lithiated iron oxides, $Li_xFe_3O_4$ and $Li_xFe_2O_3$ ($0 < x < 2$). Lithium has been inserted into Fe_3O_4 and α-Fe_2O_3 at room temperature, both chemically and electrochemically. X-ray diffraction data obtained for $Li_{1.5}Fe_3O_4$ indicated that the $[Fe_2]O_4$ subarray of the spinel structure remained intact. The A-site Fe^{3+} ions were displaced to empty octahedral positions, and the Li^+ ions in excess of $x = 1$ were located in tetrahedral positions. Lithiation of α-Fe_2O_3 caused the anion array to transform from hexagonal to cubic close packing. In this case the Li^+ ions were distributed over both 16c and 16d octahedral sites of the cubic space group $Fd3m$.

Substituted lithium ferrites (Zn, Ti, etc.) are important materials for microwave applications. For this reason a significant amount of research attention was paid to the structural properties of $LiFe_5O_8$ and the substituted lithium ferrites.

Disordered $LiFe_5O_8$ is an inverse spinel ($Fd3m$ space group) containing Fe^{3+} ion at tetrahedral A-sites and a 1:3 mixture of Li^+ and Fe^{3+} at octahedral B-sites. An order–disorder phase transformation occurs near 750°C. In the ordered $LiFe_5O_8$ ($P4_332$ or $P4_132$ space group), Fe^{3+} ions occupy tetrahedral sites 8c and octahedral sites 12d, whereas Li^+ ions are at octahedral sites 4b [19]. Cheary and Grimes [20] showed that ordered $LiFe_5O_8$ structure is built with eight domains, whose coordinates of Li atom are related to 4b positions by the $1/2 < llO >$ translation of the FCC (face-centered cell).

Dormann et al. [21] showed that the Mössbauer spectra of ordered and disordered lithium ferrites are almost identical. The spectra were fitted on the basis of two A-sites and two or three B-sites. The corresponding Mössbauer parameters are tabulated. Their interpretation is that in pure $LiFe_5O_8$, as well as in Cr- or Ti-substituted compounds, a lithium short-range order always exists.

Dormann and Renaudin [22] investigated Zn-, Ti-, and Cr-substituted lithium ferrites using Mössbauer spectroscopic measurements between 4.2 K and T_C. The hyperfine magnetic fields (HMFs) of Fe_A and Fe_B atoms were determined as a function of the temperature and the number n of nonmagnetic nearest neighbors. It was found that HMF values were practically independent on n. Mössbauer spectra of Zn-substituted lithium ferrites show relaxation for a Zn content higher than 0.3. This phenomenon is related with the localized canting structure of the spins. Hyperfine interactions in Ti-substituted lithium ferrites were also studied by Kishan et al. [23].

Nickel Ferrites

Nickel ferrite, $NiFe_2O_4$, is an inverse spinel in which the tetrahedral or A-sites are occupied by Fe^{3+} ions, and the octahedral or B-sites by Fe^{3+} and Ni^{2+} ions. At room temperature, $(Fe_{1.0}^{3+})_A [Ni_{1.0}^{2+}Fe_{1.0}^{3+}]_B O_4^{2-}$ is magnetically ordered. The Mössbauer spectrum is characterized by two sextets, which are not well resolved. The application of a 70-kG longitudinal field at 4.2 K fully resolved the A and B sublattice, while the $\Delta_m = 0$ lines were absent [24].

Uen and Yang [25] prepared $NiFe_2O_4$ using the ceramic method and studied the Mössbauer effect in this sample in the temperature range 13.4 to 890 K. Hyperfine interactions for both A- and B-site ferric ions were determined. An anomalous change in the slope of the center shift of the B-site ferric ion was clearly observed near 475 K. When the temperatures were far below the Néel temperature, the relaxation effects on the Mössbauer line shape were small. However, when the temperatures were near the Néel temperature, the fluctuation of the ferric spin affected the line shape significantly. [57]Fe Mössbauer spectra of $NiFe_2O_4$ at various temperatures are shown in Figure 7 [25]. The solid curves are the best two-site fits to the measured spectra.

Morrish et al. [26,27] used [57]Fe Mössbauer spectroscopy to study the magnetic structure of fine $NiFe_2O_4$ particles. The particle size varied between 250 and 1300 Å, and the corresponding crystallite size between 250 and 500 Å. Mössbauer spectra, recorded when a longitudinal magnetic field was applied, unambiguously established the presence of a

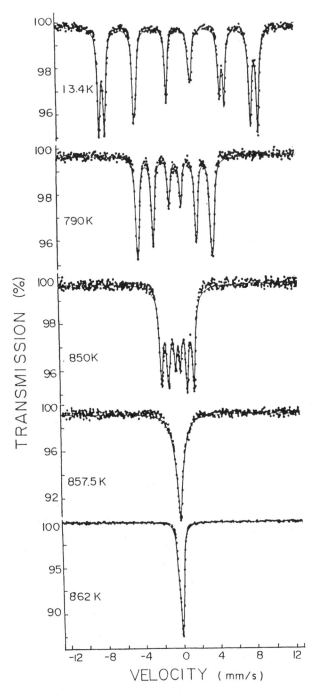

Figure 7 Typical Mössbauer spectra of $NiFe_2O_4$ at various temperatures. The solid curves are the best two-site fits to the measured spectra. (From Ref. 25: courtesy of T.-M. Uen and *Jpn. J. Appl. Phys.*)

noncollinear magnetic structure at the surface layer of $NiFe_2O_4$. This means that the $NiFe_2O_4$ particles may be visualized as having a core with the normal spin arrangement and a surface layer in which the spin arrangement takes a noncollinear structure. Morrish and Haneda [28] also investigated the magnetic structure near the surface of the particles, such as α-Fe_2O_3, [57]Fe surface-enriched α-Fe_2O_3, $NiFe_2O_4$, [57]Fe-doped CrO_2, and α-Fe. The application of Mössbauer spectroscopy in the study of the bulk and near-surface magnetic properties is very important, due to their use in magnetic inks, magnetic fluids, recording tapes, catalysts, and so on.

Peev et al. [29] used Mössbauer spectroscopy to study the thermal decomposition of coprecipitated $Fe(NH_4)_2(SO_4)_2 \cdot 6H_2O$ and $Ni(NH_4)_2(SO_4)_2 \cdot 6H_2O$ salts. They found that $NiFe_2O_4$ was the end product at 900 to 1100°C.

Raw [30] investigated the experimental conditions for the synthesis of magnetites and nickel ferrites. This investigation is actuated by the fact that nonstoichiometric nickel ferrite, $Ni_xFe_{3-x}O_4$, is the principal corrosion product formed in the primary cooling circuit of the pressurized water reactor (PWR).

Cobalt Ferrites

Mössbauer spectra of $CoFe_2O_4$ indicate that this ferrimagnetic spinel is not completely inverse and that the degree of inversion depends on the heat treatment of the material [31].

Cruset and Friedt [32] investigated the valence state of [57]Fe after [57]Co decay in $CoFe_2O_4$. It was concluded that the [57]Fe ions formed after [57]Co decay are in the trivalent state and occupy the octahedral B-sites in cobalt ferrite.

Ito [33] doped synthetic magnetite at high temperature with carrier-free [57]Co. The Mössbauer emission spectrum showed that the [57]Fe ions formed after [57]Co decay occupy B-sites in magnetite. The charge distribution of Fe ions, stabilized after the decay of [57]Co, is the same as that of Fe ions at B-sites of the host magnetite.

The coordination of Co^{2+} ions in $Co_xFe_{3-x}O_4$ spinels ($x \leq 0.04$) was followed by Mössbauer spectroscopy at room temperature both with and without external magnetic field [34]. It was shown that Co^{2+} ions occupied the octahedral B-sites.

Conversion electron Mössbauer spectroscopy (CEMS) was used to determine the cation distributions at the surface layer of Co^{3+}-substituted cobalt ferrite, $Co^{2+}Fe_{2-x}Co_x^{3+}O_4$ ($0 \leq x \leq 2$) [35]. The results indicated a shift of Co^{2+} from octahedral to tetrahedral sites when Co^{3+} ions were incorporated into the $CoFe_2O_4$ structure.

Ultrafine particles of $CoFe_2O_4$ and $MnFe_2O_4$ were prepared using coprecipitation method [36]. The particle size of these ferrites was strongly dependent on the history of ferrite preparation. [57]Fe Mössbauer spectra recorded at room temperature varied from a doublet for the particles of the superparamagnetic dimension to a nearly pure sextet for the ferrimagnetic samples with large particles.

Mössbauer spectra of small $CoFe_2O_4$ particles (10 to 100 nm size range and up) were also investigated by Haneda and Morrish [37]. The noncollinear magnetic structure of $CoFe_2O_4$ small particles was found in a similar manner as for $NiFe_2O_4$ particles. Figure 8 shows the Mössbauer spectrum of small $CoFe_2O_4$ particles recorded at 4.2 K [37]. Their particle size is 500 ± 150 Å and the crystallite size is 240 Å.

Acicular particles of γ-Fe_2O_3 and Fe_3O_4 are primary materials for magnetic recording. However, the coercivity of these particles is too low for many applications. Their magnetic properties can be improved by doping with divalent cobalt ions. One way of adding the

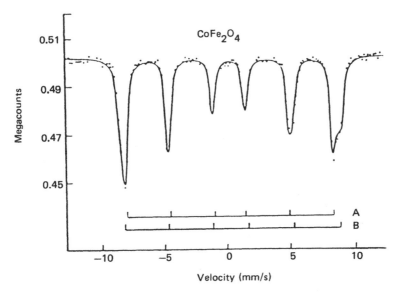

Figure 8 ^{57}Fe Mössbauer spectrum of small $CoFe_2O_4$ particles recorded at 4.2 K. (From Ref. 37; courtesy of the American Institute of Physics.)

cobalt is uniform doping, which is performed by diffusing the cobalt ions into the bulk of oxide particles. Recent investigations showed that it is better to apply the surface doping of Co^{2+}, or epitaxial deposition of ferrite layer.

Mössbauer emission spectroscopy was used to study the cobalt-surface-doped acicular magnetite particles [38]. $Co_xFe_{3-x}O_4$ layer, with x not much less than unity, is found to be responsible for the increased coercivity of the particles. $^{57}Co^{2+}$ ions were adsorbed on γ-Fe_2O_3 acicular particles (0.4 to 0.6 μm in length) [39]. The intensities of the second and fifth lines of the Mössbauer emission spectra, as a function of the applied magnetic field parallel to the gamma ray, were studied. The second and fifth lines (I_2 and I_5) did not vanish completely even under an applied magnetic field of 35 kOe.

Auric et al. [40] studied Mössbauer effect in cobalt–ferrite epitaxial magnetite. Mössbauer spectra did not show any change in the bulk properties of magnetite, due to the coating layer. Their experimental results were explained by a discontinuous surface layer including small particles magnetically coupled to the bulk magnetic moment. The coupling between the bulk magnetite and the cobalt ferrite coating has been introduced to explain coercive force results.

Copper Ferrites

Hannoyer and Lenglet [41] investigated the ^{57}Fe Mössbauer spectra of copper ferrites, $Cu_xFe_{3-x}O_4$ (x = 0, 0.05, 0.1, 0.15, 0.5, 0.85, 1) at 300 K. The spectra of samples with x = 0.15 and x = 0.5 were also recorded at 4.2 K without external magnetic field or in the 50-kOe field. The sample with x = 0.5 was investigated at 300 K in a 100-kOe magnetic field. The Mössbauer spectrum of sample $Cu_xFe_{3-x}O_4$ (x = 0) showed spectral lines typical for Fe_3O_4. Incorporation of copper ions in the Fe_3O_4 structure caused the broadening of spectral lines, due to the changes in the surrounding of the Mössbauer resonant atoms. The spectral lines corresponding to A- and B-sites were clearly resolved in the Mössbauer spectrum of sample $Cu_{0.5}Fe_{2.5}O_4$ at 300 K and in a 100-kOe longitudinal field. A-site lines were narrow and the Mössbauer parameters corresponded to Fe^{3+} ions. B-site lines were asymmetrically broadened and they were computer fitted, assuming the superposition of two field distributions ($B_I = Fe^{3+}$ and B_{II} = average iron). The Mössbauer spectra of copper ferrites are more complicated than of other spinels of divalent metals, because copper ions are present in two oxidation states, Cu^+ and Cu^{2+}, as shown by XPS measurements.

Gerard and Grandjean [42] used Mössbauer spectroscopy to investigate the structural properties of the compound $CuCr_{1.96}{}^{57}Fe_{0.04}O_4$ from the copper ferrite–chromite series, $CuCr_{2-x}Fe_xO_4$. The cations distribution in this compound can be written in the form $(Cu_{0.96}^{2+}Fe_{0.04}^{3+})(Cu_{0.02}^{2+}Cr_{0.98}^{3+})_2O_4$. However, the Mössbauer spectra indicate that in addition to the Fe^{3+} ions, Fe^{3+} ions also exist in this compound. This is in agreement with the results [43] obtained during investigation of the system $CuCr_{2-x}Fe_xO_4$ ($x = 0.02$, 0.04, 0.10, and 0.20).

The Mössbauer effect has been studied [44] in the copper–manganese ferrite series $Cu_{1-x}Mn_xFe_2O_4$ ($x = 0$ to 0.8) at 4.2 K in an external magnetic field of 4.2 T applied parallel to the direction of gamma rays. The cation distributions and the HMF values for ^{57}Fe at A- and B-sites were determined. For the Mn-rich compositions, a part of the manganese at the B-site is present in a trivalent state along with an equal amount of iron in divalent state.

Other Mössbauer Studies with the Spinel Ferrites

Mössbauer spectroscopy was used to investigate the formation of $Ni_{0.38}Zn_{0.62}Fe_2O_4$ during the ceramic fabrication procedure [45]. The spectra of samples were strongly dependent on the time and temperature of ferritization. On the basis of the experimental results, Peev et al. [45] proposed the possible mechanism of the ferritization.

Srivastava et al. [46] showed that the magnetic relaxation effects are different in $Zn_xFe_{3-x}O_4$ and $Zn_xCu_{0.1}Fe_{2.9-x}O_4$ ferrites. This has been explained on the basis of local Jahn–Teller distortion, produced due to the presence of a small amount of Cu^{2+} ions.

The cadmium–nickel ferrites have the cubic spinel structure with the unit cell containing eight formula units of the form $(Cd_x^{2+}Fe_{1-x}^{3+})_{tet}(Ni_{1-x}^{2+}Fe_{1+x}^{3+})_{oct}O_4$, where "tet" denotes the tetrahedral positions and "oct" the octahedral positions. Mössbauer study of $Cd_xNi_{1-x}Fe_2O_4$ ($x = 0.2$, 0.4, 0.6, 0.8, and 1.0) at different temperatures showed the similarity between this system and the system $Zn_xNi_{1-x}Fe_2O_4$ [47]. Figure 9 shows the Mössbauer spectra of $Cd_{0.6}Ni_{0.4}Fe_2O_4$ at different temperatures. These spectra show the presence of spin relaxation at intermediate temperatures. The absorption lines become increasingly broader, and more and more intensity is transferred to the central part of the spectrum as temperature increases.

The $Mn_{0.7}Zn_{0.3}Cr_xFe_{2-x}O_4$ ($0 \leq x \leq 1$) ferrites were also investigated by ^{57}Fe Mössbauer spectroscopy [48]. At liquid N_2 temperature, the average HMF decreased slightly with increased Cr concentration. On the other hand, at room temperature the HMF strongly decreased for $x = 0$ to 0.9 and then became zero for $x = 1$. Spectra recorded in a magnetic field of 6 T showed that the collinear character of the magnetic structure is preserved up to $x \approx 0.5$. Finally, the investigation of the site occupancy demonstrated a notable Fe^{3+} amount at tetrahedral sites, which led to an important inversion degree in these ferrites.

Manganese–zinc ferrites are important materials for high-frequency and magnetic ap-

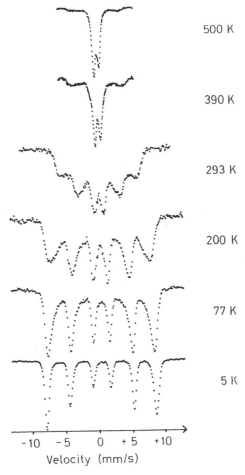

500 K

390 K

293 K

200 K

77 K

5 K

Velocity (mm/s)

Figure 9 ^{57}Fe Mössbauer spectra of $Cd_{0.6}Ni_{0.4}Fe_2O_4$ recorded at different temperatures. (From Ref. 47; courtesy of R. Wäppling.)

plications, owing to their high magnetic permeability and electric resistivity. Michalk [49] investigated the formation of Mn–Zn ferrites under low oxygen pressure. Manganese ferrite formed at temperatures below 500°C (Fe_2O_3 + MnO → $MnFe_2O_4$). With further increase in the temperature of firing, Zn^{2+} ions were incorporated in the spinel structure, forming $Mn_{1-x}Zn_xFe_2O_4$. Mn–Zn ferrite formation is completed at 800°C. The relaxation spectrum corresponding to $Mn_{0.6}Zn_{0.4}Fe_2O_4$ was observed at 750°C.

Pavlyukhin et al. [14] found that the mechanical activation of zinc or nickel ferrite also affects its physicochemical properties. For instance, the temperature of magnetic transition of mechanically activated zinc ferrite increased sharply, while that of nickel ferrite decreased. Variation in the magnetic properties of these ferrites was explained by the transfer of cation from tetrahedral to the vacant octahedral sites in the spinel structure. This transfer of cation is caused by pulses of mechanical action that produced in the ferrite structure disordering of the anion and cation sublattices.

X-ray photoelectron spectroscopy (XPS) and ^{57}Fe Mössbauer spectroscopy were used to study molybdenum-bearing ferrite [50]. The XPS data for Mo $3d$ and Fe $2p$ electrons combined with the Mössbauer data suggest that Fe_2MoO_4 is characterized with the following valence distribution, $(Fe^{2+})_{tet}[Fe^{2+}Mo^{4+}]_{oct}O_4$.

The Systems $GeFe_2O_4$ and $Ge_xZn_{1-x}Fe_2O_4$

$GeFe_2O_4$ is a normal spinel with Fe^{2+} ions at B-sites [51,52]. The Mössbauer spectrum of $GeFe_2O_4$ is characterized with a large quadrupole splitting, $\Delta = 2.9$ mm/s, at room temperature. The Mössbauer spectra of $Fe_{3-x}Ge_xO_4$ were recorded at high temperatures, with and without an applied field of 50 kOe [53]. The strong asymmetrical broadening of the Fe(B) subspectrum gave evidence of at least four different local configurations on the B-sites with increasing Fe^{2+} character, as a result of the chemical substitutional disorder in the distribution of Ge^{4+} over the A-sites. This result was explained as the averaging effect of the fast electron diffusion $Fe^{2+}-Fe^{3+}$ in regions with different local configurations.

$Ge_xFe_{3-x}O_4$ ($0 \leq x \leq 1$) has been studied by x-ray diffraction, Mössbauer spectroscopy, and dc conductivity measurements [54]. X-ray diffraction confirmed that this mixed oxide in all cases consisted of a single spinel phase. Ge^{4+} ions partially occupied the B-sites. For $x \geq 0.4$, in the Mössbauer spectra the central quadrupole doublet appeared, which can be ascribed to Fe^{2+} ions in $GeFe_2O_4$ at B-sites. For samples with $0.7 \leq x \leq 1$, the parameters of the central quadrupole doublet varied for δ_{Fe} between 1.14 and 1.11 $\pm$ 0.02 mm/s and for Δ between 2.75 and 2.80 $\pm$ 0.02 mm/s.

The Mössbauer spectra of the series $Zn_{1-x}Ge_xFe_2O_4$ with $x = 0.25$, 0.50, and 0.75 were recorded between 77 and 623 K [55]. Fe^{2+} and Fe^{3+} ions are all found at the B-sites, and the broadening of the spectra between the Curie point and 623 K is correlated with an electron hopping process via a conduction band.

Recently, Lee and Kim [56] investigated the structural properties of mixed oxide spinels, $Zn_xGe_{1-x}Fe_2O_4$ with $0 \leq x \leq 1$. The ^{57}Fe Mössbauer spectra of $Zn_xGe_{1-x}Fe_2O_4$ for $x = 0$ and $x = 1$ consist of quadrupole doublets, which correspond to $GeFe_2O_4$ and $ZnFe_2O_4$, respectively. The room-temperature spectra recorded for samples with $0.1 \leq x \leq 0.9$ were interpreted as the superposition of three quadrupole doublets, corresponding to Fe^{2+} and Fe^{3+} ions at B-sites, Fe^{2+} ions at the A-site, and a single Lorentzian line corresponding to Fe^{3+} ions at the A-site.

HEXAGONAL FERRITES

The hexagonal ferrite, $BaFe_{12}O_{19}$, is important in the production of the ceramic permanent magnets [57], due to its strong magnetic anistrocopy, caused to a great extent by the presence of an iron ion in the bipyramidal lattice site [58]. Very thin hexagonal platelike particles of $BaFe_{12}O_{19}$ were prepared by ceramic and hydrothermal methods [59]. The particles of $BaFe_{12}O_{19}$ showed small magnetization of 15.5 emu/g (at $H_{ext} = 17$ kOe) and a very small coercive force of 300 Oe at room temperature. These effects were explained by modification of the $BaFe_{12}O_{19}$ structure, in which the $2b$ sites are greatly reduced or absent.

Ast et al. [60] investigated magnetic properties of antiferromagnetic $Ba_2Fe_6O_{11}$ in the phase subsystem $BaFe_2O_4$–$BaFe_{12}O_{19}$. $BaFe_6O_{11}$ is a complex antiferromagnetic with Néel temperature $T_N = 488$ K. On the basis of the Mössbauer spectra it was concluded that the structural characteristics of $Ba_2Fe_6O_{11}$ are similar to those of antiferromagnetic perovskite-like compound, $Sr_7Fe_{10}O_{22}$, which appears as a stable composition in the phase diagram $SrO-Fe_2O_3$. The Mössbauer spectra of metastable $Ba_2Fe_6O_{11}$ at 295 or 78 K can be considered as the superposition of three sextets, corresponding to three different sublattices of iron. A central quadrupole doublet is additionally observed in the room-temperature spectrum.

Many efforts have been made to improve the magnetic properties of the hexagonal ferrite, $BaFe_{12}O_{19}$. The hexagonal unit cell of $BaFe_{12}O_{19}$ contains 10 oxygen layers, where in each fifth layer an oxygen ion is substituted with barium ions. Fe^{3+} ions occupy five crystallographically different lattice sites. In the structure of $BaFe_{12}O_{19}$, twelve Fe^{3+} ions are distributed as follows. Three different octahedral positions ($12k$, $4f_2$, $2a$) are occupied by 6, 2, and 1 Fe^{3+} ion, respectively. Two Fe^{3+} ions have tetrahedral surroundings ($4f_1$), and one Fe^{3+} ion is located in a trigonal bipyramid ($2b$) with fivefold coordination.

Kreber and Gonser [61] applied Mössbauer spectroscopy to study As-substituted barium ferrite, $BaFe_{12}O_{19}$. It was found that incorporation of As in $BaFe_{12}O_{19}$ [the composition near the formula $BaO(As_2O_3)_{0.25}(Fe_2O_3)_6$] significantly improved the properties of barium ferrite. Generally, the Mössbauer spectrum of $BaFe_{12}O_{19}$ is a superposition of five subspectra, due to the five different iron positions in the structure of this ferrite. Mössbauer spectra indicated that the As ions preferentially occupied the bipyramidal lattice site $2b$.

Hexagonal ferrites of composition $BaFe_{12-x}In_xO_{19}$ ($0 \leq x \leq 4$) and $BaFe_{12-x}Sc_xO_{19}$ ($0 \leq x \leq 3$) were prepared by a standard ceramic method and analyzed by ^{57}Fe Mössbauer spectroscopy [62,63]. It was shown that the sextet attributed to $12k$ sublattice split into the subsextets $12k^I$ and $12k^{II}$ with lower hyperfine fields. Their relative intensity increases with increasing x. The spectra were interpreted in the sense of the preferential occupation of the octahedral sites ($4f_{VI}$) by Sc^{3+} or In^{3+} ions (R block). Room-temperature Mössbauer spectra for the series $BaFe_{12-x}Sc_xO_{19}$ [$x = 0$ (a), 0.4 (b), 0.8 (c), 1.2 (d)] are shown in Figure 10 [63]. The sextets relative to the various sublattices are indicated. The distinguishable magnetically weakened $12k$ iron ions are labeled $12k^I$ and $12k^{II}$.

$SrFe_{12}O_{19}$ is hexagonal ferrite with structural properties similar to those of $BaFe_{12}O_{19}$. The structural properties of iron in $SrFe_{12}O_{19}$ are shown in Table 3, and the corresponding hyperfine parameters calculated for ^{57}Fe in $SrFe_{12}O_{19}$ are shown in Table 4.

Rao et al. [65] substituted partially Fe^{3+} ions with Cr^{3+} ions. The series $SrFe_{12-x}Cr_xO_{19}$ with $x = 0$, 3, 6, 8, 9, or 10 was prepared. It was confirmed that Cr atoms do not occupy the tetrahedral sites, $4f_1$.

Albanese et al. [66] investigated the influence of cation distribution on the magnetization of Y-type hexagonal ferrites (Mg_2-Y and Co_2-Y). The magnetizations of the various sublattices of Mg_2-Y ferrite all have the same temperature dependence, while for Co_2-Y ferrite three different behaviors of the sublattice magnetization were observed. These differences in the magnetic behavior of the ferrites investigated were explained by the preferential occupation of the octahedral sites inside the T block with divalent cations, Mg^{2+} or Co^{2+}. In particular, the presence of nonmagnetic ions, such as Mg^{2+}, at these sites is responsible for the peculiar magnetic order observed in Mg_2-Y ferrite.

Spin-order and magnetic properties of $BaZn_2Fe_{16}O_{27}$ (Zn_2-W) hexagonal ferrite were investigated using Mössbauer spectroscopy and magnetometric measurements [67]. The saturation magnetizations in Zn_2-W ferrite measured at different temperatures and extrapolated at 0 K are very high. The results were interpreted by assuming a local reversal or a weakening of the Fe^{3+} magnetic moments, due to the perturbing action of Zn^{2+} ions. Zn_2-W hexagonal ferrite is of interest for special applications as permanent-magnet material. Similar investigations with $BaZn_2Fe_{16-x}In_xO_{27}$ and $BaZn_2Fe_{16-x}Sc_xO_{27}$ were also performed [68].

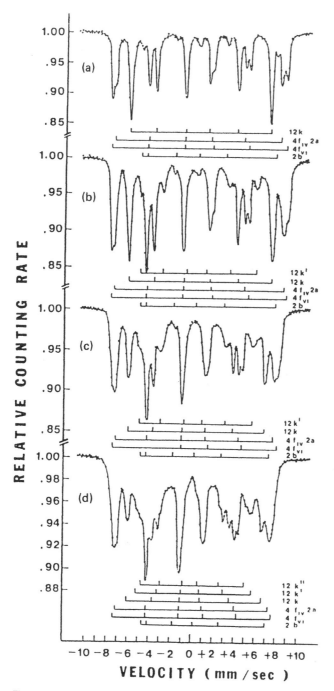

Figure 10 Room-temperature Mössbauer spectra for the series $BaFe_{12-x}Sc_xO_{19}$ with $x = 0$ (a), 0.4 (b), 0.8 (c), and 1.2 (d). The sextets relative to the various iron sublattices are indicated. The distinguishable magnetically weakened $12k$ iron ions are labeled $12k^I$ and $12k^{II}$. (From Ref. 63; courtesy of Springer-Verlag.)

Table 3 Structural Properties of Iron in $SrFe_{12}O_{19}$

Site	Coordination	Point Symmetry	Population	Spin	Block
$12k$	Octahedral	m	6	Up	S-R
$4f_1$	Tetrahedral	$3m$	2	Down	S
$4f_2$	Octahedral	$3m$	2	Down	R
$2a$	Octahedral	$3m$	1	Up	S
$2b$	Trigonal or bipyramidal	$6m2$	1	Up	R

Source: Ref. 64.

Table 4 Hyperfine Parameters Calculated for ^{57}Fe in $SrFe_{12}O_{19}$ on the Basis of Mössbauer Spectrum Recorded at 300 K

Site	δ_{Rh} (mm/s)	E_m (mm/s)	HMF (kOe)	Absorption Area (%)
$12k$	0.25	0.70	415	50
$4f_2$	0.20	0.38	493	17
$4f_1$	0.30	0.30	507	17
$2a$	0.08	0.28	508	8
$2b$	0.20	2.30	425	8

Source: Ref. 64.

^{57}Fe Mössbauer spectra of the hexagonal ferrite with W-structure, $BaFe_2^{2+}Fe_{16}^{3+}O_{27}$, were recorded at different temperatures [69]. All seven sublattices, corresponding iron ions in different structural positions, were observed. Fast electron exchange between Fe^{2+} and Fe^{3+} ions gives rise to sharp lines at 300 and 78 K and makes them indistinguishable. However, at 5 K the exchange is slow and the Fe^{2+} ions are detected from the presence of a weak subspectrum with broadened lines separated from the main spectrum of the Fe^{3+} ions. Fe^{2+} ions occupied one of the seven sublattices, together with Fe^{3+} ions, in the ratio 2:1. Similar results were obtained for $SrFe_2^{2+}Fe_{16}^{3+}O_{27}$.

IRON GARNETS

The preparation and structure, as well as the electric and magnetic properties, of yttrium iron garnet (YIG) ceramics were investigated extensively, because these ceramic materials have important applications in microwave and other techniques [70]. Eibshütz and Lines [71–73] measured the Mössbauer spectrum of amorphous yttrium iron garnet (α-YIG) at 4.2 K. The second-order quadrupole shift, the sign distribution of EFG, and line-width asymmetry of the hyperfine magnetic splitting spectrum at 4.2 K were discussed.

Mössbauer spectroscopy and magnetic measurements were applied in the study of Al-substituted YIG [74]. Garnets were prepared by presintering and sintering mixtures of Fe_2O_3, Y_2O_3, and Al_2O_3 at high temperatures. The formation of phases, such as α-Fe_2O_3, $YFeO_3$, and $Y_3Fe_5O_{12}$, was observed during the preparation of YIG. Mössbauer spectra allowed quantitative determination of incorporated Al^{3+} ions in the garnet structure. The magnetization measurement and the values of Curie temperature supported the Mössbauer spectroscopic results.

Investigation of different properties of YIG by Mössbauer spectroscopy was also performed by other researchers [75,76].

Hyperfine interactions in europium iron garnet (EuIG) were followed by [151]Eu Mössbauer spectra [77]. The hyperfine magnetic fields at the two magnetically inequivalent [151]Eu sites were measured. Their values extrapolated to O K were 630 ± 5 and 570 ± 5 kOe, respectively.

Stadnik et al. [78] studied spin reorientation in erbium iron garnet (ErIG). Single ErIG crystals were grown from the $PbO \cdot PbF_2 \cdot 0.15 V_2O_5$ flux by the low cooling method. [57]Fe Mössbauer spectra recorded from 1.5 to 297 K confirmed the occurrence of spin reorientation from *<100>* to the *<111>* direction. The spin reorientation takes place gradually between 74 and 95 K. In this work the signs of the quadrupole coupling constants at the octahedral sites were also determined.

Iron garnets of yttrium and rare earths, $R_3Fe_5O_{12}$, where R = Y, Sm, Eu, Gd, Tb, Dy, Er, or Tm, were studied by Mössbauer spectroscopy at 590 K [79]. Regular changes in the quadrupole splitting value in dependence on the ionic radius of R^{3+} were found for *a* and *d* sublattices. The changes can be ascribed to the increased local distortion of polyhedron with decreasing radius of R^{3+} cation. The absolute Mössbauer fractions for rare earth iron garnets were determined.

Stadnik used [57]Fe and [151]Eu Mössbauer spectroscopy to investigate the hyperfine interactions in $Eu_{3-y}Sc_{2+y}Fe_3O_{12}$ ($0 \le y \le 0.5$) [80,81] and $Eu_3Sc_2Fe_3Fe_3O_{12}$ [82]. [57]Fe Mössbauer spectroscopy was also used to investigate Fe^{3+} and Ga^{3+} distribution over the octahedral and tetrahedral sites in Ga-substituted iron garnets of Nd and Pr [83].

[57]Fe and [119]Sn Mössbauer spectra were studied at 4.2 K in iron garnets with the composition $Y_{3-x}Ca_x[Fe_{2-x}Sn_x]FE_3O_{12}$ ($0 \le x \le 2$) [84]. The hyperfine magnetic splittings were observed for all concentrations in these garnets, for both [57]Fe and [119]Sn. The resonance lines were broadened significantly. Mössbauer effect was related to the critical concentration $x = x_c \approx 1.2$ and the full magnetic disorder.

Spin orientations and hyperfine magnetic fields for two rare earth iron garnet films, $(Y_{1.40}Sm_{0.30}Lu_{0.46}Ca_{0.84})(Fe_{4.16}Ge_{0.84})O_{12}$ and $(Y_{0.90}Sm_{0.57}Lu_{0.98}Ca_{0.55})(Fe_{4.45}Ge_{0.55})O_{12}$, were determined as a function of depth using energy-resolve CEMS (conversion electron Mössbauer spectroscopy) [85].

RARE EARTH ORTHOFERRITES

The orthoferrites, $RFeO_3$, where R is yttrium or a rare earth element, belong to materials that exhibit weak ferromagnetism. Detailed Mössbauer studies on the rare earth orthoferrites appeared as early as the 1960s [86–90]. [57]Fe Mössbauer spectra of rare earth orthoferrites show hyperfine magnetic splitting. The hyperfine magnetic field, extrapolated to O K, decreases regularly with an atomic number of R from 564 kOe for $LaFeO_3$ to 545.5 kOe for $LuFeO_3$. This can be ascribed to the effect of the ionic radius. The Néel temperature also decreases from 740K for $LaFeO_3$ to 623 K for $LuFeO_3$. The values of hyperfine magnetic field for rare earth orthoferrites, as well as their Néel temperatures, are summarized in Table 5.

Table 5 Hyperfine Magnetic Field
and Néel Temperature of Rare Earth
Orthoferrites

Rare Earth Orthoferrite	HMF[a] (kOe)	T_N (K)
$LaFeO_3$	564	740
$PrFeO_3$	559	707
$NdFeO_3$	557	687
$SmFeO_3$	552	674
$EuFeO_3$	552	662
$GdFeO_3$	551	657
$TbFeO_3$	550	647
$DyFeO_3$	548	645
$HoFeO_3$	548	639
$ErFeO_3$	546	636
$TmFeO_3$	545	632
$YbFeO_3$	546.5	627
$LuFeO_3$	545.5	623

Source: Ref. 4.
[a] Extrapolated to O K with estimated error ±
2 kOe.

Belakhovski et al. [91] applied Mössbauer spectroscopy in the study of low-temperature spin configurations in single crystals, $ErFeO_3$ and $DyFeO_3$. Boekema et al. [92] studied the covalency effect and hyperfine interactions in the rare earth orthoferrites. Using the LCAO method, expressions are derived relating the Néel temperature, magnetic hyperfine field, and isomer shift to the various covalency parameters and to the Fe—O—Fe bond angle. Application of these expressions to the series of orthoferrites $LaFeO_3$ to $LuFeO_3$ resulted in excellent agreement with experimentally determined values for the Néel temperature, hyperfine magnetic field, and isomer shift. The LCAO (linear combination of atomic orbitals) method was also applied in the interpretation of high-pressure effects in the rare earth orthoferrites [93].

OTHER MÖSSBAUER STUDIES WITH THE PEROVSKITES

Mössbauer spectroscopy was used extensively in the study of structural and magnetic properties of other oxide compounds with the perovskite structure. The structure of $Pb(FeTa)_{1/2}O_3$ is of the disordered perovskite type, in which the Fe^{3+} and Ta^{5+} ions are randomly distributed at the B-sites in the ABO_3 lattice [94]. This compound is ferroelectric below 243 K and becomes antiferromagnetic below 180 K. The dielectric and magnetic properties of $Pb(FeTa)_{1/2}O_3$ are very similar to those of $Pb(FeNb)_{1/2}O_3$, which is ferroelectric below 387 K and antiferromagnetic below 143 K. The Mössbauer spectrum of $Pb(FeTa)_{1/2}O_3$ shows the central quadrupole doublet at room temperature, which corresponds to Fe^{3+} ions. As the temperature decreased below 160 K, the ^{57}Fe Mössbauer lines of $Pb(FeTa)_{1/2}O_3$ became extremely broad. Since the distribution of the magnetic Fe^{3+} and the nonmagnetic Ta^{5+} ions is random, the hyperfine magnetic field fluctuates from site to site in the $Pb(FeTa)_{1/2}O_3$ crystals, thus giving rise to extreme broadening of the Mössbauer lines.

^{151}Eu Mössbauer spectra of several orthorhombic perovskites, $EuMO_3$, where M = Cr, Mn, Fe, or Co, are characterized at 295 K with single but broad lines, due to unresolved hyperfine effects [95]. Induced magnetic exchange interactions, which should occur at the europium ions in the solid solution $EuFe_{0.8}Cr_{0.2}O_3$, have been found to be much smaller than predicted.

^{151}Eu and ^{57}Fe Mössbauer spectra were used to study magnetic exchange interactions in the orthorhombic perovskite solid solution, $EuFe_{1-x}Co_xO_3$ ($0 < x < 1$) [96]. Magnetic hyperfine interactions were observed in the ^{57}Fe Mössbauer spectrum at 295 K for $x \leq$ 0.4 and at 85 K for $x \leq 0.5$ with some additional spectra at other temperatures. For x = 0 the hyperfine magnetic fields of 508 kOe at 295 K and 555 kOe at 85 K were measured. The replacement of neighboring Fe^{3+} ion by diamagnetic Co^{3+} ion results in a reduction of the hyperfine magnetic field at the ^{57}Fe nucleus. The ^{151}Eu single line is considerably broadened with the increase of x. This effect is proved to be of magnetic origin.

^{119}Sn was incorporated into various perovskites to give the phases $(A_{0.95}$ $Ca_{0.05})(Fe_{0.85}M_{0.10}Sn_{0.05})O_3$ with A = La, Eu, or Lu and M = Al, Ga, Sc, Cr, Mn, Co, or Ni [97]. The observed supertransferred hyperfine field for ^{119}Sn with six nearest-neighbor Fe^{3+} cations is strongly dependent on the nature of A, but is almost independent on the M substituent.

^{151}Eu Mössbauer spectroscopy was used to study the quadrupole interactions in the orthorhombic perovskites $EuMo_3$, where M = Co, Cr, Mn, Fe, or Sc [98]. It is shown that the quadrupole coupling constant $eV_{zz}Q_g$ at room and liquid nitrogen temperatures is negative and in the range -9.3 to -5.8 mm/s. Also, it is observed that the electronic charge density at the ^{151}Eu nucleus, in the investigated series $EuMO_3$, does not depend on the type of M^{3+} ion. At the given absorber temperature, the isomer shifts were within the experimental error, practically the same. The increase in isomer shift, δ, with decreasing absorber temperature is caused by the second-order Doppler effect.

Ayyub et al. [99] observed an anomalous behavior of the hyperfine magnetic field in the microcrystalline $YFeO_3$–Fe_2O_3 mixed-phase system. The effects of superparamagnetism and the exchange coupling between the oxide microcrystals were introduced in the interpretation of their results. Different phases, such as $YFeO_3$, $Y_3Fe_5O_{12}$, Y_2O_3, α-Fe_2O_3, and γ-Fe_2O_3, were observed during the preparation of samples.

Iron oxides and many mixed oxides contain iron in the oxidation state $2+$ and/or $3+$. On the other hand, the number of oxides containing highly charged iron ions, Fe^{4+} to Fe^{6+}, is very limited, because these oxidation states are stabilized only under strongly oxidizing conditions. In Table 6 mixed oxides containing Fe^{4+} ions are presented [100]. Table 6 shows that Fe^{4+} state can be stabilized in the perovskite structure. $SrFeO_3$ and $CaFeO_3$ are typical representatives of this group of compounds.

$SrFeO_3$ is characterized by a single Mössbauer line at 298 K, as shown in Figure 11 [101]. This compound is antiferromagnetic below 134 K and gives a hyperfine magnetic field of 331 kOe at 4 K. The iron site symmetry is cubic, and for this reason there is no quadrupole splitting from the nominal $3d^4$ configuration. In nonstoichiometric strontium ferrate(IV), $SrFeO_{3-x}$, there is also the appearance of a Fe^{3+} state.

Mössbauer spectra of $CaFeO_3$ are shown in Figure 12 [100]. At room temperature, $CaFeO_3$ exhibits a single line spectrum with isomer shift close to that of $SrFeO_3$. However, at 4 K the Mössbauer spectrum of $CaFeO_3$ consists of two sextets, which is different in

Table 6 Mixed Oxides Containing Fe^{4+} Ions

Mixed Oxide	Symmetry	Structural Type
$SrFeO_3$	Cubic	Perovskite
$CaFeO_3$	Cubic tetragonal	Perovskite
$Sr_3Fe_2O_{6.9}$	Tetragonal	$Sr_3Ti_2O_7$
$Sr_2FeO_{3.7}$	Tetragonal	K_2NiF_4
$Sr_{0.5}La_{1.5}Li_{0.5}Fe_{0.5}O_4$	Tetragonal	K_2NiF_4
$BaFeO_{2.95}$	Hexagonal	$BaTiO_3$

Source: Ref. 100.

relation to the spectrum of $SrFeO_3$ recorded at 4 K (one sextet). This effect is explained by a very simple charge disproportionation model. In accordance with Takano et al. [102], at low temperatures half of the Fe^{4+} ions in $CaFeO_3$ lose one electron, giving Fe^{5+} ions, and the other half catch the electron, giving Fe^{3+} ions.

Recently, two new Fe (IV) oxides with high-spin configuration, $SrLaMg_{0.50}Fe_{0.50}O_4$

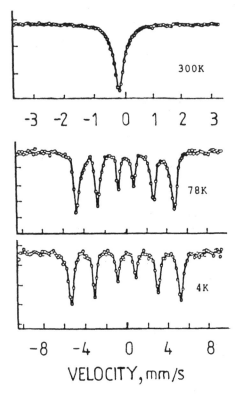

Figure 11 ^{57}Fe Mössbauer spectra of $SrFeO_3$. (From Ref. 101; courtesy of the American Institute of Physics.)

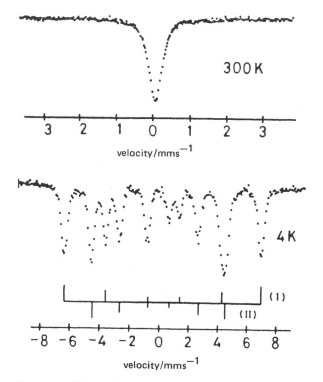

Figure 12 ^{57}Fe Mössbauer spectra of $CaFeO_3$. (From Refs. 100 and 102; courtesy of M. Takano.)

and $SrLaZn_{0.50}Fe_{0.50}O_4$, were found [103]. These compounds possess a K_2NiF_4-type derived structure.

Six-coordinated iron(V) is observed in La_2LiFeO_6 perovskite-type oxide [104]. Iron(V) is the rare oxidation state of iron. This state is also found in K_3FeO_4, Na_3FeO_4, and Rb_3FeO_4.

MIXED OXIDES AS HIGH-TEMPERATURE SUPERCONDUCTORS

After the discovery of superconducitivity above 30 K in $La_{2-x}Ba_xCuO_4$ [105] and above 90 K in $YBa_2Cu_3O_{7-\delta}$ [106], many researchers all over the world began to investigate chemical, physical and structural properties of these and similar compounds. High-temperature superconducting cuprates discovered since 1986 belong to five families [107]: (1) $La_{2-x}M_xCuO_4$ (M = Ca, Sr, or Ba) of the K_2NiF_4 structure; (2) $RBa_2Cu_3O_7$ system, where R = Y, La, Nd, Sm, Eu, Gd, Dy, Ho, Er, Tm or Yb; (3) the $Bi_2(Ca,Sr)_{n+1}Cu_nO_{2n+4}$ system; (4) the $Tl_2Ca_{n-1}Ba_2Cu_nO_{2n+4}$ system; and (5) the $Tl_2Ca_{n-1}Ba_2Cu_nO_{2n+3}$ ($n =$ 1, 2, 3, 4) system. Other oxides not containing copper ions, for instance $Ba_{1-x}K_xBiO_3$, can also exhibit high-temperature superconductivity [108]. The researchers immediately recognized the important role of mixed valence copper, oxygen content, and its ionic state, as well as the perovskite-like structure in the phenomenon of high-temperature (high-T_c) superconductivity of cuprates.

Ramakrishnan and Rao [107] reviewed recent investigations in the field of high-temperature superconductivity and briefly discussed the theoretical models developed to describe the effect. In the last two years the number of publications devoted to high-T_c superconductors has increased very rapidly. The subjects of these publications were the synthesis of known and new high-T_c superconductors, application of different techniques in the characterization of high-T_c superconductors, the relation between magnetic and structural properties of high-T_c superconductors at one side and the effect of superconductivity at the other side, investigation of the chemical bond in these compounds, and so on. These publications were published mainly in the following journals: *Physica C, Phys. Rev. B, Phys. Rev. Lett., J. Solid State Chem., Solid State Commun., J. Phys. Chem. Solids, Hyperfine Interact., J. Mater. Sci., J. Mater. Sci. Lett.,* and *J. Supercond.* For readers of this book, only publications that describe the application of Mössbauer spectroscopy in the study of high-T_c oxide superconductors will be reviewed.

The oxide superconductors, $YBa_2Cu_3O_{7-\delta}$, have a structure that can be related to that of mineral perovskite, $CaTiO_3$ [109,110]. $YBa_2Cu_3O_7$ possesses orthorhombic symmetry with the parameters $a = 3.821$ Å, $b = 3.885$ Å, and $c = 11.676$ Å [111], and the perovskite cell is tripled along the c-axis. In $YBa_2Cu_3O_7$ the copper ions occupy two inequivalent crystallographic sites. Cu(1) sites are located in planar Cu—O squares with $(CuO_2)_\infty$ chains at $z = 0$ along $<010>$, whereas Cu(2) is in square pyramidal configuration with puckered (CuO_2) planes perpendicular to $<001>$ at $z = 0.36$ and $z = 0.64$. The

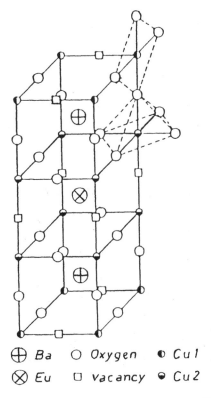

⊕ Ba   ○ Oxygen   ◐ Cu1

⊗ Eu   ▢ vacancy   ⬤ Cu2

Figure 13 Schematic crystal structure of $YB_2Cu_3O_7$. (From Ref. 112; courtesy of Springer-Verlag.)

schematic crystal structure of $YBa_2Cu_3O_7$ is shown in Figure 13 [111,112]. The structural environment of Cu(2) sites is quite well preserved over a wide range of temperatures and oxygen content. Orthorhombic $YBa_2Cu_3O_{7-\delta}$ transforms to a nonsuperconducting tetragonal structure with $\delta \geq 0.4$ on loss of oxygen at elevated temperatures. In the high-temperature tetragonal phase, Cu(1) ions form a metallic linear chain with in-plane oxygens, where Cu(1) ion is bonded to four oxygens.

To understand the mechanism of high-T_c superconductivity, many researchers have investigated the effects caused by substituting for Y with $4f$ rare earth element or for Cu by a $3d$ transition metal in the oxide superconductor $YBa_2Cu_3O_{7-\delta}$ (Y–Ba–Cu–O).

Dalmas de Réotier et al. [113] demonstrated the application of a ^{151}Eu, ^{155}Gd, or ^{170}Yb Mössbauer effect in the study of valency and electric field gradient in $RBa_2Cu_3O_{7-\delta}$ compounds, where R = Eu, Gd, or Yb. Chmist et al. [114] carried out resistivity, electron spin resonance, Mössbauer, and heat measurements for $RBa_2Cu_3O_{7-\delta}$ high-T_c superconductors, where R = Y, Eu, or Er. Figure 14 shows the electrical resistivity behavior of a $YBa_2Cu_3O_{7-\delta}$ superconductor with the superconducting transition temperature $T_c = 90$ K, and the width of superconducting transitions, $\Delta T_c = 2$ K. Figure 15 shows the ^{151}Eu Mössbauer spectra of $EuBa_2Cu_3O_{7-\delta}$ recorded at various temperatures. Mössbauer spectroscopy indicates that Eu atoms are in the Eu^{3+} form and behave as almost free ions.

Toniwaki and Sasaki [115] also found that europium is in the trivalent state in the high-T_c superconductor $EuBa_2Cu_3O_y$. The Debye temperature of ^{151}Eu is estimated at 240 K; however, the lattice vibrations could not be described well by a simple Debye model.

Taylor et al. [116] did not observe hyperfine magnetic splitting at the ^{151}Eu nucleus at temperatures down to 1.3 K. ^{151}Eu Mössbauer effect measurements showed that $Eu_{0.1}Gd_{0.9}Ba_2Cu_3O_x$ became magnetically ordered below 2.1 K, with the temperature dependence expected for spontaneous ordering of Gd^{3+} ions.

A ^{155}Gd Mössbauer effect was measured in the ceramic superconductor $GdBa_2Cu_3O_7$ [117]. It was concluded that the rare earth sites are separated from the sites that are responsible for the superconducting effect.

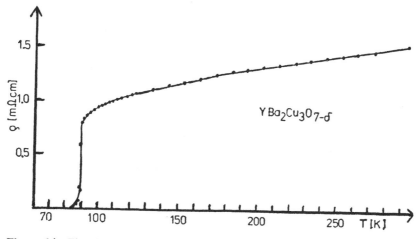

Figure 14 Electrical resistivity of $YBa_2Cu_3O_{7-\delta}$ as a function of temperature. (From Ref. 114; courtesy of *Acta Phys. Polonica.*)

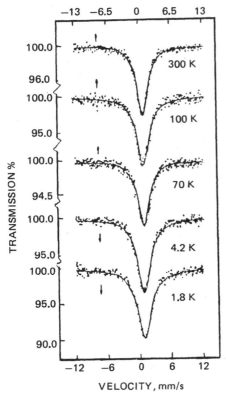

Figure 15 ^{151}Eu Mössbauer spectra of EuBa$_2$Cu$_3$O$_{7-\delta}$ recorded at different temperatures. (From Ref. 114; courtesy of *Acta Phys. Polonica*.)

The ^{155}Gd Mössbauer spectrum of GdBa$_2$Cu$_3$O$_7$ showed at 4 K the quadrupole doublet (δ = 0.47 mm/s, Δ = 2.81 mm/s, and Γ = 0.88 mm/s), and magnetic ordering of the sample was observed at 1.35 K [118].

The structural role of Er ions in ErBa$_2$Cu$_3$O$_{7-x}$ is similar to that of other rare earths, as shown by ^{166}Er Mössbauer spectroscopy [119].

^{57}Fe Mössbauer spectra of YBa$_2$Cu$_{2.9375}$Fe$_{0.0625}$O were recorded between 50 and 120 K [120]. Spectra were interpreted at the superposition of two quadrupole doublets, with similar quadrupole splittings and different isomer shifts. It was suggested that Fe atoms occupy the Cu(1) sites and behave as low-spin Fe(II) and/or Fe(III). This implies a strong bonding of iron, as well as copper atoms, in the cyrstal lattice. In accordance with Gómez et al. [120], Fe atoms in YBa$_2$Cu$_{3-x}$Fe$_x$O$_{7-\delta}$ have three possibilities: (1) to be located at Cu(2) sites, (2) to be located at Cu(I) sites, and (3) to be located at Cu(2) and Cu(1) sites. In case (1) the Mössbauer spectra should be a pure quadrupole doublet, because the environment of Cu(2) sites does not change with the oxygen content. In case (2) there is the possibility of a pair of quadrupole doublets, due to the oxygen deficiencies in the perovskite structure. In case (3) the ^{57}Fe Mössbauer spectra should consist, at least, of a pair of quadrupole doublets with a large difference in their quadrupole splittings.

Transmission Mössbauer spectra of YBa$_2$Cu$_{2.875}$Fe$_{0.125}$O$_\delta$ powder samples recorded between 12 and 300 K were interpreted as the superposition of three quadrupole doublets

[121]. The outer doublet did not change with the temperature. It was concluded that Fe atoms occupy only Cu(1) sites in three different configurations: (1) in the center of the oxygen octahedra, as Fe^{3+}; (2) in the fourfold coordinated chain, also as Fe^{3+}; and (3) in the two-coordinated axial symmetry as Fe^{2+}. The possibility of the presence of Fe^{4+} is discussed.

The temperature behavior of the asymmetries in the quadrupole doublets of the superconducting $YBa_2Cu_{3-x}Fe_xO_y$ with $x = 0.125$ suggested the existence of small magnetic fields, superimposed to an electric field gradient at the positions of Fe atoms, which are assumed to be the Cu(1) sites [122]. The effect of the superconductivity of this compound was explained by a magnetic exchange mechanism.

^{57}Fe absorption and ^{57}Co emission Mössbauer spectroscopy were used to study $YBa_2(Cu_{1-x}M_x)_3O_{7-y}$ (M = ^{57}Fe or ^{57}Co, $3.5 \times 10^{-5} \leq x \leq 0.1$, $y = 0$ or $y = 0.8$) between 4.2 and 295 K with and without an external magnetic field [123]. Four iron states, denoted as A, B, C, and D, were considered with different dominance values as a function of x and y. A preferential Cu(1) substitution by Fe or Co is suggested. Three of the four iron states, A, B, and D, have been assigned to high-spin Fe^{4+}, and C, to high-spin Fe^{3+}. For Co it is also assumed that it occurs mostly in the four-valent state. Figure 16 shows ^{57}Fe Mössbauer spectra of $YBa_2(Cu_{0.9}Fe_{0.1})_3O_7$ recorded different temperatures. With increasing temperature, the decrease in magnetic splitting for all subspectra is accompanied by relaxation line broadening.

A fully c-axis-oriented epitaxial thin film of $YBa_2(Cu_{0.97}{}^{57}Fe_{0.03})_3O_7$ was deposited on <*100*>$SrTiO_3$ [124]. The spectra obtained by CEMS at different angles between the normal of the film (i.e., the c-axis and the gamma-ray direction) show the superposition of three quadrupole doublets. This is in accordance with the powder absorption spectra, which also showed the superposition of three quadrupole doublets for the same compound; however, their intensities were different.

Suharan et al. [125] investigated ^{57}Fe Mössbauer spectra of $YBa_2Cu_{2.97}{}^{57}Fe_{0.03}O_{7-\delta}$. The best fits were obtained when the presence of Fe^{4+} ion is also supposed in the structure of $YBa_2Cu_3O_{7-\delta}$. On the other hand, Tamaki et al. [126] deconvoluted the Mössbauer spectrum of $YBa_2(Cu_{0.95}Fe_{0.05})_3O_{7-\delta}$, recorded at room temperature, into four quadrupole doublets. The authors proposed a method for analysis of the Mössbauer spectra of $YBa_2(Cu_{1-x}Fe_x)_3O_{7-\delta}$.

Mössbauer spectra of the oxide superconductor $YBa_2(Cu_{1-x}Fe_x)_3O_{7+\delta}$ ($0.0018 \leq x \leq 0.10$), were recorded at room temperature [127] and deconvoluted into three quadrupole doublets with splittings of 1.91, 1.10, and 0.55 mm/s and isomer shifts of 0.07, -0.09, and 0.03 mm/s, respectively. The relative intensities of the quadrupole doublets were dependent on the x value.

Dunlap et al. [128] investigated the structural and magnetic properties of samples having the general composition $YBa_2(Cu_{1-x}Fe_x)_3O_{7-\delta}$. The authors concluded that the substitution of Cu by Fe atoms is still unclear and that Fe appears to have a substantial substitution for Ba.

Neutron diffraction and Mössbauer effect were measured, as a function of iron concentration, in orthorhombic ($x = 0.01$, 0.02) and tetragonal ($x \geq 0.05$) $YBa_2(Cu_{1-x}Fe_x)_3O_{7+\delta}$ [129]. The Mössbauer spectra gave the relative change in oxygen configuration about Fe atoms as a function of Fe concentration. A ^{57}Fe Mössbauer effect can be used to investigate the kinetics of oxygen loss, in vacuum at high temperatures, in Fe-doped Y–Ba–Cu–O superconductors [130].

A sample of composition $YBa_2(Cu_{0.95}{}^{119}Sn_{0.05})_3O_{7-\delta}$ and samples of composition

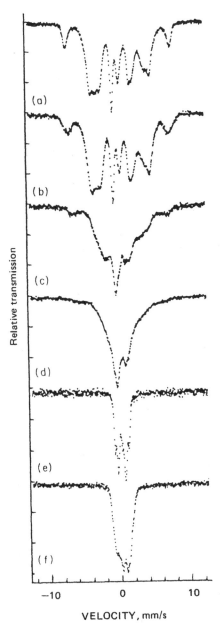

Figure 16 Mössbauer absorption spectra of $YBa_2(Cu_{0.9}Fe_{0.1})_3O_7$ at different temperatures with and without applied magnetic field: (a) $T = 4.2$ K, $H_{ext} = 0$; (b) $T = 4.2$ K, $H_{ext} = 5T$; (c) $T = 15$ K, $H_{ext} = 0$; (d) $T = 19$ K, $H_{ext} = 0$; (e) $T = 100$ K, $H_{ext} = 0$; (f) $T = 100$ K, $H_{ext} = 5T$. From Ref. 123; courtesy of the American Physical Society.)

$YBa_2(Cu_{1-x}{}^{57}Fe_x)_3O_{7-\delta}$ (where $x = 0.002$ and 0.05) were prepared under very different experimental conditions [131]. For Sn-doped $YBa_2Cu_3O_{7-\delta}$, the ^{119}Sn Mössbauer spectrum was similar to the spectrum of SnO_2, and very probably for given experimental conditions, Sn is out of the $YBa_2Cu_3O_{7-\delta}$ structure. ^{57}Fe Mössbauer spectra were interpreted as the superposition of three quadrupole doublets; however, their relative intensities were dependent on the experimental conditions for the preparation of samples.

Violet et al. [132] recorded ^{57}Fe Mössbauer spectra at room temperature for $HoBa_2(Cu_{0.95}Fe_{0.05})_3O_{7-x}$ over a range of oxygen deficiency x, in order to observe the effect of oxygen stoichiometry on local oxygen configurations relative to the Cu-substituted Fe-probe ions. ^{57}Fe Mössbauer spectra for the complete range of oxygen stoichiometry are given in Figure 17. The strong effect of oxygen content on the Mössbauer spectra is observed. For $x < 0.5$, these spectra were deconvoluted into two doublets and the third very weak doublet. The outer quadrupole doublet (lines 1 and 6) and the first inner doublet (lines 2 and 5) were ascribed to low-spin Fe^{2+}, and the second inner doublet (lines 3 and 4) was ascribed to high-spin Fe^{3+}. The authors of this work did not find experimental or theoretical justification for the Fe^{4+} state.

Coey and Donnelly [112] doped $EuBa_2Cu_3O_7$ with 1% ^{57}Fe. ^{57}Fe Mössbauer spectra of this sample were deconvoluted into two quadrupole doublets ascribed to Fe^{3+} at $Cu(1)$ and $Cu(2)$ sites and one doublet that is typical for octahedral Fe^{3+}.

^{57}Fe Mössbauer measurements with $EuBa_2Cu_3O_x$ ($6 \leq x \leq 7$) were performed with samples where 1 or 2% of the copper was replaced by iron [133]. The orthorhombic–tetragonal transition occurred on heating to 500°C in vacuum for $x = 6.6$. Mössbauer spectra were fitted using four quadrupole doublets for the orthorhombic phase and two doublets for the tetragonal phase. It was suggested that iron at $Cu(1)$ sites can also be in the Fe^{4+} state, while at $Cu(2)$ sites it is only in the Fe^{3+} state.

The ^{57}Fe Mössbauer spectrum of $GdBa_2(Cu_{1-x}F_x)_3O_4$ ($x = 0.04$) was computer deconvoluted into three quadrupole doublets [134]. Two outer doublets, I and II, were assigned to Fe^{3+} ions in an intermediate spin state $S = \frac{3}{2}$ and located at the $Cu(2)$ and $Cu(1)$ sites, respectively. The quadrupole doublet, II, was assigned to Fe^{3+} in the high-spin state $S = \frac{5}{2}$, probably coordinated octahedrally. The microstructure and superconductivity of $RBa_2Cu_{3(1-x)}Fe_{3x}O_y$, where R = Y or Gd and $0 \leq x \leq 0.4$, were investigated [135]. ^{57}Fe Mössbauer spectra were computer-fitted assuming the superposition of four quadrupole doublets, I to IV. The following Mössbauer parameters were measured (in mm/s): $\delta_{Fe}(I) = 0.05$, $\Delta(I) = 1.93$, $\delta_{Fe}(II) = 0.00$, $\Delta(II) = 1.11$, $\delta_{Fe}(III) = 0.32$, $\Delta(III) = 0.58$, $\delta_{Fe}(IV) = -0.16$, and $\Delta(IV) = 1.70$. Quadrupole doublets, I and II, were assigned to Fe^{3+} ions at the $Cu(2)$ and $Cu(1)$ sites, respectively, with spin state $\frac{3}{2}$. Quadrupole doublets, III and IV, were assigned to Fe^{3+} ($S = \frac{5}{2}$) and Fe^{4+} ions, probably coordinated octahedrally at the $Cu(1)$ sites.

The samples of $(La_{2-}Sr_y)(Cu_{1-x}{}^{57}Fe_x)O_{4-\delta}$ were prepared with y ranging between 0 and 0.20 with a constant, low ^{57}Fe concentration ($x = 0.005$) [136]. The Mössbauer and resistivity measurements performed with these samples showed the presence of three successive phases in dependence on the $y < 0.02$, semiconducting phase with two-dimensional spin-glass Cu magnetic order for $0.02 < y < 0.07$, and superconducting phase for $y > 0.07$.

The $Bi_2Sr_4Fe_3O_{12+x}$ compound isostructural with the $Bi_2Sr_2CaCu_3O_{10+x}$, the 110-K superconductor, was investigated by Mössbauer spectroscopy and magnetic measurements [137]. The Néel temperature, $T_N = 40$ K, was determined for $Bi_2Sr_4Fe_3O_{12+x}$. The analysis of Mössbauer spectra indicated the presence of Fe in two states of valence with relative

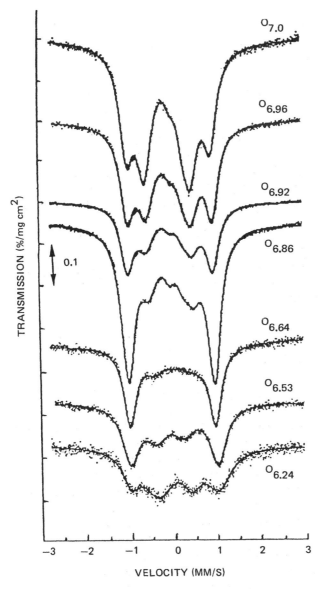

Figure 17 ^{57}Fe Mössbauer spectra of $HoBa_2(Cu_{0.95}Fe_{0.05})_3O_{7-x}$, recorded at room temperature, for different oxygen contents. The transmission is normalized to absorber thickness in mg/cm². (From Ref. 132; courtesy of Elsevier Science Publishers.)

abundance $Fe^{3+}/Fe^{4+} = 0.16$. In addition, Fe^{3+} was in a high-spin state. To understand better the phenomenon of high-T_c superconductivity, other researchers also applied Mössbauer spectroscopy in characterization of the oxide superconducting materials [138–147].

The key results of Mössbauer studies performed with high-T_c oxide superconductors can be summarized as follows. Up to date, Mössbauer studies, as well as other studies performed with high-T_c oxide superconductors, did not give all relevant parameters needed for developing a unique model of high-T_c superconductivity. It is generally accepted that the traditional Bardeen–Cooper–Schrieffer (BCS) theory [148] of superconductivity is inadequate as an explanation of high-T_c superconductivity.

Magnetic substitution on the Y site in $YBa_2Cu_3O_{7-\delta}$ superconductor has little effect on the superconductivity. It is concluded that rare earths in this type of high-T_c superconductor are extremely isolated from the conduction electrons. An important exception is Pr: $Y_{1-x}Pr_xBa_2Cu_3O_7$ for $x > 0.55$ is not superconducting oxide. On the other hand, substitutions of both magnetic and nonmagnetic ions on the Cu sites strongly affect the superconducting properties of high-T_c superconductors. It is observed that Zn has a stronger effect on the T_c value than do Fe impurities, which is unusual, since nonmagnetic ions normally have little effect on superconductivity. Magnetic ordering of the Cu ions has not been observed in high-T_c oxide superconductors. However, the magnetic hyperfine splitting for Fe impurities at Cu sites is shown in $YBa_2Cu_3O_{7-\delta}$ at low temperatures with preserved superconductivity effect. The magnetic phenomena in high-T_c oxide superconductors are very complicated, and it is now evident that this type of research must be intensified if we wish to understand the phenomenon of high-T_c superconductivity.

The presence of the 4 + state of iron impurities in $YBa_2Cu_3O_{7-\delta}$, in addition to the Fe^{2+} and Fe^{3+} states, was suggested, Different configurations of oxygens formed around Fe impurities have a strong influence on the shape of ^{57}Fe Mössbauer spectra. Some researchers claimed that Fe impurities are located only at Cu(1) sites, and others that both Cu(2) and Cu(1) sites are occupied by Fe impurities. Often, these results are difficult to compare, since the oxide samples are prepared under very different experimental conditions and with different oxygen stoichiometries.

Research in the field of high-T_c superconductivity is in a state of great expansion.

REFERENCES

1. R. L. Mössbauer, *Z. Phys.*, 151: 124 (1958).
2. R. L. Mössbauer, *Z. Naturforsch.*, 14a: 221 (1959).
3. V. I. Goldanskii and R. H. Herber, *Chemical Applications of Mössbauer Spectroscopy*, Academic Press, New York, 1968.
4. N. N. Greenwood and T. C. Gibb, *Mössbauer Spectroscopy*, Chapman & Hall, London, 1971.
5. U. Gonser, *Mössbauer Spectroscopy*, Springer-Verlag, New York, 1976.
6. T. C. Gibb, *Principles of Mössbauer Spectroscopy*, Chapman & Hall, London, 1976.
7. R. L. Cohen, *Applications of Mössbauer Spectroscopy*, Academic Press, New York, 1976.
8. A. Vértes, L. Korecz, and K. Burger, *Mössbauer Spectroscopy*, Elsevier, Amsterdam, 1980.
9. W. Kündig, H. Bömmel, G. Constabaris, and R. H. Lindquist, *Phys. Rev.*, 14: 327 (1966).
10. L. Néel, *Ann. Geophys.*, 5: 99 (1949).
11. H. Hibst, *Angew. Chem. Int. Ed. Engl.*, 21: 270 (1982).
12. H. Hibst, *J. Magn. Magn. Mater.*, 74: 193 (1988).
13. L. M. Letyuk and G. I. Zhuravlev, *Chemistry and Technology of Ferrites* (in Russian), Khimiya, Leningrad, 1983.
14. Yu. T. Pavlyukhin, Ya. Ya. Medikov, and V. V. Boldyrev, *J. Solid State Chem.*, 53: 155 (1984).
15. C. Gleitzer and J. B. Goodenough, *Mixed-Valence Iron Oxides in Structure and Bonding 61*, Springer-Verlag, Berlin, 1985.
16. G. A. El-Shobaky and A. A. Ibrahim, *Thermochim. Acta*, 118: 151 (1987).
17. J. M. Fernandez-Rodriguez, J. Morales, and J. L. Tirado, *J. Mater. Sci.*, 23: 2971 (1988).
18. M. M. Thackeray, W. I. F. David, and J. B. Goodenough, *Mater. Res. Bull.*, 17: 785 (1982).

19. P. B. Braun, *Nature*, 27: 1123 (1952).
20. R. W. Cheary and N. W. Grimes, *Acta Crystallogr.*, A34: 74 (1976).
21. J. L. Dormann, M. Nogues, and A. Tomas, *Ferrites, Proc. International Conference*, Sept.–Oct. 1980, Japan, p. 166.
22. J. L. Dormann and P. Renaudin, *Ferrites, Proc. International Conference*, Sept./Oct. 1980, Japan, p. 156.
23. P. Kishan, C. Prakash, J. S. Baijal, and K. K. Laroia, *Phys. Status Solidi (a)*, 84: 535 (1984).
24. J. Chappert and R. B. Frankel, *Phys. Rev. Lett.*, 18: 165 (1967).
25. T. M. Uen and F. H. Yang, *Jpn. J. Appl. Phys.*, 27: 955 (1988).
26. A. H. Morrish and K. Haneda, *J. Appl. Phys.*, 52: 2496 (1981).
27. K. Haneda, H. Kojima, and A. H. Morrish, *J. Magn. Magn. Mater.*, 31–34: 951 (1983).
28. A. H. Morrish and K. Haneda, *J. Magn. Magn. Mater.*, 35: 105 (1983).
29. T. M. Peev, L. Bozadjiev, T. Stoilova, and S. Nikolov, *J. Radioanal. Nucl. Chem. Lett.*, 85: 151 (1984).
30. G. Raw, *The Synthesis and Characterization of Magnetites and Nickel Ferrites*, Report AERE-R 10777, United Kingdom Atomic Energy Authority, Harwell, Oxfordshire, England, 1983.
31. G. A. Sawatzky, F. van der Woude, and A. H. Morrish, *J. Appl. Phys.*, 39: 1204 (1968).
32. A. Cruset and J. M. Friedt. *Phys. Status Solidi (b)*, 45: 189 (1971).
33. A. Ito, *J. Phys.*, 35: C6–325 (1974).
34. E. De Grave, R. Leyman, and R. Vanleerberghe, *Phys. Lett.*, 97A: 354 (1983).
35. J. W. D. Martens, W. L. Peeters, and H. M. Van Noort, *J. Phys. Chem. Solids*, 46: 411 (1985).
36. R. E. Vandenberghe, R. Vanleerberghe, and G. G. Robbrecht, *Preparation and Characterization of Some Ultra-Fine Spinel Studies in Inorganic Chemistry*, Vol. 3, *Ferrites*, R. Metselaar, H. J. M. Heijligers, and J. Schoonman (eds.), Elsevier, Amsterdam, 1983, p. 395.
37. K. Haneda and A. H. Morrish, *J. Appl. Phys.*, 63: 4258 (1988).
38. M. P. Sharrock, P. J. Picone, and A. H. Morrish, *IEEE Trans. Magn.*, MAG-19: 1466 (1983).
39. T. Okada, H. Sekizawa, F. Ambe, S. Ambe, and T. Yamada, *J. Magn. Magn. Mater.*, 31–34: 903 (1983).
40. P. Auric, G. M. Chen, H. L. Luo, D. Y. Yang, and K. Sun, *J. Magn. Magn. Mater.*, 72: 319 (1988).
41. B. Hannoyer and M. Lenglet, Mössbauer Study of the Oxidation States of Iron in $Cu_xFe_{3-x}O_4$, *Solid State Chemistry 1982, Proc. 2nd Conference*, Veldhoven, The Netherlands, June 7–9, 1982; *Studies in Inorganic Chemistry*, Vol. 3, R. Metselaar, H. J. M. Heijligers, and J. Schoonman (eds.), Elsevier, Amsterdam, 1983, p. 617.
42. A. Gerard and F. Grandjean, *Hyperfine Interact.*, 4: 460 (1978).
43. A. Gerard and F. Grandjean, *J. Phys. C Solid State Phys.*, 12: 4601 (1979).
44. A. Narayanasamy and L. Häggström, *J. Phys. C Solid State Phys.*, 16: 591 (1983).
45. T. M. Peev, T. Mitev, T. Dimova, and L. Peneva, *Bulg. J. Phys.*, 6: 651 (1979).
46. C. M. Srivastava, S. N. Shringi, A. S. Bommanavar, and M. Vijay Babu, *Ferrites, Proc. International Conference*, Sept./Oct. 1980, p. 162.
47. (a) W. Karner, R. Wäppling, and T. Nagarajan, *Mössbauer Study of the Cadmium–Nickel Ferrite System*, Report UUIP-1044, Institute of Physics, Uppsala University, Uppsala, Sweden, 1981.
    (b) W. Karner, R. Wäppling, and T. Nagarajan, *Phys. Scr.*, 36: 544 (1987).
48. J. P. Eymery, S. B. Raju, and J. F. Dinhut, *Phys. Status Solidi (a)*, 67: 89 (1981).
49. C. Michalk, *J. Magn. Magn. Mater.*, 68: 157 (1987).
50. H. Furukawa, T. Kanzaki, and T. Katsura, *J. Electron. Spectr. Relat. Phenom.*, 36: 1 (1985).

51. M. J. Rossiter, *J. Phys. Chem. Solids*, 26: 775 (1965).
52. H. B. Mathur, A. P. B. Sinha, and C. M. Yagnik, *Solid State Commun.*, 3: 401 (1965).
53. M. Rosenberg, S. Dey, and P. Deppe, *Ferrites, Proc. International Conference*, Japan, Sept./Oct. 1980, p. 101.
54. C. S. Lee and W. G. Kim, *J. Korean Phys. Soc.*, 21: 184 (1988).
55. F. Grandjean and A. Gerard, *Solid State Commun.*, 25: 679 (1978).
56. C. S. Lee and W. G. Kim, *J. Korean Phys. Soc.*, 22: 57 (1989).
57. W. D. Corner, *Phys. Technol.*, 19: 158 (1988).
58. E. Kerber, U. Gonser, A. Trautwein, and F. E. Harris, *J. Phys. Chem. Solids*, 36: 263 (1975).
59. H. Yamada, M. Takano, M. Kiyama, T. Takada, and T. Shinjo, "Study of the Mössbauer effect on the surface of BaO·6Fe$_2$O$_3$, *4th International Conference on Ferrites*, Part II, *Advances in Ceramics*, American Ceramic Society, 1986, p. 169.
60. G. Asti, M. Carbucicchio, A. Deriu, E. Lucchini, and G. Sloccari, *Mater. Res. Bull.*, 13: 163 (1978).
61. E. Kreber and U. Gonser, *Appl. Phys.*, 1: 339 (1973).
62. G. Albanese, A. Deriu, E. Lucchini, and G. Slokar, *IEEE Trans. Magn.*, MAG-17: 2639 (1981).
63. G. Albanese, A. Deriu, E. Lucchini, and G. Slokar, *Appl. Phys.*, A26: 45 (1981).
64. P. M. Rao, A. Gérard, and F. Grandjean, *Phys. Status Solidi (a)*, 54: 529 (1979).
65. P. M. Rao, A. Gérard, and F. Grandjean, *J. Magn. Magn. Mater.*, 15–18: 645 (1980).
66. G. Albanese, M. Carbucicchio, A. Deriu, G. Asti, and S. Rinaldi, *Appl. Phys.*, 7: 227 (1975).
67. G. Albanese, M. Carbucicchio, and G. Asti, *Appl. Phys.*, 11: 81 (1976).
68. G. Albanese, M. Carbucicchio, L. Pareti, and S. Rinaldi, *Phys Status Solidi (a)*, 73: K193 (1982).
69. A. M. van Diepen and F. K. Lotgering, *Solid State Commun.*, 27: 255 (1978).
70. A. Sztaniszláv, E. Sterk, L. Fetter, M. Farkas-Jahnke, and J. Lábár, *J. Magn. Magn. Mater.*, 41: 75 (1984).
71. M. Eibshütz and M. E. Lines, *Phys. Rev. B*, 25: 4256 (1982).
72. M. E. Lines and M. Eibshütz, *Phys. Rev. B*, 25: 6042 (1982).
73. M. Eibshütz and M. E. Lines, *Phys. Rev. B*, 26: 2288 (1982).
74. C. Michalk and W. Thiel, *Phys. Status Solidi (a)*, 90: 325 (1985).
75. Wang Ya-Qi and Zheng Xun-Yi, *Hyperfine Interact.*, 28: 447 (1986).
76. M. Toulemonde and F. Studer, *Philos. Mag. A*, 58: 799 (1988).
77. M. Stachel, S. Hüfner, G. Crecelius, and D. Quitmann, *Phys. Rev.*, 186: 355 (1969).
78. Z. M. Stadnik, G. H. M. Calis, and H. Lipko, *Solid State Commun.*, 38: 719 (1981).
79. V. N. Belogurov and V. A. Bilinkin, *Phys. Status Solidi (a)*, 63: 45 (1981).
80. Z. M. Stadnik and B. F. Otterloo, *J. Solid State Chem.*, 48: 133 (1983).
81. Z. M. Stadnik, *J. Magn. Magn. Mater.*, 37: 138 (1983).
82. Z. M. Stadnik, *Solid State Commun.*, 51: 79 (1984).
83. I. S. Lyubutin, B. V. Mill, R. I. Chalabov, and A. V. Butashin, *J. Phys. Chem. Solids*, 46: 363 (1985).
84. A. P. Dodokin and I. S. Lyubutin, *J. Phys.*, 40: C2-342 (1979).
85. K. Saneyoshi, T. Toriyama, J. Itoh, K. Hisatake, and S. Chikazumi, *J. Magn. Magn. Mater.*, 31–34: 705 (1983).
86. M. Eibshütz, G. Gorodetsky, S. Shtrikman, and D. Treves, *J. Appl. Phys.*, 35: 1071 (1964).
87. D. Treves, *J. Appl. Phys.*, 36: 1033 (1965).
88. M. Eibshütz, S. Shtrikman, and D. Treves, *Phys. Rev.*, 156: 562 (1967).
89. J. M. D. Coey, G. A. Sawatzky, and A. H. Morrish, *Phys. Rev.*, 184: 334 (1969).
90. L. M. Levinson, M. Luban, and S. Shtrikman, *Phys. Rev.*, 177: 864 (1969).
91. M. Belakhovsky, M. Bogé, J. Chappert, and M. Sivardiére, *Solid State Commun.*, 20: 473 (1976).

92. C. Boekema, F. Van der Woude, and G. A. Sawatzky, *Int. J. Magn.*, 3: 341 (1972).

93. C. Boekema, F. Van der Woude, and G. A. Sawatzky, *Phys. Rev. B*, 11: 2705 (1975).

94. S. Nomura, K. Kaneta, and M. Abe, *Jpn. J. Appl. Phys.*, 18: 681 (1979).

95. T. C. Gibb, *J. Chem. Soc. Dalton Trans.*, p. 2245 (1981).

96. T. C. Gibb, *J. Chem. Soc. Dalton Trans.*, p. 873 (1983).

97. T. C. Gibb, *J. Chem. Soc. Dalton Trans.*, p. 2035 (1983).

98. Z. M. Stadnik and E. De Boer, *Solid State Commun.*, 50: 335 (1984).

99. P. Ayyub, M. Multani, and R. Vijayaraghavan, *Phys. Lett. A*, 119: 95 (1986).

100. M. Takano and Y. Takeda, *Bull. Inst. Chem.* Res. Kyoto Univer., 61: 406 (1983).

101. P. K. Gallagher, J. B. MacChesney, and D. N. E. Buchanan, *J. Chem. Phys.*, 41: 2129 (1964).

102. M. Takano, N. Nakanishi, Y. Takeda, S. Naka, and T. Takeda, *Mater. Res. Bull.*, 12: 923 (1977).

103. G. Demazean, Z. Li-Ming, L. Fournés, M. Pouchard, and P. Hagenmuller, *J. Solid State Chem.*, 72: 31 (1988).

104. G. Demazean, B. Buffat, F. Ménil, L. Fournés, M. Pouchard, J. M. Dance, P. Fabritchnyi, and P. Hagenmuller, *Mater. Res. Bull.*, 16: 1465 (1981).

105. J. G. Bednorz and K. A. Muller, *Z. Phys. B*, 64: 189 (1986).

106. M. K. Wu, J. R. Ashburn, C. J. Torng, P. H. Hor, R. L. Meng, L. Gao, Z. J. Huang, Y. Z. Wang, C. W. Chu, *Phys. Rev. Lett.*, 58: 908 (1987).

107. T. V. Ramakrishnan and C. N. R. Rao, *J. Phys. Chem.*, 93: 4414 (1989).

108. R. J. Cava, B. Batlogg, J. J. Krajewski, R. Farrow, L. W. Rupp, A. E. White, K. Short, W. F. Peck, and T. Kometani, *Nature*, 332: 814 (1988).

109. Yu. T. Pavlyuhin, A. I. Rykov, and N. G. Hainovsky, *Pramana J. Phys.*, 31: L437 (1988).

110. J. Galy, R. Enjalbert, P. Millet, C. Faulmann, and P. Cassoux, *J. Solid State. Chem.*, 74: 356 (1988).

111. J. J. Capponi, C. Chaillout, A. W. Hewat, P. Lejay, M. Marezio, N. Nguyen, B. Raveau, J. L. Soubeyroux, J. L. Tholence, and R. Tournier, *Europhys. Lett.*, 3: 301 (1987).

112. J. M. D. Coey and K. Donnelly, *Z. Phys. B*, 67: 513 (1987).

113. P. Dalmas de Réotier, P. Vulliet, A. Yaouane, P. Chaudouet, S. Garcon, J. P. Sénateur, F. Weiss, L. Asch, and G. M. Kalvius, *Physica C*, 153–155: 1543 (1988).

114. J. Chmist, A. Lewicki, Z. Tarnawski, A. Kozlowski, J. Zukrowski, W. M. Woch, A. Kolodziejczyk, and K. Krop, *Acta Phys. Pol.*, A74: 757 (1988).

115. M. Taniwaki and H. Sasaki, *Physica C*, 153–155: 1549 (1988).

116. R. D. Taylor, J. O. Willis, and Z. Fisk, *Hyperfine Interact.*, 42: 1257 (1988).

117. H. H. A. Smit, M. W. Dirken, R. C. Thiel, and L. J. de Jongh, *Solid State Commun.*, 64: 695 (1987).

118. J. D. Cashion, J. R. Fraser, A. C. McGrath, R. H. Mair, and R. Driver, *Hyperfine Interact.*, 42: 1253 (1988).

119. J. A. Hodges, P. Imbert, J. B. Marimon da Cunha, and J. P. Sanchez, *Physica C*, 160: 49 (1989).

120. R. Gómez, S. Aburto, M. L. Marquina, M. Jiménez, V. Marquina, C. Quintanar, T. Akachi, R. Escudero, R. A. Barrio, and D. Rios-Jara, *Phys. Rev. B*, 36: 7226 (1987).

121. R. Gómez, S. Aburto, V. Marquina, M. L. Marquina, M. Jiménez, C. Quintanar, T. Akachi, R. Escudero, R. A. Barrio, and D. Rios-Jara, *Physica C*, 153–155: 1557 (1988).

122. R. Gómez, S. Aburto, V. Marquina, M. L. Marquina, C. Quintanar, M. Jiménez, R. A. Barrio, R. Escudero, D. Rios-Jara, and T. Akachi, *Modern Phys. Lett. B*, 3: 1127 (1989).

123. L. Bottyán, B. Molnár, D. L. Nagy, I. S. Szücs, J. Tóth, J. Dengler, G. Ritter, and J. Schober, *Phys. Rev. B*, 38: 11373 (1988).

124. J. Dengler, G. Ritter, G. Saemann-Ischenko, B. Roas, L. Schütz, B. Molnar, D. L. Nagy, and I. S. Szücs, *Physica C*, 162–164: 1297 (1989).

125. S. Suharan, C. W. Johnson, D. H. Jones, M. F. Thomas, and R. Driver, *Solid State Commun.*, 67: 125 (1988).

126. T. Tamaki, M. Nishizawa, A. Ito, and T. Fujita, *Physica C*, 162–164: 987 (1989).
127. C. Saragovi-Badler, F. Labenski de Kanter, M. T. Causa, S. M. Dutrús, C. Fainstein, L. B. Steren, M. Tovar, and R. Zysler, *Solid State Commun.*, 66: 381 (1988).
128. B. D. Dunlap, J. D. Jorgensen, W. K. Kwok, C. W. Kimball, J. L. Matykiewicz, H. Lee, and C. U. Serge, *Physica C*, 153–155: 1100 (1988).
129. B. D. Dunlap, J. D. Jorgensen, S. Segre, A. E. Dwight, J. L. Matykiewicz, H. Lee, W. Peng, and C. W. Kimball, *Physica C*, 158: 397 (1989).
130. E. Baggio-Saitovitch, R. B. Scorzelli, I. Souza Azevedo, and C. A. dos Santos, *Solid State Commun.*, 74: 27 (1990).
131. Yu. T. Pavlyuhin, N. G. Hainovsky, Y. Y. Medikov, and A. I. Rykov, *Pramana J. Phys.*, 31: L445 (1988).
132. C. E. Violet, R. G. Bedford, P. A. Hahn, N. W. Winter, and Z. Mei, *Physica C*, 162–164: 129 (1989).
133. K. Donnelly, J. M. D. Coey, S. Tomlinson, and J. M. Greneche, *Physica C*, 156: 579 (1988).
134. M. Takano, H. Mazaki, Z. Hiroi, Y. Bando, Y. Takeda, and O. Yamamoto, *J. Ceram. Soc. Jpn. Int. Ed.*, 96: 398 (1988).
135. M. Takano, Z. Hiroi, H. Mazaki, Y. Bando, Y. Takeda, and R. Kanno, *Physica C*, 153–155: 860 (1988).
136. P. Imbert, G. Jehanno, and J. A. Hodges, *Hyperfine Interact.*, 50: 599 (1989).
137. M. Pissas, V. Papaefthymiou, A. Simopoulos, A. Kostikas, and D. Niarchos, *Solid State Commun.*, 73: 767 (1990).
138. X. Z. Zhou, M. Raudsepp, Q. A. Pankhurst, A. H. Morrish, Y. L. Luo, and L. Maartense, *Phys. Rev. B*, 36: 7230 (1987).
139. Q. A. Pankhurst, A. H. Morrish, and X. Z. Zhou, *Phys. Lett. A*, 127: 231 (1988).
140. E. Mattievich, L. F. Moreira, M. F. da Silveira, R. F. R. Pereira, H. S. De Amorim, M. R. Amaral, Jr., and E. Meyer, *Hyperfine Interact.*, 42: 1254 (1988).
141. X. Z. Zhou, A. H. Morrish, Q. A. Pankhurst, and M. Raudsepp, *J. Phys.*, 49: C8-2213 (1988).
142. L. Er-Rakho, C. Michel, Ph. Lacorre, and B. Raveau, *J. Solid State Chem.*, 73: 531 (1988).
143. I. V. Zubov, A. S. Ilyushin, R. N. Kuz'min, V. S. Moisa, A. A. Novakova, and A. A. Bush, *Physica C*, 162–164: 37 (1989).
144. E. R. Bauminger, I. Felner, I. Nowik, and U. Yaron, *Physica C*, 162–164: 1281 (1989).
145. J. L. Dormann, S. Sayouri, G. T. Bhandage, S. C. Bhargava, G. Priftis, H. Pankowska, O. Gorochov, and R. Suryanarayanan, *Physica C*, 162–164: 1371 (1989).
146. S. Katsuyama, Y. Ueda, and K. Kosuge, *Physica C*, 165: 404 (1990).
147. P. Dalmas De Réotier, P. Vulliet, A. Yaouanc, P. Chaudouët, J. P. Sénateur, and F. Weiss, *Europhys. Lett.*, 11: 463 (1990).
148. J. Bardeen, L. N. Cooper, and J. R. Schrieffer, *Phys. Rev.*, 108: 1175 (1957).

12

Nuclear Magnetic Resonance Spectroscopy and Imaging of High-Performance Ceramics

Keith R. Carduner
Scientific Research Laboratories
Ford Motor Company
Dearborn, Michigan

Galen R. Hatfield
Washington Research Center
W.R. Grace & Company
Columbia, Maryland

William A. Ellingson and Stephen L. Dieckman
Materials and Components Technology Division
Argonne National Laboratory
Argonne, Illinois

INTRODUCTION

Nuclear magnetic resonance spectroscopy has recently become an established and powerful technique for the study of high-performance ceramics, specifically the high-refractory ceramics silicon carbide (SiC), silicon nitride (Si_3N_4), and related materials. These materials show great promise in high-temperature, high-strength applications of significant industrial value. NMR studies of these materials have largely fallen into two categories. For the purpose of this chapter, we have chosen to identify these two areas as "spectroscopy" and "imaging." *Spectroscopy* is typically applied to probe the structure and chemistry of ceramic materials on the atomic or molecular scale. While most of the applications of NMR spectroscopy to date have been in organic systems such as polymers and resins, this emphasis has been due primarily to historical reasons. As this chapter will show, NMR spectroscopy is finding new applications in solid inorganic systems, including the ceramics and ceramic composites. In these complex materials, NMR spectroscopy is capable of identifying both crystalline and amorphous phases, determining the structure of such phases, and studying the chemistry of these phases at the atomic level. Aside from this versatility, NMR spectroscopy provides several unique advantages, including the ability to study amorphous phase structure and the ability to observe a wide variety of nuclei.

The second area of NMR research into ceramic materials is NMR *imaging*. Imaging provides a macroscopic probe of ceramic materials that may be applied to determination of the distribution of molecular entities in a ceramic and to the mapping of porosity or voids. Generally speaking, production of reliable, high-quality ceramic components requires the development of more sophisticated diagnostic methods. Further advances in ceramics processing will require the detection of defects so that flaws can be examined and eliminated. Many high-volume processes, such as slip casting and injection molding, require detailed knowledge of the distribution of organics such as plasticizers and carriers. Thorough understanding of the defects in the green body stage of ceramics processing is desirable because subsequent processing, be it sintering, hot isostatic pressing, or machining, can be expensive and time consuming. These costs, if added to poor parts, create higher per part costs for acceptable parts. As shown in this chapter, NMR imaging is a noninvasive method that offers the potential of providing extensive diagnostic information about macroscopic spacial variations of the local chemical environment internal to both the green body and, even more exciting, to the finished ceramic itself.

The purpose of this chapter is to review many of the recent applications of NMR spectroscopy and imaging to highly refractory ceramics of significant industrial interest. Our emphasis is on applied research of high-performance ceramics and is not, therefore, a complete overview of the applications of NMR to ceramics. Although most of the examples will be drawn for published work of the authors, additional examples are selected from the literature to complete a description of the current art. Since solid-state NMR spectroscopy and NMR imaging are relative new fields, some time will be spent introducing each technique. For the imaging section, an overview of instrumentation is also provided. The review begins with a discussion of NMR spectroscopy, followed by a description of a series of examples. Following that, we present an overview of imaging, again supported by a series of examples. In the conclusion, some areas where the authors expect significant advancement over the next few years in the characterization of ceramic materials by NMR are discussed to provide readers with an outline of what to expect next.

BACKGROUND: SPECTROSCOPY

In this section we highlight many of the general concepts used when studying ceramic samples by high-resolution solid-state NMR. More complete reviews on the theory and practice of solid-state NMR are available. These include the applications-oriented book by Fyfe [1], monographs by Oldfield and Kirkpatrick [2] and Maciel [3] specifically addressing inorganic applications, a book emphasizing experimental technique by Fukushima and Roeder [4], and finally, an excellent description of the theory of NMR by Mehring [5]. Readers are referred to these texts for greater detail.

NMR Concept

The most basic description of NMR begins with the concept of a nucleus as a spinning charged particle with an associated magnetic moment. The direction of this moment is random in the absence of any magnetic field. However, if the nucleus encounters a magnetic field, B_0, its moment typically aligns itself either with or against B_0. The higher-energy state (against B_0) is less populated than the lower-energy state (with B_0), and it is possible to excite the nuclei by applying a "pulse" of energy. For a given nucleus in a given B_0, the frequency of energy that causes this transition varies slightly depending on the electronic environment of that nucleus. The local electronic environment can "shield" or "deshield" the nucleus from the applied field, creating a "shift" in frequency. These differences, or shifts, are measured in the NMR experiment and are aptly called *chemical shifts*.

Chemical Shift

The chemical shift, then, is sensitive to the local environment around the nucleus. Different environments will give rise to different chemical shifts. This environment is determined largely by the bonding network surrounding the nucleus, including the identity of neighboring atoms, the distances to those neighbors, and the angles defined by their location. As an illustration, consider two common ceramic materials, silicon carbide (SiC) and silicon nitride (Si_3N_4). In β-SiC, silicon nuclei are in a cubic site surrounded by carbon atoms. In β-Si_3N_4, however, they are in a tetrahedral site surrounded by nitrogen atoms. The different silicon environments should give rise to different ^{29}Si chemical shifts. In fact, the ^{29}Si NMR spectra of these materials reveal peaks at different shifts, as expected. As shown in Figure 1, the ^{29}Si chemical shift of β-SiC is -18ppm and the shift of β-$Si_3N/_4$ is -49 ppm. (The differences in line width is discussed below.)

Differences in chemical shift will also arise from more subtle differences in the nuclear environment. For example, there are two crystalline forms of silicon nitride, called α and β. Both structures consist of interleaved sheets of 8- and 12-membered rings of silicon and nitrogen. In the α form, each alternate sheet is inverted and offset slightly with respect to the underlying sheet, creating two unique silicon sites in the structure [6]. The ^{29}Si NMR spectrum of α-Si_3N_4 is given in Figure 2. Even though both silicon atoms are tetrahedrally surrounded by four nitrogen atoms, NMR is able to resolve them. Note that the chemical shifts are similar, reflecting the structural similarities. In the β form, the interleaved sheets of silicon and nitrogen are stacked in a regular manner. The result is that there is only one unique silicon site [7] and thus only one peak in the ^{29}Si NMR spectrum. This is also illustrated in Figure 2. Note that the chemical shifts for α and β are similar, reflecting the similarities in nuclear environment. The average Si—N bond

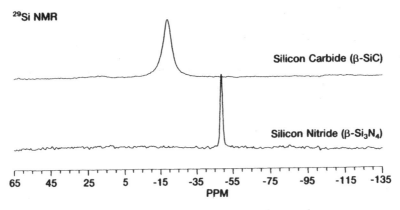

Figure 1   Silicon-29 spectra of silicon carbide (β-SiC) and silicon nitride (β-Si₃N₄).

lengths are 0.1733 nm for the β form and 0.1747 and 0.1740 nm for the α structure [6,7]. NMR is capable of distinguishing the silicon sites in each form even though the differences between them are very slight. As it turns out, the differences between these sites that influence the NMR spectra have been shown by Carduner et al. [8] to be related to differences in the configuration of the next-nearest-neighbor shell of silicon atoms. It

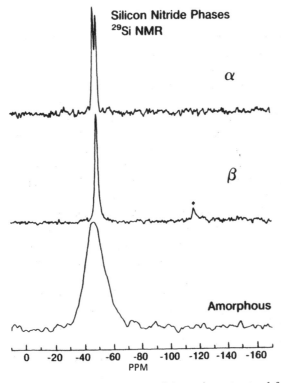

Figure 2   Silicon-29 spectra of the various structural forms of silicon nitride (Si₃N₄): α and β-crystalline and amorphorus. (Adapted with permission from Ref. 39.)

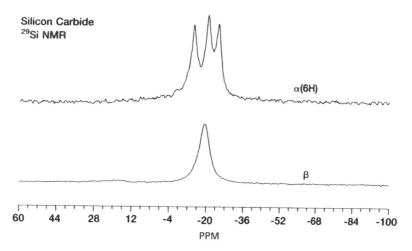

Figure 3 Silicon-29 spectra of β and α (6H) silicon carbide (SiC).

appears that the chemical shift of silicon in ceramic materials can probe the local environment out to 5 to 10 Å.

Silicon carbide (SiC), existing in the many crystalline forms discussed by Marshall et al. [9], provides another illustration. These forms (polytypes) are based on a cubic structure called β or a variety of hexagonal structures collectively called α. ^{29}Si NMR spectra of β-SiC and one α-SiC (6H) are given in Figure 3. The spectra are different, reflecting the different structural environments. Similar to α-Si_3N_4, the ^{29}Si NMR spectrum of α-SiC (6H) contains more than one line, reflecting subtle differences in the nuclear environment. As for α-Si_3N_4 (discussed above), Finlay et al. [10], Hartman et al. [11], and Guth and Petuskey [12] show that these three peaks arise from differences in the second-nearest-neighbor shell, roughly 0.5 nm away. This illustration reinforces the point that the environment is not simply the nearest-neighbor network, but may involve more distant interactions as well.

This type of sensitivity to local structure has led to many attempts [2,13–27] to correlate chemical shift with structural details such as bond angle [13–18] and type of next-nearest neighbors [19–22]. Correlations such as these as well as to other features of structure and bonding [23–27] hold considerable promise for the structural analysis of amorphous ceramic materials.

Crystallinity, Local Order, and Amorphous Phases

The examples above illustrate NMR spectroscopy of materials with known crystalline structure. However, NMR is not limited to systems with long-range crystalline order. Instead, it observes nuclei in any structural environment, crystalline or amorphous. The result is that NMR has the unique ability to obtain detailed structural information on the entire composition of a system. For example, the ^{29}Si NMR spectrum of amorphous (a) silicon nitride is given in Figure 2. The x-ray diffraction pattern of this sample revealed no crystalline features. The greater line width is related to heterogeneity in the local silicon environments of an amorphous phase. This can be explained simply by considering the nature of the chemical shift. For example, in β-Si_3N_4 there is one silicon site and one

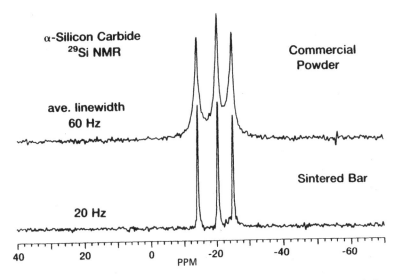

Figure 4   Silicon-29 NMR spectra of a silicon carbide (SiC) powder and sintered bar.

NMR line. In α-Si$_3$S$_4$, there are two sites and two lines. These lines have similar chemical shifts, reflecting structural similarities. Thus, in an amorphous system with many diverse but similar sites, we would expect to see many lines with similar chemical shifts. The result is a series of overlapping lines, or an increase in line width. In other words, the observed line width for a given system appears to depend on, among other things, the degree of local structural order. This is illustrated for powdered and sintered α-SiC in Figure 4. The sintered piece is expected to contain a higher degree of order at the atomic level, a fact that is reflected in the observed line width.

Quantitation in Ceramic Phases

Chemical shift and line width are not the only information obtained in an NMR experiment. The integrated intensity of each peak is also a sensitive indicator of structure. NMR is essentially a "counting" technique, and intensities depend on the concentration of the nucleus giving rise to the NMR peak. This can provide valuable information on both specific crystal structures and on samples that are mixtures. For example, in the ^{29}Si NMR spectrum of α-Si$_3$N$_4$ (Figure 2) there are two peaks, corresponding to two crystallographically distinct sites. The integrated intensities of these peaks are in a 1:1 ratio, revealing that the silicon sites are also present in a 1:1 ratio. This is confirmed in the crystal structure determination of Kato et al. [6]. NMR, then, is capable of obtaining information on the number and type of each site present within a phase. Information such as this is vital for characterizing new ceramic materials and phases that are not crystallographically well defined.

Integrated NMR intensities also reveal the concentrations of components in mixtures. This is important, among other things, for quantifying the presence of impurities. For example, the ^{29}Si NMR spectrum of amorphous Si$_3$N$_4$ with a high concentration of surface oxides is given in Figure 5. The surface oxides appear at roughly -100 ppm, the chemical shift characteristic of a tetrahedral arrangement of silicon–oxygen bonds. By integrating the areas beneath the peaks, we find that the surface oxides account for roughly 30% of

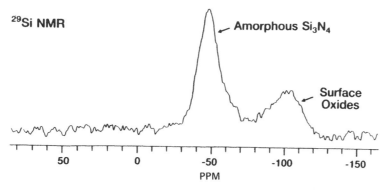

Figure 5 Silicon-29 NMR spectrum of an amorphous silicon nitride powder showing the presence of surface oxides.

the silicon in the sample. To turn the silicon headcount into a wt %, the relative formula weights of Si_3N_4 and SiO_2 must be taken into account. Correcting for the difference in the number of silicons in each structure (Si_3N_4 contributes three times as much signal as SiO_2 for equivalent numbers of formula units) indicates that the sample contains roughly 15 wt % surface oxide. This illustration emphasizes two important points: (1) NMR is capable of quantitating the various components present in a system, and (2) NMR is able to quantitate both crystalline and amorphous phases.

In closing, it should be noted that it is necessary to use proper experimental parameters when seeking purely quantitative information. Most important, care must be taken to account for differences in the NMR phenomenon called *spin-lattice relaxation*. The procedures for obtaining quantitative data have been well described elsewhere [1–5] and will not be repeated here.

Multinuclear NMR

All of the illustrations above have used ^{29}Si NMR, simply for introductory purposes. However, NMR is not limited to observing silicon but is capable of studying a wide variety of nuclei. Excellent overviews of the multinuclear capability of NMR are available [28–30]. Several examples relevant to ceramics include nitrogen (^{15}N), carbon (^{13}C), lanthanum (^{139}La), aluminum (^{27}Al), and yttrium (^{89}Y). The "observability" of these nuclei varies markedly, and most work has employed ^{29}Si, ^{13}C, ^{31}P, and ^{27}Al. The other nuclides have seen significantly less application, primarily because of experimental difficulties. Dupree et al. have employed ^{89}Y NMR to study yttria/alumina phases [31]. Recently, the same group used ^{139}La to study the La–Si–Al–O–N system [32]. In most cases, the fact that NMR is multinuclear makes it possible to obtain structural information on each nuclei in each phase present in a sample. For example, Hartman et al. [11] have obtained both ^{29}Si and ^{13}C NMR spectra of α-SiC. Their spectra are illustrated in Figure 6. As we have seen previously, the ^{29}Si NMR spectrum of the 6H polytype reveals three types of silicon in 1:1:1 ratio. It is most interesting to find that the ^{13}C NMR spectrum also reveals three types of carbon in a 1:1:1 ratio. Thus we expect to find structural similarities between the carbons and silicons. This is, in fact, the case for the 6H polytype [11]. However, NMR reveals that the structural environments are quite different in the 15R polytype. Here there are three silicon sites in a 1:2:2 ratio and at least five carbon

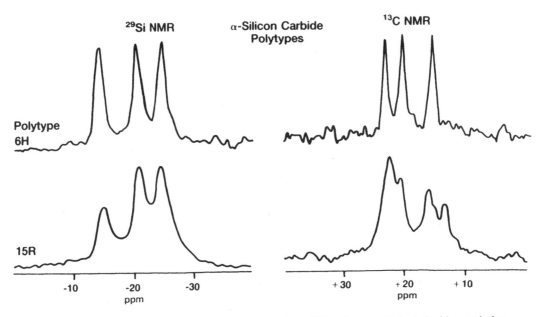

Figure 6 Silicon-29 and carbon-13 NMR spectra of two SiC polytypes. (Adapted with permission from Ref. 11.)

sites present in a complex ratio. These ratios have been used to predict [11] the structure of the 15R polytype. Carduner and Carter have used a similar multinuclear approach to study the 3C β-SiC phase [33] and have observed a single resonance in both the ^{29}Si and ^{13}C spectra, as expected on the basis of x-ray structural studies of this material. The multinuclear approach therefore allows for a more detailed characterization of the structure of SiC. The use of this multinuclear approach can be extended to many other ceramic materials as well.

Experimental Notes

One of the primary advantages of solid-state NMR is that samples can be examined ''as received.'' Minimal sample preparation is necessary. Samples can be powders, slips, fibers, or sintered pieces. The only requirement is that the sample fit into a cylindrical holder which is roughly 20 mm in length and 10 mm in diameter. Samples larger than this are typically cut or machined down to the appropriate size.

NMR is a bulk-sensitive technique. As a result, the data obtained are representative of the entire sample under study. Results obtained by traditional methods such as vibrational spectroscopy (ATR) and x-ray diffraction (XRD) are dependent on the sampling depth of the technique. The maximum sampling depth is roughly 0.5 nm for ATR and roughly 0.5 nm for XRD using typical copper radiation. In large samples where heterogeneity may exist between several microns from the surface and the interior, NMR is the method of choice. A good example of this concerns crystallization, which may occur at different rates for the surface and the interior.

For purpose of completeness, it is noted here that experiments carried out by the authors were performed on either a Chemagnetics CMX300, a Varian LX200, or a Bruker MSL300 NMR spectrometer. All spectra were acquired using standard single-pulse or cross-polarization techniques under magic-angle spinning (MAS) conditions. The details of these techniques have been well described in Refs. 1–5 and will not be repeated here. Pulse delays ranged from 2 s to 1 h. The samples investigated include fibers, sintered bars, powders, and slips. For details on the results from cited literature, readers are referred to the original texts.

APPLICATIONS: SPECTROSCOPY

In this section we outline a series of applications of NMR spectroscopy to the study of high-performance ceramics. The examples chosen are from the author's own work as well as from the work of others appearing in the literature. The examples are chosen to provide a feeling for the range of application of the technique to the study of structure and processing of ceramics and are by no means an exhaustive review of the published work.

NMR Studies of Lanthanum Silicon Nitride

Lanthanum silicon nitride ($LaSi_3N_5$) is a new type of ceramic material that is being investigated for its potential advanced heat and turbine engine applications. A structure for $LaSi_3N_5$ has been reported by Inoue et al. [34,35]. The ^{29}Si NMR spectrum of $LaSi_3N_5$ reported by Hatfield et al. [36] is reproduced as Figure 7. The spectrum consists of two peaks at -64.5 and -56.5 ppm that are present in a 2:1 ratio. This result indicates the presence of two "types" of silicon (as distinguished by NMR) in the sample. The spectrum is free of typical ^{29}Si impurity signals such as those for SiC and Si_3N_4. ^{13}C spectra were also acquired (not shown) and revealed that the sample was free of carbon signals. These results indicate that the sample is rather pure, a fact that has been confirmed by elemental analysis.

The observation of two ^{29}Si NMR peaks indicates the presence of two silicon types in the $LaSi_3N_5$ network. An understanding of the dominant influences affecting the chemical shift in $LaSi_3N_5$ will permit a detailed understanding of this and similar ceramic phases. In turn, this will enable one to probe the atomic-level changes induced by preparation, processing, and impurities.

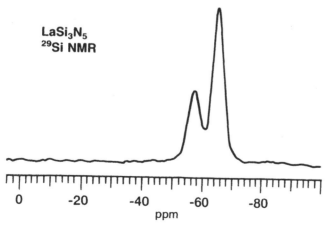

Figure 7   Silicon-29 NMR spectrum of $LaSi_3N_5$.

Hatfield et al. propose [36] that the ^{29}Si NMR peaks differ in chemical shift due to the interaction between lanthanum and silicon. The nitrogens in the network do not appear to affect the shift differences, since each silicon is tetrahedrally bound to four nitrogens [34,35]. However, within a 0.5-nm sphere about silicon, there exist two types of silicon characterized by the number of neighboring lanthanum atoms. One type, accounting for two-thirds of the total, resides in an environment with five neighboring lanthanum atoms, while the second type, accounting for the other third, has four lanthanum neighbors. These two types of silicon account for the two ^{29}Si NMR peaks. Comparing the relative concentrations of the two types of sites with the NMR spectrum, one concludes that the site with four lanthanum neighbors can be assigned to the peak at -54.5 ppm, while the site with five lanthanum neighbors can be assigned to the -64.5-ppm peak.

Studies such as this and several on SiC [10–12] and Si_3N_4 [8] suggest that a roughly 0.5-nm "sphere of influence" exists about silicon that can often be used to rationalize observed chemical shifts. This approach may prove to be valuable in other ceramic systems, as well. Changes in this "sphere," indicated by changes in chemical shift, may reveal important information on the nature of atomic-level interactions induced by processing, stresses, and so on. For example, note that for $LaSi_3N_5$ an increased Si-La interaction results in a more negative chemical shift.

Determination of Phase Composition in Silicon Nitride Powders

Silicon nitride (Si_3N_4) is an important ceramic, both for its own use and as starting material in the production of other ceramics and composites. Applications of this ceramic in autos and machine tools are discussed by Kalamasz et al. [37] and Milberg [38]. Si_3N_4 can be present in α or β crystalline forms and in an amorphous structure. In typical Si_3N_4 precursor powders, all three phases, along with impurities such as silicon oxynitrides and silicates, may be present. The physical properties of Si_3N_4 ceramics are very sensitive to the composition of the precursor powder.

For this reason it is necessary to have a rapid, reliable method for determining the composition of batches of Si_3N_4 powder before carrying out subsequent processing steps, including final product formation and sintering. Traditionally, x-ray powder diffraction (XRD) has been used. However, XRD often fails in the identification of amorphous species. Recall from Figure 2 that NMR, on the other hand, is readily capable of observing and quantifying all of the various phases of Si_3N_4.

^{29}Si NMR spectra of six Si_3N_4 powders are given in Figure 8. All of the spectra are dominated by Si_3N_4 lines at roughly -50 ppm. In some samples, other silicon species, such as oxynitrides and oxides, are observed and can readily be identified by their chemical shift. These are also identified in Figure 8, along with the results of an NMR phase composition analysis on each sample. When these values were compared with XRD analyses, Carduner et al. found [39] that as much as 30% amorphous phase was not detectable by XRD. As a result, the XRD-determined α/β ratio was typically 10% lower than that measured by NMR. The ability to characterize the phase composition of any ceramic material completely and accurately is an important link to understanding its physical properties.

Analysis of Ceramic Fibers

There has been considerable interest in the use of ceramic fibers as reinforcement in high-temperature composite materials. Rauch et al. provide an early, but still informative overview of this field [40]. Characterization of these materials, however, is often difficult

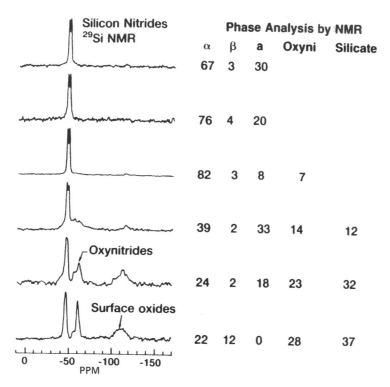

The following table summarizes the Phase Analysis by NMR shown in the figure:

α	β	a	Oxyni	Silicate
67	3	30		
76	4	20		
82	3	8	7	
39	2	33	14	12
24	2	18	23	32
22	12	0	28	37

Figure 8 Silicon-29 NMR spectra of various commercial silicon nitride powders. (Adapted with permission from Ref. 39.)

by traditional methods since they are insoluble, predominantly amorphous, and often present experimental difficulties due to their sample state. Solid-state NMR is being widely used to probe these materials. Reports on the application of NMR techniques to analysis of ceramic fibers are given by Inknott et al. [41], Lipowitz et al. [42], West and Maxka [43], and more recently by Murthy et al. [44]. Two representative examples are described below.

NMR Analysis of Fiber Structure

One method of preparing SiC fibers is by the pyrolysis of polymeric precursors. Three fibers prepared via this route have been studied by Lipowitz and Turner [45] using ^{29}Si NMR and the results are shown in Figure 9. The top two spectra are standard NICALON commercial Si—C—O fibers, with the top (SGN) one containing roughly 15 wt % oxygen and the middle (CGN) containing about 10 wt % oxygen. In the Si—C—O fibers, there are five possible tetrahedral structures for the silicon atoms if only Si—C and Si—O bonds are present (SiC_4, SiC_3O, SiC_2O_2, $SiCO_3$, and SiO_4). All appear to be present in the Nicalon fibers. Oxygenated structures are more prevalent in SGN than in CGN, reflecting the higher oxygen concentration. Note that the most intense signal in both fibers occurs close to the chemical shift of SiC (-18 ppm), showing SiC to be the dominant species. The bottom spectrum is of a Si—N—C fiber derived from a hydridopolysilazane polymer. Note that the spectrum of this sample is similar to that of amorphous Si_3N_4 (Figure 2), indicating that α-Si_3N_4 is the dominant species.

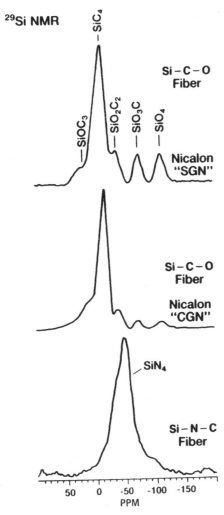

Figure 9 Silicon-29 spectra of three ceramic fibers. (Adapted with permission from Ref. 45.)

NMR Analysis of Fiber Processing

The preparation of silicon carbide fibers from a polycarbosilane precursor has been studied using NMR by Taki et al. [46,47]. Polycarbosilane (PC) consists of a skeleton of alternate carbon and silicon atoms. The PC fibers are obtained first by melt-spinning of PC, and then are cured in air at 100 to 200°C. The oxygen introduced during the oxidation curing process influences the tensile strength of the resulting SiC fibers. Thus it is important to determine the chemical structure of the oxidation-cured PC fibers and to better understand the curing process.

The ^{29}Si NMR spectrum of untreated PC fiber is given in Figure 10. There are largely two silicon types in PC, as shown below.

$$\begin{array}{ccc} & \overset{\displaystyle CH_3}{\underset{\displaystyle |}{}} & & & \overset{\displaystyle H}{\underset{\displaystyle |}{}} \\ -CH_2-\!\!\!\!&Si&\!\!\!\!-CH_2- & \qquad & -CH_2-\!\!\!\!&Si&\!\!\!\!-CH_2- \\ & \overset{\displaystyle |}{\underset{\displaystyle CH_3}{}} & & & \overset{\displaystyle |}{\underset{\displaystyle CH_3}{}} \end{array}$$

These two silicon types will hereafter be referred to as "SiC$_4$" and "SiC$_3$H." Assignments are indicated in Figure 10. The peak marked with an asterisk is due to polydimethylsilane, which is used as an external chemical shift reference. Spectra of PC fibers heated at 160 and 175°C in oxygen are also given. The major change observed upon curing is the formation of Si—O—Si bonds. The intensity for these bonds increases linearly with the rise of curing temperature. As the intensity of the Si—O—Si bonds increases, the intensity for SiC$_3$H decreases, indicating that the oxygen atoms attack the Si—H bond of the SiC$_3$H site and consequently convert it to an Si—O—Si bond through a cross-link to a neighboring chain.

As mentioned earlier, the presence of Si—O—Si bonds lowers the mechanical properties

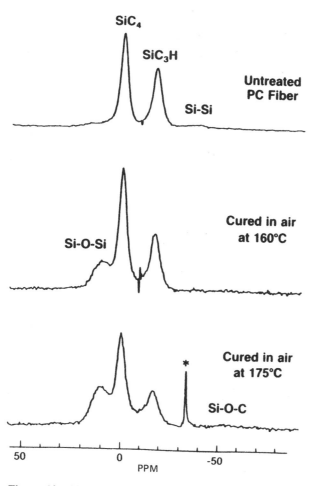

Figure 10   Silicon-29 spectra of untreated and cured PC fibers. (Adapted with permission from Ref. 47.)

of the SiC fiber at high temperatures. In an attempt to reduce the amount of silicon oxide in the SiC fiber produced, radiation curing of the PC fiber was also investigated [46]. The ^{29}Si-NMR spectra of untreated and irradiated fibers are shown in Figure 11. In the electron irradiation-cured PC fibers, Si—O—Si signals were not observed. However, the relative intensity of the SiC$_3$H site decreased linearly with an increase in dose. This indicates that the Si—H bond is broken by the electron irradiation and that either Si—C or Si—Si bonds are formed.

Following cure, the cross-linked fibers are annealed at high temperature to induce conversion to SiC. Again employing ^{29}Si NMR, Taki et al. [48] show that the conversion process involves preferential loss of the Si—H linkage beginning at 500°C with total conversion to an amorphous SiC phase by 1200°C. Similarly, nitridation of PC was also studied [49]. In a series of NMR spectra covering treatments from 500 to 1000°C, it appears that the nitridation starts with attack of the Si—H linkage with subsequent pyrolysis and nitridation of SiC$_4$ units. The nitridation reaction to form Si$_3$N$_4$ is virtually complete at 700°C.

NMR Investigation of Dispersion Aid Mechanisms

Many ceramic powders do not sinter well due to heterogeneities such as particle-size distribution and uneven dispersion. One approach in solving this problem has been to suspend the powder in a slurry containing a *dispersion aid* (DA). DAs typically contain

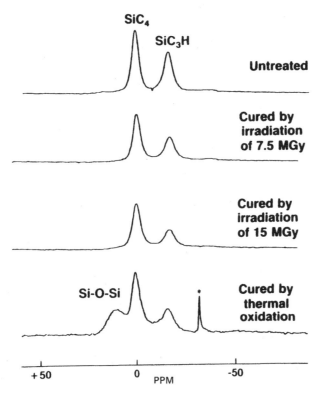

Figure 11 Silicon-29 NMR spectra of untreated and radiation-cured PC fibers. (Adapted with permission from Ref. 46.)

a long-chain hydrophobic moiety and a reactive site that can "couple" with the inorganic solid. In concept, the DA then forms a long-chain monomolecular layer around each particle, thereby improving dispersion. This is illustrated in Figure 12. When the solution is removed from the slurry, the DA acts to prevent phase separation, to promote adhesion, and to create a more uniformly dispersed ceramic. This typically results in improved physical and mechanical properties, such as rheology and impact strength of the final sintered ceramic. An understanding of the interaction between the DA and the ceramic particle is vital for optimizing these properties. NMR is uniquely capable of studying the DA in solution, as part of a slurry and as a solid within the ceramic network.

In this illustration, ^{31}P NMR is used to follow the interaction between zirconia powder and a titanatepolyphosphate DA reported to have the following structure:

$$CH_2{=}CH{-}CH_2O{-}CH_2$$
$$CH_3CH_2{-}C{-}CH_2O{-}Ti[O{-}P{-}O{-}P(OC_8H_{17})_2]_3$$
$$CH_2{=}CH{-}CH_2O{-}CH_2 \qquad OH$$

The ^{31}P NMR spectrum of the dispersion aid in water is given in Figure 13 (bottom). This spectra, acquired using standard solution-state NMR methods described by Yoder and Schaeffer [50] under conditions of proton decoupling, contains many ^{31}P peaks. The presence of more than one peak indicates the presence of many ^{31}P moieties. This result is inconsistent with the proposed single structure provided by the supplier. In fact, the spectrum is consistent with partial polymerization into a mixture of polyphosphates of linear, branching, and ring structures, as, for example, described by Villa et al. [51]. FAB-MS studies agree with the conclusion from the NMR that the solution, as received, is a mixutre of polyphosphate components.

A complete characterization of the DA is still required. However, even without this information, several interesting points can be observed from Figure 13. The spectrum of the DA is dominated by three major ^{31}P lines (A, B, C), indicating three major "types" of ^{31}P sites. When the DA is suspended in a slurry with the ceramic, two of these peaks (A, C) remain, while the third (B) is significantly reduced. Definitive assignments of the

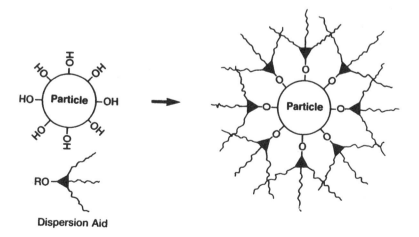

Figure 12 Schematic illustration of the dispersion aid–ceramic particle interaction.

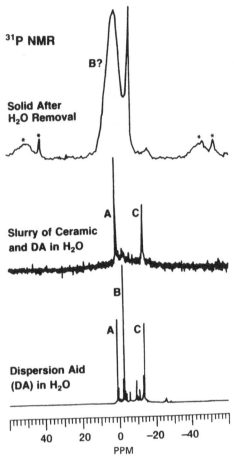

Figure 13 Phosphorus-31 NMR spectra of a dispersion aid as a solution, a slurry, and as a solid (see the text).

peaks are necessary to interpret this observation. However, one might speculate that the ^{31}P site which gives rise to peak B is now bound to the solid ceramic surface and cannot be ''seen'' by traditional solution-state NMR. When the water is removed from the slurry, a solid mixture of ceramic and DA remains that can be studied by a solid-state ^{31}P NMR (employing high-power proton decoupling [1–5]). In the spectrum of the solid material (top, Figure 13), peaks marked with an asterisk are due to spectral features called spinning sidebands and should be ignored. The ^{31}P spectrum of the solid is dominated by two signals, one sharp and one broad. Again, without a detailed structure determination and spectral assignment, it is difficult to interpret this result. However, one might speculate that the broad line is due to B bound to the ceramic and that the narrow line is due to ''free'' B in the network.

The results shown here are clearly preliminary and the interpretation speculative. However, this example demonstrates the ability of NMR to study a complex type of ceramics processing in which the material is first prepared in a solution state, partially densified in a slurry, and then compacted in a solid. Since NMR is amenable to all of these processing steps, it is possible to gain information on the dispersion aid in each of its states and, potentially, to fully characterize the DA/ceramic interaction. Another example of a similar problem studied using NMR imaging is presented in the imaging section of this chapter.

Structure of SiAlON Ceramics

''SiAlON'' ceramics are materials composed of three-dimensional arrangements of $(Si,Al)(O,N)_4$ tetrahedra and occasionally contain aluminum in a sixfold octahedral coordination. These ceramics are of interest because of their thermal and mechanical properties. Properties such as the glass transition temperature, hardness, and elastic modulus are dependent on the substitution of nitrogen for oxygen in the local structure of oxynitride glasses. However, the structure of many of these phases remains unknown. Direct evidence for the incorporation of nitrogen in the network is sparse and there is no direct evidence of aluminum–nitrogen bonding. Traditional x-ray diffraction methods have been hampered by similar x-ray scattering factors for aluminum and silicon. Solid-state NMR, however, is providing a new and powerful method for probing the structure and chemistry of these materials. Many studies of these materials have appeared, [52–59]. Two examples of this work are described below.

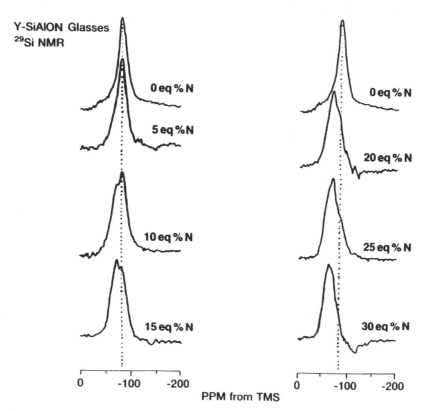

Figure 14 Silicon-29 NMR spectra of a series of Y–Si–Al(O,N) glasses with increasing substitution of nitrogen for oxygen. (Adapted with permission from Ref. 52.)

For example, consider a series of Y–Si–Al–(O,N) glasses of increasing nitrogen content. ^{29}Si NMR spectra of several SiAlON materials with constant Y:Si:Al ratio and increasing substitution of nitrogen for oxygen are given in Figure 14. NMR is capable of identifying the discrete $(Si,Al)(O,N)_4$ structural units in these materials. Each spectrum is composed of several partially resolved peaks, indicating the coexistence of a variety of $Si(O,N)_4$ structural groups. Aujla et al. present a detailed characterization of the chemical shifts [52]. For now it can be said that the observed spectra are consistent with the occurrence of $Si(O_3N)$ and $Si(O_2N_2)$ groups together in the network. Note that with increasing nitrogen substitution, there is a systematic shift in intensity that can be explained as nitrogen progressively entering the network in a threefold coordination bonding to two silicon and one aluminum atom. The data obtained also reveal a preference for Si—N bonding compared with Al—N bonding.

^{27}Al NMR has also been used to study SiAlON systems. Figure 15 shows the ^{27}Al NMR spectra of two SiAlONs, referred to as "X-phase" and as a "polytypoid" [53]. The signals attributed to octahedral AlO_6 and tetrahedral AlO_4 and to AlN_4 are indicated. Peaks marked with an asterisk are due to spinning sidebands and can be ignored. For the X-phase, the observed 1.9:1.0 intensity ratio of tetrahedral:octahedral aluminum is in complete agreement with the structure, which provides for 20 to 22 tetrahedral and 12 octahedral aluminum atoms per unit cell. The spectrum of the polytypoid indicates the presence of only two types of aluminum. The small signal at 2.3 ppm is unambiguously attributed to AlO_6 octahedra, and the intense signal at 108.2 ppm is tentatively assigned to aluminum coordinated by four nitrogen atoms. The intensity ratio of the two signals is 9:1. The results of this preliminary examination show that NMR can readily and quantitatively distinguish the various types of aluminum present in complex oxynitride ceramics.

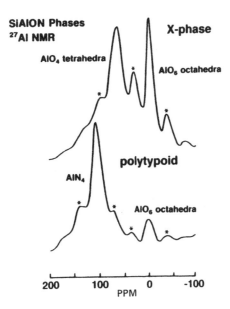

Figure 15 Aluminum-27 NMR spectra of SiAlON X-phase and polytypoid. (Adapted with permission from Ref. 53.)

Sintering of Si_3N_4 and Grain Boundary Phases

The densification process used to turn Si_3N_4 powders into structural ceramics is called *sintering*. In this process, the precursor powder is heated in the absence of oxygen and often under high pressure, converting α-Si_3N_4 into β-Si_3N_4. Since pure α-Si_3N_4 will not sinter, this process is often performed by mixing α-Si_3N_4 powder with 5 to 10% of a sintering aid. One common aid is Y_2O_3. A small amount of the powder reacts with the sintering aid to form new compounds that tend to collect on the grain boundaries between microcrystals of β-Si_3N_4. These intergranular phases have important consequences on the physical and thermal properties of Si_3N_4 components and thus need to be identified and quantified. Carduner et al. [60] have demonstrated that ^{29}Si MAS NMR is readily used to detect, identify, and quantify Y–Si–O–N phases associated with the grain boundaries. It should be noted that NMR is capable of observing these phases but cannot distinguish them from similar phases that may be present in other parts of the sample. It had previously been shown that the major source of Y–Si–O–N phase impurity is at the grain boundary of these sintered ceramics. Thus the assumption is made that the NMR signals are due to phases at the grain boundaries.

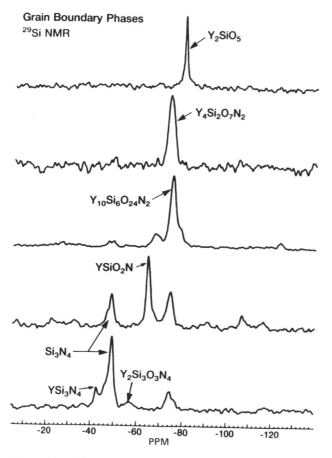

Figure 16 Silicon-29 NMR spectra of potential grain boundary phases which occur upon sintering Si_3N_4 with Y_2O_3.

The ^{29}Si MAS NMR spectra of potential grain boundary phases are given in Figure 16. Only the spectra of Y_2SiO_5 and $Y_4Si_2O_7N_2$ show pure materials. All of the other samples consist of mixtures of the phases. The resonant frequencies of these materials are well resolved, permitting ready indentification in the sintered solid. Carduner et al. [60] and Dupree et al. [61] discuss the assignment of these resonances to specific species.

^{29}Si NMR spectra of two sintered Si_3N_4 samples are given in Figure 17. In agreement with the known conversion of α to β during sintering, the similarity between the spectra of ceramics A and B with the β powder in Figure 2 is not surprising. The line width is narrowed from 150 Hz for the powder to 116 Hz for the ceramic, reflecting an increase in local order as discussed earlier. For ceramic B, a shoulder indicating the presence of some α-Si_3N_4 is just visible. Note the presence of phases associated with the grain boundaries in both ceramics. The intergranular Y–Si–O–N compounds produce less intense features, due to their lower relative concentration and are expanded in the figure. Integration of these peaks permits determination of the wt % of these phases, and the results are listed below.

Ceramic	α-Si_3N_4	β-Si_3N_4	$YSiO_2N$	$Y_{10}Si_6O_{24}N_2$
A	30.5	64.8	4.7	—
B	—	87.8	—	12.2

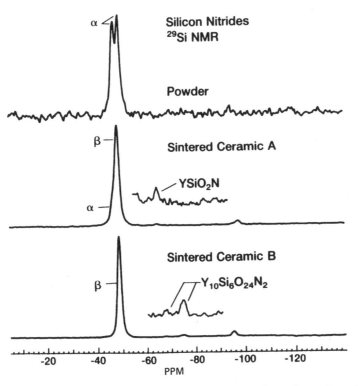

Figure 17 Silicon-29 NMR spectra of a Si_3N_4 powder and two sintered pieces.

For sample B, the determined percentage of Y–Si–O–N phase is virtually identical to the theoretical wt % of 12.9%. However, in sample A, the determined 4.7% is significantly less than the expected 14.4% based on stoichiometry. This discrepancy may simply be related to the presence of paramagnetic impurities. The effects of these on the NMR of ceramics are discussed in a later section.

In the sintering of many ceramics, subtle differences in the intergranular phases have a profound effect on the physical properties. Techniques capable of nondestructively identifying and quantifying these phases are crucial. NMR provides an important and powerful tool for probing solid-state chemistry in these systems.

Relationship of NMR to TEM and XRD Analysis of SiC Powders

Precursor β-SiC powders are routinely characterized prior to sintering for their average particle size, particle-size distribution, and percentage of α phase. Recently, sintering studies using a variety of commercially available β-SiC powders have shown that the sintering behavior depends also on the degree of local order of the initial β powder. The sensitivity of the line width of the NMR spectrum to local order, as portrayed in Figure 4 and associated discussion, makes this technique ideal for rapid determination of local order. In one study of seven commercially available β-SiC powders, Carduner et al. [62] documented the correlation between local order and its appearance in NMR spectra, x-ray diffraction (XRD) peaks, and transmission electron microscopy (TEM) images. A selection of the results for two of the powders is given in Figure 18. The narrower NMR peak is associated with more symmetrically formed powder granual as well as better microscopic order, as indicated by the sharp spots in the electron diffraction pattern. The NMR line width between these two cases changes by 600%, in comparison with approximately a 50% change in the line width of the XRD peaks.

The interpretation of the change in the line width of the NMR spectra for these two powders depends on the analysis from the TEM work. The advantage of the NMR, after suitable "calibration" by the TEM, is the ability to assay a relatively large sample of powder (ca. 1 g) rapidly and quickly determine the average local order. The NMR can then be used routinely to analyze batches of β-SiC powder as a check on consistency with TEM applied when significant variation from the desired NMR spectral line width is observed.

Effect of Paramagnetic Impurities on NMR Spectra of Ceramics

It was noted several times throughout the chapter that in some spectra, features called *spinning sidebands* (SSBs) were present that should be ignored. Adequate explanations for the appearance of SSBs require consideration of the mathematics of coordinate rotations and are beyond the scope of this section. An excellent discussion is provided by Maricq and Waugh [63]. Simply put, they are a direct result of the sample rotation during MAS data acquisition. In addition, for a given rotation speed, the relative intensity of an SSB is related to the local "anisotropy" about the given nucleus. Unfortunately, anisotropy can take on at least two meanings, including electron distribution anisotropy due to the bonding geometry around the nucleus as well as longer-range spatial inhomogeneity of the magnetic field resulting from large ferromagnetic impurities in the ceramic sample. In the former case, then, sidebands contain information about molecular structure, while in the later, information is present relating to impurities.

Generally speaking, sidebands related to electronic distribution anisotropy for [29]Si in

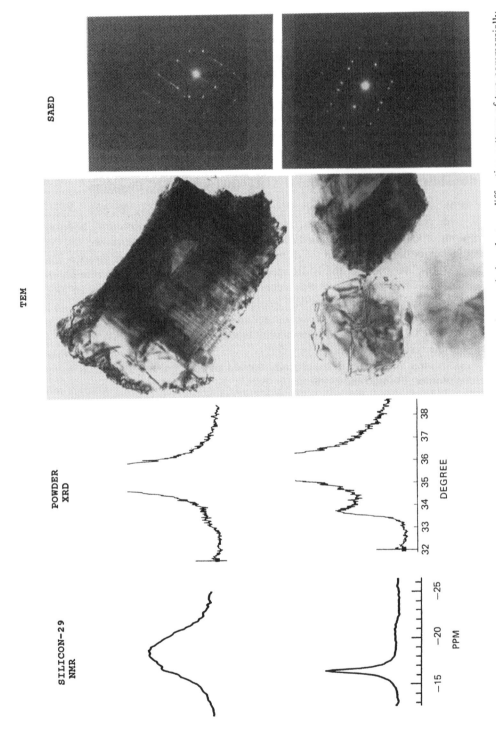

Figure 18 Comparison of ^{29}Si NMR spectra, XRD peaks, TEM images, and transmission electron diffraction patterns of two commercially available β-SiC powders.

ceramics are small, owing to the tetrahedral symmetry that typically describes the silicon atom environment. Nevertheless, we have observed that in silicon carbide (SiC) powders, for example, the appearance of SSBs is sample dependent. Typical examples of this can be seen in Figure 19 for two commercially available β-SiC powders. The peak marked with an arrow is due to the central (isotropic) resonance for β-SiC, the peak that has been discussed in this chapter, and peaks marked with an asterisk are due to SSBs. According to the manufacturer, the only difference between the samples is particle-size distribution and average particle size. This is borne out by the fact that the peaks in both the upper and lower traces have the same chemical shift. For this reason, it is thought that the SSBs are related to the presence of paramagnetic impurities. Oldfield et al. have discussed the occurrence of SSBs in the presence of ferromagnetic impurities that perturb the static magnetic field [64] and can give rise to SSBs. If the impurity is removed, one expects the sidebands to go away. In fact, rinsing of the β-SiC powder in the lower trace with concentrated acid will remove the SSBs. SSBs like these may be expected in any type of ceramic sample, owing to the prevalence of ball milling as part of ceramic powder preparation.

A further effect of paramagnetic impurities is to affect spin-lattice relaxation. For ^{29}Si in ceramics, this parameter can be as long as a few hours, with a significant negative impact on the rate of signal averaging. Paramagnetic impurities can reduce the T_1 significantly and, in the extreme, result in significant line broadening of spectral features. In ceramics it is possible that the paramagnetic impurities are not distributed uniformly through the sample. Sintered ceramics, for example, can be characterized by a higher concentration of impurities at the grain boundaries than in the grains themselves. The

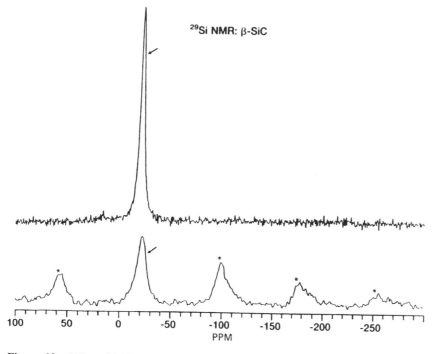

Figure 19 Silicon-29 NMR spectra of two commercially available β-SiC powders.

discrepancy between the expected concentration of a grain boundary phase and that actually observed, as discussed in one of the sections above, may result from the presence of sufficient paramagnetic impurities that broaden part of the NMR signal beyond observability.

BACKGROUND: IMAGING

For the remainder of this chapter, the application of NMR imaging to ceramics will be discussed beginning with generation of NMR images and an overview of instrumentation required for NMR imaging of ceramics.

Pulsed Magnetic Resonance Phenomenon

Nuclear magnetic resonance imaging, also called magnetic resonance imaging (MRI), can provide sectional (planar) or multidimensional (whole-body) images from three-dimensional objects. Both Foster and Hutchinson [65] and Mansfield and Morris [66] provide excellent descriptions of the principles of imaging and review a variety of imaging techniques. Unlike most x-ray computed tomography (CT) machines, in which moving parts are used, MRI has no moving parts. It operates by rapid electronic switching of magnetic coils and pulsed radio-frequency (RF) transceiver coils. In addition, unlike other conventional tomographic imaging modalities (x-ray or ultrasound), MRI can map both the chemical and physical properties throughout the sample volume. Before discussion of the imaging aspect of magnetic resonance, the basic concept of pulsed magnetic resonance phenomenon will be briefly reviewed. Although many of the concepts discussed here are relevant to NMR spectroscopy are discussed above, a through understanding is necessary to appreciate NMR imaging.

Magnetic resonance is a quantum mechanical phenomenon that is exhibited by atomic nuclei that have an odd number of either protons or neutrons. Nuclei with an odd number of protons or neutrons possess a spin that inherently gives them a magnetic field vector, albeit a very small one, like the axis of a spinning top. In a classical sense, spin-possessing nuclei are thought of as spinning charges with an associated magnetic field. If a magnetic field is externally applied to these spinning nuclei, a weak torque is exerted on them and they tend to align with the direction of the applied magnetic field, as illustrated in Figure 20a. For various reasons, not all nuclei align with the field, and therefore a resultant magnetic moment vector results. It is evident that the strength of the resultant magnetic vector is proportional to both the population of nuclei and the strength of the magnetic field.

During the alignment process, the nuclei oscillate, or process, about the externally applied magnetic field vector. The precession rate is not random but well behaved and directly proportional to the strength of the applied magnetic field:

$$\omega_L = \gamma B$$

where $\omega_L = 2\pi\nu$ and ν is the Larmor frequency (kHz), γ is the gyromagnetic ratio (kHz/G, an intrinsic physical constant of the nuclei of interest), and B is the applied magnetic field strength (G).

The bulk equilibrium magnetization contains no net oscillatory magnetization since the phases of the precessions are random. Therefore, to detect an NMR signal, the net bulk magnetization vector must be perturbed away from the equilibrium position. This step is

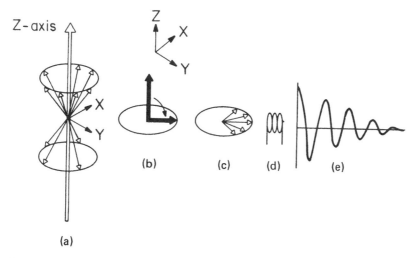

Figure 20 Schematic diagram of the pulsed NMR phenomena and detection of the NMR signal: (a) alignment of nuclei with external magnetic field B_0; (b) perturbation of the bulk magnetic vector into the *x-y* plane; (c) dephasing of the bulk magnetization vector in the *x-y* plane; (d) transceiver RF coil used to perturb the magnetic vector and receive the NMR signal; (e) typical free induction decay NMR signal detected.

achieved by applying a second, pulsed magnetic field via an RF coil (Figure 20b). Usually, this second field is much weaker (several thousandfold) than the main magnetic field, B_0 (traditionally, this is also the *z*-axis).

To achieve a perturbation at all, owing to its small magnitude this second field must possess certain unique properties. It is chosen such that the frequency of the applied RF exactly matches the Larmor frequency, ν_L. For hydrogen protons, it is 42.577 MHz at 10,000 G. Application of a carefully chosen RF at the Larmor frequency can perturb the bulk magnetization and generate an oscillatory magnetic component. The nuclei, in fact, precess at the Larmor frequency about the new net field, which is the vector sum of the static and applied RF.

If the RF field is turned off precisely when the angle of nutation reaches 90°, the magnetization is left in the transverse, or *xy*-plane (Figure 20b). Because the individual nuclei have the same evolutionary history, they are now phase coherent (i.e., they precess in exactly the same manner and in the same plane). Consequently, the net bulk magnetization vector now precesses about B_0 in a transverse plane where the *xy*-plane component can be detected. Thermal relaxation of the bulk net magnetization from coherence occurs both in the *xy*-plane (Figure 20c) and away from the *xy*-plane (back along the *z*-axis) and is characterized by two relaxation rates, known as T_2 and T_1, respectively. The spin-spin (T_2) and spin-lattice (T_1) relaxation rates depend on intra- and intermolecular processes and therefore provide important information concerning the local chemical environment of the nuclei of interest.

Experimentally, the NMR signal is excited by a tuned RF coil with an axis perpendicular to B_0 (Figure 20d). The coil used for excitation is also suitable for detection (a transceiver), as is a separate, mutually orthogonal coil. The oscillating NMR magnetization induces a signal in the RF coil. The induced signal immediately following an RF pulse, termed a *free induction decay* (FID), produces a time-domain (analog) signal decay (Figure 20e) as the nuclei relax back to equilibrium. The decay of the oscillatory signal is characterized by T_2, while T_1 determines the intensity of subsequently generated FIDs.

Producing NMR Images

To obtain an image by NMR, the information detected must be spatially encoded. This process is accomplished by applying a third magnetic field over the sample volume in the form of a magnetic field gradient. Since the Larmor frequency is proportional to the applied magnetic field B_0, if B_0 can be made to vary with spatial position, the Larmor frequency will also change with spatial position. This behavior can be seen in the following discussion. Suppose that the magnetic field varies in a direction normal to the B_0 direction (e.g., x). Then

$$B_z(x) = B_0 + xG_x$$

where B_z is the z component of the magnetic field, B_0 the static field, and G_x the magnetic field-gradient strength (G/cm). However, recalling Eq. (1), $\omega_L = \gamma B$, and substituting Eq. (2) into Eq. (1) gives

$$\omega_L = \gamma(B_0 + xG_x)$$

Equation (3) implies that the Larmor frequency is a function of spatial position, x. Thus if an RF receiver is used to detect the NMR signal and the signal is subsequently Fourier transformed, the signal strength can be determined accurately as a function of position. Note that this procedure makes clear demands on the bandwidth of the RF electronics, that is, that

$$\delta\nu = \gamma LG$$

where $\delta\nu$ is the spectral bandwidth of the receiver, γ the gyromagnetic ratio (kHz/G), L the largest sample dimension (cm), and G the magnetic field gradient (G/cm).

It is the application of magnetic field gradients for spatial encoding of the data that distinguishes MRI from NMR spectroscopy. With pulsed-field gradients in orthogonal directions and appropriate RF pulse sequences, sectional images are obtained that are similar to those generated by x-ray CT. As with x-ray CT, the image reconstruction is accomplished with various digital computer reconstruction algorithms, depending on how the data were obtained. One method of acquiring data is by back projection, exactly analogous to x-ray CT, as shown schematically in Figure 21. Representative spectra corresponding to changing the angle of the gradient through the sample are shown. A more elegant approach, and in fact the basis for most commercial NMR imaging instrumentation, is the use of switched gradients in three dimensions to uniquely encode the location of each pixel of the image. This technique, illustrated in Figure 22, is called Fourier imaging or NMR zuematography and was first described by Kumar et al. [67]. The three orthogonal gradients are switched on at separate times and the experiment is repeated with either an increase in times t_1 and t_2 or programmed stepwise increase in the respective gradient amplitudes between repetitions. The resultant data sets are then reconstructed into images using complex multidimensional Fourier Transforms.

Instrumentation for NMR Imaging: An Overview

A complete MRI system is shown schematically in Figure 23. Note that all three magnetic fields are shown: the static field B_0, the pulsed field gradients (G_x, G_y, and G_z), and the pulsed RF system. Bottomley [68] showed that resolution of the NMR image depends on

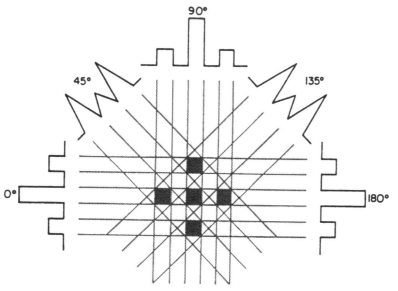

Figure 21 Schematic diagram showing the acquisition of NMR back projection data.

the strength of the field gradient, the spin-spin relaxation time T_2, as well as the gradient switching time. When imaging a liquid such as the protons in water distributed around a human body, the long T_2 of the protons in this molecule allow for the use of field gradients that are significantly smaller and switching times generally longer than can be used when imaging solids, such as ceramics, where T_2 times are much shorter. For this reason, the application of imaging to solids has followed the development of systems

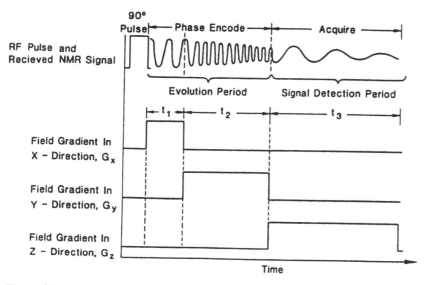

Figure 22 Idealized Fourier imaging pulse sequence as proposed by Kumar et al. [67].

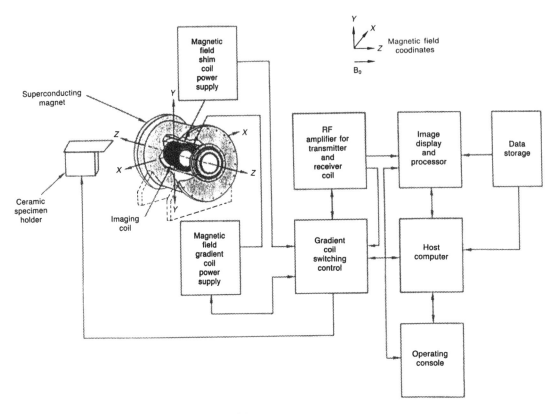

Figure 23 Schematic diagram of an MRI system.

employing higher field gradients (up to 60 G/cm) and rapid gradient switching times (under 0.5 ms). In addition to this trend in the development of imaging for solids, since resolution is roughly proportional to $1/T_2$, a variety of pulse techniques have been developed to narrow the intrinsically broader resonances of solid-state NMR lines in the hopes of increasing resolution.

APPLICATIONS: IMAGING

In this section we develop a series of applications of NMR imaging to ceramics processing. This is still very much an emerging field and it is hoped that the examples provide some insight into areas of active research.

Optimizing Slip Casting Rates

When slip casting ceramics, the casting rate of the slip and knowledge of the location of the cast surface/slip interface are critical for optimizing defect-free part formation. The efficacy of plaster molds used in slip casting depends on the amount of solids that fill the open pores as well as other parameters. The efficiency of the mold is reflected in the rate of solids buildup at the mold wall. Because most slips are water based (hence an abundance of ^1H protons), MRI provides an excellent way to observe the casting process in near-real-time.

Hayashi et al. [69] have described the use of a 1.0T MRI system to examine an Al_2O_3

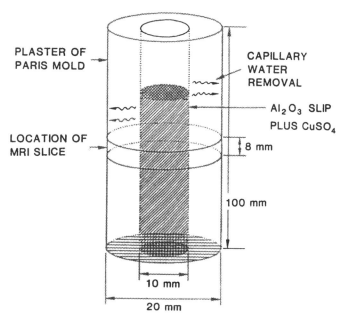

Figure 24 Schematic diagram of the slip-casting NMR experiment. (Adapted with permission from Ref. 69.)

slip in a plastic of paris cylindrical mold (20 mm OD and 10 mm ID). A schematic diagram of the mold and position of the MRI "slice" is shown in Figure 24. In this case, the water content of the slip was 30 wt % and had $CuSO_4$ added to reduce the T_1 of the water. The experiment consisted of observing the cross section of the mold noted in Figure 24 at different time intervals. The slice thickness was 8 mm and each 128×128 pixel image was obtained over 64 s.

Results of applying a spin-echo (spin-echo application enhances differences in gray level, owing to differences in T_2 depending on the environment of the water) MRI imaging technique are shown in Figure 25, where a sequence of six cross-sectional images are presented as a function of time after the slip was poured into the mold. In each image, three distinct rings of different "brightness" are apparent. The bright region in the center represents the slip; the next ring at the wall is the solids. The last ring is the mold itself. The mold is observable because water is migrating through it.

It is important to note that the signal intensity is proportional not just to the proton density, but to the proton relaxation parameters (T_2 and T_1) as well. Thus if T_2 and T_1 and proton density can be measured separately, T_2 and T_1 can be used as a way to study the local chemical environment of the ^1H protons. Since the chemical environment of the water molecules changes in the slip-casting process from the slip, to the solids, to the water migrating out through the mold, one would expect to see differences in T_1 and T_2. Such differences are observed and, for example, the values of T_2 for each location in the slip casting process are listed below.

	Al_2O_3 Slip	Deposit	Mold
T_2 (ms)	14	7	5

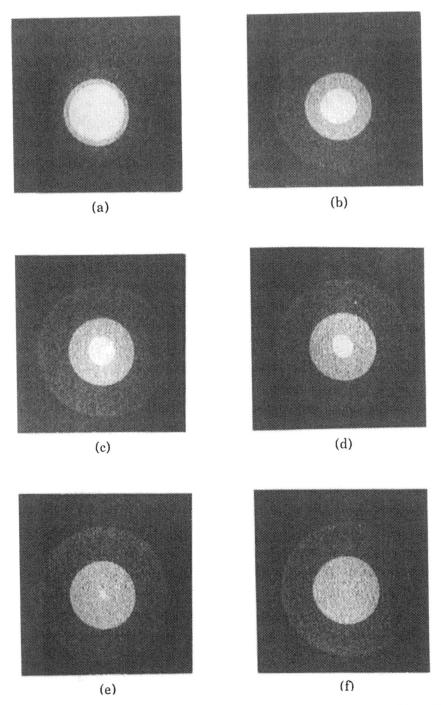

Figure 25 NMR images of the setup in Figure 24: (a) NMR image observed immediately after pouring Al_2O_3 slip into the mold; (b) NMR image observed 10 min after pouring; (c) 20 min later; (d) 30 min later; (e) 60 min after pouring; (f) 70 min later. (Adapted with permission from Ref. 69.)

The longer T_2 for the solution is consistent with general results for comparison of T_2 values for solutions and solids (or partially immobilized water in semisolids). This situation is analogous to the ^{31}P example mentioned earlier that concerned the study of dispersion aid interactions.

Since the water content difference between the slip and the solids is small, the approximately 2:1 intensity ratio between the slip signal intensity and the solids signal intensity was thought primarily to be related to the T_2 relaxation phenomenon. This result is understandable because the relative motion of ^1H in the slip is expected to be much greater than the ^1H motion in the deposit. By measuring the thickness of the deposit as a function of time, one can obtain the rate of solids buildup. A typical result is illustrated in Figure 26.

Quantifying Organic Distribution

Another efficient high-volume ceramics production technique is injection molding. In the injection-molding process, the distribution of organics (binder, plasticizers, and mold-release agents) used as the carrier for the ceramic powder is important, as this can affect homogeneity and green density, parameters that in turn will affect subsequent local densification rates and hence mechanical properties. Yeh et al. [70] and Zhang et al. [71] have compiled a list of typical defects and their causes that can form in injection-molded parts. Many of these defects can be traced to poor distribution of the organics. A partial list is provided in Table 1.

The problem of imaging solids arises when trying to image organics in injection-molded materials because the organics used are polymeric materials that behave in a manner similar to solids. As discussed above, this behavior means that the NMR signal to be detected decays rapidly (i.e., short T_2) and requires special imaging methods. To overcome this problem, special NMR instrumentation is required that combines high gradient field strengths (ca. 10 G/cm gradients) and short gradient switching times. Further details of this instrumentation are given by Ellingson et al. [72] and Gopalsami et al. [73].

With such instrumentation, spatial encoding of the information in both spatial dimen-

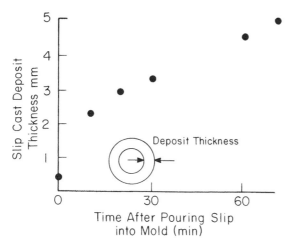

Figure 26 Change in deposit thickness with time obtained from the NMR images shown in Figure 25. (Adapted with permission from Ref. 69.)

Table 1   Injection Molding Defects and Causes

Type of Defect	Cause
Incomplete part	Improper feed material
	Poor tool design
	Improper material and/or tool temperature
Large pores	Entrapped air
	Improper material flow and consolidation during injection
	Agglomerates
	Large pockets of organic binder/plasticizer due to incomplete mixing
Knit lines	Improper tool design or feed material
	Incorrect temperatures
Cracks	Sticking during removal from tool
	Improper tool design
	Improper extraction of binder/plasticizer

sions, x-y, can be done in a fashion that compensates for the effects of spectral line broadening. Figure 27 shows a transaxial NMR image of a green, Si_3N_4 compact disk with 15 wt % organic material. In this sample, several holes were drilled, with diameters from 1.1 to 4.8 mm. Total imaging time was 29.3 min. Spatial resolution in the image is 129×128 y pixels, corresponding to approximately 640 μm in either direction.

It is possible to achieve images with higher resolution and to observe lower concentrations of NMR sensitive nuclei. The price is often a significant increase in imaging time. Figure 28 is such an example. It shows an image of a 25-mm-diameter Al_2O_3 sample containing only 2.5 wt % binder. Data acquisition required 17 h. Image resolution is

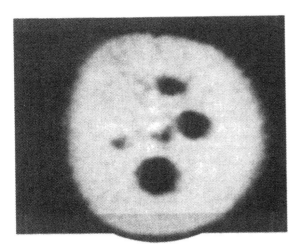

Figure 27   Transaxial ^{1}H NMR image of a green compact with 15 wt % organic binder taken with a two-dimensional spin-echo RF pulse sequence. Echo time was 4.42 ms and total imaging time was 29 min. Resolution is approximately 640 μm in either direction. Lighter areas correspond to higher binder content. Drilled holes contain no binder.

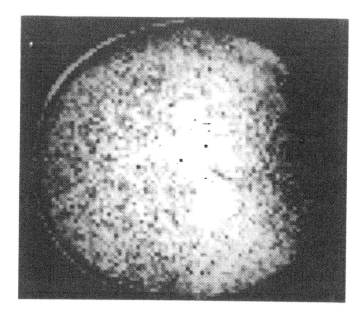

Figure 28 Two-dimensional image of a 25-mm-diameter Al_2O_3 sample with 2.5 wt % binder. Echo time is 3.2 ms and total imaging time is 17 h. Image resolution consists of 128 × 128 pixels, corresponding to ca. 320 μm of spatial resolution. Lighter areas contain more binder. Blackened area to left side is experimental artifact.

about 320 nm in either direction. If a more coarse spatial resolution would have been acceptable, the time could be reduced by 50% or more.

Signal intensity is not related to proton density alone, but also to the local chemistry, which affects T_2 and T_1 values. One NMR parameter that is not as affected by T_2 is the amplitude of the signal immediately following simple RF excitation pulse sequences: namely, free induction decay (FID). If one measures the local FID values, this is dependent only on ^1H concentration. If the images of Figures 27 and 28 had been obtained using FID and a back-projection technique (no gradient switching time and therefore no dependence on T_2), as shown in Figure 21, the gray scale values could be quantitatively related to organic concentration.

Mapping Porosity

If a ceramic material is in a state of processing in which open-connected porosity is present, MRI can also be used to map this porosity. Although mercury porosimetry has long been used to obtain porosity measurements, it has two limitations: (1) it is limited to small (<1 cm^3) specimens, and (2) it does not yield tomographic (i.e., spatial) information. On the other hand, MRI imaging has the potential to provide not only tomographic (i.e., three-dimensional) information, but also information on specimens of any size and shape.

To use MRI to study porosity in ceramics, however, filler fluids must be used to image the internal empty volume of open-connected porosity ceramics. In using this technique, it is assumed that there is adequate penetration of a compatible filler fluid into all the internal volumes of interest, including pore spaces and internal voids. Penetration is usually assisted by selecting a fluid of low viscosity and low interfacial tension with the microscopic surfaces of the sample.

Ellingson et al. [74] and Ackerman et al. [75] described the use of MRI to measure bulk porosity on a set of partially sintered Al_2O_3 disks. The disks were each 25 mm in diameter and had densities of 1.640, 1.703, and 1.720 g/cm^3. The filler fluid used was benzene, which has a proton molarity of 67.3 M. This solution provided a reasonably high proton concentration (that of water is 111 M). Paramagnetic doping to reduce spin-lattice relaxation time was accomplished by the addition of chromium acetylacetonate, $Cr(CH_3COCH_2COCH_3)_3$. By placing the specimen in an MRI facility, local signal intensities of the benzene were obtained.

Figure 29 shows the relation between MRI-measured bulk porosity and the density of the Al_2O_3 partially sintered compacts. A tomographic discrimination of the porosity in the interior and at the edges of the disks was achieved. The reduction in NMR signal with increased density is observed. The higher porosity at the edges was also expected. Recognizing that this is bulk characterization, MRI methods are also being developed that allow one to establish pore-size distribution. This method is based on measurements of local magnetic resonance relaxation phenomena, which are then related to relaxation phenomena associated with known pore sizes.

FUTURE DIRECTIONS

Spectroscopy

NMR spectroscopy of ceramics is generally hampered by low natural abundance of relevant nuclides, low spin lattice relaxation times, and in the case of quadrupolar nuclides (^{27}Al,

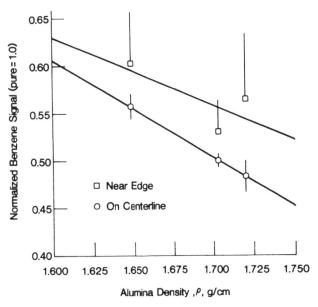

Figure 29 NMR signal intensity versus measured density for three partially sintered Al_2O_3 disks filled with benzene. Solid lines are least-squares fits. Error bars are standard deviations about the mean pixel intensity within each region of interest.

for example), the presence of SSBs and other broadening mechanisms that reduce resolution of chemically distinct sites. Although significant advances have been made to overcome each of these difficulties, one can expect some important gains in the foreseeable feature. Commercially available magnetic field strengths are increasing at a rate of roughly 25% every 2 to 3 years. Higher magnetic fields increase signal strength from a constant number of nuclei, while some solid-state broadening mechanisms fall off as the inverse of field strength. New instrumentation for magic angle spinning is available with spinning speeds up to 25 kHz and are capable of producing virtually SSB-free spectra. Other advances in spinning technology are showing promise in the removal of solid-state broadening interactions that cannot be removed by traditional MAS. Along another direction, technology for spinning solid samples of up to $5g$ will have an important impact on overcoming the T_1 problem, since a greater amount of signal will be acquired with each pulse. Significant reductions in total signal averaging time should be possible. Finally, an ever-increasing body of knowledge is being amassed on many of the nuclides of relevance to high-performance ceramics, thus providing a ''library'' of information for the analysis of new materials, with resulting increase in the depth of spectral interpretation.

Imaging

Significant effort is presently under way to achieve higher spatial resolution, faster data acquisition, and better displays for diagnostic interpretation. A target for spatial resolution for organics may be $< 100 \mu m$. To achieve this will require higher magnetic field strengths, perhaps more easily achievable with new superconducting ceramics, and resultant necessary higher gradient strengths. Faster data acquisition, such as obtaining full three-dimensional images in 5 min, may be another goal to be obtained. Use of sophisticated new parallel-processing architecture computers with specific internal architectures for MRI may allow this. Clearly, three-dimensional display of full three-dimensional images (i.e., volumetric displays obtainable with MRI) using new advanced imaging systems will also be beneficial in data interpretation.

Advanced imaging systems coupled with artificial intelligence systems will probably also make MRI more than just a process development tool, but give it the potential for ''near-on-line'' diagnostics. With the chemical sensitivity of NMR and with data properly taken, each individual pixel element can be a region for highly detailed spectroscopy. Thus in the future, one may be able to determine a rather complete ''picture'' of the chemical environment throughout the internal structure of a ceramic part. No other method being developed currently has the potential for providing this detailed local chemistry information nondestructively.

ACKNOWLEDGMENTS

For the examples taken directly from the literature, the original research groups are acknowledged. The preliminary dispersion aid/ceramic study was done in collaboration with A. Fanelli. We would like to thank R. O. Carter III and M. P. Shatlock for their continued support of this work.

REFERENCES

1. C. A. Fyfe, *Solid State NMR for Chemists*, CFC Press, Guelph, Ontario, 1983.
2. E. Oldfield and R. J. Kirkpatrick, *Science*, 227: 1537 (1985).
3. G. E. Maciel, *Science*, 226: 282 (1984).

4. E. Fukushima and S. B. W. Roeder, *Experimental Pulse NMR*, Addison-Wesley, London, 1981.
5. M. Mehring, *Principals of High Resolution NMR in Solids*, Springer-Verlag, New York, 1983.
6. K. Kato, Z. Inque, K. Kijima, I. Kawada, H. Tanaka, and T. Yamane, *J. Am. Ceram. Soc.*, 58: 90 (1975).
7. R. Grun, *Acta Crystallogr.*, B35: 800 (1979).
8. K. R. Carduner, C. S. Blackwell, W. B. Hammond, F. Reidinger, and G. R. Hatfield, *J. Am. Chem. Soc.*, 112: 4676 (1990).
9. R. C. Marshall, J. W. Faust, and C. E. Ryan, *Silicon Carbide*, University of South Carolina Press, Columbia, S.C., 1973.
10. G. R. Finlay, J. S. Hartman, M. F. Richardson, and B. L. Williams, *J. Chem. Soc. Chem. Commun.*, 159 (1985).
11. J. S. Hartman, J. F. Richardson, B. L. Sherriff, and B. G. Winsborrow, *J. Am. Chem. Soc.*, 109: 6059 (1987).
12. J. R. Guth and W. T. Petuskey, *J. Phys. Chem.*, 91: 5361 (1987).
13. J. V. Smith and C. S. Blackwell, *Nature*, 303: 223 (1983).
14. A. R. Grimmer, F. von Lampe, M. Magi, and E. Lippmaa, *Mh. Chem.*, 114: 1053 (1983), and 115: 561 (1984).
15. G. Engelhardt and R. Radeglia, *Chem. Phys. Lett.*, 108: 271 (1984).
16. R. Radeglia and G. Englehardt, *Chem. Phys. Lett.*, 114: 28 (1985).
17. S. Ramdas and J. Klinowski, *Nature*, 308: 521 (1984).
18. N. Janes and E. Oldfield, *J. Am. Chem. Soc.*, 107: 6769 (1985).
19. E. Lippmaa, M. Magi, A. Samoson, G. Engelhardt, and A. R. Grimmer, *J. Am. Chem. Soc.*, 102: 4889 (1980).
20. E. Lippmaa, M. Magi, A. Samoson, M. Tarmak, and G. Engelhardt, *J. Am. Chem. Soc.*, 103: 4992 (1981).
21. A. R. Grimmer, *Chem. Phys. Lett.*, 119: 416 (1985).
22. M. Magi, E. Lippmaa, A. Samoson, G. Engelhardt, and A. R. Grimmer, *J. Phys. Chem.*, 88: 1518 (1984).
23. B. L. Sherriff and H. D. Grundy, *Nature*, 332: 819 (1988).
24. K. A. Smith, R. J. Kirkpatrick, E. Oldfield, and D. M. Henderson, *Am. Mineral. J.*, 68: 1206 (1983).
25. A. R. Grimmer and R. Radeglia, *Chem. Phys. Lett.*, 106: 262 (1984).
26. J. M. Newsam, *J. Phys. Chem.*, 89: 2002 (1985).
27. J. B. Higgins and D. E. Woessner, *Eos*, 63: 1139 (1982).
28. J. Mason, *Multinuclear NMR*, Plenum, New York, 1984.
29. R. K. Harris and B. E. Mann, *NMR and the Periodic Table*, Academic Press, New York, 1978.
30. T. Axenrod and G. A. Webb, *Nuclear Magnetic Resonance Spectroscopy of Nuclei Other Than Protons*, Wiley, New York, 1974.
31. R. Dupree and M. E. Smith, *Chem. Phys. Lett.*, 148: 41 (1988).
32. R. Dupree, M. H. Lewis, and M. E. Smith, *J. Am. Chem. Soc.*, 111: 5125 (1989).
33. K. R. Carduner and R. O. Carter III, *Ceram. Int.*, 15: 327 (1989).
34. Z. Inoue, *J. Mater. Sci. Lett.*, 4: 656 (1985).
35. Z. Inoue, M. Mitomo, and N. Ii, *J. Mater. Sci.*, 15: 2915 (1980).
36. G. Hatfield, G. Yamanis, B. Li, W. Hammond, and F. Reidinger, *J. Mater. Sci.*, 25: 4032 (1990).
37. T. G. Kalamasz, G. Goth, R. P. Worthen, and A. E. Pasto, *Automot. Eng.*, 96: 63 (1988).
38. M. E. Milberg, *Chemtech*, 17: 552 (1987).
39. K. R. Carduner, R. O. Carter III, M. E. Milberg, and G. M. Crosbie, *Anal. Chem.*, 59: 2794 (1987).

40. H. W. Rauch, W. H. Sutton, and L. R. McCreight, *Ceramic Fibers and Fibrous Composite Materials*, Academic Press, New York, 1968.

41. K. E. Inknott, S. M. Wharry, and D. J. O'Donnell, *Mater. Res. Soc. Symp. Proc.*, 73: 165 (1986).

42. J. Lipowitz, J. A. Rahe, and T. M. Carr, *ACS Symp. Ser.* 360, p. 156 (1988).

43. R. West and J. Maxka, *ACS Symp. Ser. 360*, p. 6 (1988).

44. V. S. R. Murthy, M. H. Lewis, M. E. Smith, and R. Dupree, *Mater. Lett.*, 8: 263 (1989).

45. J. Lipowitz and G. L. Turner, *Polym. Prepr.*, 29: 74 (1988).

46. T. Taki, S. Maeda, K. Okamura, M. Sato, and T. Matsuzawa, *J. Mater. Sci. Lett.*, 7: 209 (1988).

47. T. Taki, K. Okamura, M. Sato, T. Sequchi, and S. Kawanishi, *J. Mater. Sci. Lett.*, 6: 826 (1987).

48. T. Taki, M. Inui, K. Okamura, and M. Sato, *J. Mater. Sci. Lett.*, 8: 918 (1989).

49. T. Taki, M. Inui, K. Okamura, and M. Sato, *J. Mater. Sci. Lett.*, 8: 1119 (1989).

50. C. H. Yoder and C. D. Schaeffer, *Introduction to Multinuclear NMR*, Benjamin/Cummings, Menlo Park, Calif., 1987.

51. M. Villa, K. R. Carduner, and G. Chiodelli, *J. Solid State Chem.*, 69: 19 (1987).

52. R. S. Aujla, G. Leng-Ward, M. H. Lewis, E. F. W. Seymour, G. A. Styles, and G. W. West, *Philos. Mag.*, B54: L51 (1986).

53. J. Klinowski, J. M. Thomas, D. P. Thompson, P. Korgul, K. H. Jack, C. A. Fyfe, and G. C. Gobbi, *Polyhedron*, 3: 1267 (1984).

54. N. D. Butler, R. Dupree, and M. H. Smith, *J. Mater. Sci. Lett.*, 3: 469 (1984).

55. B. C. Gerstein and A. T. Nicol, *J. Non-Cryst. Solids*, 75: 423 (1985).

56. R. Dupree, M. H. Lewis, G. Leng-Ward, and D. S. Williams, *J. Mater. Sci. Lett.*, 4: 393 (1985).

57. K. A. Smith, R. J. Kirkpatrick, E. Oldfield, and D. M. Henderson, *Am. Mineral. J.*, 68: 1206 (1983).

58. R. Dupree, M. H. Lewis, M. E. Smith, *J. Am. Chem. Soc.*, 110: 1083 (1988).

59. R. Dupree, M. H. Lewis, and M. E. Smith, *J. Appl. Crystallogr.*, 21: 109 (1988).

60. K. R. Carduner, R. O. Carter III, M. J. Rokosz, G. M. Crosbie, and E. Stiles, *Chem. Mater.*, 1: 302 (1989).

61. R. Dupree, M. H. Lewis, and M. E. Smith, *J. Am. Chem. Soc.*, 60: 249 (1988).

62. K. R. Carduner, S. S. Shinozaki, M. J. Rokosz, C. R. Peters, and T. J. Whalen, *J. Am. Ceram. Soc.*, 73: 228 (1990).

63. M. M. Maricq and J. S. Waugh, *J. Chem. Phys.*, 70: 3300 (1979).

64. E. Oldfield, R. A. Kinsey, K. A. Smith, J. A. Nichols, and R. J. Kirkpatrick, *J. Magn. Reson.*, 51: 325 (1983).

65. M. A. Foster and J. M. S. Hutchinson (eds.), *Practical NMR Imaging*, IRL Press, Oxford, 1987.

66. P. Mansfield and P. G. Morris (eds.), *NMR Imaging in Biomedicine*, Academic Press, New York, 1982.

67. A. Kumar, I. Welti, and R. R. Ernst, *J. Magn. Reson.*, 18: 69 (1975).

68. P. A. Bottomley, *Rev. Sci. Instrum.*, 53: 1319 (1982).

69. K. Hayashi, K. Kawashima, K. Kose, and T. Inouye, *J. Phys. D. Appl. Phys.*, 21: 1037 (1988).

70. H. C. Yeh, J. M. Wimmer, M. R. Huang, M. E. Rorabaugh, J. Schienle, and K. H. Styhr, *Improved Silicon Nitride for Advanced Heat Engines*, NASA Tech. Rep. NASA-CR-175006 (1985).

71. J. G. Zhang, M. J. Edirsinghe, and J. R. G. Evans, *Ind. Ceram.*, 9: 72 (1989).

72. W. A. Ellingson, J. L. Ackerman, L. Garrido, P. S. Wong, and S. Gronemeyer, *Development of Nuclear Magnetic Resonance Imaging Technology for Advanced Ceramics*, Argonne National Laboratory Rep. ANL-87-53 (1985).

73. N. Gopalsami, G. A. Forster, S. L. Dieckman, and W. A. Ellingson, ''Development of NMR imaging probes for advanced ceramics,'' in *Review of Progress in Quantitative Nondestructive Evaluation*, D. O. Thompson and D. E. Chimenti (eds.), Plenum Press, New York, 1990.

74. W. A. Ellingson, J. L. Ackerman, L. Garrido, J. D. Weyland, and R. A. Dimilia, *Ceram. Eng. Sci. Proc.*, 8: 503 (1987).

75. J. L. Ackerman, L. Garrido, W. A. Ellingson, and J. D. Weyland, ''The use of NMR imaging to measure porosity and binder distributions in green-state and partially sintered ceramics,'' *Nondestructive Testing of High Performance Ceramics*, American Ceramic Society, Westerville, Ohio, 1987, p. 88.

Index

Printed and bound by CPI Group (UK) Ltd, Croydon, CR0 4YY

17/10/2024

01775703-0006